TRANSPORTATION ISSUES, POLICIES AND R&D SERIES

WIRELESS TECHNOLOGIES IN INTELLIGENT TRANSPORTATION SYSTEMS

TRANSPORTATION ISSUES, POLICIES AND R&D SERIES

Public Transit Issues and Developments
Calvin B. Lang (Editor)
2009. ISBN: 978-1-60692-689-5

Railway Transportation: Policies, Technology and Perspectives
Nicholas P. Scott (Editor)
2009. ISBN: 978-1-60692-863-9

Yacht Modelling and Adaptive Control
Chengmo Xiao and Sing Kiong Nguang
2009. ISBN: 978-1-60741-430-8

Motorcycle Safety and Crashes
Joseph Da Corte (Editor)
2009. ISBN: 978-1-60741-884-9

Aeropolitics
Ruwantissa Abeyratne
2009. ISBN: 978-1-60876-102-9

Automotive Industry: Technical Challenges, Design Issues and Global Economic Crisis
Gordan A. Maxwell and Stuart K. Drummond (Editors)
2010. ISBN: 978-1-60876-143-2

Wireless Technologies in Intelligent Transportation Systems
Ming-Tuo Zhou, Yan Zhang and Laurence T. Yang (Editors)
2010. ISBN: 978-1-60741-588-6

Automobiles: Performance, Safety Assessment, and Energy Consumption
Matin F. Kody (Editor)
2010. ISBN: 978-1-61668-218-7
Automobiles: Performance, Safety Assessment, and Energy Consumption
Matin F. Kody (Editor)
2010. ISBN: 978-1-61668-387-0 (Online Book)

High Speed Passenger Rail: Viability, Challenges and Federal Role
Augelli Biocchetti (Editor)
2010. ISBN: 978-1-60741-985-3

Congestion Pricing in Traffic Control
Marco D. Sheehan (Editor)
2010. ISBN: 978-1-60741-963-1

Bus, Motor Carrier and Trucking Safety Issues
Samuel B. Metzler (Editor)
2010. ISBN: 978-1-60876-755-7

Head Restraints and Whiplash: The Past, Present and Future
Ediriweera Desapriya
2010. ISBN: 978-1-61668-150-0

Airport and Aviation Security
Amelia K. Voegele (Editor)
2010. ISBN: 978-1-61668-583-6

Airport and Aviation Security
Amelia K. Voegele (Editor)
2010. ISBN: 978-1-61668-724-3 (Online Book)

TRANSPORTATION ISSUES, POLICIES AND R&D SERIES

WIRELESS TECHNOLOGIES IN INTELLIGENT TRANSPORTATION SYSTEMS

MING-TUO ZHOU,
YAN ZHANG
AND
LAURENCE T. YANG
EDITORS

Nova Science Publishers, Inc.
New York

For permission to use material from this book please contact us:
Telephone 631-231-7269; Fax 631-231-8175
Web Site: http://www.novapublishers.com

LIBRARY OF CONGRESS CATALOGING-IN-PUBLICATION DATA

Wireless technologies in intelligent transportation systems / Ming-Tuo Zhou, Laurence T. Yang, [editors].
p. cm.
Includes bibliographical references and index.
ISBN 978-1-60741-588-6 (hardcover)
1. Intelligent transportation systems. 2. Wireless communication systems. I. Zhou, Ming-Tuo. II. Yang, Laurence Tianruo.
TE228.3.W57 2009
629.2'7--dc22

2009015638

Published by Nova Science Publishers, Inc. ✦ New York

CONTENTS

PREFACE

A number of wireless technologies have been developed in recent years to meet the increasing needs of high-speed wireless communications in civil and military applications. The advances include WiFi (IEEE 802.11), WiMAX (IEEE 802.16), sensor networks, wireless Mesh/Ad hoc networks, mobile IP, smart antenna, cognitive radio, and so on. These emerging technologies will significantly impact the design and operation of Intelligent Transportation Systems, which aims to effectively provide higher vehicles safety, traffic management, and communications among vehicles and transport infrastructure.

Organized into three parts, 'Wireless Technologies for Intelligent Transportation Systems' provides readers a thorough technical guide covering various wireless technologies developed in the most recent years for Intelligent Transportation Systems applications. It presents key technologies of circuits and physical layer, network protocols, system designs and applications. The broad content covers topics of radar sensor, radio channel modeling, smart antenna, medium access control, routing protocol, data dissemination, handover, security, mesh networking, road traffic estimation and monitoring, and location-based services.

This comprehensive book is a collection of basic concepts, major issues, design approaches, application examples, and future research directions of various advanced technologies developed for Intelligent Transportation Systems. With its broad coverage allowing cross reference, it serves as an essential reference for engineers, researchers, students, scientists, professors, designers and planners of Intelligent Transportation Systems.

Chapter 1 presents an overview of the radar sensor technology used in automotive applications. These applications are outlined and the performance tradeoffs of the radar sensors is discussed, including the different modulation schemes and scanning antenna types used. The test requirements of these radar sensors during production, installation on a vehicle, and during after-market service are also discussed. Different test methods used for characterizing and aligning these radar sensors are presented.

New applications for wireless communications are emerging with increasing frequency. For Intelligent Transportation Systems (ITSs), this will include applications that require communications between vehicles, and communications between vehicles and roadside entities. In Chapter 2 the authors address terrestrial ITS applications, specifically automotive applications.

Any communication system requires the following components: message source, transmitter, channel, receiver, and message destination. For wireless communication, the channel is often uncontrolled, or only very loosely controlled, by the system designer and user. Hence the channel can play an important role in communication reliability. In the case of vehicular communications for ITS, the channel will often be dynamic, lossy, and distorting. Design of effective communication systems for these vehicle-to-vehicle (V2V) and vehicle-to-roadside (V2R) channels is thus challenging, and requires good models. This chapter describes such models.

Most envisioned V2V and V2R applications will be relatively short range—on the order of a few hundred meters to a kilometer. In some cases propagation path loss may limit system performance, but the authors do not focus on this form of "large scale" channel effect, which in principle can be overcome with larger transmit power. Although obstruction by buildings ("shadowing") and other large obstacles and vehicles may occur, these effects are often of secondary importance in comparison to small scale fading, caused by multipath propagation. This chapter provides an overview of channel characterization, then describes models for the V2V and V2R channels. These models primarily address small scale fading. Treatment of "medium scale" effects is also noted, but is of lesser depth. Our coverage focuses on statistical models for small scale fading. A brief discussion of deterministic models is included for comparison and completeness.

The V2V (and to a lesser degree, the V2R) channel can be very dynamic, with time variation rates up to double those of conventional, e.g., cellular radio, channels. With the low antenna heights of V2V and V2R systems, radio line of sight (LOS) is more frequently and more thoroughly obstructed. Because of these effects, the V2V and V2R channels will generally exhibit more severe fading than conventional channels. Due to the potentially rapid time variation, V2V and V2R channels will also be best modeled as statistically *non*-stationary.

Background. Research and development on smart antennas that are recognized as a promising technique to improve the performance of mobile communications have been extensive in the recent years. Smart antennas combine multiple antenna elements with signal processing capabilities in both space and time to optimize its radiation and reception pattern automatically in response to the signal environment. This technology has significant impact on intelligent transportation systems (ITS) communications leading to increased user capacity, higher data rates and enhanced channel reliability. Furthermore, due to its capability of locating mobile units, smart antennas can facilitate vehicle location and navigation, trigger a myriad of location-based services and applications, and eventually benefit automotive telematics, public transit systems, and many other transportation systems.

Material and Methods. Chapter 3 concentrates on the signal processing aspects of smart antenna systems. Smart antennas are often classified as either switched-beam or adaptive-array systems, for which a variety of algorithms have been developed to enhance the signal of interest and reject the interference. The antenna systems need to differentiate the desired signal from the interference, and normally requires either *a priori* knowledge or the signal direction to achieve its goal. There exists a variety of methods for direction of arrival (DOA) estimation with conflicting demands of accuracy and computation. Similarly, there are many algorithms to compute array weights to direct the maximum radiation of the array pattern toward the signal and place nulls toward the interference, each with its convergence property and computational complexity. This chapter discusses some of the typical algorithms for DOA estimation and beamforming. The concept and details of each algorithm are provided.

Results. Smart antennas can significantly help in improving the performance of ITS communication systems by increasing channel capacity and spectrum efficiency, extending range coverage, multiplexing channels with spatial division multiple access (SDMA), and compensating electronically for aperture distortion. They also reduce delay spread, multipath fading, co-channel interference, system complexity, bit error rates, and outage probability. In addition, smart antennas can locate mobile units or assist the location determination through DOA and range estimation. This capability can support and benefit many location-based services including emergency assistance, tracking services, safety services, billing services, and information services such as navigation, weather, traffic, and directory assistance.

Conclusion. Smart Antennas examine nearly all aspects of array signal processing. This chapter delivers a detailed treatment of antenna array processing schemes, algorithms to adjust weighting, DOA estimation methods, and diversity-processing methods that combat fading. In general, the algorithm selection is a trade-off between complexity and performance. Smart antennas will make the future intelligent transportation systems to be less complex while providing more attractive and convenient services. The authors believe that this technology has the great potential to make our transportation systems operate more safely and efficiently, with less congestion, pollution, and environmental impact.

Intelligent Transportation Systems (ITS) using wireless and mobile ad-hoc sensor networks focus on adding information and communications to our transportation infrastructure and vehicles. ITS has inspired many autonomous applications and has transformed our day to day lives. The challenges faced by ITS are response time, data aging and communication cost caused due to rapid changes and movement of vehicles. These challenges frustrate the user because they result in inaccurate information, higher cost of using the system, and mediocre system reliability. Hence, an intelligent multi-objective approach with a sensor focus will deal with these problems by reducing costs, increasing robustness, and improving information accuracy by consistently making the optimum trade-off decisions that will benefit the user. In Chapter 4, the authors analyze routing protocols and their importance in disseminating information using distributed sensor technology while trading off the prevention of data aging with the efficient use of sensor resources.

Chapter 5 presents an impact study of wireless Medium Access Control (MAC) Protocols on Intelligent Transportation System (ITS) applications using the Dedicated Short Range Communication (DSRC) standard. After a detailed review of the existing contention-based and TDMA-based MAC protocols for vehicular communication, a novel Medium Access Control protocol for inter-vehicle communication applications is introduced. The primary contributions of this chapter are the design of a self-configuring TDMA protocol VeSOMAC, and its evaluation for inter-vehicle data transfer applications. A novel feature of the proposed VeSOMAC is its in-band control mechanism for exchanging TDMA slot information during distributed MAC scheduling. It is shown that the in-band control mechanism can be used for fast protocol convergence during topology changes due to vehicle movements. Simulated performance comparison between VeSOMAC and the DSRC-recommended 802.11 MAC protocol demonstrates that the proposed protocol can accomplish better inter-vehicle data transfer performance through enhanced UDP and TCP throughput and fewer MAC layer packet drops compared to 802.11. Experiments in transient scenarios representing VeSOMAC slot reorganization due to network topology changes demonstrate that while the application level performance with VeSOMAC degrade during topology changes, it can still outperform the contention based 802.11 for data-intensive vehicular applications.

As emerging civilian instantiates of Mobile Ad Hoc Networks (MANETs), Vehicular Ad Hoc Networks (VANETs) are critical components in Intelligent Transportation Systems

(ITS). VANETs, though special cases of MANETs, have distinct characteristics: high mobility with limitation of road topology, rapidly changing network topology, unbounded network size, and time-sensitive data exchange, to name only a few. These unique characteristics, together with the fact that many VANET applications are safety-related, make security provisioning in VANETs a challenging while indispensable task. In this chapter, the authors systematically investigate the security provisioning issues of VANETs, to give a rather complete understanding on the requirements, challenges, potential solutions, open issues, and the future research directions from different perspectives. Based on the analysis of threat models and different applications requirements, the authors categorize the VANET security requirements into three major classes (the information security, the data security, and the trust management), along with considerations about network performance constraints. In-depth investigations of these security provisioning issues and future research directions are presented. A security provisioning framework that enables interactions of different security provisioning modules is also proposed in Chapter 6.

With the recent rapid revolution in wireless network technology, wireless metropolitan area networks (WMAN) and vehicular networks (VN) have both gained significant research focuses, in academia as well as industry. VN are vital for the realization of intelligent transportation systems (ITS). The short-range communications for ITS are most likely supported by IEEE 802.11p WAVE (Wireless Access for Vehicular Environment), whereas the long-range communications would be realized by infrastructure-based WMAN such as 3G technology or IEEE 802.16 WiMAX (Worldwide Interoperability for Microwave Access). It is thus timely to study how WiMAX may be utilized in the emerging VN for ITS. Chapter 7 surveys the recent developments in mobility management of IEEE 802.16 networks, and presents a comprehensive feasibility study of 802.16 handoff schemes for the support of ITS. With the strong worldwide deployment of WiMAX networks, the authors believe that the study is a solid contribution in the practical realization of ITS

VANET applications can be classified into two categories: user applications and safety applications. Both of them must rely on efficient vehicular ad hoc routing protocols in order to deliver the information to the corresponding destination(s). Many possible VANET applications, e.g., accident avoidance or distributed traffic management applications, do not need explicit unicast data exchange, but locally aggregate and process the data broadcast by other vehicles. One-to-many communications is of great importance for VANETs since vehicles are likely to exchange data by diffusion rather than in a unicast manner. Many broadcast techniques have already been proposed for MANETs, but classicalMANET flooding techniques may not be suitable for VANETs and do not take advantage of the additional information available in vehicles. In Chapter 8 the authors investigate various broadcast techniques that may be used in VANETs.

The continuing and emerging advances in the development of wireless technologies significantly accelerate the design and operation of new intelligent transportation systems. These advances have to be combined with other fields of research and technologies in order to derive end-users' services. The research presented in this chapter proposes to combine three fundamental ITS technologies, i.e. the wireless communications, geolocalisation technologies and the geographic information systems in order to design transportation services able to monitor in real-time mobile objects (e.g. pedestrians, vehicles). Chapter 9 is illustrated by disaster response management and coordination that aims at monitoring terrestrial (indoor and outdoor) and maritime rescue operations.

The conventional maritime communications are mainly based on satellite and HF / VHF / UHF networks that are either expensive or low-speed. With the development of economy and technologies, it needs better communication means in maritime Intelligent Transportation Systems as well as in systems providing general communication services like voice call and Internet surfing. Chapter 10 introduces high-speed and low-cost wireless network technologies for maritime communications, on topics of maritime communication environment, analysis, evaluation and new developments of WiMAX mesh MAC and routing protocols for maritime communications.

Car industry is evolving in a way to provide smarter and safer cars. Nowadays, embedded electronics provide users with a safer and more enjoyable traveling experience than ever before. An inescapable step in this evolution is the connection of vehicles to the Internet which opens the way to an infinite number of enhancements. Basically, services related to Intelligent Transportation Systems (ITS) can be categorized into safety related and infotainment related services. Some of them use the Internet while others rely on car-to-car communications.

In Chapter 11 the authors explain how vehicles can take advantage from wireless communication technology diversity using IPv6 mobility protocols and standards to provide embarked applications with a full continuous IPv6 connectivity. Consequently, the Internet flexibility could simplify the development of various services (from security to infotainment). The CALM architecture, designed at ISO, is described and a focus is made on the support of multiple heterogeneous communication interfaces.

After a brief review of the work done in standardization bodies and in the academic world, this chapter provides an analysis of what should be implemented inside a vehicle. It also points out several missing features in the IETF/IEEE standards. The chapter ends, giving some insights of what a “full-featured” heterogeneous networks and mobility management framework for ITS should be.

Today, cars become more and more intelligent. They are equipped with hundreds of sensors and microcomputers that are able to measure speed, tire pressure, brake temperature, to detect raindrops on the windshield, and to help the driver to react in function of these measurements. The next step researchers started to work on was related to car-to-car communication: how could a car disseminate relevant information to other cars, in order to increase road safety. A vehicular network can be considered as a special case of a wireless ad hoc network. However, the traditional ad hoc routing algorithms might not work properly, or should at least be modified according to the specificities of such an environment, mainly the high speeds and the fast changing topology. In Chapter 12 the authors present first the shortcomings of traditional communication solutions in ITS scenarios, and describe then ITS specific routing and information spreading techniques, based on partial, directed flooding or geocasting (i.e., location-based multicasting). At the end of the chapter the authors also analyze hybrid architectures that involve road side infrastructural elements to help the communication between the mobile vehicles.

In the area of Intelligent Transportation Systems the introduction of wireless communications is reshaping the information distribution concept, and is one of the most important enabling technologies. The distribution of real-time traffic information, scheduling and route-guidance information is helping the transportation management systems in their efforts to optimize the system. The communication required to transfer all this information is rather expensive in terms of transmission power, use of the scarce resources of frequencies and also the building of an infrastructure to support the transceivers. By using information that already exists and is exchanged within the infrastructures of the GSM and UMTS networks, a lot of the resource problems are solved. The information that could be extracted from these cellular networks could be used to obtain accurate road traffic information to support real-time traffic information. In this way the cellular networks not only becomes the means to distribute information but also a source of road traffic information.

From the analysis made it is obvious that the potential of retrieving valuable road traffic information from cellular systems in a cost efficient way, i.e. by using already existing signaling data, is very high. It has however not been clear what to expect from these systems in terms of accuracy, availability and coverage. In Chapter 13 the basics for this is laid out and discussed in detail. A practical trial has also been performed and the results show clearly the potential as well as the differences in using the GSM compared to the UMTS network. The advantages and drawbacks are discussed and backed up by real measurements in two different road environments. The main advantages of using the existing signaling data, i.e., passive monitoring compared to active monitoring where the terminal sends extra data is discussed and could be summarized in three components, no user acceptance is necessary, no extra signaling is necessary and it does not drain the terminal battery.

In the future it is likely that vehicles need to communicate more frequently with each other and with some kind of traffic control centre. This traffic will also be very useful in order to estimate road traffic information using the signaling information obtained from the cellular system. However, the enhanced communication systems will also change traffic patterns in the cellular networks which will affect the potential of estimating road traffic from cellular systems. The evolvement indicates that the terminals will be in active state almost constantly, and hence the updating information will be more frequent and the information more accurate.

PART 1.
HARDWARE, IMPLEMENTATION AND PHYSICAL LAYER TECHNOLOGIES

In: Wireless Technologies in Intelligent Transportation… ISBN: 978-1-60741-588-6
Editors: Ming-Tuo Zhou et al, pp. 3-20

Chapter 1

RADAR SENSOR TECHNOLOGY AND TEST REQUIREMENTS IN AUTOMOTIVE APPLICATIONS

Ramzi Aboujaoude*
Anritsu Company, Microwave Measurements Division
Morgan Hill, California 95037, USA

Abstract

This chapter presents an overview of the radar sensor technology used in automotive applications. These applications are outlined and the performance tradeoffs of the radar sensors is discussed, including the different modulation schemes and scanning antenna types used. The test requirements of these radar sensors during production, installation on a vehicle, and during after-market service are also discussed. Different test methods used for characterizing and aligning these radar sensors are presented.

1. Introduction

Radar technology has been investigated for use on automobiles since the 1970's and has been employed for various functions on automobiles since the early 1980's [1],[2]. Initial usage of microwave radar was for collision warning applications on commercial vehicles, such as ambulances, buses, and trucks. Prometheus, the ambitious European project that started in 1986, aimed to improve vehicle safety, efficiency, and economy and was one of the main driving factors in the development of various types of sensors for automobiles, including millimeter wave radar. The advantage that radar sensors have over other types of sensors, such as optical or infrared sensors, is that they perform equally well during the day, the night, and in most weather conditions. Compared to optical or infrared sensors, radar sensors exhibit significantly less atmospheric attenuation in adverse weather conditions, such as fog, rain, or snow. In addition, radar sensors do not suffer degradation due to dirt or road

* E-mail address: ramzi.aboujaoude@anritsu.com. Phone: +1-408-778-2000; Address correspondence to: Ramzi Aboujaoude., Ph.D., Anritsu Company, Morgan Hill, CA 95037 USA.

grime. Radar can also be used for target identification and for detecting road conditions by making use of scattering signature information.

Table I. The road to collision avoidance - automotive applications that can utilize radar sensors

1) Collision Warning (forward, backup, side-looking)
2) Adaptive Cruise Control (ACC)
3) Stop and Go Function or ACC II (City Driving)
4) Air-bag pre-crash Trigger
5) Lane change, Backup aid, Parking aid
6) Radar imaging, Target identification
7) Complete sensor belt for collision avoidance

Table II. Sensor fusion that may be required for collision avoidance applications

1) Radar for medium to long range applications (76-77 GHz)
2) Radar for short range applications (24 GHz, 77-81 GHz)
3) Vision sensors
4) Ultra-sonic sensors
5) Laser sensors
6) Infra-red sensors
7) GPS receivers
8) μWave and mm-wave transceivers for inter-vehicle and roadside communication (5.8GHz and 61GHz)

After several years of implementation in the trucking industry as part of collision warning/avoidance systems (CW/A) [3], radar sensors transitioned to the broader automobile market. One of the first commercial applications implemented is Adaptive Cruise Control (ACC) for highway driving. Other applications being implemented include ACC for city driving, lane change, parking aid, and back-up aids. Table I lists the different automotive applications that can utilize radar sensors. The ultimate goal is for the automobile to have a sensor belt that provides 360° coverage around the vehicle. This sensor belt, which can be used to give the automobile autonomy and to provide complete collision avoidance, will require the use of multiple types of sensors. Using such sensor fusion helps achieve the accuracy and safety levels required of such a system [4], [5]. Table II lists some of the types of sensors that could be used in these systems. Radar sensors will undoubtedly be an integral part of any sensor fusion system.

There is an element of risk associated with using radar on automobiles, whether they are used in comfort and convenience options, such as ACC, or in safety options. Due to the critical impact of the radar sensor on the these systems, it is important to conduct accurate verification and calibration of the radar module and system at various stages of development, production, and installation.

Section 2 will give an overview of the current radar technology used in automotive applications including ACC. A discussion of the different modulation schemes and antennas used will be presented. Section 3 will discuss the test requirements of ACC radar during sensor production, during installation, and during after-market service. The test methods and equipment that can be used for these various requirements will be presented, including built-in tests that are incorporated into the radar sensors to insure proper operation in the field.

2. Automotive Radar Technology

Applications Overview

Early vehicular radar development was in the microwave frequency range of 10-24 GHz. For forward looking CW/A applications, the physical size of these radar sensors was too large for automobile use and so their usage was restricted to trucks and buses [3]. To reduce the physical size of the radar sensors for use on automobiles, the frequency of operation had to be increased to the millimeter wave range. At those frequencies, the radar antenna can be made electrically large to produce the required high directivity and narrow beamwidth, while keeping the physical size small enough for installation on an automobile. With the exception of some early development work at the 60 GHz oxygen absorption band and at 94 GHz, most current millimeter-wave radar systems operate in the 76-77 GHz frequency range [6], [7].

First generation ACC systems were designed for highway driving only, where the speed of the automobile remains at highway driving levels. ACC radar sensors provide range and closing-rate information to the cruise control system, which can control the brakes, throttle, and the automatic transmission of the vehicle. The ACC system adapts the speed of the vehicle according to the speed of the vehicle ahead, in order to maintain a separation time-interval between the two vehicles. The driver sets the maximum speed and the minimum separation desired. The radar system can also be used to locate and track multiple targets on the road ahead in order to anticipate traffic conditions in the driving lane. In order to restrict its use to highway driving, the ACC system will only apply the brakes up to a specified level. If additional braking is required to slow the automobile, the ACC function is disabled and the driver must assume control of the automobile.

Second generation systems make use of multiple radar sensors to extend the ACC system to city driving or stop-and-go traffic. The same radar sensors used for ACC can also be used in pre-crash sensing, parking aid, and CW/A systems. When used in a CW/A application, the radar sensors allow for warning signals to be given and for air bags to be activated. CW/A systems can also be designed to take control of a vehicle when a collision is anticipated. For most applications beyond highway ACC, multiple radar sensors must be used in conjunction with the long-range forward-looking 76-77 GHz ACC radar. These sensors are typically short-range and are used to monitor the traffic in front, on the sides, and behind the vehicle. Various sensor types can be used for these near-range applications, including infrared, vision, ultrasonic, and microwave radar.

The unlicensed Industrial Scientific Medical (ISM) band at 24.125 GHz is presently being used for short-range radar. In addition to the standard ISM band, government agencies in the U.S.A., Europe, Australia, and New Zealand have permitted the use of ultra wide band (UWB) technology in the 22-29 GHz frequency range for automotive radar. This lower frequency can be used for these short-range applications because the radar does not require high gain antennas. These radar sensors do not need to detect objects further than 30 m away and their angular coverage is broad. Therefore, lower gain antennas with wider beamwidths can be used, resulting in radar sensors that are physically small enough for automobiles. In addition, compared to 76 GHz, the technology at 24 GHz is more mature and is lower in cost. This is due in part to the build up of the infrastructure for broadband wireless communication systems in that frequency range.

The widespread use of automotive radar hinges on the ability of the manufacturers to achieve the required targets for performance, size, and cost. With the most recent developments and product introductions, the performance and size targets appear to have been mostly met. Meeting the cost targets may prove to be more challenging, with long term cost targets for the 76 GHz sensors set at approximately $100 and for the 24 GHz sensors even lower than that. These are extremely difficult targets to meet, even with the expected higher production volumes. Therefore, next generation sensors are being developed with lower cost designs. Early versions of ACC radar used dielectric lens antennas, waveguides, and discrete components including Gunn oscillators. Current and future designs include lower cost scanning antennas, millimeter wave front-ends using flip-chip Microwave Monolithic Integrated Circuit (MMIC) technology, low-cost oscillator designs, and automated assembly, packaging and testing [8].

ACC Radar System Requirements

The ACC radar must be able to detect targets of varying sizes (motercycles, cars, trucks, etc.) up to 200 m ahead. These targets could be moving relative to the radar at speeds of up to 250 km/h. The radar must have an angular coverage of at least ±8° in the azimuth plane. This coverage is needed to track the targets (other vehicles as well as fixed objects such as guard rails or sign posts) in the lanes ahead especially when approaching curves in the road. In addition, the angular coverage is needed for early detection of vehicles cutting into the driver's lane. In the elevation plane, the radar angular coverage must be narrow to reduce the effect of ground bounce signals and reflections from overhead structures, such as bridges.

Table III summarizes the typical performance specifications of ACC radar modules. Some of the limitations of current systems are their inability to detect very close-in targets and their limited angular coverage.

Table III. Typical performance specifications of ACC radar sensors

<table>
<tr><td colspan="2">Transmit Frequency</td><td>76-77 GHz</td></tr>
<tr><td colspan="2">Transmit Power</td><td>>10 dBm</td></tr>
<tr><td rowspan="2">Target Detection Distance</td><td>Range</td><td>2 to 150 m</td></tr>
<tr><td>Accuracy</td><td><± 1 m or ± 5%</td></tr>
<tr><td rowspan="2">Relative Velocity</td><td>Range</td><td>± 250 km/h</td></tr>
<tr><td>Accuracy</td><td><± 1 km/h</td></tr>
<tr><td rowspan="2">Angular Coverage</td><td>Azimuth</td><td>± 8° wide coverage with 3° minimum resolution</td></tr>
<tr><td>Elevation</td><td>3° to 4° single beam</td></tr>
<tr><td colspan="2">Antenna Gain</td><td>26 – 34 dBi</td></tr>
<tr><td colspan="2">Antenna sidelobe level</td><td>> 20 dB</td></tr>
<tr><td colspan="2">Update rate</td><td>>10 Hz</td></tr>
</table>

ACC Radar Antenna Types

The direction of arrival (DOA) of a radar signal can be obtained using various analog and digital techniques. The most common techniques used in the current ACC radar sensors are analog or digital beamforming techniques. These include electrically switching multi-beam antennas, continuous scanning single beam antennas, and amplitude or phase monopulse. Figure 1 shows these different antenna beam configurations.

For radar with multi-beam antennas in the azimuth (horizontal) plane, the ACC system will continuously switch between the three to seven overlapping transmit/receive beams. Each beam has a width of approximately 3°.

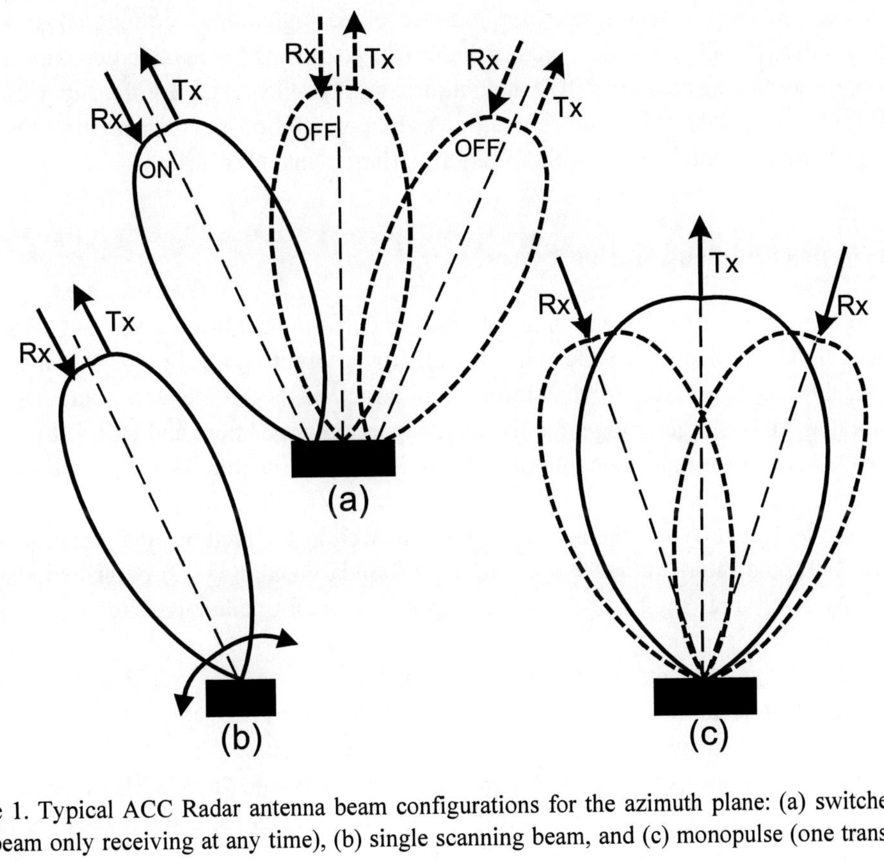

Figure 1. Typical ACC Radar antenna beam configurations for the azimuth plane: (a) switched beam (one beam only receiving at any time), (b) single scanning beam, and (c) monopulse (one transmit and two receive beams).

Alternatively, a mechanically or electrically scanning antenna is used to sweep across the desired coverage angle with a narrow 3° single beam antenna. Some radar sensors will transmit a single wide-beam, approximately 12° in width, and use monopulse receive techniques to locate vehicles off boresight on the left and right of the sensor [28]. To achieve the narrow beamwidths, electrically large antennas with high directive gains of 26–34 dBi must be used. (Electrically large antennas are antennas with radiating apertures that have large dimensions in terms of wavelength.) The most common antennas that meet these gain levels are planar microstrip arrays, dielectric lens, and parabolic reflector antennas.

In the elevation (vertical) plane, a single beam antenna with sufficiently narrow beamwidth is typically used to meet the required specifications.

Most antenna research on ACC sensors, at 76 GHz and 24 GHz, has recently been in the area of antenna scanning. To insure that the ACC system can track all the vehicles to the sides of the driver's vehicle, especially at close range, a very wide field of view is required. Also, there is interest in using the radar sensors to generate an image of the automobile's surrounding area. A high resolution (<2°) wide field-of-view (>40°) radar is required for such applications [8], [9].

In general, a narrower beamwidth (physically larger) antenna is required to obtain higher resolution using the standard techniques discussed above. To overcome these limitations, digital parameter-estimation techniques have recently been investigated for use in automotive radar. These techniques decompose the radar-received signal into orthogonal subspaces: signal and noise. The DOA of the signal is then calculated using eigenvalue decomposition of the autocorrelation matrix. These DOA techniques, such as those that use the superresolution MUSIC [10] and ESPRIT [11] algorithms, have the potential of meeting the high-resolution requirements of automotive radar without requiring larger antennas [8,12].

Radar Types and Modulation Schemes

Radar (short for Radio Detection and Ranging) consists of a transceiver that transmits a modulated signal and listens for the echo reflected by a target. The reflected signal collected at the receiver is processed to determine if a target has been detected, and to extract information about the target: range, relative velocity, angular position, and RCS [28].

In 1998, the European Telecommunications Standards Institute (ETSI) published what was the first detailed standard for the 76-77 GHz radar (ETSI EN 301 091). This standard specified the technical characteristics of the radar as well as the methods that must be used to test it. In 2002, the International Organization of Standardization (ISO) published standards (ISO 15622:2002 and 15623:2002) which specified performance requirements and test procedures for ACC systems.

In addition, government regulations also dictate the radar's allowed frequency of operation. Europe decided early on to use the 76-77 GHz band for vehicular radar. The FCC in the U.S.A. allocated two separate bands, the 46.7-46.9 GHz and the 76-77 GHz for such use. In Japan and Asia pacific, the two bands of 60-61 GHz and 76-77 GHz were allocated. The 76-77 GHz band is common to all existing regulation standards, and has thus become the de facto standard for ACC applications worldwide.

Most of the ACC radar manufacturers relied upon their experiences in building military radar to develop the technologies for ACC. The major effort in applying these technologies for automotive applications is in meeting the cost and size targets required of these sensors. All the 76-77 GHz ACC systems in production or development use either one or a combination of the following: (1) frequency modulated continuous wave (FM-CW), (2) frequency shift keying (FSK), or (3) pulse modulation schemes [13]-[19]. These techniques are the only ones specified by the ETSI standard. Of the three, FM-CW is the most commonly used technique. In addition, there is continuing research on using spread spectrum transceivers that have a combined use of inter-vehicle communications as well as radar ranging [20], [21].

FM-CW

The FM-CW radar, shown in Figure 2a, transmits a CW signal whose frequency is modulated as a function of time with a periodic waveform, such as a sawtooth waveform. Typically, the frequency deviation is on the order of 150-300 MHz with a period of approximately 1 ms. The signal reflected by the target will be delayed in time and demodulated in the radar receiver, along with the transmitted signal. The demodulation output is the intermediate frequency (IF) or beat frequency. For a moving target, the beat frequency is different for the positive and negative slopes of the modulation waveform, as shown in Figure 2b. The average of the beat frequencies determines the range to the target, and the difference contains the Doppler (or velocity) information.

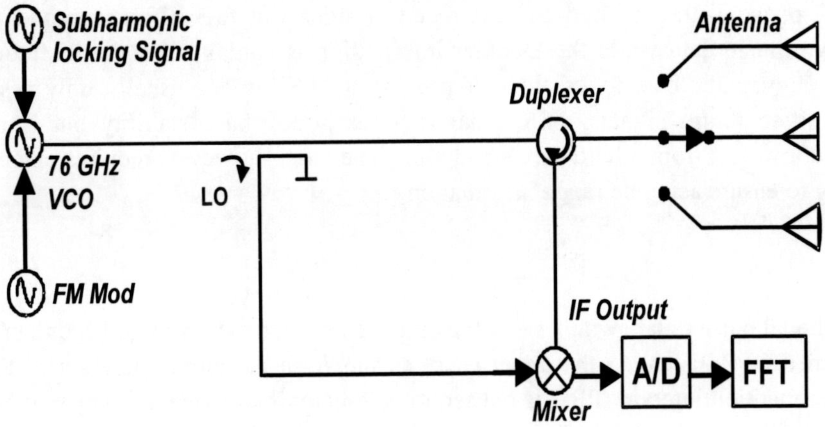

a)

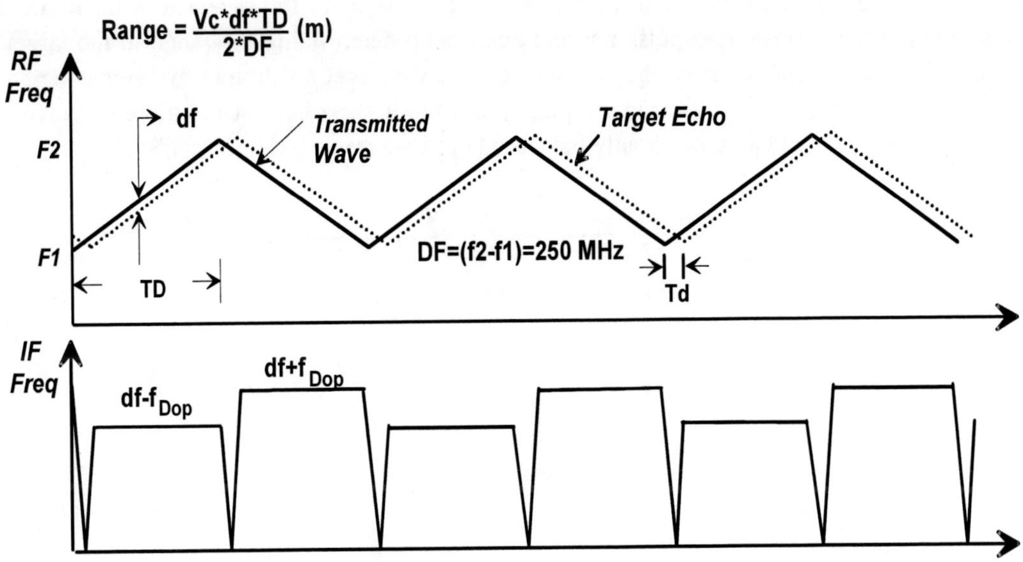

b)

Figure 2. Typical circuit diagram for FMCW Radar (a) with saw-tooth waveform and IF beat frequency for range and Doppler measurements (b).

Ensuring the linearity of the frequency sweep of the 76.5 GHz source is a critical step in FM-CW radar. A Gunn Oscillator is usually used with a frequency locked loop, although newer designs make use of MMIC-based Voltage Controlled Oscillator (VCO) technology. The post processing of the FM-CW signals is usually done using standard Fourier Transform DSP techniques providing simple and fast data analysis. FM-CW is the most commonly used technique for ACC because of its potentially high performance-to-cost ratio.

FSK

The FSK radar, shown in Figure 3a, is a narrow-band variant of the FM-CW radar. The radar transmits a CW signal whose frequency is typically changed in multiple steps of 150-1000 kHz every microsecond. The IF output is processed in a similar manner to the FM-CW radar. The phase difference between the received signals at the different frequency points contains the range data while the Doppler information is contained in the IF frequency, as shown in Figure 3b. Due to its type of processing, FSK radar usually only responds to Doppler-shifted return signals. FSK radar requires good phase stability but is otherwise simple and low cost from a hardware standpoint. The radar, however, requires extensive post processing to ensure accurate range information.

Pulse

Traditional pulse radar switches ON the output signal for a short period and then, with the output turned OFF, listens to the target echo, as shown in the circuit diagram of Figure 4a. The pulse repetition interval (PRI) is chosen to be greater than twice the propagation time of the transmitted pulse to the furthest potential target. The PRI may be staggered between subsequent pulses. The frequency of each transmitted pulse may also be changed, as shown in Figure 4b. The delayed echo from the target is demodulated in the receiver, with the time delay between the transmitted pulse and the pulse echo determining the range to the target. Velocity can be determined from the rate of change of the target position or by using coherent pulse-Doppler techniques. Pulse radar typically has the highest cost of the three techniques used for ACC radar and may eventually be phased out in favor of FM-CW or FSK.

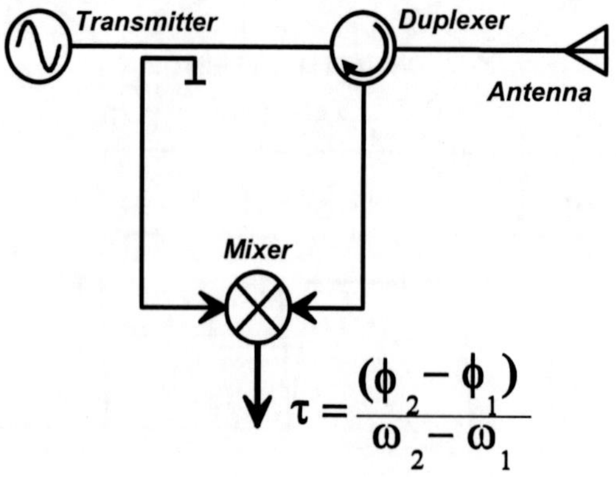

a)

Figure 3. Continued on next page.

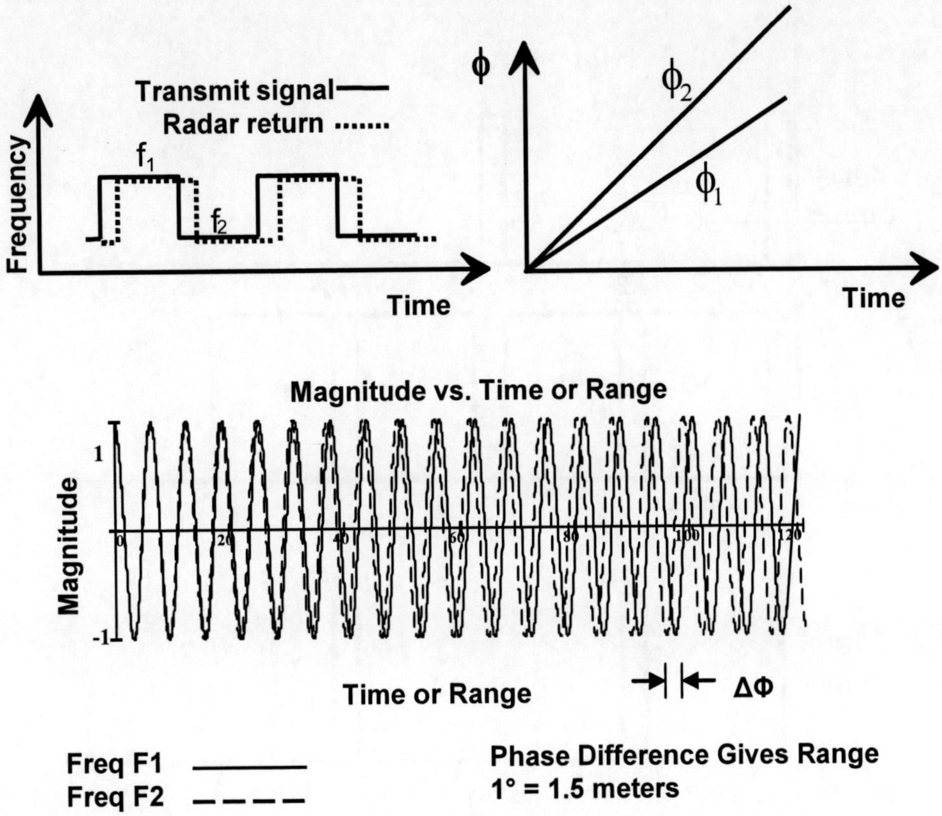

b)

Figure 3. Typical circuit diagram for FSK Radar (a) which uses phase offsets to measure range and the IF frequency to measure Doppler (b).

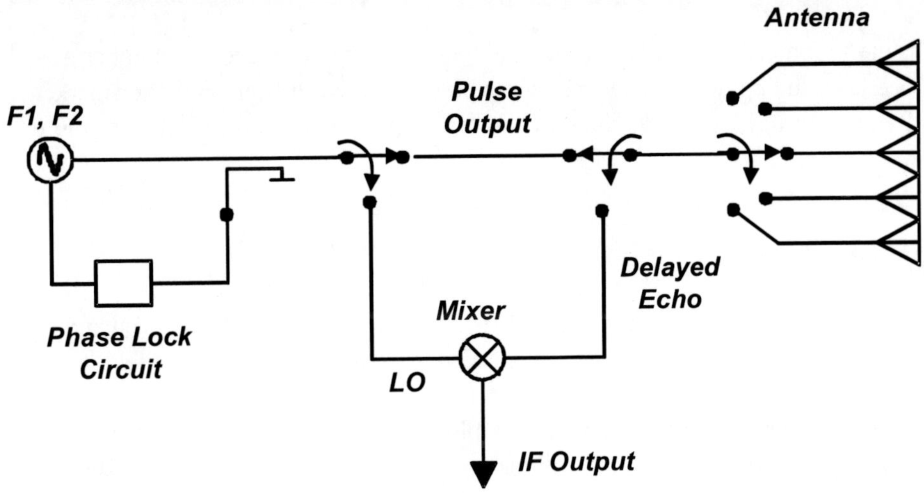

a)

Figure 4. Continued on next page.

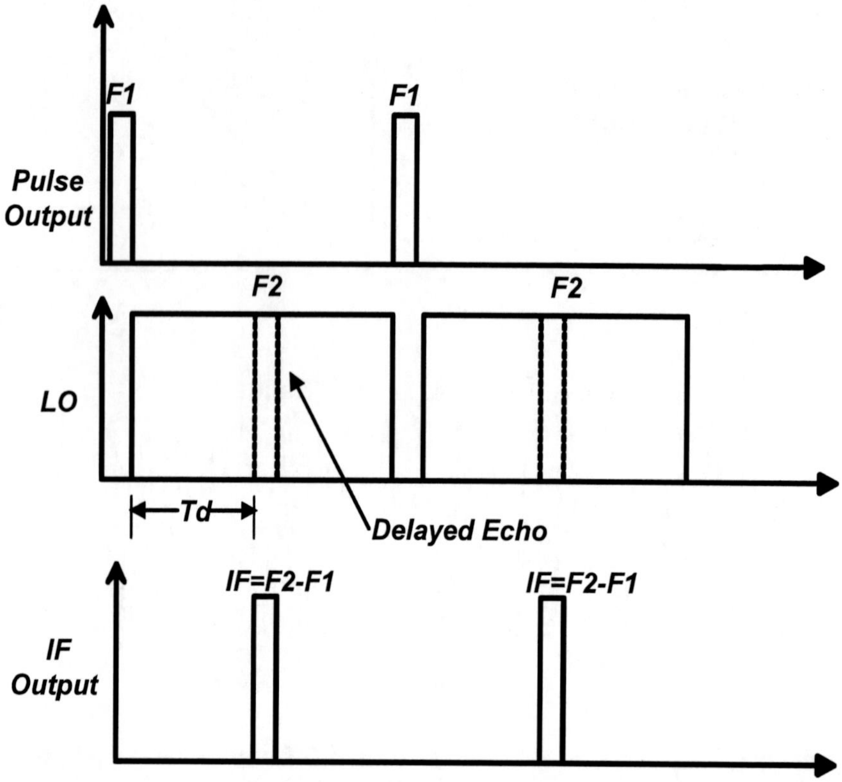

b)

Figure. 4. Typical Pulse Doppler Radar Modulation circuit (a) and pulse timing diagram (b).

3. ACC Radar Test Requirements

Due to the safety aspect of ACC radar, detailed testing is needed at a number of stages in the process, including research and development (R&D), production, and after-market service [33]. Below is a description of some of the test requirements and solutions at the different stages.

Component Level

As a first step, discrete devices and MMICs must be checked at high speed and on-wafer, if applicable. This testing is especially needed for MMICs with low yield to minimize potential rework at the module level. The main radar components that need testing are oscillators, frequency multipliers, mixers, amplifiers and switches. These tests can be carried out using standard microwave and millimeter wave instruments, including vector or scalar network analyzers, spectrum analyzers and power meters.

Sensor Functional Testing

During R&D and low-volume pre-production, full characterization of the radar is needed. With high-volume mass-production, it is expected that a subset of the full characterization testing will still be performed on each radar module being produced, and that a statistical sample of the modules will be fully characterized. Determining which subset of the testing will be required for mass-production depends on the yield and reliability performance of the radar modules currently in pre-production.

The transmitter portion of the radar can be tested by analyzing the transmitted signal in terms of power and spectral characteristics. Power measurements may include peak or average power, pulse timing/jitter, and duty cycle measurements. Spectral measurements include center frequency, modulation bandwidth, spurious content, and noise measurements. In addition, the antenna must be tested for beamwidth, directivity, gain, and sidelobe level. Equally important, the alignment of the antenna within the radar module housing must be checked. If the offset angle between the peak of the antenna beam (the beam-maximum) and the reference on the housing is not within tolerance, then compensation could be made either in software or during installation of the sensor on the automobile.

For testing the transmitted signal and antenna of the radar, standard spectrum analyzers and power meters can be used. Most radar modules have the antennas integrated with the millimeter-wave front-end and so this testing must be done with the radar radiating into free space. The most common setup is to place a calibrated test antenna facing the radar antenna and connect its output to the test equipment. A down-converter module is usually used to translate the radar signal at the test antenna to a lower frequency at which standard test equipment operates. For accurate measurements, especially for antenna patterns, the test antenna must be set up in the far field region of the radar and in an electromagnetically quiet area, such as an anechoic chamber. The distance to this far field region is a function of the size of the radar antenna. For the most common ACC radar, this distance is usually 1.5-7 m.

The best approach for testing the receiver of the radar is to have a known target reflect the transmitted signal back to the radar. For ACC systems, the ability of the radar to detect moving targets, with known radar cross-section (RCS), at ranges up to 150 m must be tested. One approach is to set up well-characterized targets, such as metallic spheres or trihedral reflectors, in an anechoic chamber at specified distances from the radar. In most R&D and production facilities, however, it is not feasible to set up a large chamber to test targets 150 m away. In addition, it is very difficult to test moving targets in such chambers. Road tests using actual automobiles may be conducted, but such testing is costly and may not provide the accuracy or consistency in the measurements.

A repeatable and accurate solution that allows this testing to be done in a confined space uses electrically simulated targets instead of physical targets. A target simulator receives the signal transmitted by the radar, delays it in time, modifies its amplitude and frequency, and transmits it back to the radar. When this modified signal is received at the radar, it will appear as if it had come from a target at a specific distance, with a specific speed, and of a specific size or RCS. The specifications of the radar and its performance limits can be fully characterized, and its parameters can be calibrated by using this technique. The receive and transmit antennas of the target simulator must be in the far field region of the radar, which could be as close as 1.5 m. Therefore, with a target simulator, a 150-meter moving target of variable RCS can be simulated in a confined space of only a few meters.

Traditionally, target simulation required the use of a rack and stack of microwave equipment and accessories, which was costly and impractical for the high-volume production lines of ACC radar. In 1997, dedicated target simulation equipment that works with the different types of ACC radar was developed by Anritsu Company [22]. This equipment was integrated, along with signal analysis equipment, into the radar assembly lines by several radar manufacturers enabling them to characterize and calibrate their radar sensors during production. In addition to using anechoic chambers for this type of testing, such as depicted in Figure 5, compact range technology was also developed to reduce the distance between the ACC radar and the target simulator even further [23].

To reduce the cost of the necessary equipment for full ACC radar testing, Anritsu Company further integrated, in 2001, the functions of signal analysis and target simulation into a single instrument [24]. This equipment, which is based on MMIC and Surface Acoustic Wave (SAW) delay-line technology, is also smaller and lighter than previous versions, allowing for easier bench top testing in R&D settings as well as more compact production line testing.

In addition to bench-top and radar production testing, on-vehicle system testing at the R&D stage is also required. At that stage, evaluation of different radar types and mounting locations is performed. Testing is also needed during the stages of installation, EMC verification, or servicing of the radar modules on the automobiles. During those stages, the test system may need to provide functional testing as well as aid in the positional and angular alignment of the radar. The functional testing could be performed using the RTS equipment described above. On-vehicle alignment is discussed in the next section.

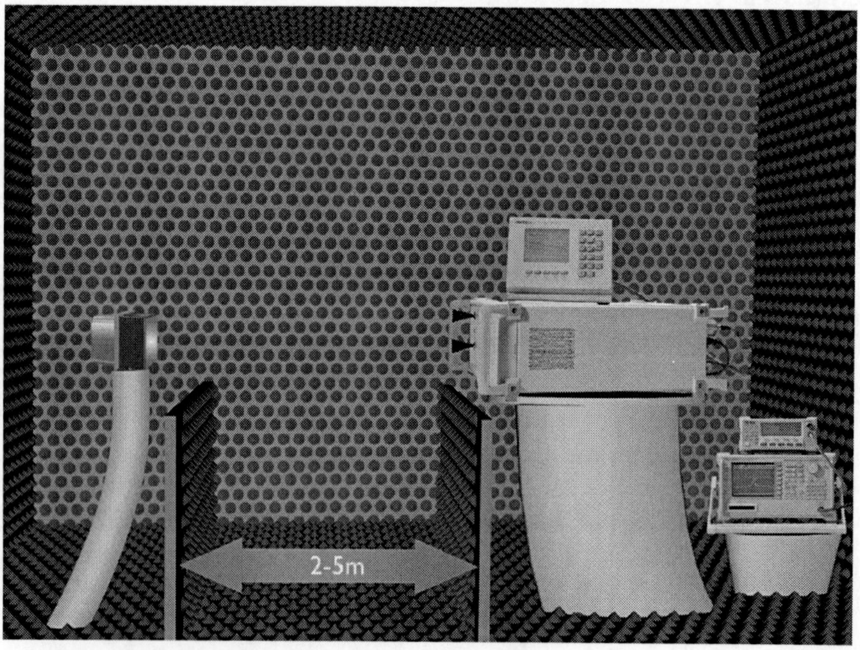

Figure 5. Setup of radar test equipment in an anechoic chamber for radar module testing in R&D or production environments. The test equipment consists of a Radar Test System (RTS), a power meter, and a spectrum analyzer. The radar module is shown physically aligned with the RTS antennas and located 2-5 m away.

Sensor Alignment on Vehicle

Vehicle manufacturers must perform positional and angular alignment of the sensor on the automobile to insure accurate system operation. In addition, they are responsible for instructing their service centers on how to service and install these radar systems in the field during regular maintenance or after an accident. Therefore, vehicle manufacturers must get involved in producing full test plans for the radar systems that they use.

Radar misalignment is potentially one of the most significant problems for these radar systems. To illustrate the need for alignment, consider the fact that an angular error of 2° can cause a displacement of greater than 4 m at a distance of 120 m ahead. This will cause the radar to report that a target is straight ahead, when in fact it is in the next lane.

In most cases, the radar must be installed on the vehicle such that the beam maximum of the radar receive-antenna is pointing in the direction of vehicle motion. This direction can be represented by the thrust vector of the automobile. The total offset angle or misalignment error is the sum of the misalignment between the radar housing-reference and the antenna beam, and between the radar housing and the vehicle thrust vector. These angular errors are shown in Figure 6.

There are three general methods of performing alignment on the vehicle, as illustrated below. These techniques can be used during installation of the radar during vehicle production as well as during after-market service.

Optical Mechanical Alignement

A laser is usually mounted on the radar housing and positioned to point perpendicular to the housing reference. A receiver is placed in front of the laser and the orientation of the radar mounted on the vehicle is adjusted until the laser beam is pointing in the direction of the thrust vector. Alternatively, a mirror is mounted on the radar housing and a laser is pointed towards it. When the radar is aligned, the laser beam would reflect upon itself [25]. This technique is simple and low cost but is limited in accuracy. First, there is a potentially high angular error when mechanically mounting the laser or mirror to the radar housing. Second, this technique only compensates for the mechanical offset angle (α) and not for the squint angle (θ) (see Figure 6). To address this issue, the radar manufacturers could imbed the value of the radar's squint angle (which they measure during module testing) within the module itself. The vehicle manufacturers could then use that information to compensate for the squint angle during assembly.

Using Internal Angle Measurements

For radar systems that can measure angle of arrival, placing a physical or simulated target in the direction of the thrust vector allows the radar to measure the angle of the signal reflected from that target. Using that information, the orientation of the radar sensor mounted on the vehicle is adjusted until the internal angle (total offset angle, Φ) is at 0°. Because the radar is radiating/receiving in the presence of the automobile, any RF alignment errors associated with module housing and radome cover distortions, or distortions caused by secondary surfaces installed in front of the radar, are accounted for. The accuracy of this technique is limited by the accuracy of the radar's internal angle measurement as well as the mechanical alignment of the centerline of the reflector to the center of the radar antenna. A planar reflector with a sharp angular response, such as a flat plate, may reduce or eliminate the potential error due to any misalignment between its centerline and the radar antenna center point [26]. Alternatively, the simulated targets of an RTS can be used for aligning such a radar sensor to the vehicle thrust vector [27]. The advantage of using an RTS instead of a physical reflector is that functional testing of the radar could be performed simultaneously with alignment. These functional tests could include transmit power and frequency measurements as well as target measuring accuracy. In addition, using a simulated target with a far-range and high-speed setting allows the radar to perform range or Doppler gating on the received signal. For example, if the RTS is set to a target with a range of 120 m and a speed of 120 km/h, then the radar can focus on this target only, ignoring all other undesired reflections coming from clutter around the vehicle. When using physical targets instead of an RTS, it may be difficult to distinguish between the desired and undesired signal reflections. Therefore, in that case, an anechoic-type chamber may be required to remove the clutter and all unwanted signal reflections that may introduce measurement errors.

RF Alignement

Most ACC radar systems have some angle measurement capability in the azimuth plane, but not in the elevation plane. When no internal angle measurement capability is available, or when the accuracy of this internal angle measurement is not sufficient, RF alignment techniques can be used for measuring the beam-maximum of the transmit antenna. One classical technique measures the power of the transmit beam and determines its maximum by adjusting the radar angle to get a peak power reading, or to determine symmetry points along the beam pattern (such as the 3 dB beamwidth) [26], [28], [29]. This technique is simple in that it only requires one millimeter-wave receiver. The accuracy of this technique, however, is limited by the beamwidth of the radar transmit-antenna. Depending on the sensitivity of the receiver, the accuracy for typical automotive radar may be only ±1°, which may be unacceptable. An alternative approach is to use a passive interferometer system. Such a system replaces the one receiver by four receivers, two in the azimuth plane and two in the elevation plane. These receivers are located a fixed distance apart from each other with the radar positioned at the centerline of each antenna pair. Figure 7 shows the configuration of the two antennas in any single plane. The separation distance, D, between the antennas in each pair is chosen so that their beams intersect at the 6 dB or 10 dB points (i.e. the points where the signal from the antenna beam is 6 dB or 10 dB below the peak signal). Consequently, the ratio between the received signals at the two antennas in a plane produces a sharp null in the direction of the beam maximum, as illustrated in Figure 7. A sharp null usually requires less sensitive receivers to measure accurately than a broad peak. Using four receivers, Anritsu developed an RF radar alignment system (RAS) with a demonstrated accuracy of ±0.25° for both the azimuth and elevation angles [30]. The separation distance, R, between the radar and RAS is chosen to be the shortest possible distance that maintains the symmetry of the amplitude of the radar main beam. This distance is usually much shorter than the far-field distance of the radar and could be as short as 0.75 m for typical ACC radar. In addition to aligning the radar, the RAS, which inherently measures the received signal power, can test the functionality of the radar by measuring its average radiated power.

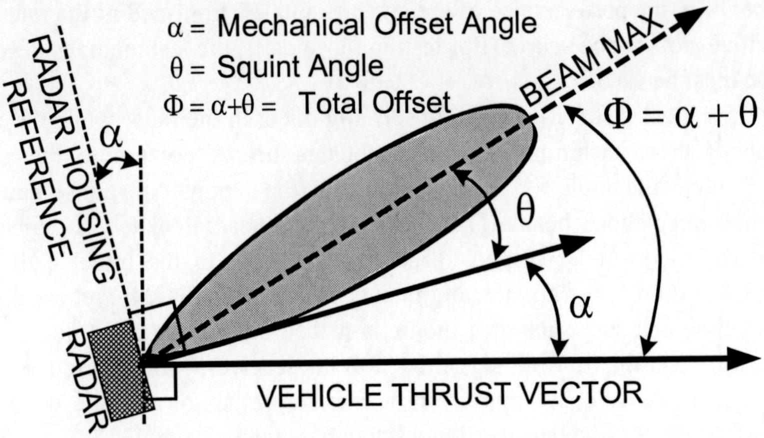

Figure 6. Possible angular errors when installing radar on vehicle.

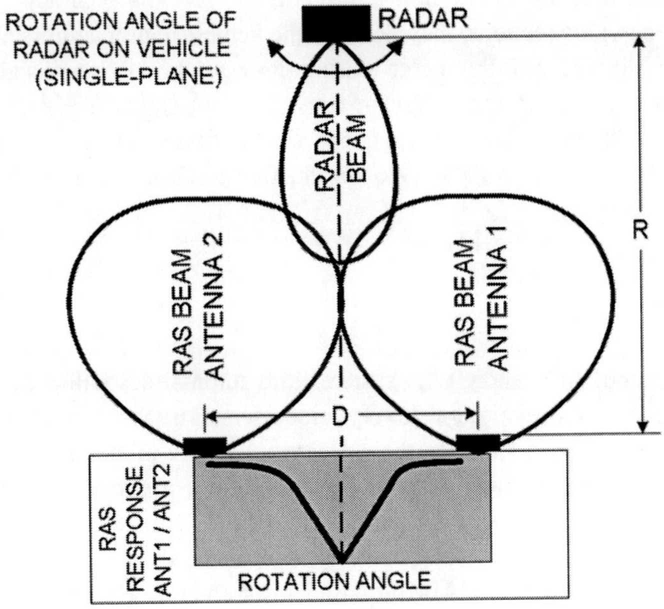

Figure 7. Radar Alignment System (RAS) showing two of the four-interferometer receivers. The radar-transmitted beam is detected by each antenna and the ratio of the response is used to calculate the offset angle of the radar (azimuth and elevation). A typical response from the RAS is a sharp null at the radar boresight angle.

Built-in Testing and Alignment

Most radar systems contain built-in test capabilities. For example, injecting an IF signal into the IF radar loop will simulate the presence of a target and will allow the radar functionality to be tested during development, production, or even during operation on a vehicle. Such a test can be useful in detecting DC power or DSP processing problems and can be made to run at continuous intervals while the radar is in operation. However, such a built-in test will not check the performance of the antenna and RF front-end of the radar, which are the most sensitive parts of the sensor. For testing those parts, the test methods described in the sections above must be used.

Several techniques exist for performing self-alignment in the radar while the vehicle is in motion. Some of these techniques use two antenna beams pointed to the ground in a symmetrical fashion and look for the difference in the Doppler response caused by any misalignment between those beams [13]. Other techniques calculate the trajectory of the vehicle ahead (using the radar response) and compare that to the travel path of the host vehicle, in order to correct for any misalignments of the radar [31]. In yet another proposed method, a reflecting element, such as a diode, is placed directly on the radome cover of the radar in order to reflect the transmit signal back to the receiver. Modulating the diode causes the reflected signal to have a Doppler shift. Such a reflection can be used to test the functionality of the radar as well as to align a scanning antenna beam [32].

Two advantages of using built-in testing are the ability to test the radar without any external test equipment, and the ability to continuously test and align the radar while it is in operation on the vehicle. Although built-in tests do not have the accuracy of the test methods described in the pervious sections, they do offer the vehicle manufacturers a simple method to monitor the performance of the radar during its operation. One disadvantage of these methods, however, is that in most built-in tests the front-end of the radar is not tested. In addition, built-in tests do not use independent and calibrated test equipment, and therefore, they cannot be used for any regulatory or production testing, where measurement accuracy and traceability are critical.

4. Conclusion

The integration of ACC and CW/A systems into automobiles will continue at a fast pace. Originally designed exclusively for luxury models and trucks, the popularity and safety benefits of these systems will lead to their design into more moderately priced automobiles. Innovative modulation and antenna scanning techniques are being used to design radar sensors that meet the size and performance specification targets. What will accelerate the expanded use of these sensors on automobiles is lowering their cost. Low cost designs that utilize printed antennas, flip-chip MMIC based front-ends, low-cost oscillator designs, and automated assembly and packaging are being implemented. Lowering the cost of testing is another essential step for reducing the overall system cost. The use of target simulation systems allows accurate and repeatable testing of the critical radar module specifications in a small confined space, and in a short period of time. In addition to lowering production costs, it is important to keep final installation and after-market-service costs low. RF alignment test systems ensure that the radar is properly mounted on a vehicle during installation, reducing costly off-line adjustments and after-market warranty repair. As a complement to independent testing, the use of built-in tests will further insure that the radar performance continues to meet the minimum requirements while the radar is installed and operational on a vehicle.

References

[1] Belohoubek, E. F. (1982). Radar control for automotive collision mitigation and headway spacing. *IEEE Trans. Vehicular Technology*, Vol. 31 (No. 2), pp. 89-99.

[2] Brus, E. (1987). Vehicular Radar: The ultimate aid for defensive driving? *Microwaves and RF*, *Sept,* pp. 53-58.

[3] Woll, J.D. (1995). VORAD collision warning Radar. *IEEE International Radar Conference Digest*, pp. 369-372.

[4] Polychronopoulos, A. (2006). Research activities on Sensor Data Fusion by ProFusion2 project, *13th ITS World Congress, London, UK.*

[5] Park, S. B. (2005). ProFusion 1and2 - Sensor data fusion for reliable environment sensing. *IV'05 - The Intelligent Vehicles symposium, Las Vegas, USA.*

[6] Wegner, J. (1998). Automotive mm-wave radar: status and trends in system design and technology. *IEE Colloquium on Automotive Radar and Navigation Techniques*, *Ref. No. 1998/230*, pp. 1-7.

[7] Dixit, R., and Rafaelli, L. (1997) Radar requirements and architecture trades for automotive applications. *IEEE MTT-S Digest*, vol.3, pp. 1253 -1256

[8] Automotive radars and prospective circuits/antenna technologies from "car collision avoidance" to "autonomous driving". (2002). *IEEE Microwave Symposium Workshop Notes, workshop No. WMA*

[9] Zelubowski, S. (1994). Low cost antenna alternatives for automotive radars. *Microwave Journal*, pp. 55-62

[10] Schmidt, R. (1986). Multiple emitter location and signal parameter estimation. *IEEE Trans. Antennas and Propag.*, Vol. AP-34 (No. 3), pp. 276-280.

[11] Roy, R. and Kailath, T. (1989). ESPRIT – Estimation of signal parameters via rotational invariance techniques. *IEEE Trans. on Acoustics. Speech, and Signal processing,* Vol. 37 (No. 7), pp. 984-995.

[12] Asano, Y., Ohshima, S., Harada, T., Ogawa, M., and Nishikawa, K. (2001). Proposal of millimeter-wave holographic radar with antenna switching. *IEEE MTT-S Digest*, pp. 1111-1114.

[13] Russel, M.E., Crain, A., Curran, A., Campbell, R. A., Drubin, C.A., and Miccioli, W.F. (1997). Millimeter-wave radar sensor for automotive intelligent cruise control (ICC). *IEEE Trans. Microw. Theor Techn., MTT-45*, pp. 2444-2453.

[14] Olbrich, H., Beez, T., Lucas, B., Mayer, H., and Winter, K. (1998). A small, light radar sensor and control unit for adaptive cruise control. *SAE Proceedings,* Vol 980607, pp. 25-30.

[15] Menzel, W., Pilz, D., and Leberer, R. (1999). A 77GHz FM/CW radar frontend with a low-profile, low-loss printed antenna, *IEEE MTT-S Digest*, pp. 1485-1488.

[16] Camiade, M., Domnesque, D., Alleaume, P.F., Mallet, A., Pons, D., and Dämbkes, H. (1999). Full MMIC millimeter-wave front-end for a 76.5GHz adaptive cruise control car radar. *IEEE MTT-S Digest*, pp. 1489-1492.

[17] Matsumura, T., Hirao, M., Sato, T., Saryo, T., Ohmuro, M., Mizutani, H., and Sida, N. (1999). 76GHz FM-CW radar with electrically scanned seven beams for automotive applications. *6th World Congress on ITS*, *Toronto Canada.*

[18] Fujimura, K., Hitotsuya, M., Yamano, S., and Higashida, H. (1999). 76GHz automotive radar for ACC. *6th World Congress on ITS, Toronto Canada.*

[19] Gresham, I., Jain, N., Budka, T., Alexanian, A., Kinayman, N., Ziegner, B., Brown, S., and Staecker, P. (2001). A compact manufacturable 76-77-GHz radar module for commercial ACC applications. *IEEE Trans. Microw. Theor Techn, MTT-49*, pp. 44-58.

[20] Mizutani, K., and Kohno, R. (2001). Inter-vehicle spread spectrum communication and ranging system with concatenated EOE sequence. *IEEE Trans. Intelligent Transportation Systems*, Vol. 2 (No. 4), pp. 180-191.

[21] Nishikawa, S., and Endo, H. (1999). Application of millimeter-wave sensors in ITS. *Furukawa Review*, No. 18, pp. 1-5.

[22] Abou-Jaoude, R., and Grace, M. (2000). Test systems for automotive radar, *IEEE Vehicular Technology Conference (VTC2000-Spring), Tokyo, Japan.*

[23] Flacke, J., and Boumans, M. (1999). Compact range measurement system for automotive radars. *Proceedings of AMTA '99, 21st Annual Meeting and Symposium*, pp. 50-55.

[24] Abou-Jaoude, R., Grace, M., Geller, D., Noujeim, K., Bradley, D., and Oldfield, W. (2001). Low cost 76GHz radar target simulator and test system. *Microwave Engineering Europe*, June, pp. 25-30.

[25] Säger, P., Landsiedel, T., and Neugärtner, J. (2002). *Method and apparatus for aligning a beam path for a beam emitting sensor,* United States Patent, No. 6418775.

[26] Schirmer, G., Adolph, D., Winter, K., Mayer, H., Lucas, B., Beez, T., Winner, H., and Olbrich, H. (2002). *Method and device for adjusting a distance sensor*. United States Patent, No. 6363619.

[27] Grace, M., and Bradley, D. (2001). *Automobile radar antenna alignment system using transponder and lasers*. United States Patent, No. 6329952.

[28] Skolnik, M. L. (1980). *Introduction to Radar Systems* (2nd ed), McGraw-Hill.

[29] Henrio, J.F., and Artis, J.P. (2002). *Method and device for the alignment of an automobile radar*. United States Patent, No. 6437731

[30] Grace, M., Abou-Jaoude, R., Noujeim, K., and Bradley, D. (2000). 76GHz radar antenna alignment system. *Proceedings of the 30th European Microwave Conference, Paris, France,* vol. 3, pp. 175-178.

[31] Alland, S., and Searcy, J. (1999). Automatic sensor azimuth alignment. *United States Patent,* No. 5964822.

[32] Thordarson, G., and Båck, I. (2002). *Car radar testing*. United States Patent, No. 6392586.

[33] Aboujaoude, R. (2003). ACC Radar Sensor Technology, Test Requirements, and Test Solutions. *IEEE Trans. Intellient Transportation Systems*, Vol 4 (No 3), pp. 115-122.

In: Wireless Technologies in Intelligent Transportation… ISBN: 978-1-60741-588-6
Editors: Ming-Tuo Zhou et al, pp. 21-49

Chapter 2

RADIO CHANNEL MODELING FOR VEHICLE-TO-VEHICLE/ROAD COMMUNICATIONS

David W. Matolak*

School of Electrical Engineering and Computer Science
Ohio University, Athens, OH, 45701

Abstract

New applications for wireless communications are emerging with increasing frequency. For Intelligent Transportation Systems (ITSs), this will include applications that require communications between vehicles, and communications between vehicles and roadside entities. In this chapter we address terrestrial ITS applications, specifically automotive applications.

Any communication system requires the following components: message source, transmitter, channel, receiver, and message destination. For wireless communication, the channel is often uncontrolled, or only very loosely controlled, by the system designer and user. Hence the channel can play an important role in communication reliability. In the case of vehicular communications for ITS, the channel will often be dynamic, lossy, and distorting. Design of effective communication systems for these vehicle-to-vehicle (V2V) and vehicle-to-roadside (V2R) channels is thus challenging, and requires good models. This chapter describes such models.

Most envisioned V2V and V2R applications will be relatively short range—on the order of a few hundred meters to a kilometer. In some cases propagation path loss may limit system performance, but we do not focus on this form of "large scale" channel effect, which in principle can be overcome with larger transmit power. Although obstruction by buildings ("shadowing") and other large obstacles and vehicles may occur, these effects are often of secondary importance in comparison to small scale fading, caused by multipath propagation. This chapter provides an overview of channel characterization, then describes models for the V2V and V2R channels. These models primarily address small scale fading. Treatment of "medium scale" effects is also noted, but is of lesser depth. Our coverage focuses on statistical

* E-mail matolak@ohiou.edu. Tel 740 593 1241, Fax 740 593 0007. Address correspondence to: David W. Matolak, Ph.D., School of Electrical Engineering and Computer Science, Ohio University, 322E Stocker Center, Athens, Ohio, 45701,

models for small scale fading. A brief discussion of deterministic models is included for comparison and completeness.

The V2V (and to a lesser degree, the V2R) channel can be very dynamic, with time variation rates up to double those of conventional, e.g., cellular radio, channels. With the low antenna heights of V2V and V2R systems, radio line of sight (LOS) is more frequently and more thoroughly obstructed. Because of these effects, the V2V and V2R channels will generally exhibit more severe fading than conventional channels. Due to the potentially rapid time variation, V2V and V2R channels will also be best modeled as statistically *non*-stationary.

Abbreviations

BER	Bit Error Ratio
CIR	Channel Impulse Response
IEEE	Institute of Electrical and Electronics Engineers
ITS	Intelligent Transportation System(s)
LOS	Line of Sight
MAC	Medium Access Control
PDP	Power Delay Profile
PHY	Physical Layer
RMS-DS	Root-Mean-Square Delay Spread
Rx	Receiver
Tx	Transmitter
US	Uncorrelated Scattering
WLAN	Wireless Local Area Network
WSS	Wide-Sense Stationary
V2R	Vehicle to Roadside
V2V	Vehicle to Vehicle

1. Introduction

This chapter describes characteristics of the mobile physical wireless channel for vehicle-to-vehicle (V2V) and vehicle-to-roadside (V2R) settings. As a relatively new area of study, the abbreviations VTV and VTR are also used for V2V and V2R, respectively. We employ the latter. The defining local environments for these cases are first described. A brief motivation for accurate channel characterization and the channel's effect upon communication system performance is also provided. Distinguishing features of the V2V and V2R channels are identified, and comparisons are drawn between these channels and more traditional ones such as those for cellular radio. Our focus is on statistical models for small scale fading, but a brief description of large and medium scale fading and of work on deterministic small scale fading modeling is also given. Basic statistical channel characterization functions are described, in the delay, frequency, Doppler, and time domains. We review some of the basic theoretical findings and models for V2V/R channels, and also review some recent experimental work that has led to development of empirical models. Both the theoretical and experimental models are cast in terms of the statistical channel characterization functions. The newest of these empirical models address channel characteristics not typically accounted for; this includes correlated scattering, and statistical non-stationarity. In the remainder of this section we define these channels and highlight the importance of their characterization.

1.1. Defining the V2V and V2R Channels

The V2V and V2R channels are almost self explanatory. In this section we provide some basic definitions as background for the remainder of the chapter. Our definitions can not be all-encompassing (e.g., vehicle-to-vehicle communication between Martian rovers is not addressed), but are intended to focus on what are viewed as major new applications, primarily in the area of ITS [1]. Some of our discussion will be applicable to a range of environments.

1.2. The V2V Channel

The V2V channel represents the wireless channel that exists between two terrestrial vehicles, such as two automobiles or other types of vehicles such as trucks, vans, buses, etc. In many settings, networks of vehicles may be communicating with one another. The most common environments for these communications will be on established roadways in cities, suburbs, and on highways and throughways through and between such locales. We do not consider "off-road" settings explicitly, but note that in open off-road areas, the V2V channel will exhibit characteristics similar to those of highways in open areas. The "off-road" forest or mountainous area will exhibit differences in the form of greater attenuation and obstruction. We leave these unique V2V settings for future work.

For an urban area, Figure 1 depicts a plan view of a V2V setting. The various lines indicate conceptual radio propagation paths, and will be described subsequently. One transmitter (Tx) and one receiver (Rx) are indicated, but in general all vehicles will have both Tx and Rx. In suburban areas, the density of vehicles would generally be smaller. On highways, vehicle density is highly time-dependent, particularly near cities, where peak vehicle density values occur during morning and evening rush hours, with substantially smaller densities during the remainder of the day.

In any of these settings the V2V channel is the wireless link between the Tx and Rx of any two vehicles. The channel may include a line-of-sight (LOS) path, or it may be an obstructed, or non-LOS (NLOS) link. Either or both vehicles may be in motion. We define the channel as the complete set of communication system parameters for all electromagnetic wave components that travel from transmitter to receiver in the frequency band of interest. The "communication system parameters" are defined subsequently.

1.3. The V2R Channel

The V2R channel is the channel between a vehicle and another transmitter/receiver (transceiver) located on the roadside.

A V2R connection between a lamp post transceiver and a vehicular Rx is indicated in Figure 1. In the V2R case, most roadside transceivers will be mounted above automobile height, on poles, buildings, overpasses, etc. The V2R channel has only one platform (Tx or

Rx) in motion. The V2R channel definition follows that of the V2V channel, specifically: the channel is the complete set of communication system parameters for all electromagnetic wave components that travel from transmitter to receiver in the frequency band of interest. With only one platform in motion, the V2R channel dynamics are generally not as rapid as those of the V2V channel.

Figure 1. Diagrammatic depiction of V2V and V2R channels, illustrating multipath propagation.

1.4. V2V/V2R Communication Frequency Bands, and the DSRC Standard

In principle, V2V communication[1] could take place at any frequency band that is available and convenient for use. In practice, based upon some desired system characteristics (e.g., range, and cost of devices for the given band), and upon external factors (e.g., regulatory constraints), limitations arise, which lead system designs to specific bands. Given the rapid growth of wireless communications in the past decade, many desirable VHF and UHF frequency bands are dedicated to other services and are hence unavailable for V2V use.

The history of V2V communication is interwoven with that of land mobile radio, and we do not address that here other than to mention a few examples [2]. There have been some past V2V communication systems that are ad hoc, e.g., the "citizen's band" (CB) radios made popular for the trucking industry in the USA. These employ a small (~440 kHz) band around 27 MHz, and these communications are almost exclusively voice, for personal use. Similarly,

[1] The term "V2V communication" is used henceforth for brevity to include both V2V and V2R communication; when discussing channels, the two are distinguished.

there are multiple military systems that can operate in V2V modes, e.g., the US Dept. of Defense's "Speakeasy" radios [3]. These radios were designed to operate in several bands. Herein we focus on more contemporary V2V bands aimed for ITS.

One band that is being planned for use for V2V communication is the 5.9 GHz band. This is the band that is the focus of the Dedicated Short Range Communication (DSRC) standard [4], originated by the US Dept. of Transportation. This standard has moved to the IEEE under the 802.11p group [5], and is also known as Wireless Access for Vehicular Environments (WAVE). Another band that is currently being considered for V2V communication is the 700 MHz public safety band [6]. Details on this band for V2V applications are less well developed than for DSRC.

For DSRC, as explicitly indicated, the expected range is short—less than 1 km. The available bandwidth is 75 MHz, from 5.85-5.925 GHz. The band is part of the "unlicensed national information infrastructure" (UNII) in the USA, and as such is a shared band, but transportation use is deemed primary by the Federal Communications Commission (FCC). This band is divided into seven, 10-MHz channels, one of which is reserved for priority messages (so called "high-availability, low-latency" or HALL channel) [7]. This channel plan initially reserves two blocks of two concatenated 10-MHz channels, i.e., two 20-MHz blocks within this 75-MHz band are reserved. These reserved blocks could be used with the 20 MHz bandwidth, or subdivided in the future into smaller bands. Based upon this, and upon other emerging wireless technologies that may be used in V2V communications [8], we consider channel bandwidths of 10 MHz or less.

Since our intent here is description of the V2V channel, we address transmission schemes for use on the V2V channel mainly for illustration. The DSRC standard is a modified version of the IEEE 802.11 standard [9] for wireless local area networks (WLANs). This standard specifies the transmission scheme at the lowest two layers of the communications protocol stack, the physical (PHY) and the medium access control (MAC). Modulation is orthogonal frequency division multiplexing (OFDM) with various signaling alphabets, and multiple user access is controlled by time-division [10]. More detail can be found in [4], [5], [9]. In the case of public safety systems, the "P34" system standard is also an OFDM system [11].

1.5. V2V/V2R Channels vs. Traditional Mobile Channels

Traditional mobile channels are, in the terrestrial case, often referred to as land mobile radio channels [12]. In this setting, one end of the communication link is at a non-mobile "base station" (BS). For cellular radio and most other land mobile systems such as those used by public safety organizations, base stations are carefully engineered in terms of location and facilities. They often include backup power systems, provisions for multiple antennas, and the availability of powerful signal processing. In terms of physical features, a prominent feature is the presence of an elevated antenna, often on a tower, possibly a tower on a hilltop. These towers can be tens of meters above all surrounding obstacles. This is one feature that distinguishes traditional mobile channels from the V2V channel, where both Tx and Rx antennas are attached to vehicles, with heights of at most a few meters. The V2R channel does provide an elevated antenna for one end of the link, but these heights are expected to be less than ten meters, and mostly well below surrounding building heights, particularly in urban areas. In the V2V case, there may be significant scatterers around *both* Tx and Rx, not only around one or the other, as in the traditional land mobile case.

Referring to Figure 1, in the V2V channel, the primary propagation obstacles may be vehicles themselves, but signals can also be obstructed by buildings or other civil structures, terrain, or groves of plants. Because of such obstructions, and the limitations on transmitter power levels, link distances in such urban areas are anticipated to be short, from a few to a few tens of meters. As the environment "opens up" to suburban and rural/highway settings, obstacles become less dense and achievable link distances can also increase. Yet even in these cases, link distances are not planned to be as large as those in cellular radio, which may reach several tens of kilometers.

Another difference between traditional land mobile channels and the V2V channel is that in the V2V case—but *not* the V2R case—both Tx and Rx may be in motion. Thus rates of V2V channel time variation can be as much as double that of the traditional case. Because of this more rapid time variation, traditional statistical models that assume wide-sense stationarity (WSS)—the invariance of channel statistics over some moderate time period—may not be applicable, or will only be applicable for a shorter duration than in the traditional land mobile case.

For the purpose of characterizing V2V and V2R channels, it is convenient to use multiple channel classes, as is commonly done in traditional land mobile radio. In the cellular case, where each class aims to represent a particular type of physical situation, one typically sees the rural, suburban, and urban classes. We propose several V2V classes subsequently. Some characterizations (within or outside such classes) also explicitly identify the presence of an LOS component, and divide into LOS and NLOS cases.

1.6. Importance of Channel Modeling

Whatever transmission[2] technique or standard is employed, wireless channel knowledge is essential to the design and performance of any V2V communication system [13]. Specifically, *mathematical* channel characterization results are critical for all communication system physical layer waveform design and analysis. One example way in which channel models are used is in comparison ("tradeoff") studies that evaluate contending transmission schemes. Waveform features such as bandwidths and packet durations can be optimized, for example, using good knowledge of the channel, and various transmission schemes can be compared on an equal basis with a common model.

Good channel models can also provide some estimates of expected system performance (bit error ratio, latency, etc.) for any waveform used across the channel. Appropriate design of measures to counteract channel effects also requires accurate channel knowledge. Such measures include the use of equalizers, or multiple diversity antennas. Whether used in analysis or computer simulations, the channel models are in effect used as blocks in a cascade of models that includes the other components in a wireless communication system. Simulations are a natural way to use wholly empirical models (those that "re-play" stored measured channel data).

Modern communication systems are highly adaptive. Yet even with such an adaptive (or, re-configurable) communication system, impairments caused by the channel can severely degrade performance if they are not taken into account. Such degradations have been widely studied, e.g., [14], [15]. Two well-known results of inadequately accounting for channel characteristics include

(1) a bit error probability (or bit error ratio, BER) "floor," in which error probability reaches a lower limit regardless of received power level; and
(2) a large latency, which in the case of some protocols would translate to a link outage.

One potential cause of a large latency is channel fading that causes multiple packet errors and forces the system to employ re-transmissions. These retransmissions also reduce the achievable throughput, and this can significantly degrade both objective and subjective performance for many applications. Thus, channel fading characteristics should be modeled as accurately as possible, since they affect both PHY (BER) and protocol performance. In particular, for any V2V/V2R applications involving safety, channel effects must be carefully considered and the system designed to ameliorate or properly counteract them.

2. Statistical Channel Characteristics

In this section we describe important characteristics of wireless channels. Beginning with basic concepts applicable to all wireless channels, we define important terms and functions used to characterize and analyze these channels. Some specification to V2V/R channels is then done, to introduce the existing and new V2V/R channel models subsequently presented.

2.1. Basics

For communication systems purposes, we generally model the wireless channel as a linear, time-varying filter. This means that the channel is completely described by the channel impulse response (CIR) $h(\tau)$, or equivalently, the Fourier transform of this, the channel transfer function (CTF) $H(f)$. The variable τ is the delay variable, commonly denoted t in the study of time-invariant (TI) systems, but necessarily distinct from t in time-varying systems. The variable f is frequency in Hz, and the transform relationship is

$$H(f) = \int h(\tau) e^{-j2\pi f\tau} d\tau \,. \, (h(\tau) \leftrightarrow H(f)) \tag{1}$$

The inverse transform relationship is similar to (1): it exchanges $h(\tau)$ and $H(f)$, replaces $-j$ with j, and integration[3] is with respect to f.

In the case with time variation, we generalize h and H to be $h(\tau,t)$ and $H(f,t)$. Here $h(\tau,t)$ is the channel output at time t due to an impulse input at time $t-\tau$. For TI cases, it can be shown that $h(\tau,t)$ reduces to $h(\tau)$ [16]. Note that the notion of linearity relies on the rate of

[2] We use the terms "transmission technique" or "transmission scheme" to collectively denote both transmission *and* reception schemes.

[3] Unless otherwise noted, all integrals are from $-\infty$ to ∞.

channel time variation being slow enough to allow the usual convolution relation between input and output signals. Specifically, if the input signal is $v(t)$ and the output $y(t)$, we have

$$y(t) = \int v(t-\tau)h(\tau,t)d\tau = \int v(\tau)h(t-\tau,t)d\tau \tag{2}$$

For this relation to be valid for digital modulations, $h(\tau,t)$ must remain essentially constant during at least one symbol interval T_s of $v(t)$. This holds true for terrestrial channels with most symbol rates R_s (=1/T_s) of current interest. Also, since physical channels must be causal, $h(\tau,t)$=0 for τ<0. Similarly, physical channels have finite-duration CIRs, so if this duration is denoted T_h, then (2) becomes

$$y(t) = \int_0^{T_h} v(t-\tau)h(\tau,t)d\tau \tag{3}$$

Note also that the CIR $h(\tau,t)$ we refer to is typically complex—it is the "complex envelope," or "lowpass equivalent" response, from which the actual bandpass channel response $h_B(\tau,t)$ is obtained via the formula $h_B(\tau,t)=2Re\{h(\tau,t)e^{j\omega_c t}\}$, with $Re(x)$ denoting the real part of x, and $\omega_c=2\pi f_c$, with f_c the carrier, or "center" frequency [14]. Most often for simulation and analysis, the complex envelope is used, and unless otherwise noted, our discussion in this chapter employs the complex envelope.

For a distortionless channel, the CIR is

$$h(\tau,t) = A(t)\delta(\tau-\tau_0(t)), \tag{4}$$

where $\delta(t)$ is the Dirac delta, and $A(t)$ can be complex, i.e., $A(t)=\alpha(t)exp[j\theta(t)]$, $\alpha(t)=|A(t)|$, $\theta(t)=\angle A(t)$. This yields the transfer function

$$H(f,t) = A(t)e^{-j2\pi f\tau_0(t)} = \alpha(t)e^{-j[2\pi f\tau_0(t)-\theta(t)]}. \tag{5}$$

The distortionless condition means the channel only scales and delays the time domain signal, and this is equivalent to a constant gain and a linear phase shift across the signal spectrum. Tacit again is that variations of $A(t)$ and $\tau_0(t)$ must be *slow* with respect to the rate of input signal variation. For a TI case, the distortionless channel relationships become

$$h(\tau,t) = h(\tau) = A\delta(\tau-\tau_0) \leftrightarrow H(f,t) = H(f) = \alpha e^{j\theta}e^{-j2\pi f\tau_0}.$$

Implicit in this discussion is the dependence of our channel characterization on the signal characteristics themselves—for example, $h(\tau,t)$ might change significantly only when observed over some large number of symbols, say N symbols. This change occurs over NT_s seconds, which may constitute a very large number of symbols for the signal $v(t)$ in question, hence the channel is essentially distortionless for $v(t)$ over any given symbol interval. If the input signal were changed so that its new symbol rate (and bandwidth) decreased substantially to R_{sn}, such that the new symbol time $T_{sn}>>T_s$, the new signal $v_n(t)$ could experience significant channel variation over a few symbols or even a single symbol, making the exact same physical wireless channel a very distorting channel for $v_n(t)$. This dependence on signal characteristics (symbol time and bandwidth) underlies the characterization of all types of wireless channels.

Broadly speaking, the term "fading" refers to variation of the channel's characteristics over time (or space). Typically this is of most interest with regard to the channel amplitude $\alpha(t)$, as variation of amplitude is usually more significant to communication system performance than variation of phase θ. Variation of delay τ_0 is typically very slow with respect to signaling rates for terrestrial communications. A fading channel can be distortionless, and this is termed "frequency non-selective," or "frequency flat" fading. For most terrestrial communication systems, fading is slow in that channel variations occur only over many (hundreds or thousands of) T_s, but for the largest velocities and higher carrier frequencies, channel variations may not be classifiable as slow (although the rate of variation is almost always slow enough so that (2) still holds). Classification of the fading rate as fast or slow for any given input signal is somewhat arbitrary.

Also implicit in the previous definitions of the CIR is transmission over a single path from Tx to Rx. This is the simplest type of channel, not often encountered in terrestrial settings in the presence of obstacles between Tx and Rx.

Finally in this section we note that the functions that describe variation of amplitude, phase, and delay are often very well modeled as random. For the simple single-path channel this may not offer much advantage over a deterministic model, but for more complicated V2V channels, stochastic models will generally be preferable to deterministic ones. In principle, given all the electrical, geometric, and kinematic parameters of objects in the environment, we could compute the resulting electromagnetic field at any point in space distant from the transmitting antenna. In many practical situations this knowledge is unavailable or insufficiently accurate. In addition, even if it is available and accurate, we may require significant computational resources to solve the electromagnetic field equations, which could constrain how fast we could estimate electric field strengths and received powers, and thus severely limit the velocities for which our calculations apply. Finally worth noting is that in complicated environments with mobility, we are most often not interested in the exact value of field strength (or its square, proportional to received power density) at a specific point, but in some average value over a small spatial extent.

Deterministic channel models are mostly site specific, and to be accurate, they are computationally intensive in comparison to statistical models. This is one reason that statistical models are more attractive. Statistical models do not aim to provide exact estimation of a channel's small scale fading characteristics at points in space at any particular time. Instead, they attempt to faithfully emulate the *variation* in these channel effects.

2.2. Small Scale vs. Large Scale Fading

In this chapter, we concern ourselves only with small scale fading, which will most often arise due to the destructive interference from multiple replicas of the transmitted signal arriving at the receiver with different delays. This results from multipath propagation, as depicted schematically in Figure 1—the signal from Tx to Rx travels multiple paths, each of which can have a different amplitude, delay, and phase. This small scale fading is observed on spatial scales on the order of one-half wavelength—the distance required for a propagating wave's phase to change by π radians. In contrast, for frequency bands of current interest for V2V/R communication, (VHF and higher), large scale fading occurs on scales of many (e.g., 20 or more [13]) wavelengths. Large scale fading is also often termed shadowing, obstruction, or blockage [17].

Propagation path loss, or attenuation, may also be grouped with large scale fading, but strictly, this loss—which is frequency dependent, but for most signal bandwidths still essentially frequency non-selective—is not a fading phenomenon. This loss is due to signal spatial spreading as the electromagnetic wave propagates (and hence is sometimes called spreading loss). Figure 2 illustrates the components of signal variation in a plot of received power versus distance. This figure is a generalization of a similar one in [18]. Path loss does not vary that rapidly with distance, and generally can be accounted for by proper specification of transmit power and other link budget parameters. Shadowing typically varies over moderate distances (e.g., on the size of buildings) so that fading margin and/or power control can compensate. For countering small scale fading, other types of transceiver processing, in addition to those applied for large scale fading, may be needed. For V2V/R cases, when either or both Tx and Rx are moving, all types of fading can occur, and for complicated environments such as urban or suburban, these fading effects are most compactly modeled stochastically.

As channel characterization research continues, other classifications also arise. The idea of "medium-scale" or "meso-scale" fading has been given some attention [19], [20]. This classification or scale of fading is intended to be between small scale and shadowing. More will be said about this in a subsequent section.

2.3. The Multipath Channel Impulse Response

Generalizing (4) to allow for multiple transmission paths, the multipath CIR can be expressed as follows:

$$h^{(e)}(\tau,t)=\sum_{k=0}^{L(t)-1} z_k(t)\alpha_k(t)exp\{j[\omega_{D,k}(t)(t-\tau_k(t))-\omega_c(t)\tau_k(t)]\}\delta[\tau-\tau_k(t)] \quad (6)$$

where again, $h(\tau,t)$ represents the response of the channel at time t to an impulse input at time $t-\tau$. Equations (4) and (6) appear in the form of "discrete impulses" via the Dirac deltas. This can be interpreted as the channel imposing specific discrete attenuations, phase shifts, and delays upon any signal transmitted. This approximation is very good for signal bandwidths of tens of MHz or more, but may not be adequate for very wideband signals such as ultrawideband (UWB) [21]. In the UWB case, each path may impose filtering via its own "individual path IR," expressed in (6) by replacing $\delta(\tau-\tau_k)$ with $h_k(\tau-\tau_k)$. In some channels, such as HF troposcatter channels, this discreteness of impulses may not be appropriate, and the baseband CIR is a continuous function of both τ and t [14].

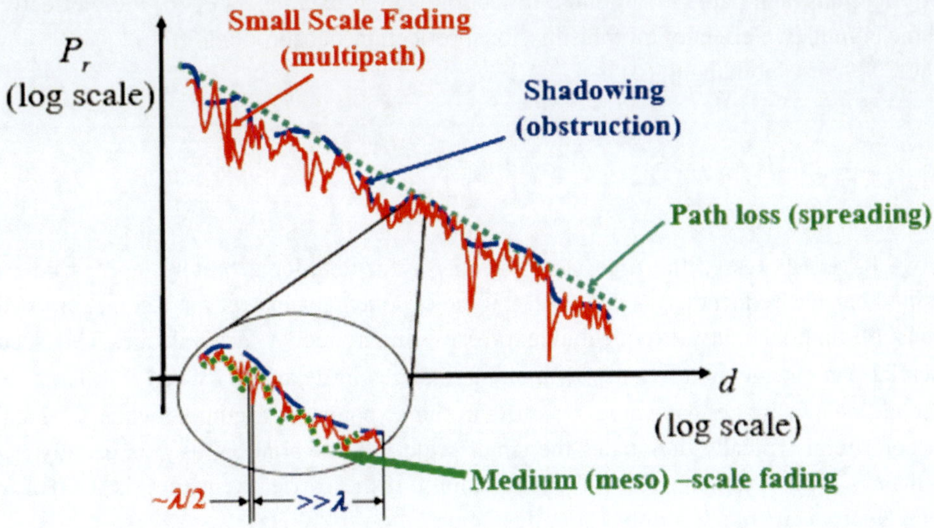

Figure 2. Illustration of received power vs. distance, showing relative scales of signal fading.

The variable α_k in (6) is the amplitude of the kth path, and the argument of the exponential is the kth resolved phase, and τ_k is the delay of the kth path. The carrier frequency is $f_c=\omega_c/(2\pi)$, and the kth resolved Doppler frequency is $f_{D,k}=\omega_{D,k}/(2\pi)$ where $f_{D,k}(t)=v(t)f_c cos[\theta_k(t)]/c$, where $v(t)$ is relative velocity between Tx and Rx, $\theta_k(t)$ is the aggregate phase angle of all components arriving in the kth delay "bin," and c is the speed of light. The delay bin width is approximately equal to the reciprocal of the signal bandwidth e.g., for a 10 MHz signal, the bin width is 100 nanoseconds—components separated in delay by an amount smaller than the bin width are "unresolvable." The kth resolved component in (6) thus often consists of multiple terms ("subcomponents") from potentially different spatial angles $\theta_{k,i}$ received in the kth delay bin. We do not address "spatial" channels in this work, e.g., [22], [23], primarily because at least initially, V2V communications will employ omni-directional antennas (except possibly for some stationary V2R antennas mounted at roadsides). Omni-directional V2V/R antennas are assumed henceforth.

The form of (6) is generalized from that typically seen in texts [15] and allows for

- an "environment" classification (superscript "e" on h); this can be used to denote CIRs for the various channel classes;
- a time-varying number of transmission paths $L(t)$;
- a "persistence process" $z(t)$ accounting for the finite "lifetime" of propagation paths;
- the explicit time variation of carrier frequency $\omega_c(t)$ to account for transmitter oscillator variations and/or carrier frequency hopping.

For our purposes, we make use of the first and third of these generalizations. The third generalization (persistence process) actually imposes the second generalization (time-varying number of paths).

Note that the amplitudes (α's) and phases represent aggregate channel effects. Referring to Figure 1, the path numbered "0" is the direct, possibly LOS path, paths "1-4" denote single-reflections, and path “5” a double reflection, so the parameters of (6) must effectively model the cumulative effect of all transmissions, reflections, absorptions, etc.

The CTF corresponding to (6) is

$$H^{(e)}(f,t) = \sum_{k=0}^{L(t)-1} z_k(t)\alpha_k(t)e^{j2\pi f_{D,k}(t-\tau_k(t))}e^{-j2\pi f_c\tau_k(t)}e^{-j2\pi f\tau_k(t)} \tag{7}$$

where we have suppressed the time variation in $f_{D,k}$ for notational brevity, and assume the transmitted carrier frequency f_c is constant. All the exponentials except the last one are strong functions of time; the last exponential expresses the frequency dependence. The second exponential can change significantly with small changes in delay $\tau_k(t)$ when f_c is large, e.g., nanosecond delay changes can cause 2π shifts in this exponential argument when f_c =1 GHz. This second term typically dominates the small scale fading variation, as f_c is usually *much* larger than $f_{D,k}$. For example, for V2V applications, if the carrier frequency is 5 GHz, and relative velocity is 63 m/s (roughly 140 miles/hour), $f_{D,max}$=1050 Hz<<f_c.

The CIR form in (6) indicates variation in both delay τ and time t. This function was termed the input delay spread function by Bello in his classic treatment of random channels [24]. Figure 3 illustrates a conceptual CIR, showing both fading in time, and variation of impulse energy (proportional to α_k^2) with delay. This type of CIR is implementable in analysis, simulations, or hardware in terms of the popular tapped-delay line (TDL) model. The TDL is a linear, finite IR filter, as shown in Figure 4. Implicit in Figure 3 is the notion of multiple time scales—the short-term “delay” scale (τ), and the longer term “time” scale over which the random processes evolve. For most channels considered to be “slowly fading,” the CIR is viewed as “decaying” or ending in delay τ long before any of the components change in time t appreciably. This will be true even for V2V channels.

2.4. CIR and CTF Correlation Functions, and Doppler

For channels modeled as random, a complete specification would contain all the probability density functions for all parameters, including joint densities of any order. In practice, this is never available, and fortunately, not necessary. Our brief discussion here employs the forms of Bello [24], and is necessarily concise, with an aim to those features of interest for V2V channels. As in [24], we limit the discussion to 2nd-order statistics. For a comprehensive treatment, see [24]; a good summary appears in [13]. The CIR correlation function is[4]

$$R_{hh}(\tau,\tau_1,t,t_1) = E[h(\tau,t)h^*(\tau_1,t_1)]. \tag{8}$$

[4] In most cases, a factor of ½ multiplies the expectation in the correlation function, to account for the complex envelope scaling [14]. We neglect that factor here, as it does not change the shape of the function, which is of ultimate interest.

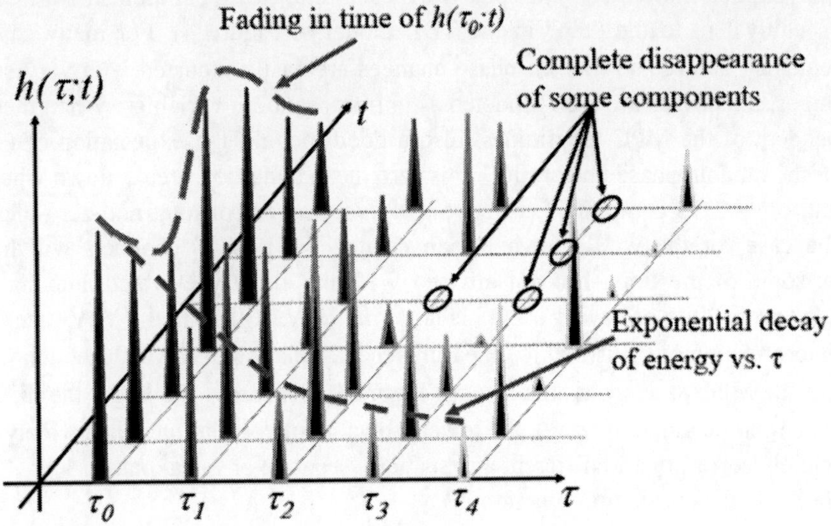

Figure 3. Conceptual illustration of time-varying CIR. (From D. W. Matolak, "Channel Modeling for Vehicle-to-Vehicle Communications," *IEEE Communications Magazine,* vol. 46, no. 5, pp. 76-83, May 2008. © 2008 IEEE, reprinted with permission.)

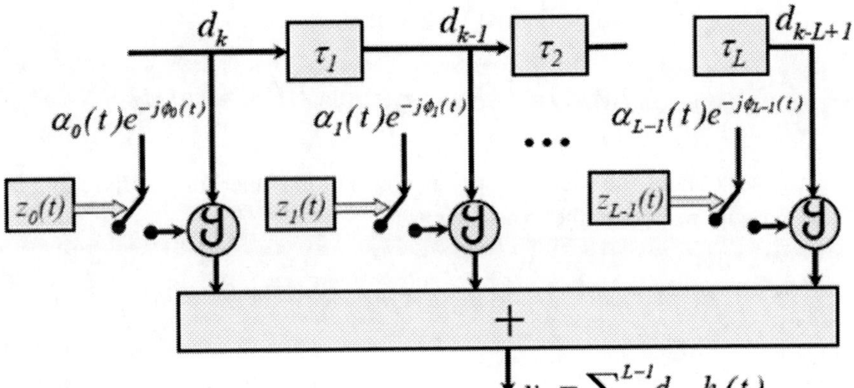

Figure 4. Tapped-delay line model for time-varying CIR. (From D. W. Matolak, "Channel Modeling for Vehicle-to-Vehicle Communications," *IEEE Communications Magazine,* vol. 46, no. 5, pp. 76-83, May 2008. © 2008 IEEE, reprinted with permission.)

Bello focused on correlation and not covariance functions because for many cases of interest, the channel is considered to be zero mean. Also, for stationary or wide-sense stationary assumptions, the correlation and covariance differ only by a constant. In regard to the zero-mean condition specifically, re-writing (6) as

$$h^{(e)}(\tau,t) = \sum_{k=0}^{L-1} \beta_k(t) e^{j\phi_k(t)} \delta[\tau - \tau_k(t)] \tag{9}$$

we see the explicit amplitude and phase of each multipath component (MPC, for our discussion equivalent to the "tap" in the TDL model of Figure 4). For many cases where carrier frequency is large so that 2π phase changes are easily incurred with very small path delay changes, the phases are well modeled as uniform random variables. When these phases are independent of the MPC amplitudes (also a good model), the expectation of (9) is zero because of the random phase uniformity. This zero mean condition breaks down when there is a dominant, often LOS component, with a slowly varying or constant non-zero-mean phase. This is the case for the well known Rician channel [25]. V2V channels will have LOS conditions some of the time, but not always. With the slightly elevated antennas of V2R channels, the probability of having a LOS is larger in the V2R than in the V2V case.

The interpretation of (8) is that R_{hh} measures the correlation of the MPC at delay τ, time t, with that of the MPC at delay τ_1, time t_1; see Figure 3. Generally, the larger the difference in delays (the larger the value of $|\tau-\tau_1|$), the lower the value of correlation, since widely differing delays typically correspond to different and spatially separate physical paths.

For the CTF, the correlation function is

$$R_{HH}(f,f_1,t,t_1)=E[H(f,t)H^*(f_1,t_1)] \tag{10}$$

with an interpretation analogous to that for R_{hh}, i.e., R_{HH} measures the correlation between the CTF evaluated at frequency f, time t, and that at frequency f_1, time t_1. The relationship between R_{hh} and R_{HH} is the double Fourier transform, i.e.,

$$R_{HH}(f,f_1,t,t_1)=\int\int R_{hh}(\tau,\tau_1,t,t_1)e^{-j2\pi f\tau}e^{-j2\pi f_1\tau_1}d\tau d\tau_1 \tag{11}$$

We can also obtain the delay-Doppler correlation function as the double Fourier transform of R_{hh} with respect to the t variables:

$$R_{SS}(\tau,\tau_1,\nu,\nu_1)=\int\int R_{hh}(\tau,\tau_1,t,t_1)e^{-j2\pi\nu t}e^{-j2\pi\nu_1 t_1}dtdt_1 \tag{12}$$

The function R_{SS} describes the correlation between the MPC at delay τ and Doppler shift ν with the MPC at delay τ_1 and Doppler shift ν_1. The final correlation function of Bello, here denoted $R_{DD}(f,f_1,\nu,\nu_1)$ is a double Fourier transform of R_{SS} with respect to the delay variables, and also a double Fourier transform of R_{HH} with respect to the time variables. This function R_{DD} is the correlation function of the output Doppler-spread function $H_O(f,\nu)$. Since H_O and R_{DD} are generally used less than the other functions, we do not mention them further.

2.5. Uncorrelated Scattering

When the MPCs at different delays are uncorrelated, the channel is termed an uncorrelated scattering (US) channel. In terms of (9), assuming phases are independent of amplitudes, the amplitude functions $\beta_k(t)$ and $\beta_j(t)$ satisfy

$$R_{\beta\beta}(t,\Delta t)=E[\beta_k(t)\beta_j(t+\Delta t)]=E[\beta_k(t)]E[\beta_j(t+\Delta t)] \quad (13)$$

If we let $e^{j\phi_k(t)}=\chi_k(t)$, then US also implies

$$R_{\chi\chi}(t,\Delta t)=E[\chi_k(t)\chi_j(t+\Delta t)]=E[\chi_k(t)]E[\chi_j(t+\Delta t)] \quad (14)$$

and when the random phases ϕ_k are uniform, $R_{\chi\chi}$=0 since $E(\chi_k(t))$=0. In the US case, (8) becomes

$$R_{hh}(\tau,\tau_1,t,t_1)=E[h(\tau,t)h^*(\tau_1,t_1)]=R_{hh}(\tau,t,t_1)\delta(\tau-\tau_1) \quad (15)$$

that is, R_{hh} is non-zero only when $\tau=\tau_1$. It can be shown that (15) implies that the CTF correlation function becomes

$$R_{HH}(f,f_1,t,t_1)=E[H(f,t)H^*(f_1,t_1)]=R_{HH}(\Delta f,t,t_1) \quad (16)$$

where $\Delta f=|f-f_1|$. This can be interpreted as meaning that US implies that the CTF correlation function does not depend upon absolute values of f and f_1, but only on their difference. This condition is termed wide-sense stationarity (WSS) in frequency. Thus we can express the relationship US(delay)⇔WSS(frequency). Similar to (15), with US, the delay-Doppler correlation function of (12) becomes $R_{SS}(\tau,\nu,\nu_1)\delta(\tau-\tau_1)$.

2.6. Wide-Sense Stationarity

In the more traditional use of stochastic processes as functions of time, we have the concept of wide-sense stationarity (WSS): a process is WSS if its mean is constant, and its autocorrelation depends only on the time difference, and not on absolute times [26]. For our CIR, WSS means that

$$R_{hh}(\tau,\tau_1,t,t_1)=E[h(\tau,t)h^*(\tau_1,t_1)]=R_{hh}(\tau,\tau_1,\Delta t) \quad (17)$$

where $\Delta t=|t-t_1|$. When the CIR is WSS in time, then the CTF autocorrelation becomes

$$R_{HH}(f,f_1,t,t_1)=E[H(f,t)H^*(f_1,t_1)]=R_{HH}(f,f_1,\Delta t) \quad (18)$$

and the delay-Doppler correlation function becomes

$$R_{SS}(\tau,\tau_1,\nu,\nu_1)=\delta(\nu-\nu_1)\int R_{hh}(\tau,\tau_1,\Delta t)e^{-j2\pi\nu_1\Delta t}d\Delta t=\delta(\nu-\nu_1)P_S(\tau,\tau_1,\nu) \quad (19)$$

where $P_S(\tau,\tau_1,\upsilon)$ is the delay-Doppler cross power spectral density. The meaning of (19) is that if each scatterer in the channel produces a delay and Doppler shift, the contributions to the CIR from different scatterers are uncorrelated if the scatterer Doppler shifts are different; this relationship is expressed as WSS(time)⇔US(Doppler).

2.7. Wide-Sense Stationarity, Uncorrelated Scattering

When both US(delay) and WSS(time) apply, we have the WSSUS channel, and the correlation functions reduce to

$$R_{hh}(\tau,\tau_1,t,t_1)=E[h(\tau,t)h^*(\tau_1,t_1)]=R_{hh}(\tau,\Delta t)\delta(\tau-\tau_1) \quad (20)$$

$$R_{HH}(f,f_1,t,t_1)=E[H(f,t)H^*(f_1,t_1)]=R_{HH}(\Delta f,\Delta t) \quad (21)$$

$$R_{SS}(\tau,\tau_1,\nu,\nu_1)=\delta(\tau-\tau_1)\delta(\nu-\nu_1)P_S(\tau,\nu) \quad (22)$$

In this case, R_{HH} is called the spaced-frequency, spaced-time (SFST) correlation function, and $P_S(\tau,\upsilon)$ is the scattering function, expressing the channel's delay-Doppler characteristics.

For the CIR correlation function, when we set Δt=0, $R_{hh}(\tau,0)=\psi_h(\tau)$, the power delay profile (PDP), the channel's average power output versus delay. The Fourier transform of this is $R_{HH}(\Delta f,0)= \psi_{Hf}(\Delta f)$, the spaced-frequency correlation function, which measures the correlation between channel effects at frequencies separated by Δf. As per the usual Fourier relationships, the width of $\psi_h(\tau)$, the multipath delay spread T_M, is reciprocally related to the width of $\psi_{Hf}(\Delta f)$, the coherence, or correlation bandwidth B_c.

Similarly, setting Δf=0 in R_{HH}, we obtain $R_{HH}(0,\Delta t)= \psi_{Ht}(\Delta t)$, the spaced-time correlation function, which measures the correlation between channel effects separated in time by Δt. The scattering function is the double Fourier transform of the SFST correlation function, and $P_S(0,\nu)=P_S(\nu)$ is the Doppler spectrum, whose width, the Doppler spread f_D, is reciprocally related to that of the width of $\psi_{Ht}(\Delta t)$, which is the coherence, or correlation time t_c.

In summary, T_M=width of $\psi_h(\tau)\sim 1/B_c$, and f_D=width of $P_S(\nu)\sim 1/t_c$. The delay spread T_M measures the amount the channel spreads an input impulse in delay, and analogously, the Doppler spread f_D measures the amount the channel spreads an input tone in frequency. The coherence bandwidth B_c measures the channel's frequency selectivity, and the coherence time t_c measures the channel's time selectivity (or its time rate of variation). Common measures for T_M are the root-mean-square delay spread (RMS-DS) and the delay window [15].

2.8. Non-stationary Channels and Correlated Scattering

All real channels can be statistically stationary for a limited time only. Thus statistical models for dynamic wireless channels can be at best only stationary in the wide sense [24]. One reason for the widespread use of the WSS (and WSSUS) assumption(s) is their relative mathematical simplicity. Yet while this simplicity is appealing to researchers and design engineers, it has inherently limited application. For example, almost all wireless standards, including cellular [27], [28], LAN [29], or V2V [30], [31] do indeed model the *distribution* of the channel delay spreads, which is a tacit way of implying non-stationarity without explicitly modeling it.

For modeling channels over long terms, which in mobile settings can correspond to modeling over wide and diverse areas, the channel's changing statistics must be used if the channel is to be modeled realistically and accurately [32]. Very few NS models for terrestrial communications exist at present; examples are the COST 259 model [33], and the recent UMTS model [34], which employ a simple set of two distinct models that are "switched between" as is in [35]. Another set of NS models is represented by the author's work, e.g., [36], [37], and work on the general topic of NS modeling is growing.

In [38], Matz developed an approach to characterize the non-WSSUS channel using a framework analogous to that of Bello. The term "non-WSSUS" means both non-WSS in time (yielding time varying Doppler spectra) and "non-US" via correlated scattering among taps (yielding non-WSS in frequency). As with Bello's functions, the system functions in [38] characterize the channel with four-dimensional (4D) functions, so that the traditional WSSUS model is a special case of this generalized non-WSSUS model. For small time/frequency intervals, the non-WSSUS model can be approximated as WSSUS. Yet due to mathematical subtleties associated with NS processes [39], representation of physical channel effects such as the scattering function require added "processing" of the 4D functions, and hence although these non-WSSUS functions are elegant, they may be difficult to use in practice.

The newer NS models previously noted ([33]-[37]) attempt to extend WSSUS models to incorporate nonstationarity, in a constructive way, using for example, time-varying Doppler spectra, or multipath component persistence processes, or propagation region transitions. The challenge is to incorporate "just enough non-stationarity" to capture the primary effects, without unduly complicating the model. NS models that go above the physical layer are also being developed [40]. Given the growth of mobile ad hoc networks, research on the NS channel topic will continue to grow.

As for correlated scattering, that too has been documented to some degree, e.g., [41]. Yet for most cases, this phenomenon appears to have a smaller potential for detrimental effects on communication system performance than does non-stationarity. Nonetheless, for high-fidelity channel modeling, correlation among MPCs must be incorporated.

2.9. Remarks on V2V Channel Statistics

With both Tx and Rx in motion, V2V channels will have more rapid variation than traditional channels. Thus, for a given environment, V2V coherence times will be smaller, implying larger Doppler spreads. In principle, t_c(V2V) can be as small as half that of t_c(cellular) or t_c(V2R), so that Doppler spread is double, i.e., f_D(V2V)~$2f_D$(cellular). With smaller coherence times, the duration for which a V2V channel can be considered WSS is smaller than that for cellular channels.

Delay spreads for V2V channels will be comparable to those for V2R channels, since the positions of scatterers in the local environment will be largely the same for these two cases. The V2V channel delay spreads will typically be smaller than those for cellular channels in urban environments, simply because urban V2V link distances will be smaller than cellular link distances. For open or highway V2V/R conditions, delay spreads can be larger than in the urban setting, but they will still generally be smaller than delay spreads for cellular channels, since cellular base stations typically have elevated antennas that can radiate energy to farther distances than can low-height V2V antennas. Thus for a given environment we have the relations, T_M(V2V)~T_M(V2R)<T_M(cellular), and B_c(V2V)~B_c(V2R)>B_c(cellular).

Uncorrelated scattering is a reasonable assumption when scattering emanates from distinct physical objects in the environment, but may not always be true for V2V cases. For example, scattering from two separate cars moving uniformly at the same velocity can yield *fully* correlated effects. This is indicated in Figure 5.

Even though V2V channels may be NS, and exhibit correlated scattering, PDPs and CIRs can still be measured, and delay spreads and coherence bandwidths quantified for environments of interest. Thus even though full knowledge of the channel correlation functions R_{hh} or R_{HH} may not be available, representative CIRs and CTFs can be obtained, and these representative functions can be analyzed statistically. Similarly, Doppler spreads can be measured for V2V/R environments, and corresponding coherence times estimated, even though complete SFST correlation functions and scattering functions will not generally be available.

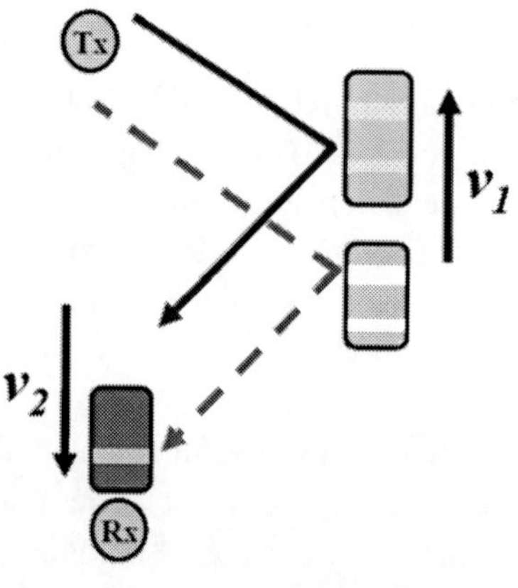

Figure 5. Illustration of correlated scattering condition.

To completely specify the TDL V2V channel model, we must specify the number of taps (obtainable from T_M), the taps' time rate of change (obtainable from f_D), and a statistical model for the tap amplitude random processes.[5] Last, we also require the relative energy of each tap. Typically the longer the delay, the weaker the MPC, so an exponentially-decaying energy vs. delay characteristic is often employed e.g., $E(k) = c_1 e^{-c_2 k} - c_3$, k=1,2, ...(see Figure 3).

[5] The tap phases are typically modeled as uniformly distributed on [0, 2π), except when there is a dominant (often LOS) component, in which case that tap's phase is modeled as having a narrow distribution about the mean phase.

3. Existing Work on V2V/V2R Channels

In this section we describe past and more-recent work on V2V channels. We review some deterministic models, established theoretical statistical models, and empirical statistical models. Early V2V channel models were often based upon models for similar settings, the most obvious of which is the cellular channel. As previously noted though, the V2V channel has significant differences from the cellular channel.

3.1. Deterministic Models

For the simplest of situations, deterministic models can be used effectively. One of these models is the "plane-earth" model [13], which assumes a single LOS path and a single ground-reflected path between Tx and Rx. This can be a very good model for the V2V channel between two cars on a level open highway [42]. Unfortunately, as the vehicle velocities and inter-vehicle distances vary, keeping track of the channel parameters becomes complicated, i.e., it is difficult to create an accurate time-varying deterministic model. In addition, uncertainties in model parameters (distance, ground reflection coefficients) will reduce the model accuracy, and ultimately, the "two isolated cars" situation is only one special case among many V2V conditions of interest.

More commonly, deterministic models use ray tracing. In this approach, local environment data is used to construct a model for all objects in the physical setting, for which electrical parameters (e.g., conductivities) are also needed as input. High frequency approximations allow plane waves to be modeled as rays, and each ray from the transmitter is traced along its path to the receiver, with reflections, diffractions, and other attenuations accounted for, yielding a composite received signal at the receiver. For accuracy, typically a large number of rays must be used, and all calculations are done by computer.

In [43], V2V channels were modeled via a ray-tracing approach. The authors modeled typical roadside structures and vehicles as the scattering objects, with parameters set by the program user. These same authors also published additional papers that provided more detailed models, e.g., [44]. As is true for other applications of ray tracing (e.g., indoor channels), reasonably accurate ray-tracing channel models can be obtained for site-specific cases, and these cases can be generalized somewhat by random generation of object properties (hence making these models a hybrid of statistical and deterministic). Nonetheless, these models are inherently site-specific, and the ray-tracing approach is generally computationally intensive, and sacrifices accuracy if computational complexity is reduced.

3.2. Theoretical Statistical Models

An early theoretical V2V channel reference is [45]. In this paper, the authors generalize the work of Jakes and Clarke [46], and use some simplifying assumptions to obtain the spaced-time correlation function $\psi_{Ht}(\Delta t)$ and the Doppler spectrum $P_S(\nu)$ for the V2V channel. The assumptions they used are, (1) scattering about both mobiles is isotropic; (2) fading amplitude statistics are Rayleigh; (3) the channel is WSSUS. The spaced-time correlation function for this case is a product of two Bessel functions—a nice extension of the cellular channel result of a single Bessel function:

$$\psi_{Ht}(\Delta t) = J_0(2\pi f_1 \Delta t) J_0(2\pi f_2 \Delta t) \tag{23}$$

where f_i is the maximum Doppler shift due to motion of mobile i, given by $f_{Di}=v_i/\lambda$, with v_i the velocity, and λ the wavelength (note the WSS assumption means a constant velocity). The Doppler spectrum for this case is expressed in terms of the complete elliptic integral of the first kind, and given its complex form, is not reproduced here. For the same assumptions, one of these same authors also derived other second-order statistical functions—the envelope level crossing rate and average fade duration—in [47]. The authors of [48] describe an effective means of simulating the isotropic scattering Rayleigh channel of [45] using the sum of sinusoids approach.

In [49] the authors generalize the two-dimensional scattering model and obtain integral equations for the spaced-time autocorrelation function and Doppler spectrum for a three-dimensional scattering case. In general, these functions depend upon antenna patterns and assumed distributions of angles of arrival, and even for the simplest of cases, must be evaluated numerically.

One other theoretical result of note is [50], in which non-isotropic scattering is modeled. Using a von Mises probability density model for angle of arrival, a closed-form expression for the spaced-time autocorrelation function is obtained:

$$\psi_{Ht}(\Delta t) = \prod_{i=1}^{2} \frac{I_0[\sqrt{\kappa_i^2 - 4\pi^2 f_i^2 \Delta t^2 + j4\pi\kappa_i f_i \Delta t \cos(\mu_i)}]}{I_0(\kappa_i)} \tag{24}$$

where f_i is the maximum Doppler shift due to motion of mobile i, μ_i is the mean angle of arrival of plane waves for vehicle i, and κ_i is a parameter of the von Mises density analogous to a standard deviation of the angle of arrival. For this model, the Doppler spectrum must be found numerically.

We note that no well established theoretical models exist for the V2V channel PDP $\psi_h(\tau)$ or its transform, the SF correlation function $\psi_{Hf}(\Delta f)$. This observation is also generally true for the cellular channel as well, simply because a wide range of PDP shapes with a range of delay spreads is known to exist in practice. For the purpose of analysis and simulations, two forms of PDP are most often used, a uniform and an exponentially decaying, with the latter being more realistic, as it is in better agreement with measurements of the V2V channel. Finally, a recent paper for multiple antenna schemes [51] provides a new analytical model for V2V ST correlation functions in three spatial dimensions.

3.3. Empirical Statistical Models

Empirical statistical models exist for a number of V2V/R cases, and here we describe several. For frequency non-selective fading channels, the authors of [52] found that the Ricean distribution fit best to their data, taken in urban, suburban, and highway environments in the 5 GHz band. The Ricean parameters for these cases were not provided, hence these models are incomplete. These authors also found that Doppler spectra best fit a Gaussian shape, with standard deviations of approximately 85 Hz in the urban environment, and 120 Hz in the highway environment.

In [53], the authors conducted measurements using a 20-MHz bandwidth signal in the 2.4 GHz band, for a highway setting. A 10-tap TDL model was provided, with Rice factors for each tap. The maximum delay of the model is 450 nanoseconds, with an RMS-DS of less than 100 nanoseconds. Doppler spectra were also provided, but only for a few of the taps, in graphical form. General observations are that the first tap has a very narrow, lowpass Doppler spectrum (20 dB bandwidth less than 100 Hz), and that the last tap has a wider bandwidth Doppler spectrum (20 dB bandwidth nearly 500 Hz). None of the spectra show the classic "Clarke" shape for isotropic scattering, which hence implies that scattering is *non*-isotropic, not surprising for a highway environment. In addition, multiple Doppler spectra were shown for a given tap, illustrating time-varying Doppler, an indication of statistical non-stationarity.

The authors of [54] performed narrowband (frequency non-selective) measurements in a suburban environment in the 5 GHz band. The single-tap models for amplitude fading employ Nakagami statistics. Some of the Nakagami parameters were less than unity, indicating fading more severe than Rayleigh, commonly assumed to be a "worst-case" amplitude fading model. In [55] and [56] these authors provide measured results on multipath statistics and Doppler for the 5 GHz band. These authors also provide a nice analysis of performance limits for DSRC based schemes over V2V channels in [57].

Finally in our survey of empirical models, the authors of [53] presented several sets of models for the 5 GHz band in [58]. The model bandwidths are 10 MHz, and three models for V2V (expressway-oncoming, urban, and expressway with wall) and three for V2R settings (expressway, suburban street, and urban canyon) are included. The models contain up to eight taps, with some taps a combination of multiple "paths." Each tap has an associated energy level, Ricean K-factor, frequency shift, and Doppler spectrum. The value of eight taps was chosen based upon the limits of a specific hardware emulator. Similarly, the set of four possible Doppler spectrum shapes and the Ricean amplitude statistics are restrictions of the hardware emulator. Maximum Doppler spreads are on the order of 1500 Hz for the highway cases, and 900 Hz for the urban cases. Values of RMS-DS were not provided. As with these authors' measurements at 2.4 GHz [53], Doppler spectral shapes largely imply non-isotropic scattering, and statistical non-stationarity is evident from time-varying Doppler spectra.

4. New Non-stationary V2V Channel Models

In this section we describe our recent work on new V2V channel models. Explicit empirical NS TDL models for several values of channel bandwidth are provided, based upon 50 MHz bandwidth measurements made in the 5 GHz band.

4.1. Modeling Multipath Component Persistence

From (6), the "appearance" and "disappearance" of MPCs can be modeled by the persistence processes, $z_k(t)$. This is illustrated in Figure 3. From our measurements in both airports [36] and V2V settings [37], we have found that these non-stationary MPC persistence effects are common. Phenomenologically, our persistence models are "medium scale" fading effects, which for a given velocity are slower than small scale multipath fading, but faster than large scale shadowing. Quantifying precisely the rates of MPC persistence is a current research topic, but from our measurements we have developed models that enable the model user to select these rates.

Regarding the need for the persistence processes, as with many modeling issues, this is an engineering judgement. Clearly, using persistence processes improves the model fidelity, at the expense of some increase in model complexity. In terms of RMS-DS statistics, we have found that inclusion of persistence in the models yields better agreement with measured data than if persistence is ignored [37]. The effects of persistence on transmission scheme performance are also of interest, and this too is an area of current research. Example results showing that NS models typically yield worse system performance than stationary models are provided in [59] and [60].

Our initial models for persistence employ 1st-order (homogeneous) Markov chains for each tap. These models are described by two matrices, the transition *(TS)* matrix and the steady state *(SS)* matrix:

$$TS = \begin{bmatrix} P_{00} & P_{01} \\ P_{10} & P_{11} \end{bmatrix}, \quad SS = \begin{bmatrix} P_0 \\ P_1 \end{bmatrix}. \tag{25}$$

where P_{ij} denotes the probability of going from state i to state j, and P_i denotes the long term probability of being in state i. State "0" denotes MPC "off," and state "1" denotes MPC "on." At any time t, $z_k(t)$ is either 0 or 1. We also have $P_0=1-P_1$; $P_{01}=1-P_{00}$; $P_{10}=1-P_{11}$.

Figure 6 shows example persistence processes for the 5th and 8th taps for some urban V2V measurement data. The *TS* and *SS* matrices for both taps are included in the figure.

4.2. Modeling Propagation Region Transitions and Time-Varying Doppler Spectra

As also noted in [36] and [37], in mobile settings, vehicles often move from one type of region to another, e.g., from urban to suburban in a V2V case. (This is of course also true for traditional radio systems.) Since these regions typically have different channel models associated with them (subscript "e" on $h(\tau,t)$ in (6)), for a comprehensive or long-term model, it is only natural to allow transitions between models for these different regions. Velocities should be user-selectable. The challenge in modeling such transitions lies in smoothing the change from one model to the next, since real channels do not change abruptly. This is an area for future work.

In real V2V/R settings, velocities change, thus yielding changing Doppler spectra. This has the potential to greatly affect communication system performance, and so models for time varying Doppler are needed. The simplest method is to have a set of Doppler spectra to "switch between," e.g., [61], yet as with propagation region transitions, changes between models must be smoothed.

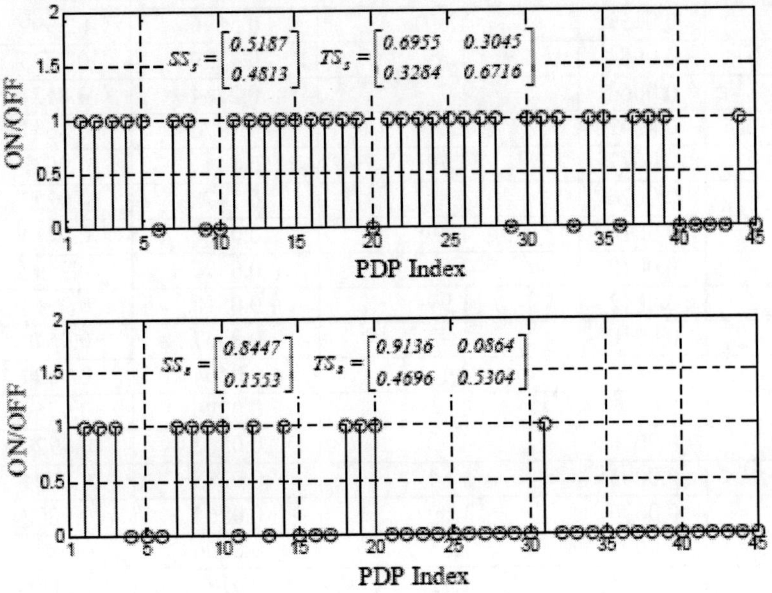

Figure 6. Example persistence processes for taps 5 and 8 for segment of travel in urban case, 10 MHz bandwidth (From I. Sen, D. W. Matolak, "Vehicle-Vehicle Channel Models for the 5 GHz Band," *IEEE Trans. on Intelligent Transportation Systems,* , vol. 9, no. 2, pp. 235-245, June 2008. © 2008 IEEE, reprinted with permission.)

4.3. Representative NS V2V Models

In this section we provide some of our empirical NS models for V2V channels. The models take the form of Figure 4, with persistence parameters as defined in (25). For our measurements, we used a 50 MHz bandwidth in the 5 GHz band, and combined CIR samples to form models of smaller bandwidths. We present here models for a 10 MHz bandwidth in Tables 1-5. Our V2V regions are defined as follows: Urban-Antenna Outside Car (UOC), Urban-Antenna Inside Car (UIC), Small City (S), Open Area–Low Traffic Density (OLT), and Open Area—High Traffic Density (OHT). The "open" areas are highways. These models were developed in [62]. Amplitude fading is modeled using the Weibull probability density function [63], which is equal to the Rayleigh distribution when the Weibull β parameter is 2; when β <2, fading is more severe than Rayleigh (as also found in [54]). Models for 5 MHz bandwidths appear in [37], along with correlation coefficients for inter-tap correlations—for many cases, taps are highly correlated, so the uncorrelated scattering assumption does not hold. Table 6 summarizes RMS-DS values for these V2V regions.

Table 1. Channel model for 10 MHz OHT channel

Tap Index k	Energy	Weibull Factor (β_k)	P_1	$P_{00,k}$	$P_{11,k}$
1	0.8982	4.30	1.0000	na	1.0000
2	0.0527	1.64	0.7960	0.3625	0.8366
3	0.0159	1.89	0.5696	0.5999	0.6973
4	0.0084	1.85	0.3739	0.7325	0.5514
5	0.0064	1.72	0.2764	0.8121	0.5084
6	0.0042	1.67	0.1976	0.8641	0.4486
7	0.0027	1.70	0.1444	0.9037	0.4297
8	0.0026	1.84	0.1178	0.9222	0.4172
9	0.0018	1.79	0.0856	0.9414	0.3750
10	0.0010	2.01	0.0576	0.9592	0.3333
11	0.0012	1.99	0.0598	0.9544	0.2844
12	0.0010	1.92	0.0467	0.9640	0.2647
13	0.0006	2.16	0.0404	0.9694	0.2721
14	0.0008	2.06	0.0296	0.9759	0.2130
15	0.0006	4.30	0.0362	0.9729	0.2803
16	0.0003	1.64	0.0285	0.9794	0.2981
17	0.0006	1.89	0.0264	0.9808	0.2917
18	0.0005	1.85	0.0206	0.9829	0.1867
19	0.0005	1.72	0.0178	0.9869	0.2769

Table 2. Channel model for 10 MHz OLT channel

Tap Index k	Energy	Weibull Factor (β_k)	$P_{00,k}$	$P_{11,k}$	P_1
1	0.949	5.15	na	1.0000	1.0000
2	0.040	1.63	0.3836	0.8525	0.8073
3	0.007	1.7070	0.5771	0.6469	0.5443
4	0.003	1.6996	0.7411	0.5212	0.3507
5	0.001	1.8066	0.8603	0.4618	0.2060

Table 3. Channel model for 10 MHz UOC channel

Tap Index k	Energy	Weibull Factor (β_k)	$P_{00,k}$	$P_{11,k}$	P_1
1	0.8319	3.19	na	1.0000	1.0000
2	0.0817	1.61	0.2717	0.9150	0.8956
3	0.0322	1.63	0.4401	0.8171	0.7538
4	0.0186	1.73	0.5571	0.7488	0.6382
5	0.0109	1.81	0.6955	0.6716	0.4813
6	0.0059	1.95	0.7843	0.5737	0.3362
7	0.0038	1.85	0.8693	0.5417	0.2218
8	0.0026	1.70	0.9136	0.5304	0.1553
9	0.0024	1.59	0.9322	0.4796	0.1156
10	0.0019	1.55	0.9436	0.4941	0.1003
11	0.0019	1.35	0.9549	0.5105	0.0843
12	0.0024	1.34	0.9655	0.5510	0.0714

Table 3. Continued

Tap Index k	Energy	Weibull Factor (β_k)	$P_{00,k}$	$P_{11,k}$	P_1
13	0.0022	1.34	0.9707	0.5364	0.0594
14	0.0008	1.33	0.9772	0.5635	0.0495
15	0.0007	1.29	0.9778	0.5179	0.0440

Table 4. Channel model for 10 MHz UIC channel

Tap Index k	Energy	Weibull Factor (β_k)	$P_{00,k}$	$P_{11,k}$	P_1
1	0.0033	2.51	0.9460	0.6098	0.1196
2	0.3764	1.76	na	1.0000	1
3	0.3274	1.69	0.1818	0.9853	0.9827
4	0.1337	1.73	0.1379	0.9213	0.9164
5	0.0614	1.65	0.2500	0.8735	0.8559
6	0.0367	1.59	0.4294	0.8245	0.7651
7	0.0255	1.71	0.5837	0.7723	0.647
8	0.0163	1.92	0.6644	0.7494	0.5706
9	0.0080	2.04	0.7201	0.6800	0.4683
10	0.0049	1.84	0.7850	0.6528	0.3833
11	0.0024	2.10	0.8460	0.6359	0.2968
12	0.0020	2.00	0.8205	0.6267	0.3242
13	0.0014	1.60	0.8708	0.6632	0.2752
14	0.0010	2.45	0.8640	0.5848	0.2464

Table 5. Channel model for 10 MHz small city channel

Tap Index k	Energy	Weibull Factor (β_k)	$P_{00,k}$	$P_{11,k}$	P_1
1	0.893	3.95	na	1.0000	1.0000
2	0.0807	1.91	0.4839	0.9446	0.9034
3	0.019	2.02	0.3452	0.7712	0.7383
4	0.00631	1.72	0.5985	0.6995	0.5701
5	0.00128	2.11	0.7217	0.4537	0.3364

Table 6. Summary of RMS-DS values for the five V2V regions (From I. Sen, D. W. Matolak, "Vehicle-Vehicle Channel Models for the 5 GHz Band," *IEEE Trans. on Intelligent Transportation Systems*, vol. 9, no. 2, pp. 235-245, June 2008. © 2008 IEEE, reprinted with permission)

Region	RMS-DS (nanosec) [min; mean; max]
UIC	[2.7; 236; 1210]
UOC	[3; 125.8; 1328.6]
OLT	[0.3; 53.2; 1113.4]
OHT	[0.5; 126.8; 1773.4]
Small	[0.7; 160; 1276]

Conclusion

In this chapter we defined the V2V/R channels, and discussed their importance to future ITS applications. After reviewing statistical channel characterization, we described how these channels differ from conventional mobile channels. Rates of time variation (and hence Doppler spreads) can be larger, WSS conditions pertain for a shorter amount of time, and with low-height antennas, fading can often be more severe in V2V channels than in conventional channels. Non-stationary channel modeling is required for high-fidelity models.

Acknowledgments

The author thanks his former students, Dr. Indranil Sen (Apple, Inc.), Dr. Beibei Wang (Motorola), and Dr. Wenhui Xiong (Qualcomm) for their past assistance in V2V channel measurements and modeling. Also deserving thanks are the editors and reviewers whose suggestions improved this work.

References

[1] US Dept. of Transportation (2008), ITS project website, http://www.its.dot.gov /index.htm

[2] Backof, C. (2006), “History of Vehicular Technology,” *IEEE Vehicular Tech. Magazine*, vol. 1, no. 1, pp. 3-7.

[3] Lackey, R. J., Upmal, D. W. (1995), “Speakeasy: The Military Software Radio,” *IEEE Communications Mag.*, vol. 33, no. 5, pp. 56-61.

[4] American Society for Testing and Materials (2008), ASTM E2213, “Standard Specification for Telecommunications and Information Exchange Between Roadside and Vehicle Systems—5GHz Band Dedicated Short Range Communications (DSRC) Medium Access Control (MAC) and Physical Layer (PHY) Specifications, http://www.astm.org

[5] Institute of Electrical and Electronics Engineers (2008), IEEE Wireless Access in Vehicular Environments website, http://grouper.ieee.org/groups/802/11/Reports/ tgp_update.htm

[6] Doumi, T. L. (2006), “Spectrum Considerations for Public Safety in the United States,” *IEEE Communications Mag.*, vol. 44, no. 1, pp. 30-37.

[7] Jones, W. (2008), US Dept. of Transportation, ITS Joint Program Office Presentation, Intelligent Transportation Society of America website, http://www.itsa.org/

[8] Institute of Electrical and Electronics Engineers (2008), IEEE Broadband Wireless Access Working Group website, http://grouper.ieee.org/groups/802/16/

[9] Institute of Electrical and Electronics Engineers (2008), IEEE Wireless Local Area Networks Working Group website, http://ieee802.org/11/ , April 2008.

[10] Aggelou, G. (2005), *Mobile Ad Hoc Networks: From Wireless LANS to 4G Networks*, McGraw-Hill, New York, NY.

[11] ITT Industries, Inc. (2005), “Technology Assessment for the Future Aeronautical Communications System,” *NASA Report CR-2005-213587*, TR 04055.

[12] Bodson, D., McClure, G. F., McConoughey, S. R., eds. (1984), Land Mobile Communications Engineering, *IEEE Press (Selected Reprint Series)*, New York, NY.

[13] Parsons, J. D. (2000), *The Mobile Radio Propagation Channel,* 2nd ed., John Wiley and Sons, New York, NY.

[14] Proakis, J. G. (2001)*Digital Communications*, 4th ed., McGraw-Hill, Boston, MA.

[15] Stuber, G. L. (2001), *Principles of Mobile Communication*, 2nd ed., Kluwer Academic Publishing, Boston, MA.

[16] Goldsmith, A. (2005), *Wireless Communications*, Cambridge University Press, New York, NY.

[17] Lutz, E. et. al. (1991), “The Land Mobile Satellite Communication Channel-Recording, Statistics, and Channel Model,” *IEEE Trans. Vehicular Tech*., vol. 40, no. 2, pp. 375-385.

[18] Sklar, B. (2001), *Digital Communications: Fundamentals and Applications*, 2nd ed., Prentice-Hall, Upper Saddle River, NJ.

[19] Sen, I., Matolak, D. W., Xiong, W. (2006), “Wireless Channels that Exhibit “Worse than Rayleigh” Fading: Analytical and Measurement Results,” *Proc. MILCOM 2006*, Washington, DC.

[20] Calcev, G., Chizhik, D., Göransson, B., Howard, S., Huang, H., Kogiantis, A., Molisch, A., Moustakas, L., Reed, D., Xu, H. (2007), “A Wideband Spatial Channel Model for System-Wide Simulations,” *IEEE Trans. Vehicular Tech*., vol. 56, no.2, pp. 389-403.

[21] Qiu, R. C. (2002), “A Study of the Ultra-Wideband Wireless Propagation Channel and Optimum UWB Receiver Design,” *IEEE Journ. Selected Areas Comm*., vol. 20, no. 9, pp. 1628-1637.

[22] Boche, H., Bourdoux, A, Fonollosa, J. R., Kaiser, T., Molisch, A., Utschick, W. Q. (2006) “Smart Antennas: State of the Art,” *IEEE Vehicular Tech. Magazine*, vol. 1, no. 1, pp. 8-17.

[23] Gesbert, D., Bolcskei, H., Gore, D. A., Paulraj, A. J. (2002), “Outdoor MIMO Wireless Channels: Models and Performance Prediction,” *IEEE Trans. Comm*., vol. 50, no. 12, pp. 1926-1934.

[24] Bello, P. (1963), “Characterization of Random Time-Variant Linear Channels,” *IEEE Trans. Comm*., vol. 11, pp. 360-393.

[25] Rice, S. O. (1948), “Statistical Properties of a Sine Wave Plus Random Noise,” *Bell System Tech. Journal*, vol. 27, pp. 109-157.

[26] Papoulis, A., Pillai, S. U. (2002), *Probability, Random Variables, and Stochastic Processes*, 4th ed., McGraw-Hill, Boston, MA.

[27] COST 231 Final Report (2007), http://www.lx.it.pt/cost231/final_report.htm

[28] Institute of Electrical and Electronics Engineers (2007), IEEE 802.16.3c-01/29r4, “Channel Models for Fixed Wireless Applications”

[29] Medbo, J., Hallenberg, H., Berg, J. E. (1999), “Propagation Characteristics at 5 GHz in Typical Radio-LAN Scenarios,” *Proc. IEEE Vehicular Tech. Conf.*, vol. 1, pp. 185-189.

[30] Institute of Electrical and Electronics Engineers (2005), IEEE 802.20-03/48, “Channel Models for IEEE 802.20 MBWA System Simulations,” http://grouper.ieee.org/groups/802//20/Contribs/C802.20-03-48.pdf

[31] Institute of Electrical and Electronics Engineers (2005), IEEE 802.20 Channel Models (V1.0), IEEE 802.20 PD-08, http://grouper.ieee.org/groups/802//20/P_Docs/IEEE_802.20-PD-08.doc

[32] Middleton, D. (2005), "Reflections and Reminiscences," *IEEE Information Theory Society Newsletter*, vol. 55, no. 3, pp. 3, 4-6.

[33] Correia, L. M. (ed) (2001), Wireless Flexible Personalised Communications (COST 259 Final Report), L. M. Correia, ed., Chichester, UK, Wiley.

[34] European Telecommunications Standards Institute (ETSI) (1997), Universal Mobile Telecommunications System (UMTS): Selection procedures for the choice of radio transmission technologies of the UMTS (UMTS 30.03, ver. 3.1.0), TR 101 112, v3.1.0, Section B.1.4.2.

[35] Schafhuber, D., Matz, G. (2005), "MMSE and Adaptive Prediction of Time-Varying Channels for OFDM Systems," *IEEE Trans. Wireless Comm.*, vol. 4, no. 2, pp. 593-602.

[36] [36] Channel: Part I, Measurement and Modeling Results for Large Airports," *IEEE Trans. Vehicular Tech.*, vol. 57, no. 4, pp. 2014-2026.

[37] Sen, I., Matolak, D. W. (2008), "Vehicle-Vehicle Channel Models for the 5 GHz Band," *IEEE Trans. on Intelligent Transportation Systems*, vol. 9, no. 2, pp. 235-245.

[38] Matz, G. (2005), "On non-WSSUS Wireless Fading Channels," *IEEE Trans. Wireless Comm.*, vol. 4, no. 5, pp. 2465 – 2478.

[39] Cohen, L. (2005), *Time-Frequency Analysis,* Prentice-Hall, Upper Saddle River, NJ.

[40] Konrad, A., Zhao, B. Y., Joseph, A. D., Ludwig, R. (2003), "A Markov-Based Model Algorithm for Wireless Networks," *Wireless Networks* (Kluwer), no. 9, pp. 189-199.

[41] Matolak, D. W., Sen, I., Xiong, W., Apaza, R. D. (2007), "Channel Measurement/Modeling for Airport Surface Communications: Mobile and Fixed Platform Results," *IEEE Aerospace and Electronics Mag.*, vol. 22, no. 10, pp. 25-30.

[42] Gallagher, B., Akatsuka, H., Suzuki, H. (2006), "Wireless Communications for Vehicle Safety: Radio Link Performance and Wireless Connectivity Methods," *IEEE Vehicular Tech. Mag.*, vol. 1, no. 4, pp. 4-16.

[43] Maurer, J., Schafer, T., Wiesbeck, W. (2001), "A Realistic Description of the Environment for Inter-Vehicle Wave Propagation Modelling," *Proc. IEEE Vehicular Technol. Conf.*, vol. 1, pp. 1437-1441.

[44] Maurer, J., Fugen, T., Schafer, T., Wiesbeck, W. (2004), "A New Inter-Vehicle Communications (IVC) Channel Model," *Proc. IEEE Vehicular Tech. Conf.*, vol. 1, pp. 9-13.

[45] Akki, A. S., F. Haber, F. (1986), "A Statistical Model of Mobile-to-Mobile Land Communication Channel," *IEEE Trans. Vehicular Tech.*, vol. VT-35, no. 1, pp. 2-7.

[46] Jakes, W. C. (1993), *Microwave Mobile Communication*, IEEE Press, New York, NY.

[47] Akki, A. (1994), "Statistical Properties of Mobile-to-Mobile Land Communication Channels," *IEEE Trans. Veh. Tech.*, vol. 43, no. 4, pp. 826-831.

[48] Patel, C. S., Stuber, G. L., Pratt, T. G. (2005), "Simulation of Rayleigh Faded Mobile-to-Mobile Communication Channels," *IEEE Trans. Comm.*, vol. 53, no. 11, pp. 1876-1884.

[49] Vatalaro, F., Forcella, A. (1997), "Doppler Spectrum in Mobile-to-Mobile Communications in the Presence of Three-Dimensional Multipath Scattering," *IEEE Trans. Vehicular Tech.*, vol. 46, no. 1, pp. 213-219.

[50] Zheng, Y. R. (2006), "A Non-Isotropic Model for Mobile to Mobile Fading Channel Simulations," *Proc. MILCOM '06*, Washington, DC.

[51] Zajić, A. G., Stüber, G. L. (2008), "Three-Dimensional Modeling, Simulation, and Capacity Analysis of Space-Time Correlated Mobile-to-Mobile Channels," *IEEE Trans. Vehicular Tech.*, vol. 57, no. 4, pp. 2042-2054.

[52] Maurer, J., Fugen, T., Wiesbeck, W. (2002), "Narrow-Band Measurement and Analysis of the Inter-Vehicle Transmission Channel at 5.2 GHz," *Proc. IEEE Vehicular Tech. Conf.*

[53] Acosta-Marum, G., Ingram, M. A. (2006), "A BER-Based Partitioned Model for a 2.4 GHz Vehicle-to-Vehicle Expressway Channel," *Wireless Personal Communications*, vol. 37, pp. 421-443.

[54] Cheng, L, Henty, B. E., Stancil, D. D., Bai, F., Mudalige, P. (2007), "Mobile Vehicle-to-Vehicle Narrow-band Channel Measurement and Characterization of the 5.9 GHz Dedicated Short Range Communication (DSRC) Frequency Band," *IEEE Journ. Selected Areas Communications*, vol. 25, no. 8, pp. 1501-1516.

[55] Cheng, L., Henty, B. E., Stancil, D. D., Bai, F. (2008) "Doppler Component Analysis of the Suburban Vehicle-to-Vehicle DSRC Propagation Channel at 5.9 GHz," *IEEE Radio and Wireless Symposium*, Orlando, FL.

[56] Cheng, L., Henty, B. E., Stancil, D. D., Bai, F. (2008), "Multipath Measurements for Vehicular Networks at 5.9 GHz," *IEEE Wireless Communications and Networking Conference*, Las Vegas, NE.

[57] Cheng, L., Henty, B. E., Bai, F., Stancil, D. D. (2008), "Performance of Time-Scaled 802.11a Waveforms over the Mobile-to-Mobile Vehicular Channel at 5.9 GHz," *IEEE Communications Magazine*, vol. 46, no. 5, pp. 84-91.

[58] Acosta-Marum, G., Ingram, M. A. (2007) "Six Time- and Frequency-Selective Empirical Channel Models for Vehicular Wireless LANs," *Proc. IEEE 1st Int. Symp. on Wireless Vehicular Communications*, Baltimore, MD.

[59] Matolak, D. W. (2008), "Channel Modeling for Vehicle-to-Vehicle Communications," *IEEE Communications Magazine, special section on Automotive Networking*, vol. 46, no. 5, pp. 76-83.

[60] Sen, I., Matolak, D. W. (2007), "V2V Channels and Performance of Multi-user Spread Spectrum Modulation," *Proc. IEEE 1st Int. Symp. on Wireless Vehicular Communications*, Baltimore, MD.

[61] Wang, R., Cox, D. (2002), "Channel modeling for ad hoc mobile wireless networks," *Proc. IEEE Veh. Technol. Conf.*, vol. 1, Birmingham, AL pp. 21-25.

[62] Sen, I. (2007), "5 GHz Channel Characterization for Airport Surface Areas and Vehicle-Vehicle Communication Systems," Ph.D. Dissertation, School of Electrical Engineering and Computer Science, Ohio University.

[63] Matolak, D. W., Sen, I., Xiong, W. (2008), "On the Generation of Multivariate Weibull Random Variates," *IET Communications Journal*, vol. 2, no. 4, pp. 523-527.

In: Wireless Technologies in Intelligent Transportation… ISBN: 978-1-60741-588-6
Editors: Ming-Tuo Zhou et al, pp. 51-84

Chapter 3

SMART ANTENNAS IN INTELLIGENT TRANSPORTATION SYSTEMS

Minghui Li[1,*], Bang Wang[2], Yilong Lu[2], Ming-Tuo Zhou[3] and I-Ming Chen[2]

[1] Centre for Ultrasonic Engineering
Department of Electronic and Electrical Engineering
University of Strathclyde, 204 George Street, Glasgow G1 1XW, United Kingdom

[2] Intelligent Systems Centre Nanyang Technological University
Nanyang Avenue, Singapore 639798

[3] Wireless Communications Laboratory, National Institute of Information and Communications Technology Singapore, 20 Science Park Road, #01-09A/10 TeleTech Park, Singapore 117674

Abstract

Background

Research and development on smart antennas that are recognized as a promising technique to improve the performance of mobile communications have been extensive in the recent years. Smart antennas combine multiple antenna elements with signal processing capabilities in both space and time to optimize its radiation and reception pattern automatically in response to the signal environment. This technology has significant impact on intelligent transportation systems (ITS) communications leading to increased user capacity, higher data rates and enhanced channel reliability. Furthermore, due to its capability of locating mobile units, smart antennas can facilitate vehicle location and navigation, trigger a myriad of location-based services and applications, and eventually benefit automotive telematics, public transit systems, and many other transportation systems.

* E-mail address: Minghui.Li@ieee.org. Tel: +44 (0)141 548 2463, Fax: +44 (0)141 548 2950. Address correspondence to: Dr. Minghui Li, Department of Electronic & Electrical Engineering, University of Strathclyde, Royal College Building, 204 George Street, Glasgow G1 1XW, United Kingdom,

Material and Methods

This chapter concentrates on the signal processing aspects of smart antenna systems. Smart antennas are often classified as either switched-beam or adaptive-array systems, for which a variety of algorithms have been developed to enhance the signal of interest and reject the interference. The antenna systems need to differentiate the desired signal from the interference, and normally requires either *a priori* knowledge or the signal direction to achieve its goal. There exists a variety of methods for direction of arrival (DOA) estimation with conflicting demands of accuracy and computation. Similarly, there are many algorithms to compute array weights to direct the maximum radiation of the array pattern toward the signal and place nulls toward the interference, each with its convergence property and computational complexity. This chapter discusses some of the typical algorithms for DOA estimation and beamforming. The concept and details of each algorithm are provided.

Results

Smart antennas can significantly help in improving the performance of ITS communication systems by increasing channel capacity and spectrum efficiency, extending range coverage, multiplexing channels with spatial division multiple access (SDMA), and compensating electronically for aperture distortion. They also reduce delay spread, multipath fading, co-channel interference, system complexity, bit error rates, and outage probability. In addition, smart antennas can locate mobile units or assist the location determination through DOA and range estimation. This capability can support and benefit many location-based services including emergency assistance, tracking services, safety services, billing services, and information services such as navigation, weather, traffic, and directory assistance.

Conclusion

Smart Antennas examine nearly all aspects of array signal processing. This chapter delivers a detailed treatment of antenna array processing schemes, algorithms to adjust weighting, DOA estimation methods, and diversity-processing methods that combat fading. In general, the algorithm selection is a trade-off between complexity and performance. Smart antennas will make the future intelligent transportation systems to be less complex while providing more attractive and convenient services. We believe that this technology has the great potential to make our transportation systems operate more safely and efficiently, with less congestion, pollution, and environmental impact.

Abbreviations

BER	Bit Error Rate
CDMA	Code Division Multiple Access
CMA	Constant Modulus Algorithm
CML	Conditional Maximum Likelihood
CRB	Cramer-Rao Bound
DOA	Direction of Arrival
ESPRIT	Estimation of Signal Parameters via Rotational Invariance Technique
FDMA	Frequency Division Multiple Access
GPS	Global Positioning System
LMS	Least Mean Square
ML	Maximum Likelihood
MMSE	Minimum Mean Square Error
MSE	Mean Square Error
MUSIC	Multiple Signal Classification
MVDR	Minimum Variance Distortionless Response

RMSE	Root Mean Squared Error
RSS	Received Signal Strength
SDMA	Space Division Multiple Access
SLL	Sidelobe Level
SMI	Sample Matrix Inversion
SNR	Signal to Noise Ratio
TDMA	Time Division Multiple Access
TDOA	Time Difference of Arrival
TOA	Time of Arrival
ULA	Uniform Linear Array
UML	Unconditional Maximum Likelihood
WSN	Wireless Sensor Network

1. Introduction

The demand for improved performance and increased capacity in mobile communications motivated recent research toward wireless systems that exploit space selectivity and diversity [1]. As a result, there are many efforts on the design of smart antenna arrays. This technology has significant impact on future intelligent transportation systems (ITS). Besides enhancing user capacity, data rates and channel reliability of ITS communications, smart antennas have the capability of locating a vehicle or assisting the location determination through positioning the on-board communication device or a mobile phone [2]. This can support and benefit various transportation applications including automotive telematics and public transit systems, and trigger many location-based services.

Smart antennas have emerged as a promising technique for mobile communication systems to overcome the problem of limited channel bandwidth and satisfy a growing demand for a larger number of mobile terminals on communication channels. The benefits of using smart antennas are that the sender can focus the transmission energy towards the desired user and in a narrower region while minimizing the effect of interference, and the receiver can form a directed beam towards the sender while simultaneously placing nulls in the directions of other transmitters. Furthermore, smart antennas can provide virtual channels in an angle domain by using spatially selective transmission and reception. This is referred to as spatial division multiple access (SDMA), which means that it is possible to multiplex channels in the spatial dimension just as in the frequency and time dimensions. All these help to reduce multipath reflections, delay spread, and co-channel interference, leading to increased network capacity, reduced energy consumption, lower bit error rates (BER), larger range coverage, lower outage probability, and decreased system complexity and infrastructure cost [3].

A myriad of transportation applications can be supported by centralized location and navigation systems, which utilize mobile communication networks and on-board vehicle radio equipment or phones to locate and navigate [4]. Smart antenna arrays can locate mobile units using such estimation models as direction of arrival (DOA), time of arrival (TOA), time difference of arrival (TDOA), and received signal strength (RSS). Various DOA-alone and hybrid positioning systems have been developed in the context of wireless networks. Compared with global positioning system (GPS) based solutions, the network-based schemes save an extra location device, overcome the GPS weakness encountered in urban canyons and indoor environments, and reduce the system complexity, cost and energy consumption. This technology can trigger different location-based services concerning vehicle safety, emergency assistance, vehicle tracking and fleet management, traffic information and directory assistance, and remote vehicle monitoring and services. Moreover, it can aid in the development of vehicle crash avoidance and antitheft systems, and improve the intelligent traffic management and control systems potentially reducing traffic congestion and air pollution [5].

This chapter concentrates on the signal processing aspects of smart antenna systems. Any smart antenna design is a trade-off between complexity and performance. Smart antennas are often classified as either switched-beam systems or adaptive-array systems. The relative merits of both classes of smart antennas, and their application and benefits for ITS are discussed in Section 2. Section 3 describes the array data model pivoting on the narrowband assumption, which is a fundamental concept in wireless DOA estimation and beamforming that allows a small time delay seen by a signal as it propagates across an array to be modeled as a simple phase shift. Once an architecture is chosen for a smart antenna system, an algorithm must be utilized to exploit it. For both switched-beam and adaptive-array systems, a variety of algorithms have been developed to enhance the desired signal while rejecting the interference. Section 4 discusses different beamforming algorithms that exploit the spatial and/or temporal characteristics of a reference signal. There exists a myriad of methods for DOA estimation with conflicting demands of accuracy and computation cost. Section 5 discusses some of the high resolution algorithms, their performance is analyzed and compared, and the pros and cons of each algorithm are discussed.

2. Smart Antennas and Their Benefits for Intelligent Transportation Systems

ITS communications can benefit from antenna array technology through exploiting either spatial selectivity or spatial diversity. Spatial selectivity separates spectrally and temporally overlapping signals through beamforming while minimizing interference, and diversity processing takes advantages of the different fading observed by antennas that are widely separated. In the literature, the term of “smart antennas” is often used to describe beamforming based technology and systems, on which we put the focus of this chapter.

2.a. Types of Smart Antennas

Smart antennas are usually categorized as either switched-beam or adaptive-array systems according to how they produce their response [6]-[7].

A *switched-beam system* is a system that can choose from many predefined patterns, in order to enhance the received signal. It is considered as an extension of the cell-sectoring scheme, in which a cell is composed of three 120-degree macro-sectors. The switched-beam method further subdivides the macro-sectors into several micro-sectors. Each micro-sector contains a predetermined fixed beam pattern. As a mobile user moves throughout the cell, the switched-beam system detects the signal strength, chooses the micro-sector containing the strongest signal, and continually switches the beams as necessary. Switched-beam systems provide range extension and enhanced coverage. Compared to conventional sectored cells, switched-beam systems may increase the range of a base station from 20% to 200% [3]. Less complexity in signal processing leads to its cost efficiency. At the same time, this technique has some apparent disadvantages. Since the beams are predefined and fixed, the signal strength varies as the user moves through the micro-sector. In addition, a switched-beam system does not discriminate between the desired signal and interference. If an interferer is near the center of the selected beam, it may be enhanced more than the desired user.

Adaptive-array systems take a very different approach and have the ability to adapt the radiation pattern to the signal environment in real time. The adaptive-array system continuously distinguishes among desired signals, multipath, and interfering signals according to their spatial and/or temporal characteristics, and updates its beam pattern based on changes in both the desired and interfering signal parameters. Adaptive-array systems can customize an appropriate radiation pattern for each individual user, which is far superior to the performance of a switched-beam system. Adaptive-array systems locate and smoothly track the desired user with main lobe and interferers with nulls, and dynamically adjust the beam pattern to enhance reception while minimizing interference.

Figure 1 illustrates the beam patterns that might be chosen by a switch-beam system and an adaptive-array system, in a scenario involving a desired user and an interferer. Both systems direct their main lobe in the general direction of the desired user, but the adaptive-array system makes a more accurate placement. On the contrary, the switched-beam system is not only unable to place the desired signal at the maximum of the main lobe, but also unable to fully reject the interference.

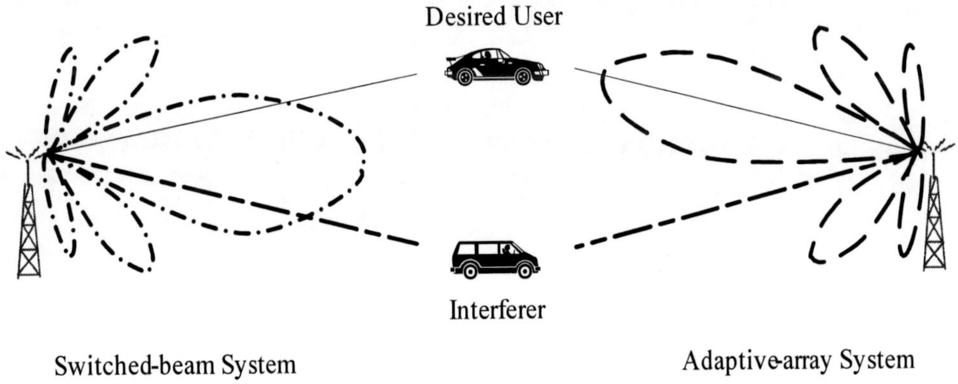

Figure 1. Beamforming lobes and nulls that switched-beam and adaptive-array systems might have for identical user-interferer scenarios.

Although providing lower gain and limited interference suppression, switched-beam systems are simple and can be easily deployed. This makes it attractive in some ITS applications, for example ITS sensor networks. Wireless sensor network (WSN) is a flexible and scalable paradigm that is drawing increasing attention from ITS community [8]-[10]. A key point to increase the reliability of a WSN is to improve the efficiency of wireless links. Adaptive arrays are generally too complex due to the size and cost constraints typical of sensor nodes, but the simpler switch-beam counterpart seems feasible. A system consisting of four fixed beam antennas is presented in [11], in which the beam selection is implemented through two digital lines based on the received signal strength. The simple system demonstrates substantial benefits: it extends the communication range by more than 250% when used outdoors, and it suppresses multipath fading by more than 70% while indoors.

Adaptive-array systems, considered by many to be "smarter" than switched-beam systems, are growing in popularity in mobile communications. A functional block diagram of adaptive arrays is shown in Figure 2. Rather than choosing a beam from predefined patterns, adaptive-array systems synthesize patterns by dynamically changing the weights (amplitudes and phases of the signals). The central part of an adaptive array is the beamformer, which applies a complex weight to each signal and sums the weighted signals. The weights are computed using the direction of the desired signal (spatial reference) or a training sequence (temporal reference). Section 4 deals with some of the beamforming algorithms. DOA estimators measure the signal directions by computing the time delays among the antenna elements, which are discussed in Section 5.

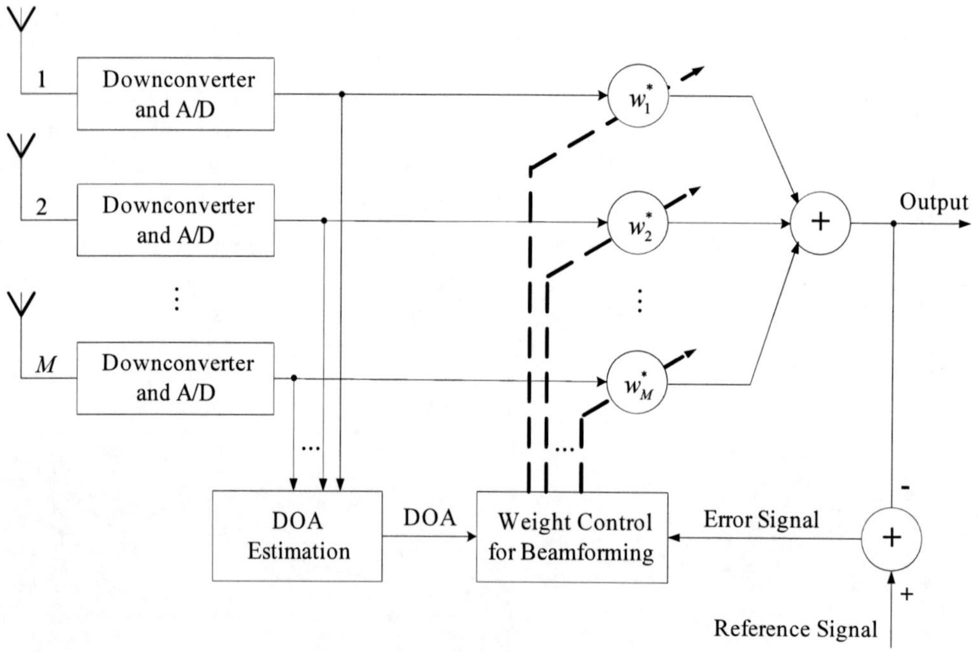

Figure 2. Functional block diagram of adaptive array systems.

Adaptive-array systems are most effective when the communication standard is designed to accommodate the adaptive array technology. As a result, medium access control (MAC) protocols that support adaptive-array and switched-beam systems in wireless networks have been developed [12]-[14].

2.b. Benefits of Smart Antennas for ITS

Smart antennas, when used appropriately, help to enhance ITS communications by increasing channel capacity and spectrum efficiency, extending range coverage, multiplexing channels with space division multiple access (SDMA), and compensating electronically for aperture distortion. Moreover, smart antennas can locate on-board vehicle mobile units, and thus enable numerous location-based service and applications.

2.b.1. Spatial Filtering for Interference Reduction

As shown in Figure 1, smart antennas attempt to enhance the received signal and reject all interfering signals through spatial filtering, and improve the signal to interference and noise ratio (SINR). This can reduce delay spread, multipath fading, co-channel interference, bit error rate, system complexity, and outage probability.

In noise or interference limited environments, the antenna array gain can be exchanged for higher detection probability and lower BER. The probability of detection in the case where a matched filter is employed for each array element is given by [15]

$$P_D = Q\left(Q^{-1}(P_F) - \sqrt{M \cdot SNR}\right), \tag{1}$$

where $Q(\cdot)$ is the Q function defined by $Q(\alpha) = \frac{1}{\sqrt{2\pi}} \int_{\alpha}^{\infty} e^{-\frac{x^2}{2}} dx$, P_F is the threshold false alarm probability, and M is the number of elements. The approximate formula for the BER in a CDMA system is given by [16]

$$P_b = Q\left(\sqrt{3 \cdot SF \cdot SIR_{omni} \cdot k^{-1}}\right), \tag{2}$$

where SF is the spreading factor, SIR_{omni} is the signal to interference ratio with an omnidirectional antenna, and

$$k = \frac{BW}{360} + SLL\left(1 - \frac{BW}{360}\right), \tag{3}$$

with BW and SLL the effective beamwidth and average sidelobe level, respectively. Smart antennas can increase the network coverage through antenna directivity and interference reduction. With small angular spread and the Hata path loss model, the range extension factor, *REF*, can be calculated as [17]

$$REF = \frac{r_2}{r_1} = M^{0.3}, \tag{4}$$

where r_1 and r_2 are the ranges with a single antenna and array, respectively. The range extension with smart antennas can lead to reduction in the number of base stations needed to serve the same area, and consequently reduce the infrastructure costs. The base station array gain can also be exploited to reduce the power transmitted by the mobile, which may relax the battery requirements, reduce the handset size and weight, and mitigate interference to other users. Smart antennas can increase the capacity through SNR improvement and co-channel interference rejection, as shown by the Shannon's expression for the capacity of a channel with bandwidth W and with additive Gaussian noise [16]

$$C = W \log_2 (1 + SNR) \quad (bit/s). \tag{5}$$

Smart antennas can also exploit spatial diversity [18], multiple-input multiple-output (MIMO) technology [19], and SDMA to enhance the user capacity.

2.b.2. Space Division Multiple Access (SDMA)

Smart antennas can be used to implement SDMA, which is among the most sophisticated utilization of smart antenna technology. It uses independently steered beams at the same frequency to support multiple users within a cell, as shown in Figure 3. This means that multiple users within the same cell can operate on the same time and frequency channel by exploiting the spatial separation of the users. With SDMA, it is possible to multiplex channels in the spatial dimension just as in the frequency and time dimensions, and this can be exploited as a hybrid multiple access technique complementing TDMA, FDMA and CDMA [20]. This concept can be seen as a dynamic (as opposed to fixed) sectoring approach in which each user defines its own sector as it moves. Through providing virtual channels in an angle domain, SDMA dramatically increases frequency reuse while it greatly improves the interference suppression capability, resulting in increased capacity and reduced infrastructure cost. SDMA enables capacity increases not only through inter-cell frequency reuse, but also through intra-cell frequency reuse [1].

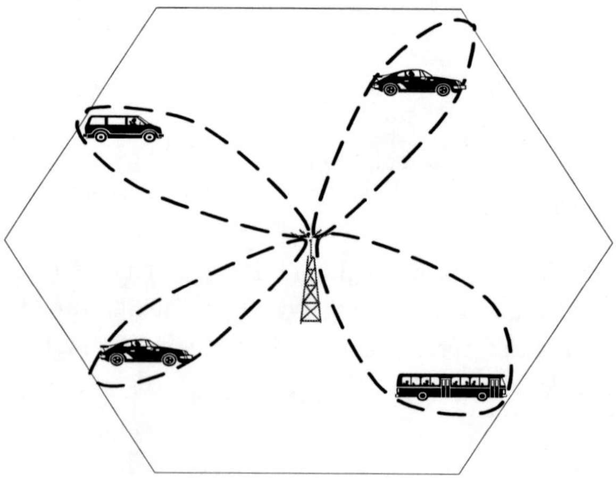

Figure 3. Base station employing SDMA. Four independent beams at the same frequency support four users within the same cell.

2.b.3. Location Positioning of Mobile Units

Vehicle location and navigation using communication networks and the on-board mobile units have received considerable attention around the world, which will benefit many applications such as automotive telematics and modern public transit systems [4]. This technology will have significant impact on future intelligent transportation systems. Besides emergency assistance, mobile unit localization can trigger many location-based services for intelligent transportation applications, which can be categorized into safety, information, tracking, remote, and billing services. Safety services, especially personal safety, are often the top priority in ITS. Information services include weather, traffic, navigation, and directory assistance. Tracking services can monitor continuously the location of the vehicle, asset, and people, which facilitate vehicle and fleet management. Remote services can provide further convenience as unlocking the car, monitoring the engine, and collecting tolls. Location-sensitive billing will be able to differentiate a variety of customer services. Furthermore, the location service can aid in the development of vehicle crash avoidance and antitheft systems, and improve the intelligent traffic management and control systems potentially reducing traffic congestion and air pollution. In addition, location of mobile users can be exploited to increase communications capacity through dynamic reuse planning. With user location information, channels can be assigned dynamically to areas of high user density and hence better capacity can be achieved [3].

Smart antennas can determine the mobile unit location or assist the location determination based on the wireless network infrastructure using estimation models, such as DOA, TOA and TDOA [21]-[24]. Those network-based location schemes are of considerable interest in ITS, and often outperform the schemes depending on GPS, because the network-based methods save an extra location device, overcome the GPS weakness encountered in urban canyons and indoor environments, and reduce the system complexity, cost and energy consumption [4]. DOA and range estimation is a basic capability of an antenna array, and a myriad of high resolution algorithms have been proposed. Various DOA-alone [25]-[27] and hybrid positioning systems [28]-[30] using communications infrastructure have been developed, including hybrid DOA-TOA systems, DOA-TDOA systems, and etc. It is demonstrated that the combination of disparate data through data fusion produces more accurate location estimates than any of the individual measurement [31].

3. Array Data Model and Problem Formulation

To understand how multiple antennas can be exploited for beamforming and DOA estimation, we must first model the signals received at the array. Consider a general scenario of an array of M elements arranged in an arbitrary geometry immersed in the far field (planar wave) of N point sources at unknown locations. To simplify the exposition, our discussion is confined to azimuth-only systems, i.e., the sensors and signals are assumed to be co-planar. However, the data model and algorithms presented here are general and the extension to azimuth-and-elevation systems is straightforward.

We start our discussion by considering the situation of a single source. Let $x(t)$ denote the signal as measured at a reference point, for instance the first sensor. The output of sensor k ($k=1,\ldots,M$) can be written as

$$\overline{y}_k(t)=\overline{h}_k(t)*x(t-\tau_k)+\overline{n}_k(t) \tag{6}$$

where $\bar{h}_k(t)$ is the impulse response of the kth sensor, $*$ denotes the convolution operation, τ_k denotes the time needed for the wave to travel from the reference point to sensor k, and $\bar{n}_k(t)$ is an additive noise. The array is assumed to be calibrated, i.e., the sensors are modeled as linear time invariant systems with known $\bar{h}_k(t)$ and known locations. The Fourier transform of the model (6) is given by

$$\bar{Y}_k(\omega) = \bar{H}_k(\omega) X(\omega) e^{-j\omega\tau_k} + \bar{N}_k(\omega) \tag{7}$$

where $\bar{Y}_k(\omega)$, $\bar{H}_k(\omega)$, $X(\omega)$ and $\bar{N}_k(\omega)$ denote the Fourier transforms of $\bar{y}_k(t)$, $\bar{h}_k(t)$, $x(t)$ and $\bar{n}_k(t)$, respectively, which are defined as, e.g., $X(\omega) = \int_{-\infty}^{\infty} x(t) e^{-j\omega t} dt$.

For a general class of physical signals, such as the carrier modulated signals encountered in communications, the signal is bandpass and its spectrum is limited to a certain band of frequencies centered about the carrier frequency ω_c. Assume that $x(t)$ is a bandpass signal, and $s(t)$ denotes the baseband signal associated with $x(t)$, which is often termed the complex envelope. $x(t)$ is assumed to be obtained by real modulation process, that is

$$x(t) = s(t) e^{j\omega_c t} + \left[s(t) e^{j\omega_c t} \right]^* = 2\,\mathrm{Re}\left[s(t) e^{j\omega_c t} \right], \tag{8}$$

where $(\bullet)^*$ denotes complex conjugate, and $\mathrm{Re}[\bullet]$ represents the real part. In the frequency domain,

$$X(\omega) = S(\omega - \omega_c) + S^*\left(-(\omega + \omega_c)\right) \tag{9}$$

where $S(\omega) = \int_{-\infty}^{\infty} s(t) e^{-j\omega t} dt$. When insert $X(\omega)$ (9) into (7), we obtain

$$\bar{Y}_k(\omega) = \bar{H}_k(\omega) \left[S(\omega - \omega_c) + S^*\left(-(\omega + \omega_c)\right) \right] e^{-j\omega\tau_k} + \bar{N}_k(\omega). \tag{10}$$

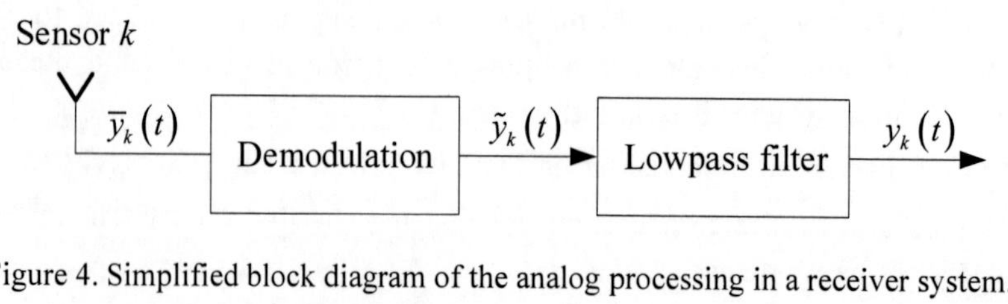

Figure 4. Simplified block diagram of the analog processing in a receiver system.

Consider a classic receiver model depicted in Figure 4: to obtain the baseband signal $y_k(t)$, the received signal $\bar{y}_k(t)$ is demodulated and filtered with a baseband (lowpass) filter. Let $\tilde{y}_k(t)$ denote the demodulated signal $\tilde{y}_k(t) = \bar{y}_k(t)e^{-j\omega_c t}$, whose Fourier transform is given by

$$\tilde{Y}_k(\omega) = \bar{H}_k(\omega+\omega_c)\left[S(\omega) + S^*(-(\omega+2\omega_c))\right]e^{-j(\omega+\omega_c)\tau_k} + \bar{N}_k(\omega+\omega_c). \tag{11}$$

When $\tilde{y}_k(t)$ is passed through a lowpass filter with bandwidth matched to $S(\omega)$, the components centered at $\omega = -2\omega_c$ is eliminated. Hence, the output signal can be expressed as

$$Y_k(\omega) = H_k(\omega+\omega_c)S(\omega)e^{-j(\omega+\omega_c)\tau_k} + N_k(\omega+\omega_c) \tag{12}$$

where $H_k(\omega+\omega_c)$ and $N_k(\omega+\omega_c)$ denote the parts of $\bar{H}_k(\omega+\omega_c)$ and $\bar{N}_k(\omega+\omega_c)$ that fall within the passband of the lowpass filter.

Now, we make the narrowband assumption: the received signals are narrowband, so that $|S(\omega)|$ decreases rapidly with increasing $|\omega|$. (12) can be reduced to the following equation in an approximate way,

$$Y_k(\omega) = H_k(\omega_c)S(\omega)e^{-j\omega_c\tau_k} + N_k(\omega+\omega_c), \tag{13}$$

whose time-domain counterpart is the following

$$y_k(t) = H_k(\omega_c)e^{-j\omega_c\tau_k}s(t) + n_k(t), \tag{14}$$

where the baseband signals $y_k(t)$ and $n_k(t)$ are the inverse Fourier transforms of $Y_k(\omega)$ and $N_k(\omega+\omega_c)$, respectively. Under the narrowband assumption, the array model (6) is simplified in that the time delay τ_k is modeled as a simple phase shift, and the sensor frequency response is flat over the signal bandwidth (but may vary with the source direction).

We introduce the so-called array steering vector

$$\mathbf{a}(\theta) = \left[H_1(\omega_c)e^{-j\omega_c\tau_1} \ldots H_M(\omega_c)e^{-j\omega_c\tau_M}\right]^T \tag{15}$$

where θ is the source DOA, and $[\bullet]^T$ denotes the transpose. By making use of (15), the array model can be expressed as

$$\mathbf{y}(t) = \mathbf{a}(\theta)s(t) + \mathbf{n}(t) \tag{16}$$

where

$$\mathbf{y}(t) = \left[y_1(t) \ldots y_M(t)\right]^T$$

$$\mathbf{n}(t) = [n_1(t) \ldots n_M(t)]^T$$

denote the array output vector and additive noise vector, respectively. In most mobile communication applications, the first sensor is selected as the reference point, and the sensors are assumed to be omnidirectional over the DOA range of interest with unit gain, the expression (15) can be simplified to the following form

$$\mathbf{a}(\theta) = \begin{bmatrix} 1 & e^{-j\omega_c\tau_2} \ldots e^{-j\omega_c\tau_M} \end{bmatrix}^T \tag{17}$$

The extension of (16) to multiple sources is straightforward. Based on the assumption of linear sensors and superposition principle, the array data model for N sources is given by

$$\mathbf{y}(t) = \mathbf{A}(\boldsymbol{\theta})\mathbf{s}(t) + \mathbf{n}(t) \tag{18}$$

where

$$\mathbf{A}(\boldsymbol{\theta}) = [\mathbf{a}(\theta_1) \ldots \mathbf{a}(\theta_N)],$$

$$\mathbf{s}(t) = [s_1(t) \ldots s_N(t)]^T,$$

denote the array transfer matrix and source signal vector, respectively.

The dependence of τ_k in (17) as a function of source direction θ can be derived using the planar wave assumption. As an example, we consider a uniform linear array (ULA) of M sensors with inter-element spacing d and a signal illuminating the array from θ as depicted in Figure 5. The first element is chosen as the reference point, we find that

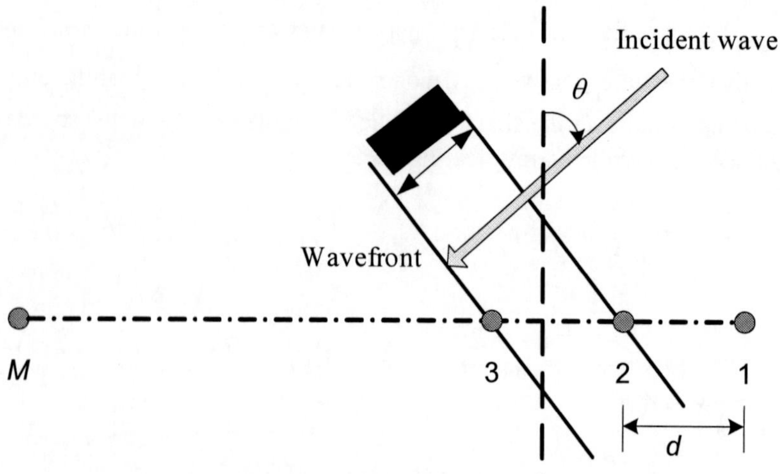

Figure 5. Uniform linear array scenario.

$$\tau_k = \frac{(k-1)d\sin\theta}{c}, \quad k = 1,\ldots,M \tag{19}$$

where c is the propagation velocity of the incident wave.

An array can be considered to sample the wavefield spatially, and the sensor outputs can be viewed as a group of discrete time samples of the incident signal with sampling interval $\tau = d\sin\theta / c$, and sampling frequency is given by

$$f_s = \frac{1}{\tau} = \frac{c}{d\sin\theta}. \tag{20}$$

The minimum sampling rate occurs when $\sin\theta = 1$

$$f_{s\min} = \frac{c}{d}. \tag{21}$$

From the Nyquist sampling theorem for a narrowband signal at ω_c, $f_s \geq 2f_c$, therefore, we have

$$d \leq \frac{\lambda}{2} \tag{22}$$

where λ is the signal wavelength. The equation (22) can be interpreted as a spatial Nyquist sampling theorem, which states that the inter-element space should be smaller than half of the signal wavelength to avoid spatial aliasing.

The steering vector $\mathbf{a}(\theta)$ describes a mapping between DOA and array response, and provides a framework for describing more complex phenomena including multipath, fading, and angle spread. In theory, $\mathbf{a}(\theta)$ is completely defined by the array geometry and the gain patterns of individual antenna elements, as in (15). However, in practice, the actual array response can deviate significantly from its analytic form, due to the effect of mutual coupling, uncertainty in antenna locations, gain and phase errors, and etc. The process of determining $\mathbf{a}(\theta)$ as a function of θ is known as array calibration, which is accomplished by placing a transmitter at a known angle and estimating $\mathbf{a}(\theta)$ from the received data [32].

The vectors of signals and noise are assumed to be stationary, temporally white, zero-mean complex Gaussian random processes with second-order moments given by

$$\begin{aligned} E\{\mathbf{s}(t)\mathbf{s}^H(s)\} &= \mathbf{P}\delta_{ts} \\ E\{\mathbf{s}(t)\mathbf{s}^T(s)\} &= 0 \\ E\{\mathbf{n}(t)\mathbf{n}^H(s)\} &= \sigma_n^2\mathbf{I}\delta_{ts} \\ E\{\mathbf{n}(t)\mathbf{n}^T(s)\} &= 0 \end{aligned} \tag{23}$$

where δ_{ts} is the Kronecker delta, $(\bullet)^H$ denotes complex conjugate transpose, $E\{\bullet\}$ stands for expectation, and $\mathbf{P}$ and $\sigma_n^2\mathbf{I}$ are the signal and noise covariance matrices, respectively. Assuming that the noise and signals are independent, the data covariance matrix is given by

$$\mathbf{R} = E\{\mathbf{y}(t)\mathbf{y}^H(t)\} = \mathbf{APA}^H + \sigma_n^2\mathbf{I} \tag{24}$$

An unbiased estimate of $\mathbf{R}$ using L data samples can be obtained using an averaging scheme

$$\hat{\mathbf{R}} = \frac{1}{L}\sum_{t=1}^{L}\mathbf{y}(t)\mathbf{y}^H(t) \tag{25}$$

4. Beamforming Algorithms

A beamformer is a device that applies a complex weight w_i, $i = 1, \cdots, M$ to the signal $y_i(t)$ of each antenna element and sums the weighted signals, as illustrated in Figure 6. The array gain pattern is defined by

$$F(\theta) = \left|\mathbf{w}^H\mathbf{a}(\theta)\right|^2, \tag{26}$$

where $\mathbf{w} = [w_1 \ldots w_M]^T$. The gain pattern describes the power in the beamformer output due to a signal impinging on the array from a particular direction θ. A judicious choice of the weights will result in a desirable beam pattern such as that of Figure 1. The beamformer usually requires either the source directions (spatial characteristics) or reference signals (temporal characteristics) to compute the weights. In this section, we discuss three techniques using spatial reference, one technique using temporal reference, and the adaptive implementations of these algorithms that adjust the weights in real time to track the signal environment changes. Many wireless standards specify that users should periodically transmit a known signal, often called a training sequence, over a specified interval for channel estimation. These known signals can be exploited as the reference signal. Source bearings can be measured using DOA estimation algorithms.

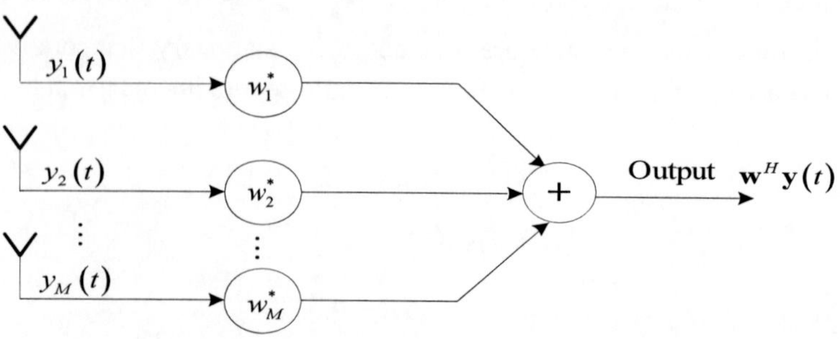

Figure 6. Narrowband beamforming structure.

4.a. Conventional Beamformer

Conventional beamformer, also known as delay-and-sum beamformer, is one of the oldest and simplest array processing algorithms [33]. The underlying idea is quite simple: when a propagating signal is present in an array's aperture, if the sensor outputs are delayed by appropriate amounts and added together, the signal will be reinforced with respect to noise or waves propagating in different directions [34]. For conventional beamformer, all the weights have equal magnitudes, and the phases (time delays) are selected based on the length of time it takes for the signal to propagate between sensors. To steer the array in a particular direction θ_d, the weights are selected as

$$\mathbf{w} = \frac{1}{M}\mathbf{a}(\theta_d) \tag{27}$$

The process is similar to steering the array mechanically in the look direction except that it is done electronically by adjusting the phases. In an environment consisting of only one source of power p_s in the array look direction θ_d and uncorrelated noise, using the data model (18) and (24), the output power of the beamformer can be measured as

$$\begin{aligned} P(\theta_d) &= E\left\{\left|\mathbf{w}^H \mathbf{y}(t)\right|^2\right\} \\ &= \frac{\mathbf{a}^H(\theta_d)\mathbf{R}\mathbf{a}(\theta_d)}{M^2} \\ &= \frac{\mathbf{a}^H(\theta_d)\left[\mathbf{a}(\theta_d)p_s\mathbf{a}^H(\theta_d) + \sigma_n^2\mathbf{I}\right]\mathbf{a}(\theta_d)}{M^2} \\ &= p_s + \frac{\sigma_n^2}{M}. \end{aligned} \tag{28}$$

The first part represents the output signal power, which is the same as the source power, and the second part represents the output noise power. Therefore, the array with these weights (27) has unity response in the look direction, and the array gain, which is defined as the ratio of the output SNR to the input SNR, equals to M, the number of elements. Though the conventional beamformer provides maximum SNR when there is no directional interference, it is not effective in the presence of directional jammers, intentional or unintentional.

4.b. Null-steering Beamformer

A null-steering beamformer can cancel a plane wave arriving from a known direction by producing a null in the response pattern in that direction. A beam with unity response in the desired direction θ_d and nulls in k interference directions $\theta_1, \cdots, \theta_k$ may be formed by estimating the weights of a beamformer using suitable constraints [35]. The desired weight vector is the solution to following simultaneous equations:

$$\begin{aligned} \mathbf{w}^H \mathbf{a}(\theta_d) &= 1 \\ \mathbf{w}^H \mathbf{a}(\theta_i) &= 0, \quad i = 1, \cdots, k. \end{aligned} \tag{29}$$

Using matrix notation, this becomes

$$\mathbf{w}^H \mathbf{C} = \boldsymbol{\delta}_1^T, \tag{30}$$

where

$$\mathbf{C} = \left[\mathbf{a}(\theta_d) \ \mathbf{a}(\theta_1) \cdots \ \mathbf{a}(\theta_k) \right],$$
$$\boldsymbol{\delta}_1 = \left[1 \ 0 \cdots \ 0 \right]^T.$$

In a general scenario where **C** is not a square matrix, a suitable estimate of the weight vector may be computed using

$$\mathbf{w} = \left(\boldsymbol{\delta}_1^T \mathbf{C}^H \left(\mathbf{C}\mathbf{C}^H \right)^{-1} \right)^H. \tag{31}$$

Although the beam pattern produced by this beamformer has nulls in the directions of interferences, it is not designed to minimize the noise at the array output. A performance analysis of the null-steering algorithm is presented in [36].

Another class of approaches for null-steering beamforming is array pattern synthesis, which is to determine a set of weights that result in a prescribed and somewhat arbitrary beam pattern response. In general, array pattern synthesis is first formulated as an optimization problem with the goal of sidelobe level (SLL) suppression and/or null placement in certain directions, while preserving the gain of the main beam toward the desired direction, and then solved by a global search algorithm such as simulated annealing [37], genetic algorithms [38], and particle swarm optimization algorithms [39] for the optimal weight vector. Compared with the constraint-based method, this approach has some obvious advantages: a) the SLL and the null location, width and depth can be fully controlled; b) wide null placement is possible for nulling interference with angle spread; c) flat sidelobe and nulls can be produced, which facilitates minimizing interference over a large area [38], [40]. All these features make this technique extremely attractive for beam pattern design of a transmitting array.

4.c. Optimal Beamformer

The optimal beamformer, also known as the minimum variance distortionless response (MVDR) beamformer [41], does not require either the knowledge of the directions and power levels of the interferers, or the level of the background noise power, to maximize the output SNR. It only requires the DOA of the desired signal. This beamformer is also known as the maximum likelihood (ML) filter [41], since it finds the ML estimate of the power of the signal source assuming all sources as interference, and the weights are the solution to the following constrained optimization problem

$$\min_{\mathbf{w}} E\left\{ \left| \mathbf{w}^H \mathbf{y}(t) \right|^2 \right\} \quad \text{subject to } \mathbf{w}^H \mathbf{a}(\theta_d) = 1 \tag{32}$$

The constraint $\mathbf{w}^H \mathbf{a}(\theta_d) = 1$ ensures that the desired signal coming from the DOA θ_d passes to the beamformer's output undistorted. This process is to minimize the mean output power $E\left\{ \left| \mathbf{w}^H \mathbf{y}(t) \right|^2 \right\}$ while maintaining unity response in the look direction. Solving the optimization problem amounts to finding the set of weights that result in the lowest-power array output subject to the directional constraint, while minimizing power presumably reduces the deleterious effects of noise and unwanted signals. To minimize the total output noise while keeping the output signal constant is the same as maximizing the output SNR.

Using the Lagrange method, we can solve the constraint problem, and the expression for the weights is given by

$$\mathbf{w} = \frac{\mathbf{R}^{-1}\mathbf{a}(\theta_d)}{\mathbf{a}^H(\theta_d)\mathbf{R}^{-1}\mathbf{a}(\theta_d)}, \tag{33}$$

where θ_d is the DOA of the desired signal. This scheme requires the number of interferers to be less than or equal to $M-2$, since an array with M elements has $M-1$ degrees of freedom, and one has been utilized by the constraint in the look direction. Multipath arrivals in wireless communication environment may violate this condition, and the beamformer may not be able to achieve the maximization of the output SNR by suppressing all interferers. However, as argued in [42], the beamformer does not have to fully suppress interference, since a few decibels increase in the output SNR can make a large improvement in the channel capacity. The applications of optimal beamformer in mobile communications for interference cancelation and performance improvement are addressed in [42]-[44].

4.d. Minimum Mean Square Error (MMSE) Beamformer

This beamformer requires a reference signal instead of the desired signal direction [45]. As shown in Figure 2, the array output $\mathbf{w}^H \mathbf{y}(t)$ is subtracted from a reference signal $d(t)$ to generate an error signal $e(t) = d(t) - \mathbf{w}^H \mathbf{y}(t)$, which is used to estimate the weights. The beamformer that extracts the desired signal should minimize the error signal in some sense, for example the mean square error (MSE), and this is the motivation for the minimum mean square error (MMSE) beamformer. The MSE is defined as

$$\begin{aligned} E\left\{ \left| e(t) \right|^2 \right\} &= E\left\{ \left| d(t) - \mathbf{w}^H \mathbf{y}(t) \right|^2 \right\} \\ &= E\left\{ \left| d(t) \right|^2 \right\} - \mathbf{r}_{yd}^H \mathbf{w} - \mathbf{w}^H \mathbf{r}_{yd} + \mathbf{w}^H \mathbf{R} \mathbf{w}, \end{aligned} \tag{34}$$

where $\mathbf{r}_{yd} = E\{\mathbf{y}(t)d^*(t)\}$, and $\mathbf{R} = E\{\mathbf{y}(t)\mathbf{y}^H(t)\}$. Setting the gradient of $E\{|e(t)|^2\}$ with respect to $\mathbf{w}^*$ equal to zero yields the well-known Wiener-Hoff solution for the optimal weight vector

$$\mathbf{w} = \mathbf{R}^{-1}\mathbf{r}_{yd}, \tag{35}$$

and the corresponding MMSE of the beamformer is given by

$$MMSE = E\{|d(t)|^2\} - \mathbf{r}_{yd}^H\mathbf{R}^{-1}\mathbf{r}_{yd}. \tag{36}$$

In general, the MMSE beamformer provides higher output SNR compared to the MVDR beamformer in the presence of a weak signal source. As the input signal power becomes large compared to the background noise, the two beamformers produce almost the same results [45]. The enhanced SNR by MMSE beamformer is achieved at the cost of signal distortion by the processor, while the MVDR beamformer is distortionless. Use of reference signals for computing array weights in mobile communications is reported in [46]-[47].

4.e. Adaptive Beamforming Algorithms

The methods presented in the previous paragraphs all require knowledge of the channel statistics, for example the data covariance matrix **R**. In conventional and null-steering beamformers, **R** is required for source DOA estimation; in MVDR beamformer, **R** is used for DOA estimation and weights computation; and in MMSE beamformer, **R** and $\mathbf{r}_{yd}$ are needed for weights estimation. In a practical situation when **R** is unknown, adaptive algorithms can be used to adjust the weights in a recursive manner and track changes in the user signals and interference.

4.e.1. Sample Matrix Inversion (SMI) Algorithm

This method computes the array weights by replacing **R** with its estimates. An unbiased estimate of **R** using L samples $\mathbf{y}(n)$, $n = 1,\ldots,L$, is given by $\hat{\mathbf{R}} = \frac{1}{L}\sum_{n=1}^{L}\mathbf{y}(n)\mathbf{y}^H(n)$. When new samples arrive, $\hat{\mathbf{R}}$ can be updated using

$$\hat{\mathbf{R}}(n+1) = \frac{n\hat{\mathbf{R}}(n) + \mathbf{y}(n+1)\mathbf{y}^H(n+1)}{n+1}, \tag{37}$$

and $\hat{\mathbf{R}}^{-1}$ can be obtained using the matrix inversion lemma as follows

$$\hat{\mathbf{R}}^{-1}(n+1) = \hat{\mathbf{R}}^{-1}(n) - \frac{\hat{\mathbf{R}}^{-1}(n)\mathbf{y}(n+1)\mathbf{y}^H(n+1)\hat{\mathbf{R}}^{-1}(n)}{1+\mathbf{y}^H(n+1)\hat{\mathbf{R}}^{-1}(n)\mathbf{y}(n+1)}, \tag{38}$$

with

$$\hat{\mathbf{R}}^{-1}(1) = \frac{1}{\varepsilon}\mathbf{I}, \quad \varepsilon > 0. \tag{39}$$

A new estimate of the weights $\mathbf{w}_{n+1}$ at time instant $n+1$ can be made using $\hat{\mathbf{R}}(n+1)$ and $\hat{\mathbf{R}}^{-1}(n+1)$. As the number of samples grows, the estimates $\hat{\mathbf{R}}$ and $\hat{\mathbf{R}}^{-1}$ approach their true values $\mathbf{R}$ and $\mathbf{R}^{-1}$, and thus the estimated weights approach the desired weight. Application of SMI for weights estimation in wireless communications has been investigated in [20], [47]-[49].

4.e.2. Least Mean Square (LMS) Algorithm

LMS is one of the most well-known adaptive algorithms [50]. Consider the cost function (34) of MMSE beamformer using a reference signal. This function is an M dimensional quadratic function. A simple recursive procedure for determining the minimum of the quadratic is to update the estimate by an amount proportional to the negative gradient of the error, that is

$$\mathbf{w}_{n+1} = \mathbf{w}_n - \mu \nabla_{\mathbf{w}_n^*} E\left\{\left|e_n\right|^2\right\}, \tag{40}$$

where n is the iteration index, μ is a positive scalar that regulates the step size and controls the convergence characteristic of the algorithm, $e_n = d(n) - \mathbf{w}_n^H \mathbf{y}(n)$, and the gradient is given by

$$\nabla_{\mathbf{w}_n^*} E\left\{\left|e_n\right|^2\right\} = 2E\left\{e_n \cdot \nabla_{\mathbf{w}_n^*}\left(e_n^*\right)\right\} = -2E\left\{e_n \mathbf{y}_n^*\right\} \tag{41}$$

Therefore, the weights update equation can be rewritten as

$$\mathbf{w}_{n+1} = \mathbf{w}_n + \mu E\left\{e_n \mathbf{y}_n^*\right\} \tag{42}$$

To ease the notion, μ is still used in (42). In its standard form, the LMS algorithm uses an instantaneous, and hence, noisy estimate of the gradient to replace the statistical gradient estimate (41). In general, the LMS algorithm iteratively finds the weight vector that minimizes the MSE cost function $E\left\{\left|e_n\right|^2\right\}$ by estimating $E\left\{e_n \mathbf{y}_n^*\right\}$ with $e_n \mathbf{y}_n^*$, a product of the array signals and the error between the reference signal and the array output, and the weight update equation is given by

$$\mathbf{w}_{n+1} = \mathbf{w}_n + \mu e_n \mathbf{y}_n^*, \tag{43}$$

where $e_n = d_n - \mathbf{w}_n^H \mathbf{y}_n$. The most attractive feature of the LMS algorithm is its low computational complexity, and the main drawback of this technique is its slow convergence rate if μ is too small, and potential for instability if μ is too large. There are ways to adjust μ to make a prudent trade-off between stability and speed of convergence [51]. Its use for mobile communication systems has been studied in [42]-[43], [52].

When both source directions and training sequences are not available, blind adaptive algorithms, such as the constant modulus algorithm (CMA) [1], [53], can be developed to exploit a known property of the desired signal, for instance the constant-modulated signal envelope.

5. Direction of Arrival Estimation

Source DOA estimation is a principle function of sensor array processing, and the problem has received considerable attention in the literature [59]-[91]. In this section, we discuss four typical techniques: a) MVDR, also known as an optimal beamformer; b) MUSIC, one of the most well-known and studied schemes; c) ESPRIT, a computation efficient technique; and d) ML, a statistically optimal algorithm representing the best performance achievable. The discussion gives the reader a good foundation to understand other DOA estimators. The application of the algorithms for mobile communications has been investigated in [26]-[30], [32] and [89]-[91].

Consider the data model (18). In most DOA literature, the number of sources N is assumed to be known (given or estimated). The estimation of N is often referred to as the detection problem [54]-[55]. The Akaike's information criterion (AIC) [56] and minimum description length (MDL) [57] algorithms are among the most referred techniques. We also assume that the number of sensors is greater than the number of sources, $M > N$, to guarantee the uniqueness of DOA estimation [58].

5.a. MVDR Estimator

The MVDR beamformer can estimate the source DOA by computing the power spectrum (the mean output power of the array as a function of the direction) and then determining the local maxima. The result estimator is also known as the Capon's algorithm [41]. If we steer the array in direction θ, the weights can be obtained by minimizing the mean output power subject to unity constraint in the look direction as discussed in Section 4.c, and $\mathbf{w}(\theta) = \dfrac{\mathbf{R}^{-1}\mathbf{a}(\theta)}{\mathbf{a}^H(\theta)\mathbf{R}^{-1}\mathbf{a}(\theta)}$. An expression for the power spectrum is given by

$$P(\theta) = E\left\{\left|\mathbf{w}^H(\theta)\mathbf{y}(t)\right|^2\right\} = \left[\mathbf{a}^H(\theta)\mathbf{R}^{-1}\mathbf{a}(\theta)\right]^{-1}. \tag{44}$$

The DOA estimates are obtained by viewing the peaks in the spectrum. Similarly, the conventional beamformer can also be used for DOA estimation by computing the power spectrum

$$P(\theta) = E\left\{\left|\mathbf{w}^H(\theta)\mathbf{y}(t)\right|^2\right\} = \frac{\mathbf{a}^H(\theta)\mathbf{R}\,\mathbf{a}(\theta)}{M^2}, \tag{45}$$

where the weights $\mathbf{w}(\theta) = \frac{1}{M}\mathbf{a}(\theta)$ steer the array in direction θ. The resolving power of the conventional beamformer depends on the beamwidth of the main lobe, which is known as the conventional resolution limit. As expected, the MVDR estimator demonstrates better resolution properties than the conventional beamformer [59].

5.b. MUltiple SIgnal Classification (MUSIC) Estimator

MUSIC [60] exploits the eigenstructure of the data covariance matrix (24), and can be classified as an eigenstructure method. The rationale for this class of algorithms is related to division of information in the data covariance matrix into two vector subspaces, namely, the signal subspace and the noise subspace. These algorithms assume that signals of interest lie in a lower dimensional signal space than the full dimensional space spanned by the data samples [61].

From (24), the data covariance matrix is given by $\mathbf{R} = \mathbf{A}\mathbf{P}\mathbf{A}^H + \sigma_n^2\mathbf{I}$. Assume the N sources to be uncorrelated, thus $\mathbf{P}$ is a diagonal matrix. Now, we can apply eigenvalue decomposition to $\mathbf{R}$. Under ideal conditions, the eigenvalues satisfy

$$\underbrace{\lambda_1 \geq \lambda_2 \geq \cdots \geq \lambda_N}_{N} > \underbrace{\sigma_n^2 = \cdots = \sigma_n^2}_{M-N} \tag{46}$$

where λ_i ($i = 1, \cdots, N$) are principal eigenvalues, all noise eigenvalues equal to σ_n^2, and the number of multiplicity is $M - N$. Assume that $\mathbf{q}_i$ ($i = 1, \cdots, M$) are eigenvectors associated with ordered λ_i. $\mathbf{q}_i$ ($i = 1, \cdots, N$) are principal eigenvectors, and $\mathbf{q}_i$ ($i = N+1, \cdots, M$) are noise eigenvectors. The representation of $\mathbf{R}$ in terms of its eigenvalues and corresponding eigenvectors is as follows:

$$\mathbf{R} = \sum_{i=1}^{N}\left(\lambda_i - \sigma_n^2\right)\mathbf{q}_i\mathbf{q}_i^H + \sigma_n^2\mathbf{I}. \tag{47}$$

Since the signals are uncorrelated, $\mathbf{R}$ can also be expressed as

$$\mathbf{R} = \sum_{i=1}^{N}\left|p_i\right|^2 \mathbf{a}(\theta_i)\ \mathbf{a}^H(\theta_i) + \sigma_n^2\mathbf{I}, \tag{48}$$

where $\left|p_i\right|^2$ is the average power of the ith source. Therefore,

$$\mathbf{A}\mathbf{P}\mathbf{A}^H = \sum_{i=1}^{N}\left(\lambda_i - \sigma_n^2\right)\mathbf{q}_i\mathbf{q}_i^H = \sum_{i=1}^{N}\left|p_i\right|^2 \mathbf{a}(\theta_i)\mathbf{a}^H(\theta_i). \tag{49}$$

It means that the principal eigenvectors $\mathbf{q}_i$ ($i=1,\cdots,N$) are linear combinations of array steering vectors $\mathbf{a}(\theta_i)$ associated with N sources, and vice versa. Equivalently, the vectors $\mathbf{q}_i$ ($i=1,\cdots,N$) and $\mathbf{a}(\theta_i)$ span the same vector subspace – the signal subspace. And the noise eigenvectors $\mathbf{q}_i$ ($i=N+1,\cdots,M$) span the noise subspace, which is orthogonal to the signal subspace.

Once the signal and noise subspaces have been estimated, a search for N directions is made by looking for steering vectors that are as orthogonal to the noise subspace as possible. Let

$$\mathbf{U}_S=\begin{bmatrix}\mathbf{q}_1 & \mathbf{q}_2 & \cdots & \mathbf{q}_N\end{bmatrix}$$

$$\mathbf{U}_N=\begin{bmatrix}\mathbf{q}_{N+1} & \mathbf{q}_{N+2} & \cdots & \mathbf{q}_M\end{bmatrix} \tag{50}$$

the MUSIC algorithm can be written as [60]

$$P(\theta)=\frac{1}{\mathbf{a}^H(\theta)\mathbf{U}_N\mathbf{U}_N^H\mathbf{a}(\theta)} \tag{51}$$

Instead of using the noise subspace, one may use the signal subspace and search for directions with steering vectors contained in this space. This amounts to searching for peaks in [62]

$$P(\theta)=\mathbf{a}^H(\theta)\mathbf{U}_S\mathbf{U}_S^H\mathbf{a}(\theta) \tag{52}$$

It is advantageous to use the one with smaller dimensions in (51) and (52). In practice, the estimates $\hat{\mathbf{U}}_S$ and $\hat{\mathbf{U}}_N$ are used, which are computed from the eigen-decomposition of the sample covariance matrix $\hat{\mathbf{R}}$ (25). The equations in (51) and (52) are the standard forms, known as spectral MUSIC. For ULAs, root-MUSIC [62], a polynomial rooting version of MUSIC, provides higher resolution capabilities. A weighted version of MUSIC, WMUSIC [63], also gives an extension in the resolution capabilities compared to spectral MUSIC.

5.c. Estimation of Signal Parameters via Rotational Invariance Technique (ESPRIT)

ESPRIT [64] is a computationally efficient eigenstructure method, which is free from searching peaks in pseudo spectrum as in previous methods. It requires that all sensors are present in pair-wise matched and co-directional doublets.

To describe mathematically the effect of the translation invariance of the sensor array, it is convenient to describe the array as being composed of two sub-arrays, identical in every aspect although physically displaced from each other by a known displacement vector $\mathbf{\Delta}$ of magnitude Δ measured in wavelengths. Following the data model (18), the outputs of two sub-arrays can be written as

$$\mathbf{x}(t) = \mathbf{A}\mathbf{s}(t) + \mathbf{n}_x(t)$$
$$\mathbf{y}(t) = \mathbf{A}\mathbf{\Phi}\mathbf{s}(t) + \mathbf{n}_y(t) \tag{53}$$

where $\mathbf{s}(t)$ is the vector of signals observed at the reference sensor of the sub-array corresponding to $\mathbf{x}(t)$, and $\mathbf{\Phi}$ is a diagonal matrix of phase delays between doublet sensors for N wavefronts, which is given by

$$\mathbf{\Phi} = diag\ \left\{e^{j2\pi\Delta\cos\theta_1},\ \ldots,\ e^{j2\pi\Delta\cos\theta_N}\right\} \tag{54}$$

and θ_i ($i = 1, \cdots, N$) is measured relative to the direction of $\mathbf{\Delta}$. The sub-array outputs can be combined to yield the total array output vector $\mathbf{z}(t)$ as

$$\mathbf{z}(t) = \begin{bmatrix} \mathbf{x}(t) \\ \mathbf{y}(t) \end{bmatrix} = \overline{\mathbf{A}}\mathbf{s}(t) + \mathbf{n}_z(t), \tag{55}$$

where

$$\overline{\mathbf{A}} = \begin{bmatrix} \mathbf{A} \\ \mathbf{A\Phi} \end{bmatrix},\ \mathbf{n}_z(t) = \begin{bmatrix} \mathbf{n}_x(t) \\ \mathbf{n}_y(t) \end{bmatrix}.$$

Let $\mathbf{E}_X$ and $\mathbf{E}_Y$ be two sets of vectors, which span the signal subspace associated with $\mathbf{x}(t)$ and $\mathbf{y}(t)$ as in (50), respectively. If the output of sub-arrays is sampled simultaneously, $\mathbf{E}_X$ and $\mathbf{E}_Y$ span the same signal subspace. The basic idea behind ESPRIT is to exploit the rotational invariance of the underlying signal subspace.

The signal subspace can also be obtained from the covariance matrix $\mathbf{R}_z$ of $\mathbf{z}(t)$ as in (50), and denote it $\mathbf{E}_S$, where

$$\mathbf{R}_\mathbf{z} = E\left\{\mathbf{z}(t)\mathbf{z}^H(t)\right\} = \overline{\mathbf{A}}\mathbf{P}\overline{\mathbf{A}}^H + \sigma_n^2\mathbf{I}. \tag{56}$$

At the same time, the signal subspace can be spanned by the columns of $\overline{\mathbf{A}}$, namely $\Re\{\mathbf{E}_S\} = \Re\{\overline{\mathbf{A}}\}$. Therefore, there must exist a unique, nonsingular matrix $\mathbf{T}$ such that

$$\mathbf{E}_S = \overline{\mathbf{A}}\mathbf{T}. \tag{57}$$

Furthermore, the invariance structure of the array implies $\mathbf{E}_S$ can be decomposed into $\mathbf{E}_X$ and $\mathbf{E}_Y$ such that

$$\mathbf{E}_S = \begin{bmatrix} \mathbf{E}_X \\ \mathbf{E}_Y \end{bmatrix} = \begin{bmatrix} \mathbf{AT} \\ \mathbf{A\Phi T} \end{bmatrix}, \tag{58}$$

from which it is easily seen that

$$\Re\{\mathbf{E}_X\} = \Re\{\mathbf{E}_Y\} = \Re\{\mathbf{A}\}. \tag{59}$$

Therefore, there must exist a unique, nonsingular matrix $\mathbf{\Psi}$, such that $\mathbf{E}_X\mathbf{\Psi} = \mathbf{E}_Y$. Since $\mathbf{E}_X$ and $\mathbf{E}_Y$ share a common column space, the rank of $\mathbf{E}_{XY} = [\mathbf{E}_X \mid \mathbf{E}_Y]$ is *N*, which implies that there exists a unique rank *N* matrix $\mathbf{F}$ such that

$$\mathbf{0} = [\mathbf{E}_X \mid \mathbf{E}_Y]\mathbf{F} = \mathbf{E}_X\mathbf{F}_X + \mathbf{E}_Y\mathbf{F}_Y = \mathbf{ATF}_X + \mathbf{A\Phi TF}_Y \tag{60}$$

where $\mathbf{F} = \begin{bmatrix} \mathbf{F}_X \\ \mathbf{F}_Y \end{bmatrix}$ spans the null-space of $[\mathbf{E}_X \mid \mathbf{E}_Y]$. Defining $\mathbf{\Psi} \overset{\text{def}}{=} -\mathbf{F}_X[\mathbf{F}_Y]^{-1}$, equation (60) can be rearranged to yield $\mathbf{AT\Psi T}^{-1} = \mathbf{A\Phi}$. Assuming $\mathbf{A}$ to be full rank implies

$$\mathbf{T\Psi T}^{-1} = \mathbf{\Phi}. \tag{61}$$

Therefore, eigenvalues of $\mathbf{\Psi}$ must be equal to diagonal elements of $\mathbf{\Phi}$, and DOA estimates can be obtained using

$$\theta_i = \cos^{-1}\left\{\frac{\arg(\lambda_i)}{2\pi\Delta}\right\}, \quad i = 1,\cdots,N, \tag{62}$$

where λ_i ($i = 1,\cdots,N$) are principal eigenvalues of $\mathbf{\Psi}$, and $\arg(\bullet)$ denotes the phase angle. ESPRIT also has some important variations, including multiple invariance ESPRIT [65], virtual interpolated array ESPRIT [66], resolution-enhanced ESPRIT [67], and higher order ESPRIT [68].

5.d. Maximum Likelihood (ML) Estimator

The ML technique estimates source DOAs by maximizing the log-likelihood function, which signifies that signals from those directions are most likely to cause the occurrence of given samples. The ML method produces superior estimates compared to other methods, especially in unfavorable conditions involving low SNR, short data samples, highly correlated or coherent sources, and small array apertures, and thus is of practical interest. It can be used as a standard to compare the performance of other methods. Regarding the source signals, there are two types of models in current use: conditional model, which assumes the signals to be deterministic and unknown sequences; and unconditional model, which assumes the signals to be random. These two models lead to different ML methods, termed CML and UML respectively [69].

5.d.1. Conditional Maximum Likelihood (CML) Estimator

Assume the signals $\mathbf{s}(t)$ to be deterministic and unknown sequences, and the noise $\mathbf{n}(t)$ to be stationary zero-mean white Gaussian process, $E\{\mathbf{n}(t)\mathbf{n}^H(t)\} = \sigma_n^2\mathbf{I}$, therefore $\mathbf{y}(t) \sim G(\mathbf{As}(t), \sigma_n^2\mathbf{I})$, where $G(\bullet)$ denotes Gaussian distribution.

The likelihood function of the snapshots $\mathbf{y}(1)$, …, $\mathbf{y}(L)$ is given by

$$L(\mathbf{y}(1), \cdots, \mathbf{y}(L)) = \prod_{t=1}^{L} \frac{1}{\pi \det[\sigma_n^2 \mathbf{I}]} \exp\left(-\frac{1}{\sigma_n^2} |\mathbf{y}(t) - \mathbf{As}(t)|^2 \right), \tag{63}$$

where, $\det[\bullet]$ stands for the determinant. Thus, the log-likelihood function is

$$\ln L = -L \ln \pi - ML \ln \sigma_n^2 - \frac{1}{\sigma_n^2} \sum_{t=1}^{L} |\mathbf{y}(t) - \mathbf{As}(t)|^2 \tag{64}$$

In (64), σ_n^2, $\boldsymbol{\theta}$ (in $\mathbf{A}(\boldsymbol{\theta})$) and $\mathbf{s}(t)$ are unknown parameters.

Firstly, we fix $\boldsymbol{\theta}$ and $\mathbf{s}(t)$, and calculate the derivatives of (64) with respect to σ_n^2,

$$\frac{\partial \ln L}{\partial \sigma_n^2} = -\frac{ML}{\sigma_n^2} + \frac{1}{\sigma_n^4} \sum_{t=1}^{L} |\mathbf{y}(t) - \mathbf{As}(t)|^2 . \tag{65}$$

We then get,

$$\sigma_n^2 = \frac{1}{ML} \sum_{t=1}^{L} |\mathbf{y}(t) - \mathbf{As}(t)|^2 \tag{66}$$

Secondly, fixing σ_n^2 and $\boldsymbol{\theta}$, we calculate the derivatives of (64) with respect to $\mathbf{s}(t)$,

$$\frac{\partial \ln L}{\partial \mathbf{s}(t)} = \frac{2}{\sigma_n^2} \mathbf{A}^H [\mathbf{y}(t) - \mathbf{As}(t)], \tag{67}$$

it arrives at

$$\mathbf{s}(t) = (\mathbf{A}^H \mathbf{A})^{-1} \mathbf{A}^H \mathbf{y}(t) \tag{68}$$

Finally, substituting (66) and (68) back into (64), we obtain the following maximization problem,

$$\arg\max_{\boldsymbol{\theta}}\left\{-ML\ln\frac{1}{ML}\sum_{t=1}^{L}\left|\mathbf{y}(t)-\mathbf{A}\left(\mathbf{A}^H\mathbf{A}\right)^{-1}\mathbf{A}^H\mathbf{y}(t)\right|^2\right\}, \tag{69}$$

which is equivalent to the following minimization problem

$$f_{CML}(\boldsymbol{\theta})=\arg\min_{\boldsymbol{\theta}}\left\{tr\left[(\mathbf{I}-\mathbf{A}\left(\mathbf{A}^H\mathbf{A}\right)^{-1}\mathbf{A}^H)\hat{\mathbf{R}}\right]\right\} \tag{70}$$

where $tr[\bullet]$ denotes the trace, and $\hat{\mathbf{R}}$ is the sample covariance matrix (25).

5.d.2. Unconditional Maximum Likelihood (UML) Estimator

If we assume that both the signals and the noise are stationary, temporally white, zero-mean complex Gaussian random processes with second-order moments satisfying (23), following a similar derivation procedure, we can conclude that the UML estimator is given by minimizing (71)

$$f_{UML}(\boldsymbol{\theta})=\ln\det\left[\mathbf{A}\tilde{\mathbf{P}}\mathbf{A}^H+\tilde{q}\mathbf{I}\right] \tag{71}$$

where

$$\tilde{\mathbf{P}}=\left(\mathbf{A}^H\mathbf{A}\right)^{-1}\mathbf{A}^H\hat{\mathbf{R}}\mathbf{A}\left(\mathbf{A}^H\mathbf{A}\right)^{-1}-\tilde{q}\left(\mathbf{A}^H\mathbf{A}\right)^{-1},$$

$$\tilde{q}=\frac{1}{M-N}tr\left\{\left[\mathbf{I}-\mathbf{A}\left(\mathbf{A}^H\mathbf{A}\right)^{-1}\mathbf{A}^H\right]\hat{\mathbf{R}}\right\},$$

and $\ln[\bullet]$ denotes the natural logarithm.

Literature [69] demonstrates that for uncorrelated sources, the statistical performances of CML and UML are similar; while for highly correlated or coherent sources, UML is significantly superior. For UML, the stochastic Cramer-Rao bound (CRB) can be achieved as $N\to\infty$ or SNR $\to\infty$; while for CML, the corresponding bound cannot be attained if $M<\infty$, even though $N\to\infty$ or SNR $\to\infty$.

The optimization of likelihood functions (70) and (71) is a nonlinear optimization problem. In the absence of a closed form solution, it requires iterative schemes for solutions. Many such schemes have been proposed for solving this problem, such as the alternating projection method [70], the simulated annealing method [71], the expectation maximization algorithm [72], and the data supported grid search technique [73]. A fast algorithm [74]-[76] based on a modified genetic algorithm is presented for solutions in general cases.

5.e. Performance Evaluation and Comparison

The main performance measures for a DOA estimator include bias, variance, and probability of resolution, which are complicated functions of source SNR, data sample size, number and directions of sources, and array geometry. A poor estimate generally results from using shorter snapshots and sources with lower SNR. Even though bias and variance both play important roles in direction estimation, the effect of bias is more critical in the threshold region. The variance is often evaluated against the CRB, which provides a lower bound on the covariance matrix of any unbiased estimator and is expected to be a good performance predictor for large samples.

The stochastic CRB associated with DOA estimation in white Gaussian noise is given by [77]

$$\mathrm{CRB}(\boldsymbol{\theta}) = \frac{\sigma_n^2}{2L}\left[\mathrm{Re}\left\{\left[\mathbf{D}^H\mathbf{P}_A^{\perp}\mathbf{D}\right]\odot\left[\mathbf{P}\mathbf{A}^H\mathbf{R}^{-1}\mathbf{A}\mathbf{P}\right]^T\right\}\right]^{-1} \tag{72}$$

where $\mathbf{X}\odot\mathbf{Y}$ denotes the Hadamard product of matrices $\mathbf{X}$ and $\mathbf{Y}$, i.e., $(\mathbf{X}\odot\mathbf{Y})_{ij} = \mathbf{X}_{ij}\mathbf{Y}_{ij}$, $\mathbf{P}$ is the signal covariance matrix, and

$$\mathbf{P}_A^{\perp} = \mathbf{I} - \mathbf{A}\left(\mathbf{A}^H\mathbf{A}\right)^{-1}\mathbf{A}^H,$$

$$\mathbf{D} = \left[\left.\frac{\partial\mathbf{a}(\theta)}{\partial\theta}\right|_{\theta=\theta_1} \cdots \left.\frac{\partial\mathbf{a}(\theta)}{\partial\theta}\right|_{\theta=\theta_N}\right] \tag{73}$$

In the follows, we present a numerical example to compare the performance of UML, ESPRIT, MUSIC and MVDR, and evaluate those estimators against the CRB. Each simulated point is calculated based on 500 independent Monte-Carlo trials. The performances of those methods are compared at two aspects: a) DOA estimation root-mean-squared error (RMSE), which is calculated as

$$RMSE = \sqrt{\frac{1}{NN_{runs}}\sum_{l=1}^{N_{runs}}\sum_{i=1}^{N}[\hat{\theta}_i(l) - \theta_i]^2}\,, \tag{74}$$

where N is the number of sources, $\hat{\theta}_i(l)$ is the estimate of the ith DOA achieved in the lth run, θ_i is the true DOA of the ith source; b) the ability to resolve closely spaced sources. By definition, two sources are said to be resolved in a given run if $\left|\hat{\theta}_i - \theta_i\right| < \left|\theta_1 - \theta_2\right|/2$, $i = 1,2$ are satisfied.

Since ESPRIT requires two displaced subarrays of matched doublets and is only applicable to certain geometry such as ULA, an 8-element ULA with half-wavelength element spacing is considered. Because correlated sources due to, for example multipath propagation are of practical interest in mobile communications, two equal-power correlated signals with the correlation factor $\gamma = 0.8$ are assumed to illuminate the array from 61° and 64° relative to the end-fire. The number of snapshots is 40, and SNR is varied.

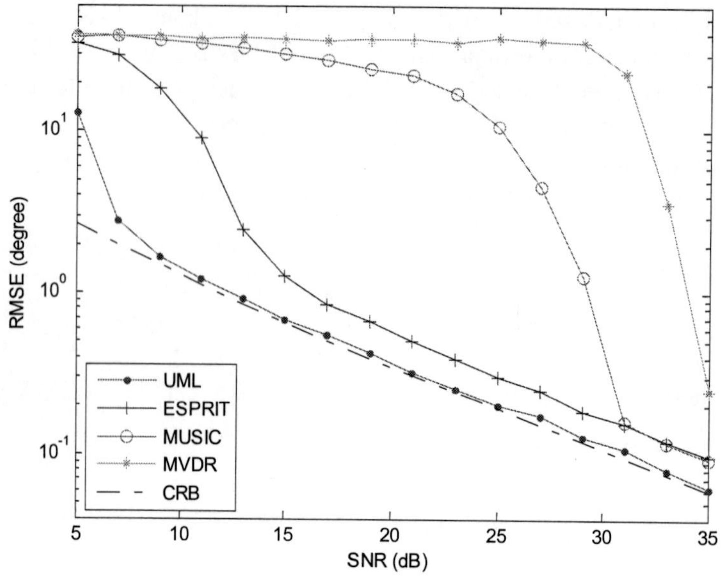

Figure 7. DOA estimation RMSE of UML, ESPRIT, MUSIC and MVDR versus SNR.

Figure 7 depicts DOA estimation RMSE obtained by UML, ESPRIT, MUSIC and MVDR, and compares them with the CRB. Figure 8 shows the resolution probabilities for the same methods. As can be seen from Figure 7 – Figure 8, UML demonstrates much better performance than the other techniques as a whole, producing more accurate estimates in terms of RMSE, and better source resolving power in terms of resolution probabilities. UML asymptotically attains the CRB when SNR gets higher. ESPRIT performs better than MUSIC and MVDR in the cases of correlated sources, and MVDR demonstrates the strongest threshold effect when SNR is small. Based on the simulation results and performance analysis carried out in the literature, we make some comments on pros and cons of each algorithm for performance evaluation.

1. UML is statistically optimal and yields superior estimates. It can tackle correlated or even coherent arrivals. For Gaussian signals, the estimates of UML are unbiased and efficient. However, UML is computation intensive due to optimization of the multi-dimensional multimodal likelihood function. [69], [77]-[78].
2. MUSIC estimates are the large sample realization of ML estimates in the presence of uncorrelated arrivals, and asymptotically efficient for large arrays. A rigorous bias analysis shows that MUSIC estimates are biased. MUSIC makes a good compromise between computation and accuracy, and is one of the most studied schemes. [77], [79]-[81].
3. ESPRIT is computation-efficient with advantages over other algorithms such as speed, storage, and indifference to array calibration. ESPRIT and MUSIC are two important eigenstructure methods: ESPRIT may yield unbiased estimates with variance less than MUSIC; ESPRIT is more robust than MUSIC and can handle

correlated arrivals; but ESPRIT requires specific array geometry. [64], [80], [82]-[83].

4. MVDR is suitable for both DOA estimation and optimal beamforming, which is less sensitive to perturbations and model errors than ML, MUSIC and ESPRIT. [59], [84]-[85].

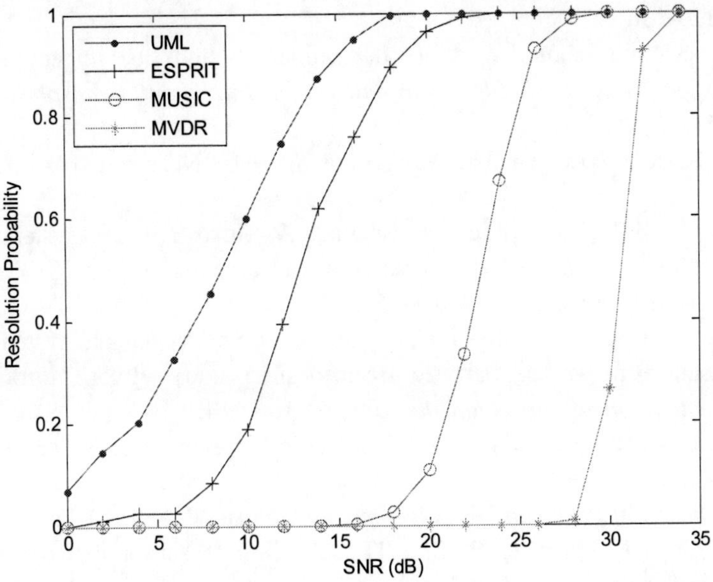

Figure 8. Resolution probabilities of UML, ESPRIT, MUSIC and MVDR versus SNR.

6. Conclusion

Smart antennas are often classified as either switched-beam or adaptive-array systems. Each class of smart antennas has its own inherent cost-complexity trade off. The relative merits of both implementations for ITS have been discussed. This chapter deals with the signal processing aspects of smart antenna systems. In general, smart antenna system design and algorithm selection are a trade-off between complexity and performance. Various beamforming algorithms using spatial and temporal reference, and the adaptive implementations are described. DOA estimation is an essential function of antenna arrays, which can be utilized to orient the beamformer or be exploited for location positioning. There exists many DOA estimation algorithms with conflicting demands of accuracy and computation. Some of the typical algorithms are discussed and analyzed. Smart antennas technology will have significant impact on future intelligent transportation systems. It will support various transportation applications, trigger many new services, and make the future ITS to be less complex while providing more attractive and convenient services. It is believed that this technology has great potential to make transportation systems operate more safely and efficiently, with less congestion, pollution, and environmental impact.

References

[1] Liberti J, Rappaport T. *Smart Antennas for Wireless Communications: IS-95 and Third Generation CDMA Applications*. Upper Saddle River, NJ: Prentice-Hall, 1999.

[2] Drane C, Rizons C. *Positioning Systems in Intelligent Transportation Systems*. Norwood, MA: Artech House, 1998.

[3] Chryssomallis M. Smart antennas. *IEEE Antennas and Propagation Magazine*. 2000;42:129-136.

[4] Zhao Y. Mobile phone location determination and its impact on intelligent transportation systems. *IEEE Transactions on Intelligent Transportation Systems*. 2000;1:55-64.

[5] Zhao Y. *Vehicle Location and Navigation Systems*. Norwood, MA: Artech House, 1997.

[6] Bellofiore S, Balanis C, Foutz J, Spanias A. Smart-antenna systems for mobile communication networks. Part 1. Overview and antenna design. *IEEE Antennas and Propagation Magazine*, 2002;44:145-154.

[7] Bellofiore S, Foutz J, Balanis C, Spanias A. Smart-antenna system for mobile communication networks. Part 2. Beamforming and network throughput. *IEEE Antennas and Propagation Magazine*. 2002;44:106-114.

[8] Li B, Wang H, Yan B, Zhang C. The research of applying wireless sensor networks to intelligent transportation system (ITS) based on IEEE 802.15.4. in *Proceedings of 6th International Conference on ITS Telecommunications*. 2006; 939-942.

[9] Chen W, Chen L, Chen Z, Tu S. WITS: A wireless sensor network for intelligent transportation system. in *Proceedings of First International Multi-Symposiums on Computer and Computational Sciences*. 2006;2:635-641.

[10] Sawant H, Tan J, Yang Q. A sensor networked approach for intelligent transportation systems. in *Proceedings IEEE/RSJ International Conference on Intelligent Robots and Systems (IROS 2004)*. 2004;2:1796-1801.

[11] Giorgetti G, Cidronali A, Gupta S, Manes G. Exploiting low-cost directional antennas in 2.4 GHz IEEE 802.15.4 wireless sensor networks. in *Proceedings 10th European Conference on Wireless Technologies*. 2007; 217-220.

[12] Bellofiore S, Foutz J, Govindarajula R, Bahceci I, Balanis C, Spanias A, Capone J, Duman T. Smart antenna system analysis, integration, and performance for mobile ad-hoc networks (MANETs). *IEEE Transactions on Antennas and Propagation*. 2002;50: 571-581.

[13] Koubaa H. Reflections on smart antennas for MAC protocols in multihop ad hoc networks. in *Proceedings European Wireless'02*. 2002; 25-28.

[14] Singh H, Singh S. Tone based MAC protocol for use with adaptive array antennas. in *Proceedings IEEE WCNC 2004*. 2004; 1246-1251.

[15] Johnson D, Dudgeon D. *Array signal processing*. USA, Prentice-Hall, 1993.

[16] Tsoulos G. Approximate SIR and BER formulas for DS-CDMA based on radiation pattern characteristics of adaptive antennas. *Electronics Letters*. 1998;34:1802-1804.

[17] Hata M. Empirical formula for propagation loss in land mobile radio services. *IEEE Transactions on Vehicular Technology*. 1980;29:317-325.

[18] Yacoub M. *Foundations of Mobile Radio Engineering*. CRC Press, 1993.

[19] Dakdouki A, Banket V, Mykhaylov N, Skopa A. Downlink processing algorithms for multi-antenna wireless communications. *IEEE Communications Magazine*. 2005;43: 122-127.

[20] Winters J, Salz J, Gitlin R. The impact of antenna diversity on the capacity of wireless communications systems. *IEEE Transactions on Communications*. 1994;42:1740-1751.

[21] Klukas R, Lachapelle G, Fattouche M. Field tests of a cellular telephone positioning system. in *Proceedings 47th IEEE Vehicular Technologies Conference*. 1997;2:470-474.

[22] Rappaport T, Reed J, Woerner B. Position location using wireless communications on highways of the future. *IEEE Communications Magazine*. 1996;34:33-41.

[23] O'Connor J, Alexander B, Schorman E. CDMA infrastructure-based location finding for E911. in *Proceedings 49th IEEE Vehicular Technology Conference*. 1999;3:1973-1978.

[24] Gutowski G, Jalloul L, Golovin E, Nakhjiri M, Yousef N, DeClerck P. Simulation results of CDMA location finding systems. in *Proceedings 49th IEEE Vehicular Technology Conference*. 1999;3:2124-2128.

[25] Niculescu D, Nath B. Ad hoc positioning system (APS) using AOA. in *Proceedings IEEE INFOCOM 2003*, 2003; 1734-1743.

[26] Xiong L. A selective model to suppress NLOS signals in angle-of-arrival (AOA) location estimation. in *Proceedings 9th IEEE International Symposium Personal, Indoor and Mobile Radio Communications (PIMRC'98)*. 1998;1:461-465.

[27] Owen R, Lopes L. Experimental analysis of the use of angle of arrival at an adaptive antenna array for location estimation. in *Proceedings 9th IEEE International Symposium Personal, Indoor and Mobile Radio Communications (PIMRC'98)*, 1998;2:607-611.

[28] Qi Y, Kobayashi H, Suda H. Analysis of wireless geolocation in a non-line-of-sight environment. *IEEE Transactions on Wireless Communications*. 2006;5:672-681.

[29] Gu Z, Gunawan E. Radiolocation in CDMA cellular system based on joint angle and delay estimation. *Wireless Personal Communications*. 2002;23:297-309.

[30] Cong L, Zhuang W. Hybrid TDOA/AOA mobile user location for wideband CDMA cellular systems. *IEEE Transactions on Wireless Communications*. 2002;1:439-447.

[31] Kleine-Ostmann T, Bell A. A data fusion architecture for enhanced position estimation in wireless networks. *IEEE Communications Letters*. 2001;5:343-345.

[32] Litva J, Lo T. *Digital Beamforming in Wireless Communications*. Artech, 1996.

[33] Bartlett M. *An Introduction to Stochastic Process*. New York: Cambridge University Press, 1956.

[34] Johnson D. The application of spectral estimation methods to bearing estimation problems. *Proceedings IEEE*. 1982;70:1018-1028.

[35] Anderson V. DICANNE, a realizable adaptive process. *Journal of the Acoustical Society of America*. 1969;45:398-405.

[36] Friedlander B, Porat B. Performance analysis of a null-steering algorithm based on direction-of-arrival estimation. *IEEE Transactions on Acoustics, Speech, and Signal Processing*. 1989;37:461-466.

[37] Murino V, Trucco A, Regazzoni C. Synthesis of equally spaced arrays by simulated annealing. *IEEE Transactions on Signal Processing*. 1996;44:119-122.

[38] Yan K K, Lu Y L. Sidelobe reduction in array-pattern synthesis using genetic algorithm. *IEEE Transactions on Antennas and Propagation*. 1997;45:1117-1122.

[39] Khodier M, Christodoulou C. Linear array geometry synthesis with minimum sidelobe level and null control using particle swarm optimization. *IEEE Transactions on Antennas and Propagation*. 2005;53:2674-2679.

[40] Li M H, Lu Y L. Dimension reduction for array processing with robust interference cancellation. *IEEE Transactions on Aerospace and Electronic Systems*. 2006;42:103-112.

[41] Capon J. High-resolution frequency-wave number spectrum analysis. *Proceedings IEEE*. 1969;57:1408-1418.

[42] Winters J. Optimum combining in digital mobile radio with cochannel interference. *IEEE Journal on Selected Areas in Communications*. 1984;2:528-539.

[43] Winters J. Optimum combining for indoor radio systems with multiple users. *IEEE Transactions on Communications*. 1987;35:1222-1230.

[44] Suard B, Naguib A, Xu G, Paulraj A. Performance of CDMA mobile communication systems using antenna arrays. in *Proceedings ICASSP-93*. 1993; 153-156.

[45] Griffiths L. A comparison of multidimensional Weiner and maximum-likelihood filters for antenna arrays. *Proceedings IEEE*. 1967;55:2045-2047.

[46] Anderson S, Millnert M, Viberg M, Wahlberg B. An adaptive array for mobile communication systems. *Transactions on Vehicular Technology*. 1991;40:230-236.

[47] Gebauer T, Gockler H. Channel-individual adaptive beamforming for mobile satellite communications. *IEEE Journal on Selected Areas in Communications*. 1995;13:439-448.

[48] Lindskog E. Making SMI-beamforming insensitive to the sampling timing for GSM signals. in *Proceedings 6th IEEE International Symposium Personal, Indoor and Mobile Radio Communications (PIMRC'95)*. 1995;2:664-668.

[49] Passerini C, Missiroli M, Riva G, Frullone M. Adaptive antenna arrays for reducing the delay spread in indoor radio channels. *Electronics Letters*. 1996;32:280-281.

[50] Widrow B, Mantey P, Griffiths L, Goode B. Adaptive antenna systems. *Proceedings IEEE*. 1967;55:2143-2158.

[51] Haykin S. *Adaptive Filter Theory*. Prentice Hall, 3rd ed., 1995.

[52] Jones M, Wickert M. Direct sequence spread spectrum using directionally constrained adaptive beamforming to null interference. *IEEE Journal on Selected Areas in Communications*. 1995;13:71-79.

[53] Treichler J, Agee B. A new approach to multipath correction of constant modulus signals. *IEEE Transactions on Acoustics, Speech, and Signal Processing*. 1983;31:459-472.

[54] Wax M, Kailath T. Detection of signals by information theoretic criteria. *IEEE Transactions on Acoustics, Speech, and Signal Processing*. 1985;33:387-392.

[55] Fuchs J. Estimating the number of sinusoids in additive white noise. *IEEE Transactions on Acoustics, Speech, and Signal Processing*. 1988;36:1846-1854.

[56] Akaike H. A new look at the statistical model identification. *IEEE Transactions on Automatic Control*. 1974;19:716-723.

[57] Rissanen J. Modeling by shortest data description. *Automatica*. 1978;14:465-471.

[58] Wax M, Ziskind I. On unique localization of multiple sources by passive sensor arrays. *IEEE Transactions on Acoustics, Speech, and Signal Processing*. 1989;37:996-1000.

[59] Cox H. Resolving power and sensitivity to mismatch of optimum array processors. *Journal of the Acoustical Society of America*. 1973;54:771-785.

[60] Schmidt R. Multiple emitter location and signal parameter estimation, *IEEE Transactions on Antennas and Propagation*. 1986;34:276-280.

[61] Tufts D, Melissinos C. Simple, effective computation of principal eigenvectors and their eigenvalues and application to high-resolution estimation of frequencies. *IEEE Transactions on Acoustics, Speech, and Signal Processing*. 1986;34:1046-1053.

[62] Barabell A. Improving the resolution of eigenstructure based direction finding algorithm. in *Proceedings ICASSP-83*. 1983; 336-339.

[63] Kaveh M, Bassias A. Threshold extension based on a new paradigm for MUSIC-type estimation. in *Proceedings ICASSP-90*. 1990;5:2535-2538.

[64] Roy R, Kailath T. ESPRIT – Estimation of signal parameters via rotational invariance techniques. *IEEE Transactions on Acoustics, Speech, and Signal Processing*. 1989;37: 984-995.

[65] Swindlehurst A, Ottersten B, Roy R, Kailath T. Multiple invariance ESPRIT. *IEEE Transactions on Signal Processing*. 1992;40:867-881.

[66] Gavish M, Weiss A. Performance analysis of the VIA-ESPRIT algorithm. *IEE Proceedings -F*, 1993;140:123-128.

[67] Hamza R, Buckley K. Resolution enhanced ESPRIT. *IEEE Transactions on Signal Processing*. 1994;42:688-691.

[68] Yuen N, Friedlander B. Asymptotic performance analysis of ESPRIT, higher order ESPRIT, and virtual ESPRIT algorithms. *IEEE Transactions on Signal Processing*. 1996;44:2537-2550.

[69] Stoica P, Nehorai A. Performance study of conditional and unconditional direction-of-arrival estimation. *IEEE Transactions on Acoustics, Speech, and Signal Processing*. 1990;38:1783-1795.

[70] Ziskind I, Wax M. Maximum likelihood localization of multiple sources by alternating projection. *IEEE Transactions on Acoustics, Speech, and Signal Processing*. 1988;36: 1553-1560.

[71] Sharman K. Maximum likelihood estimation by simulated annealing. in *Proceedings ICASSP-88*. 1988; 2741-2744.

[72] Miller M, Fuhrmann D. Maximum likelihood narrow-band direction finding and the EM algorithm. *IEEE Transactions on Acoustics, Speech, and Signal Processing*. 1990; 38:1560-1577.

[73] Stoica P, Gershman A. Maximum-likelihood DOA estimation by data-supported grid search. *IEEE Signal Processing Letters*. 1999;6:273-275.

[74] Li M H, Lu Y L. Genetic algorithm based maximum likelihood DOA estimation. in *Proceedings Radar 2002*. 2002; 502-506.

[75] Li M H, Lu Y L. Improving the performance of GA-ML DOA estimator with a resampling scheme. *Signal Processing*. 2004;84:1813-1822.

[76] Li M H, Lu Y L. Accurate direction-of-arrival estimation of multiple sources using a genetic approach. *Wireless Communications and Mobile Computing*. 2005;5:343-353.

[77] Stoica P, Nehorai A. MUSIC, maximum likelihood, and Cramer-Rao bound. *IEEE Transactions on Acoustics, Speech, and Signal Processing*. 1989;37:720-741.

[78] Viberg M, Ottersten B, Nehorai A. Performance analysis of direction finding with large arrays and finite data. *IEEE Transactions on Signal Processing*. 1995;43:469-477.

[79] Xu X, Buckley K. Bias analysis of the MUSIC location estimator. *IEEE Transactions on Signal Processing*. 1992;40:2559-2569.

[80] Stoica P, Nehorai A. Performance comparison of subspace rotation and MUSIC methods of direction estimation. *IEEE Transactions on Signal Processing*. 1991;39:446-453.

[81] Weiss A, Friedlander B. Effect of modeling errors on the resolution threshold of the MUSIC algorithm. *IEEE Transactions on Signal Processing*. 1994;42:1519-1526.

[82] Ottersten B, Viberg M, Kailath T. Performance analysis of the total least squares ESPRIT algorithm. *IEEE Transactions on Signal Processing*. 1991;39:1122-1135.

[83] Soon V, Huans Y. An analysis of ESPRIT under random sensor uncertainties. *IEEE Transactions on Signal Processing*. 1992;40:2353-2358.

[84] Lacoss R. Data adaptive spectral analysis method. *Geophysics*. 1971;36:661-675.

[85] Nuttall A, Carter G, Montaron E. Estimation of two-dimensional spectrum of the space-time noise field for a sparse line array. *Journal of the Acoustical Society of America*. 1974;55:1034-1041.

[86] Li M H, Lu Y L. Maximum likelihood DOA estimation in unknown colored noise fields. *IEEE Transactions on Aerospace and Electronic Systems*. 2008;44:1079-1090.

[87] Li M H, Lu Y L. Source bearing and steering-vector estimation using partially calibrated arrays. *IEEE Transactions on Aerospace and Electronic Systems*. 2009;45(4).

[88] Li M H, Lu Y L, Chen I M, Chen H H. High resolution direction-of-arrival estimation and location determination in wireless networks. *International Journal of Computer Research*. 2008;16:341-370.

[89] Biedka T, Reed J, Woerner B. Direction finding methods for CDMA systems. in *Proceedings 13th Asilomar Conference on Signals, Systems and Computers*. 1996;637-641.

[90] Wigren T, Eriksson A. Accuracy aspects of DOA and angular velocity estimation in sensor array processing. *IEEE Signal Processing Letters*. 1995;2:60-62.

[91] Klukas R, Fattouche M. Line-of-sight angle of arrival estimation in the outdoor multipath environment. *IEEE Transactions on Vehicular Technology*. 1998;47:342-351.

PART 2.
PROTOCOLS

In: Wireless Technologies in Intelligent Transportation… ISBN: 978-1-60741-588-6
Editors: Ming-Tuo Zhou et al, pp. 87-110

Chapter 4

COGNITIVE ROUTING PROTOCOL FOR SENSOR-BASED INTELLIGENT TRANSPORTATION SYSTEM

Rajani Muraleedharan and Lisa Ann Osadciw
Department of Electrical and Computer Engineering
Syracuse University, Syracuse, New York 13210

Abstract

Intelligent Transportation Systems (ITS) using wireless and mobile ad-hoc sensor networks focus on adding information and communications to our transportation infrastructure and vehicles. ITS has inspired many autonomous applications and has transformed our day to day lives. The challenges faced by ITS are response time, data aging and communication cost caused due to rapid changes and movement of vehicles. These challenges frustrate the user because they result in inaccurate information, higher cost of using the system, and mediocre system reliability. Hence, an intelligent multi-objective approach with a sensor focus will deal with these problems by reducing costs, increasing robustness, and improving information accuracy by consistently making the optimum trade-off decisions that will benefit the user. In this chapter, we analyze routing protocols and their importance in disseminating information using distributed sensor technology while trading off the prevention of data aging with the efficient use of sensor resources.

Abbreviations

ACK	Acknowledgement
ACO	Ant Colony Optimization
AEADMRA	Ant-based Energy-Aware Disjoint Multipath Routing Algorithm
AS	Ant System
BER	Bit Error Rate
BPSK	Binary Phase Shift Keying
BS	Base Station

CBRP	Cluster Based Routing Protocol
CCLS)	Cognitive Cross-Layered System
CEDAR	Core Extraction Distributed Ad-hoc Routing
CRP	Cognitive Routing Protocol
DoS	Denial of Service
DREAM	Distance Routing Effect Algorithm for Mobility
DSDV	Destination-Sequenced Distance-Vector
DSR	Dynamic Source Routing
EA	Evolutionary Algorithm
ExOR	Extremely Opportunistic Routing
GFSK	Gaussian Frequency Shift Keying
GPS	Global Position System
GSR	Geographical Source Routing
H-LANMAR	Hierarchical Landmark Routing Protocol
IIS	Intelligent Inference System
ITS	Intelligent Transportation System
IPS	Intelligent Parking System
MH	Mobile Host
NP	Non-deterministic Polynomial-time
POSets	Partially Ordered Sets
QoS	Quality-of-Service
QPSK	Quadrature Phase Shift Keying
SI	Swarm Intelligence
TORA	Temporally Ordered Routing Algorithm
WRP	Wireless Routing Protocol
WSN	Wireless Sensor Network
UWB	Ultra-wideband
YATI	Yosemite Area Traveler Information System

1. Introduction

Recent advances in computer, micro, nano devices and intelligence algorithms for software as well as other emerging technologies has led to fourth generation networks. Intelligent transportation systems' (ITS) primary task consists of distributing information over a vast geographic area. Tremendous progress has been made in the available information for ITS applications that is non-traffic related such as, a "yellow pages" for local businesses that dynamically provides business information and operating hours, directions, information on the modes of transportation locally available, road/weather conditions, special events, camping information, etc. Various systems are under development from computers for the car to airport kiosks. Unfortunately, existing systems, GM Onstar, Yosemite Area Traveler Information System (YATI), Orlando TravTek Project, SmartTraveler, Seattle SWIFT, and 511 Traveler Information System, and currently planned systems have many shortcomings such as mobility, scalability, adaptability and cost. Hence the need for a system that can communicate over a wireless medium with improved efficiency, reduced delay and robustness creates a demand for innovation of sensor based ITS.

In ITS, information is distributed based on resource availability, accuracy, and message prioritization. The tiny sensors have limited resources such as battery, bandwidth and memory requiring very controlled exploitation of the sensors. Thus, the trade-off decisions must focus on getting the most information in quantity and quality from the sensors in the network to give users relevant information using the minimum in resources. Unlike other networks, WSNs have several salient features that make them attractive and cost effective in a variety of real-time applications. WSN may be routinely used in commercial products, such as health monitoring, emergency responder, directions, locating resources, and habitat monitoring, to secure military applications, such as remote monitoring with Unmanned Aerial Vehicle (UAV) or Unmanned Ground Vehicles (UGV), border control, battlefield monitoring including seismic, electric and magnetic field sensors, to name a few. Data communicated in sensor based ITS is time sensitive and requires optimal, real-time decisions, which may require automation without human intervention. Hence the need for an intelligent algorithm that self-adapts to varied scenarios is crucial. One key area, routing, is most affected by the sensor topology, situation, environment, and system goals.

Real-time sensor based ITS applications require mission-driven, robust algorithms that adapt to the current situation and environment in real-time. A cognitive routing protocol (CRP) that communicates collaboratively with heterogeneous networks attains an optimal, reliable, adaptable and cost-effective protocol solution.

ITS using wireless and mobile ad-hoc sensor networks has inspired many autonomous applications. Sensor based ITS promises new products in intelligent parking system (IPS), advanced traffic management and safety, advanced travel information management, and incident management tailored for an individual or group of travelers. Wireless sensor network (WSN) based ITS consists of tiny sensors embedded within vehicles that have the ability to communicate to neighboring sensors (vehicles) and surrounding nodes (sensors or databases) (parking, traveler center etc.). The challenges faced by current ITSs are response time, data aging and communication cost in terms of energy and additional delay [1]. Due to limited resource constraints of the sensors, protocols that require heavy computations such as encryption for security cannot be supported . Thus the need for an emergent algorithm that has the capability of adapting with existing protocol technology while avoiding parts consuming too much energyis a necessity.

In this chapter, the evolution of routing including distributive, decentralized and classical with a focus on sensor based ITS are discussed in the second, and third sections respectively. Ant colony optimization is analyzed in the fourth section. The background on evolutionary algorithms, ant colony optimization, is explained with its characteristics. The cognitive routing protocol that communicates and adapts autonomously to the network's requirements in a de-centralized manner to obtain a multi-objective optimal solution is described in detail in the fifth section. The fifth section presents simulation results and discusses the advantages of incorporating cognitive routing into ITS in terms of energy efficiency, quality-of-service (QoS), scalability and robustness.

2. Distributed and De-Centralized Based ITS Approach

Unlike other networks, WSN has several salient features that make it very attractive and cost effective in many real world applications. WSN is routinely used in many applications ranging from mundane commercial to data sensitive military applications such as border control, battlefield monitoring, health monitoring, emergency responder and habitat monitoring to name a few. In this section, the communication and architecture aspect of the WSN is explored, and its impact in ITS is explained.

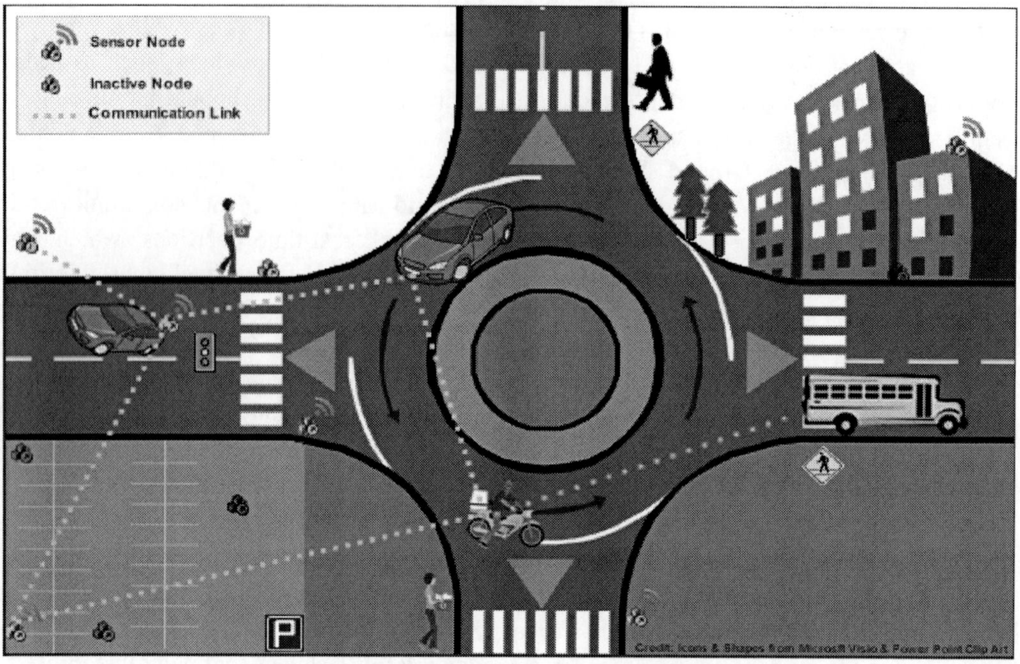

Figure 1. Sensor Based Intelligent Transportation System.

Figure 1 illustrates the system model of a cooperative, distributive and de-centralized Sensor based ITS. The sensors are embedded in the vehicles and surrounding. The nodes state can vary based on its energy level and environmental conditions. A cooperative network benefits the users with different wireless technologies and QoS. The model shown below redirects the traffic to a parking lot using intelligent parking system feature. Thus a collaborative approach is advantageous in situation like emergency evacuation or traffic congestion.

2.a. Sensor Nodes

WSN are comprised with many sensors that are tiny, inexpensive, and battery driven. The sensors have limited lifetime, bandwidth and memory due to non-renewable power supply and size. The function may include sensing, collecting and distributing data in a multi-hop fashion to the destination or sink node. Each sensor has a sensing coverage of radius r, which is primarily dependent on its available sensing energy as shown in Figure 2. The inactive node leads to lost coverage, and any distress event could be left undetectable. A single node is incapable of computing heavy measurements or relaying long messages in one route. Hence collaboration between multiple nodes is required to achieve a common goal of communicating messages using minimal resources.

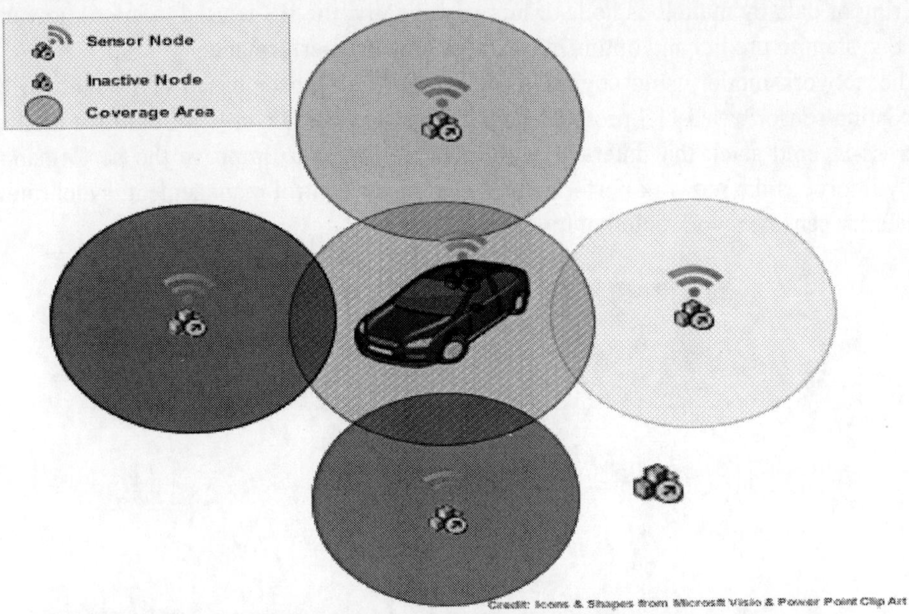

Figure 2. Sensor node and coverage for Intelligent Transportation System.

There are different types of sensors, such as temperature, heat, pressure, mechanical, humidity, biometric, motion, etc. Each sensor has its own functionalities; apart from them they also participate in routing to relay messages. There are several challenges in maintaining the lifetime of the network,

1) Location awareness
2) Quality of service
3) Energy consumption
4) Scalability
5) Adaptability
6) Security

In sensor based ITS, the sensors are embedded to vehicles, and communicate collaboratively with neighboring vehicles and base stations (BS) or mobile hosts (MH) that may have different platforms. Since each sensor is tailored for a particular mission, a network architecture that works cognitively using a cross-layer approach is preferable.

2.b. Network Architecture

Sensor based ITS improves the performance of existing ITS by co-existing with current technologies and adapting to new ones. Hence the architecture of sensor based ITS comprises of three sub-systems: cognitive cross-layered system (CCLS), security system, and intelligent inference system (IIS). The CCLS adapts to different signals, traffic, communication and system availability to attain targeted performance. The security system identifies threats and tampering of data by malicious node or humans. Finally, the IIS is used to infer knowledge of the ITS system to predict and optimize its future network performance.

The network model using cognitive intelligence for sensor based ITS is shown below. The existing sensor based ITS protocol stack uses monolithic layer [2]. Cognitive intelligence is a cross-layered stack that interacts with multiple layers to improve the performance and directly improves the network performance. The power control management, synchronization and security schemes work collaboratively.

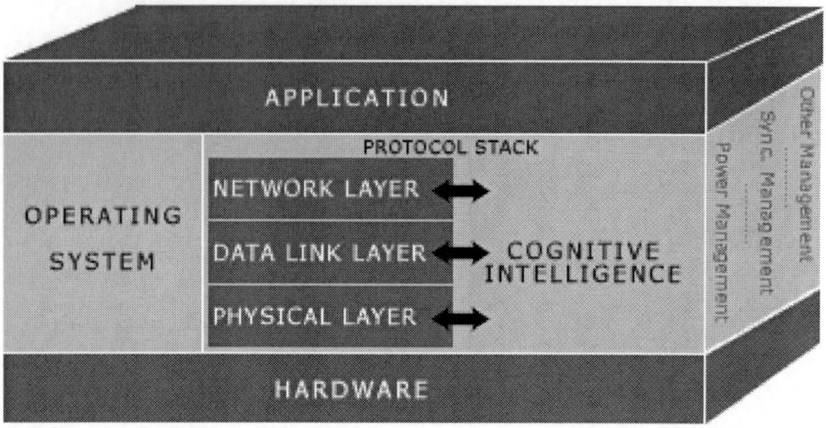

Figure 3. Network Model: Cognitive Intelligent Transportation System.

2.c. Data Communication and Processing

The communication links between nodes are asymmetric, due to multipath fading, shadowing, environmental conditions, bandwidth availability, energy, etc., making communication challenging. Sensor based ITS requires adaptive coverage, communication and processing. The power control scheme in WSN helps in setting different modes for the sensor to sense, transmit and idle; each with different energy levels. Sensor communication is half-duplex; it can either transmit or receive but cannot achieve both simultaneously.

Data communications between sensors and BS/MH is triggered by a query. These queries can be generated continuously, based on priority, time based or on an as needed basis. A continuous query is deterministic and is targeted towards data from a location or region of interest. During time sensitive or critical scenarios such as distress signals or accidents, an event is triggered and given higher priority to be queried to BS/MH or nearby sensors to alert of oncoming traffic congestion or nearest healthcare facilities. Such events are non-deterministic and are classified as priority queries. Timely queries are made based on an individual's interest during any pre-set time. For example, the busy traffic hours and delays due to construction are obvious times when a critical message may need to break through the heavy communication traffic. The need-based queries are sent when any node or BS requires additional information for validation purposes. Upon sensing any event, nodes are triggered to collect, fuse, and relay the messages to the destination.

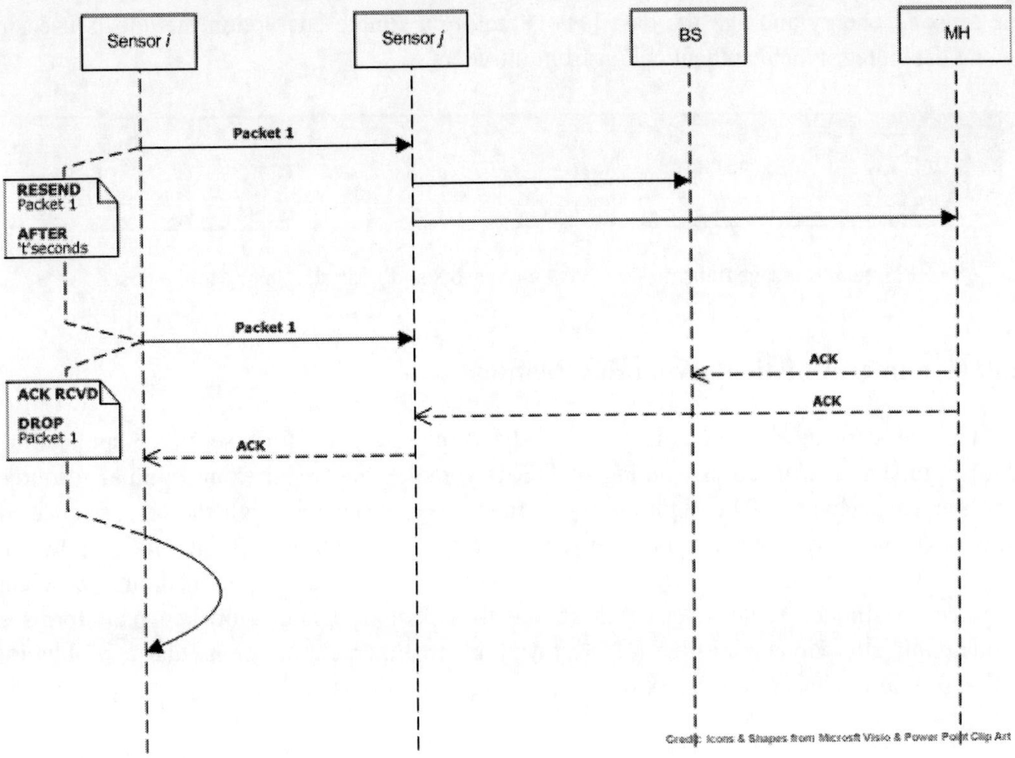

Figure 4. Message transmission between Sensors and Base station or Mobile host.

Figure 4 illustrates message transmission between sensor's i, j, BS or MH. When a query is sensed by a sensor 'i', it transmits the information to it's neighboring node, and the intermediates 'j' nodes acts as relays and finally the destination node 'd' (not shown). Upon receiving the message, the BS/MH sends back an acknowledgement (ACK) to the source. After a periodic interval of 't' seconds, the source again re-sends the data packet if no 'ACK' message is received. If the ACK message is received before re-transmission is performed, the source 'i' alerts its neighbor to drop the message, by referring to the packet's message identifier.

Data packets contain preamble and data information as shown in Figure 5. Sensor nodes cannot handle long preamble messages, as it may lead to high energy depletion during routing. Each node is associated with a unique identification, which relates to its location. The location information can be achieved using GPS, triangulation or multi-lateration methods. Unfortunately, clock synchronization is one of the major issues with sensors, due to thermal noise and other conditions. Applying modulation schemes such as an ultra-wideband modulation promises more accurate location estimation using short pulses [3].

Due to the magnitude of sensors involved in an ITS application, data fusion would reduce the energy at relay nodes caused by redundant message transmission. But data fusion would improve the quality of information received. Fusion of multiple inputs converging at one node is not a feasible solution in sensors, thus only neighbors with correlated data is fused. There is centralized and distributed fusion available for sensor networks; due to ITS distributive characteristics, binary distributive fusion best suits the application. Also, traditional encryption schemes cannot be directly applied due to heavy computation, which will spend the sensor's energy and age the data [4]. Therefore a simple encryption method is used to secure data integrity and authentication from intruders.

Figure 5. Message transmission between Sensors and Base station or Mobile host.

2.d. Nondeterministic Polynomial Problem

ITS infrastructure is dynamic, de-centralized and supports time sensitive application. Unlike conventional methods, routing in ITS is complex and requires an optimal resource management technique. The algorithms designed for ITS should weigh the performance of individual nodes by balancing load, energy, distance, bit error rate (BER) and bandwidth distribution, which together forms a multi-objective function. An optimal and efficient resource constrained route taken from source to sink node using multi-hop path forms a nondeterministic polynomial-time (NP) hard problem. Emergent solutions that can obtain a global solution to improve network throughput and reliability is preferred.

2.e. Optimization

Routing protocols for ITS require self adaptation, self organizing, distributive and de-centralized features to solve multi-objective functions. The optimality of a solution is achieved based on its local and global minima or maxima. The local minima or maxima are defined as the minimum or maximum value obtained from a local search. A global minima or maxima is defined as the minimum or maximum obtained from a global search. It is also considered as the most optimal and feasible solution for the given problem. A local minima or maxima solution can be a global minima or maxima but not vice versa. Since the sensor has varied constraints, network performance is based on the parameters chosen. The energy, distance, BER and bandwidth required to transmit messages needs to be minimal, whereas successful packet delivery and data rate should be maxima. A solution that satisfies these criterions is difficult to attain by conventional methods, hence evolutionary algorithms that is capable of adapting to real time dynamic environment is suited.

3. Classical ITS Routing Approach

There are several strategies used in solving routing issues in sensor network, each having their own advantages and disadvantages. The protocols for sensor networks emerge from the traditional wired network, link state and distance-vector algorithms [5]. In the link state protocol, which is based on link availability or weight, the path is any feasible route for transmitting data linking the source to destination. Whereas in a distance vector algorithm, the route taken is the path with the shortest distance. Due to scalability and limited resource constraints in WSN, the traditional algorithms are modified or extended. The ITS routing protocols are primarily classified as topology, which uses location or position, and extended performance based protocols, which combines multiple performance parameters of interest.

Topology routing protocols depend on the information on existing links in the network for data forwarding. They are further classified as Proactive, Reactive and Hybrid protocols. Proactive protocols use the classical routing schemes discussed previously, distance-vector or link-state and then use this to periodically update the routing table. Reactive protocols find the route based on a request at the source. Whereas, a hybrid protocol is a combination of proactive and reactive routing that is designed to be more scalable and efficient than either technique alone. Location or position based protocols use location information to route data, which reduces the overhead by eliminating the need to create another routing technique but requires location information. This can be obtained either using a global position system (GPS), multi-lateration methods, triangulation, or any other positioning scheme. In performance based protocols, one or more parameters directly influence the efficiency, QoS and reliability of applications. Obtaining sensor position may also be important to a performance protocol as it is to the position based protocols.

3.a. Topology Based Protocols

Topology based protocols are often used with wireless sensor networks due to the importance of minimizing energy usage. These protocols maintain their routes to accomplish a particular goal. One of the goals is minimizing energy usage but others such as minimizing hops, which reduces latency, as well as reducing bandwidth usage may also be included in the routing. The routes are then publicized and maintained in a variety of ways within the network.

In a destination-sequenced distance-vector (DSDV) protocol, a routing table that consist of all the hops for all routes from source to destination is maintained [6]. This table is periodically broadcasted by each node to maintain consistency, which increases the energy loss and bandwidth during table transmission.

The wireless routing protocol (WRP) in [7] maintains four tables: distance table, routing table, link-cost table and message retransmission list (MRL) table. The link status is maintained using the mobile host, which helps in eliminating looping and faster route convergence.

Dynamic source routing (DSR) in [8] is a demand based protocol where the routes discovered with hop counts are cached and continuously updated. If any route error occurs, the table is updated and the hop count is reduced. DSR also supports multipath, which helps with route failure. The ad-hoc on-demand routing protocol (AODV) in [9] combines DSR and DSDV,but it supports only unicast routing i.e., symmetric links.

Temporally ordered routing algorithm (TORA) by Park and Corson in [10] performs three functions: route discovery, maintenance and route erasure. The only disadvantage is that TORA assumes that all clocks at the nodes are synchronized, which is never the case in reality.

Hierarchical landmark routing protocol (H-LANMAR) in [11] is also a hybrid routing protocol. It consists of super-nodes equipped with high computational power, which serve as the backbone network to reduce long hop counts. Thus, the H-LANMAR protocol assumes that super-nodes are networked for the high computational power needed to provide the route and security.

Cluster-based routing protocol (CBRP) in [12] is another hybrid protocol, where the network is partitioned into clusters to reduce the overhead. Each node knows its neighbors by joining a cluster but does not have to know the nodes in the other clusters. If a destination node is outside the level, then the cluster head (CH) is used for relaying the message. The overhead needed in electing new cluster heads for dynamically changing environments is computationally high. Belding-Royer in [13] proposed the adaptive routing using clusters (ARC) protocol which is another cluster based hybrid protocol that adapts the cluster head election based on subset changes. The scalability and dynamics of the environment are designed into the protocol by supporting additional mobiles nodes using multi-hierarchical protocol (ARCH).

3.b. Location Based Protocols

Location based protocols focus on the routing algorithm in terms of location and distance. These are important factors since energy is used in proportion to distance. Thus shorter routes will use less energy to pass the message. The condition of the channel and sensors comprising the route can be used by the protocol to select the stronger, more robust route. In wireless sensor networks, the route is the major contributing factor to the lifetime of the sensors in the network. Two such protocols are introduced in this section.

Distance routing effect algorithm for mobility (DREAM) tracks the sensor nodes and updates their location if they move in a table [14]. Hence stationary nodes do not have to update their location frequently reducing the bandwidth cost. Unfortunately, DREAM is vulnerable to denial-of-service attacks on stationery nodes by intruders. In [15] geographical source routing (GSR), the protocol uses location information supported by maps for correlated routing. Whereas spatially-aware routing (SAR) in [16], the message source includes a list of intermediate geographical location in its packet header and selects a minimum distance route.

The mobility-centric data dissemination algorithm (MDDV [17]) uses digital maps and GPS knowledge to forward message along a predefined routes. The geographic protocol may be constrained to a particular sensor node location making it infeasible for dynamic environments. Roger and Toh in [18] proposed a node-centric routing protocol that relied on routing states. Thus maintaining these states for large networks like ITS incurs a large overhead. Also, the protocol must be scalable for ITS due to its dynamics. An opportunistic model was proposed by Wolfson and Xu in [19], where the data is disseminated based on geographic location. In [20] the extremely opportunistic routing (ExOR) data is forwarded to the next hop based on some probabilistic model.

3.c. Performance Based Protocols

Performance based protocols allow the network system designer to focus on achieving very specific goals within the system. The challenge is in selecting the appropriate performance parameters. The protocol performance based solely on distance or link availability is not suitable for WSN using battery-driven sensors due to the criticality of having enough transmit energy to communicate as well as sense and compute. There are other key factors to consider as well. Hence a QoS based protocol [21] is applied to network where traffic quality, time and other factors like packet delivery are characterized. In [22] Core extraction distributed ad-hoc routing (CEDAR) protocol is proposed for small and medium networks. CEDAR consists of three main functions; core extraction, link state propagation and route computation. The core extraction function provides the infrastructure; state propagation provides link status, which is incorporated into the route computation.

Intelligent protocols have been designed to address multiple objectives. Routing based on evolutionary computation is proposed in [23] where distributed system is achieved at the price of computation. Genetic algorithm (GA) is capable of solving multi-objective problems, but it falls short in finding the global optima [23].

In [24] Sun proposed the GVgrid protocol, which utilizes link information for stable and longer route lifetimes with fixed sources, but cannot be applied to a realistic environment where traffic is dynamic, unstable and multiple requests need to be processed simultaneously. The ant-based energy-aware disjoint multipath routing algorithm (AEADMRA) is proposed in [25]. This protocol uses swarm intelligence based on ant colony optimization (ACO) technique to discover multiple routes with less energy depletion and requires low routing overhead.

The link, distance and QoS based protocols have paved the way for the evolution of cross-layered protocols [26]. A detailed survey on the different types of protocols is given in [27, 28]. The criterion used to select the protocol primarily depends on the application requirements and performance.

4. Background: Ant Colony Optimization

There are many algorithms available for routing optimization such as genetic, simulated annealing, tabu search, and others. Each approach possesses trade-offs depending on the application or use. A critical selection criterion for an algorithm is reduced delay and the probability of obtaining an optimal and reliable solution. The Ant Colony Optimization (ACO) algorithm [35] is a learning algorithm with characteristics such as robustness and versatility solves any NP hard problem. In this section, the background the evolution of ACO and its characteristics is summarized.

4.a. Evolution of ANT System

The natural adaptability of insects and other creatures to survive and thrive through cooperative behavior has astounded scientist and researchers enough to analyze the swarm behavior of ants, bees, and birds and attempt to artificially replicate it. In nature, ants set out to find food from their home cooperatively. A special group of ants are appointed with this task. Stumbling upon an obstacle, the ants choose alternate routes to reach their destination. Over time, the ants find and share the best routes to food. If anything changes in the environment to affect these routes, the ants simply adapt. The swarm behavior of the ants when implemented in engineering problems are capable of solving many complex, dynamic optimization issues.

Swarm intelligence (SI) [36] evolved as the collective behavior from a group of simple autonomous agents that work collaboratively in achieving a common goal. The ACO algorithm utilizes the features of the ants in a real time environment. The agents (ants) in the system communicate interactively either directly or indirectly in a distributed problem-solving manner.

4.b. Characeristics of Ant Colony Optimization

The three main characteristics of ACO are pheromone deposition, state transition probability and tabu-list. Figure 6 shows the agent traversing a maze interactively using low and high pheromone deposition. Upon finding an obstacle, the agents deposit low pheromones along the path warning others of the problem and move forward in their quest to find the destination.

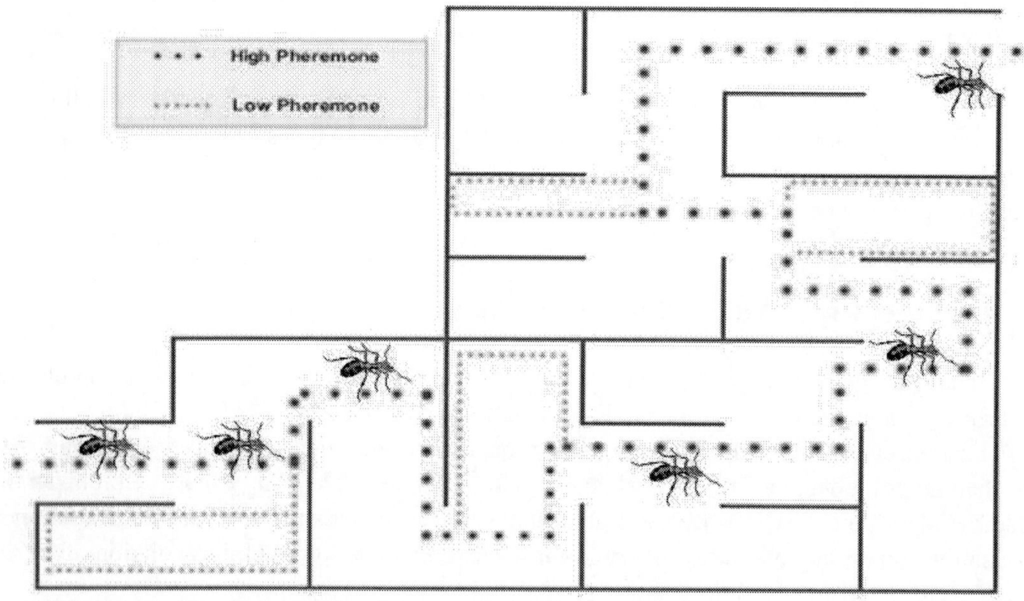

Figure 6. Agents traversing a maze.

Pheromone Deposition

Ants traveling from source to destination interact with other ants using a chemical substance called 'pheromones'. Similarly in the artificial ACO algorithm, pheromone deposition is implemented depending on the visibility of the ant agent. An agent is an entity capable of performing tasks autonomously using predefined behaviors or a set of criterion. In this application, agents mimic the swarm behavior of ants or other creatures. The pheromone deposition along a route increases as more agents traverse through the route. Also, the attractiveness of a route increases so that the probability that an agent takes a particular route increases. Agents move toward an optimal solution by sharing their knowledge with their neighbors. The initial set of agents deployed in the network traverse through all the nodes in a random manner. Each agent leaves trails by depositing pheromones on the route.

The pheromone depositions on the paths serve as a means of communication between the agents. The agents use the pheromones to help select the best route through the network. The most popular route is one that has the highest level of pheromone deposition. Energy and reduced delay are two important criteria in ITS and are incorporated in the pheromone deposition function ψ by

$$\psi_{ij}(t) = \rho\psi_{ij}(t-1) + \frac{Q}{\eta_t} \tag{1}$$

where ψ is the pheromone deposition; i and j are the source and destination; (t-1) is the previous pheromone value; η_t is the visibility of the ant for an entire tour; ρ is the memory; and Q is an arbitrary value.

The pheromones on all paths are updated at the end of a tour. Thus the route taken by the agent is both energy aware and time sensitive. A key feature of the ACO is that the pheromone deposition equation can be modified based on the performance requirements making it adaptive.

State Transition Probability

The transition probability of an agent is defined as the probability taken by the agent to choose a particular node as its next hop towards its destination. Thus, the probability function (P) includes the performance parameters that impact the network, such as distance and energy,

$$P_{ij} = \frac{(\psi_{ij}(t))^{\alpha}.(\eta_{ij}(t))^{\beta}}{\sum_{k}(\psi_{ik}(t))^{\alpha}.(\eta_{ik}(t))^{\beta}} \tag{2}$$

where η is the link cost function, α and β are the power on the pheromone deposition and cost function. The link cost function, η, is also based on the performance parameters. Hence if the distance and energy are of primary concern, then

$$\eta_{ij}=\text{cost}\ (D, E). \tag{3}$$

This cost function can be modified based on maximization or minimization forming a multi-objective min-max function.

Tabu List

A Tabu-list serves as memory tool listing the set of nodes that a single agent has visited. The agent's goal is to visit nodes in the network depending on the number of hops. This list helps agents to avoid looping that can occur during routing. The tabu list eliminates any redundant route traversals thereby conserving energy and consists of estimated energy availability in the nodes for the particular sub-optimal route with high reachability. Thus the tabu-list supports energy usage prediction and decisions concerning situation assessment.

The pheromone deposition, tabu-list, and energy monitoring help this novel ACO algorithm to obtain an optimal solution and adapt during node degradation. These three characteristics can be tailored to any specific application making ACO robust and versatile.

Success of an algorithm is defined as attaining the optimal solution. The computation time is defined as the amount of time the algorithm takes to obtain an optimal solution. In some cases, a combination of artificial intelligence such as a Bayesian network [29], and any of the evolutionary algorithms (EA) can achieve better results. A trade-off between factors affecting the overall performance of a system is primarily application dependent. The advantage of using EA is primarily because the population is searched in a parallel manner and not in a one-to-one basis. EAs do not require any derivative information or any auxiliary knowledge. EAs require only the objective function and the fitness level for performing a directive search. It uses only probabilistic transition rules rather than deterministic ones. Therefore, EAs are generally more straightforward to apply. EAs can provide a number of potential solutions to a given problem if all solutions produce identical performance. There exist possibility that EAs could lead to a pre-mature solution, but it's upon the user to define the fitness function in a way to obtain global optima rather than a local optima. In the following section, the constraints of WSN been applied to cognitive routing protocol, an improved version of ACO is discussed.

5. Proposed Approach: Cognitive Routing Protocol

Data communicated in sensor based ITS is time sensitive and requires logical decisions without human intervention. Hence the need for an intelligent algorithm that self-adapts to varied scenarios is crucial. A cognitive routing protocol (CRP) that communicates collaboratively with heterogeneous network to attain optimal, reliable, adaptable and cost-effective is necessary. CRP is derived from the biological aspect of ACO to form a cross layer approach. The CRP successfully finds the global optima rather than the local optima thereby avoiding any stagnation behavior. In addition, the learning rate of the proposed algorithm evolves over time, thus making the CRP adaptive to its environment.

There are tremendous advantages in deploying a sensor-based CRP, which will include self-adaptive, self-organizing algorithms, such as swarm intelligence. The robustness and de-centralized features of the CRP attains a feasible solution even during node degradation. The CRP's adaptiveness is based on the flexible cost function that can be modified to suite any application making it a generic solution.

WSN is faced with many constraints, namely wireless channel, quality of service, and energy efficiency issues.

5.a. Wireless Channel Constraint

The sensor communication through a wireless medium affects the quality of data transmission. The quality of information is affected in sensor based ITS based on the following:

1) The nodes are half-duplex, thus cannot receive packets if node is in transmit mode leading to traffic congestion.
2) Wireless channel has noise, fading and shadowing that could deteriorate transmission.

3) Interference from neighboring nodes, can lead to packet drop.
4) Limited bandwidth restrict transmitting heavy data i.e., biometric
5) Propagation delay
6) Error in clock synchronization affects location information.
7) Error control strategies lead to heavy computation.

The protocol design should consider these constraints to improve the quality and quantity of information.

5.b. Quality of Service Constraint

Quality of service is an important factor in wireless application as it directly affects the performance of the network. Traditional wired networks provide high reliable information without user convenience. Hence in wireless networks a trade-off between different parameters has to be attained in providing good QoS and reliable service. Also, the QoS varies dynamically based on the user's request. Hence a weighing scheme tailored to each individual request would increase the QoS. The common parameters that are used in QoS are distance, energy, and packet delivery.

5.c. Energy Efficiency

Energy efficiency impacts the lifetime of the sensors, which directly reflects the network lifetime. The energy level at each node needs to be tracked, and should be set with different threshold levels to avoid simultaneous node failure. Also, the sensors can communicate using different power control schemes, which allow them to use high energy level to transmit during priority events signal and switch to lower energy level for normal traffic. The use of thresholds at the node also helps in identifying the node's lifetime and its coverage ability.

5.d. Cross-Layer Approach

Every layer of a sensor network reflects the performance of the application. Hence a cross-layer approach that can work collectively to improve the QoS is preferred. There are also some potential risks in cross-layer design as discussed in [30] that leads to a 'spaghetti effect'. Hence, careful parameter consideration is a necessity. Thus an intelligent approach that has the learning ability to improve the reliability of network using cross-layer approach is designed and presented here using cognitive intelligence.

Salient Features

The cognitive routing protocol (CRP) overcomes the constraints of WSN by extending the ACO algorithm using partially ordered sets (POSets). POSets is a weighing scheme that helps in attaining multiple objective solutions using reinforcement.

There are different types of reinforcement namely positive, negative, penalizing and extinction. The positive and negative reinforcement helps to increase the strength of response, whereas the penalizing and extinction characterizes a weaker behavior. A reinforcement parameter is incorporated in CRP to achieve global solution. In applications like Sensor ITS, it is best to improve the learning behavior of the agents rather than penalizing or extinguishing the solution.

The performance of a system is affected due to improper and inflexible design structure. The salient features of the CRP helps in achieving the following

1. Adaptive Modulation Scheme
2. Adaptive Error Correction Scheme
3. Adaptive Power Control Scheme
4. Adaptive Traffic
5. Adaptive QoS
6. Message Prioritization

The performance parameters such as energy, distance, successful packet delivery, packet loss rate and data traffic are weighed to achieve an energy efficient, secure and mission oriented network. The meta-heuristic cross layer protocol combines the factors that affect the performance in physical, MAC and routing layer to achieve maximized performance for using minimal resource thus achieving a multi-objective solution. CRP is a hybrid algorithm that self-adapts, self-organizes and evolves to the network's needs. The salient features of the agents are combined with POSets which helps in attaining an application driven performance.

POSets

One of the major challenges in sensor management is providing a mathematical framework that can consistently represent the complex multidimensional optimization problem. POSets has been used in queuing theory, networking, and lately sensor management [25, 26]. POSets provides a graphical mathematical framework for representing relationships between a finite numbers of elements [31].

POSets formulates weights at each graphical level to flow down the importance of the performance parameter measuring the expected system goal. In this system, the current mission is locating and directing the vehicle or ITS to attain its goal. The performance is broken down to speed of alert, and downtime. Downtime is the percentage of time the system is not functional because a sensor node is depleted of energy. The speed of alert is the response of the system based on the hops, distance and BER in receiving the message.

In Figure 7, the four levels to the mission POSet are illustrated. The performance parameters are weighed based on its importance. The weights are graphically entered on the POSet as shown in Figure 7. The nodes are the items described in circles of Figure 7.

The flow down of weights i.e., the lower case variables on the arrows should sum to one from their source. The entire POSet begins with a value of 1 at the top. Then the arrows exiting that node should sum to 1 or $\sum_{k=1}^{2} x_k = 1$

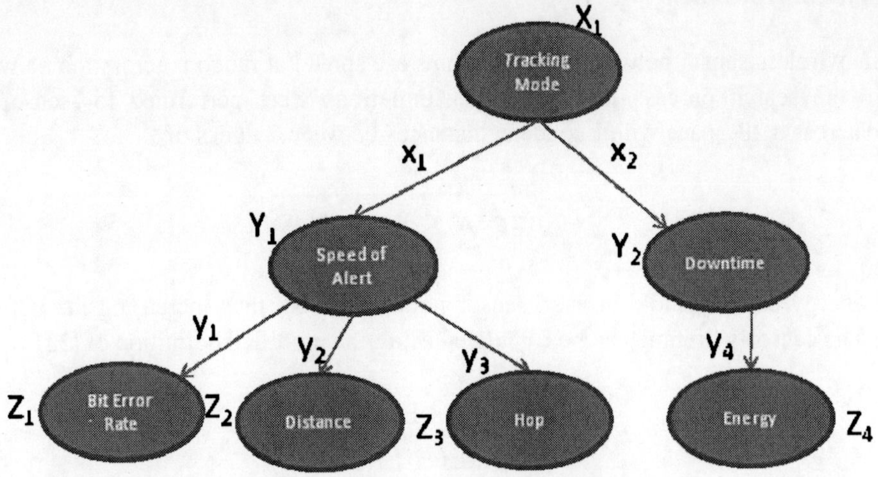

Figure 7. Illustration to support POSets computation.

The values of the next row of nodes is computed by multiplying the arrow's value by the preceding node or

$$X_1 x_k = Y_k, k = 1, 2 \tag{4}$$

The computation of the remaining POSet structure can precede in the same manner. The POSet provides a weighting scheme to guide the creation of a single global performance parameter so that sensor parameter decisions can be made by the sensor agents. For example, distance and the number of hops need to be emphasized if the sensor network needs to quickly send messages during priority queries or when distress events are sensed. Saving energy to prolong the life of the sensors is less important at that particular point in the system's lifetime. The weights are then computed from

$$W_k = Y_1 y_k, k = 1, 2, 3, 4, i = 1, 2 \tag{5}$$

The total performance is computed in for N parameters (N=4 here)

$$\sigma_{global} = \sum_{i=1}^{N} W_i \left[\frac{\left(\phi_{actual_i} - \phi_{required_i} \right)}{\phi_{required_i}} \right] \tag{6}$$

where σ is the global network performance parameter, and W_i is the weighting from the POSets structure in (5). The operator may make new decisions at this point as to the weighting applied in the POSet. These weights are then applied to the modified ACO algorithm to obtain the desired performance on the next system update.

Mathematical Approach

In the wireless sensor network, the ant agents are spread at random across the network to speed up the search process. Monte Carlo simulations were performed for sensor node scattered across a 2D space with Euclidean distances between 2 nodes of

$$D_{12} = \sqrt{(x_1 - x_2)^2 + (y_1 - y_2)^2} \quad (7)$$

The energy and threshold at each sensor node is varied, thus increasing the network's lifetime. The sensor's lifetime can be calculated using the modified definition of [32]

$$E(l) = \frac{\varepsilon_0 + \mathrm{E}(E_w)}{P_c + \lambda.\mathrm{E}(E_r) + \xi.\mathrm{E}(E_s)} \quad (8)$$

where ε_0 is the non-rechargable initial energy, $E(E_w)$, expected wasted energy, P_c, the constant power consumption, λ, the number of received messages, $E(E_r)$, the expected reporting energy,. ζ the sensing rate, and $E(E_s)$, the expected sensing energy.

The agents are energy aware and know the energy status of each sensor node. As the agent moves from node to node, energy is lost through communication. The agents do not traverse nodes with depleted energy. New paths are set up that avoid such nodes so that communication remains functional without the degraded sensor. The agents do not require initial solutions in the system during initial setup. This allows the system to be more flexible, robust, decentralized and intelligent. These agents ensure the optimal route to the destination using limited resources and also learn the network environment. Initial computational cost and time is high but this drops drastically once the agents adapt to the network and environment.

The concept of sensor based ITS application is that nodes participate in routing if their energy is above a certain threshold. The sensor's whose energy fall below a set threshold is considered down or depleted. Thus, these nodes are neglected and an alternate route is taken. This keeps the network functional even during individual node failure. Only for a distress signal (priority message) will the sensors even with low energy levels participate in routing.

The agents accumulate pheromones as they traverse from source i to sink j. Hence distance travelled, energy, BER and hops are the critical parameters that need to be considered while depositing the trails. Hence (1) is modified for this cognitive protocol as

$$\psi_{ij}(t) = \rho\psi_{ij}(t-1) + \frac{Q.E_t}{D_t.H_t.BER_t} \quad (9)$$

where D_t, H_t, E_t and BER_t are the total distance, hop, energy and bit error rate performance, respectively, of the current agent in a tour. The link cost function is modified based on the QoS parameters as

$$\eta_{ij}(t) = \min\{[W_1.(\frac{H_a - H_r}{H_a})] + [W_2.(\frac{D_a - D_r}{D_a})] + [W_3.(\frac{E_a - E_r}{E_a})] + [W_4.(\frac{BER_a - BER_r}{BER_a})]\} \quad (10)$$

In (9) 'a' is the actual and 'r' is the required value for data transmitting. The minimization cost function for the normalized value of performance parameters is applied. Thus increasing the QoS and also incorporating the cross-layered meta-heuristic approach.

The transition probability for normal traffic determines the routes chosen by the ants and is calculated as in (2) using values from both (9) and (10). Whereas for handling priority events and random queries (distress signals), the priority of the message needs to be incorporated into the transition probability as

$$P_{ij} = \frac{(\psi_{ij}(t).\Gamma_{ij}(t))^{\alpha}.(\eta_{ij}(t))^{\beta}}{\sum_{k}(\psi_{ik}(t))^{\alpha}.(\eta_{ik}(t))^{\beta}} \quad (11)$$

where Γ is the message priority, and η is the global performance from the swarm agents

As previously discussed, the tabu-list is a main feature of the ACO algorithm, which keeps track of the nodes visited by the agents. In our algorithm, since we modified the pheromone and transition probability, the routes reflect energy efficiency and resource optimized solutions. The meta-heuristic algorithm ensures the optimal route to the destination using limited resources and also learning the network environment.

5.f. Experimental Simulations and Results

A sensor network with 16 nodes is evenly distributed along the 2D network and 10 agents are randomly placed on the nodes. Since the nodes move in a preset path, an accident is randomly imposed on a node where a distress signal is triggered. The agents upon sensing the distress signal work interactively in disseminating the information to all nearby 'key' locations (BS and MH).

The parameter settings of the agents play a key role in network adaptation and global optimal solution. The values are assumed to be $\alpha = 4$, $\beta = 7$, $\rho = .7$, $Q = 9$, and initial pheromone value ψ as 10, which is from our previous work in [33]. The stability of the algorithm is analyzed by iterating all scenarios for 100 runs. The robustness of the algorithm is analyzed in an additive white gaussian noise (AWGN) and rayleigh fading channel. The three modulation techniques available for selection in the protocol are binary phase shift keying (BPSK), quadrature phase shift keying (QPSK) and gaussian frequency shift keying (GFSK).

Figure 8 illustrates the prioritized event queried by a sensor and the optimal route taken by CRP. The dash dot line represents the local optimal route taken by CRP, which was rejected in time to select the globally optimum solution represented in solid line. The reactivity of the cognitive algorithm depends on the learning rate and the reinforcement parameter setting on the agent.

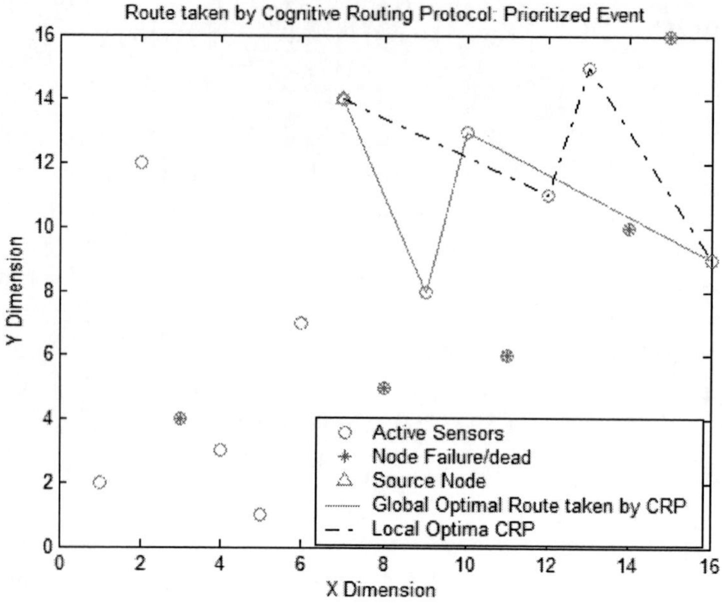

Figure 8. Route taken by CRP during Prioritized Event.

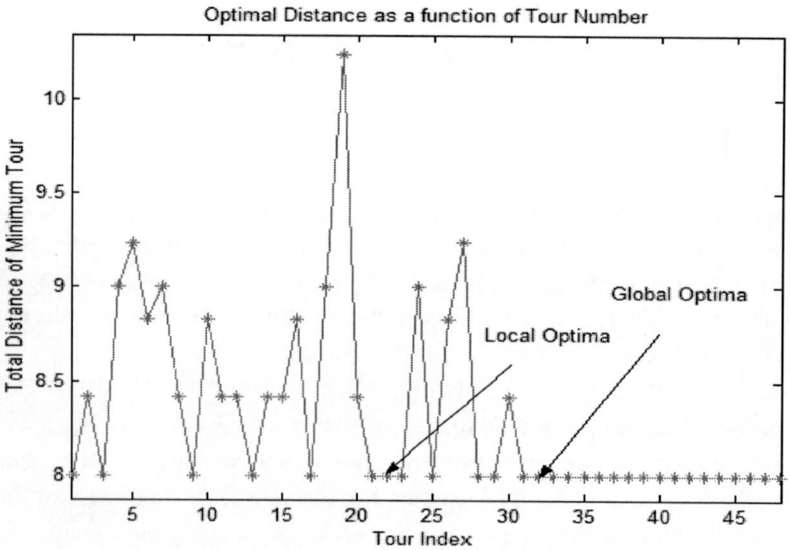

Figure 9. Optimal distance taken by CRP.

Figure 9 illustrates the agent behavior towards attaining the global solution as the number of tours taken over time. The pheromone deposition on the local optima evaporates, which leads the agents towards the global optima. POSets weigh the performance parameters based on system requirements, hence if distance is weighted more, then shortest path is chosen than energy or any other resource management parameters.

Figure 10 below shows CRP using each type of service request, where the image coefficient with high, medium and low prioritization is transmitted. Since errors in a distress signal could cause a delayed response, the route taken for high prioritized data should have low BER. The data packet has the information that provides the proper type of query. Interestingly the collaborative approach of CRP helps in attaining not only a global optimal, but versatile, feasible and reliable solutions.

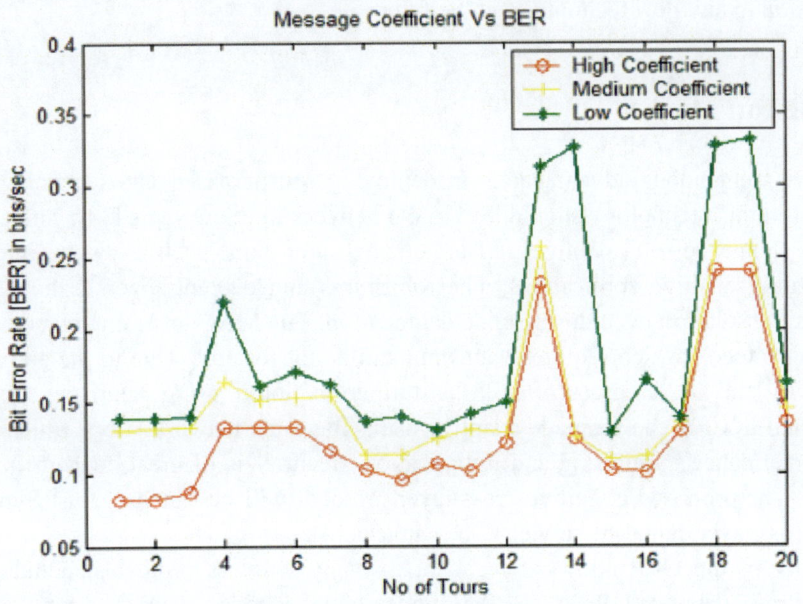

Figure 10. Message Transmission Using CRP.

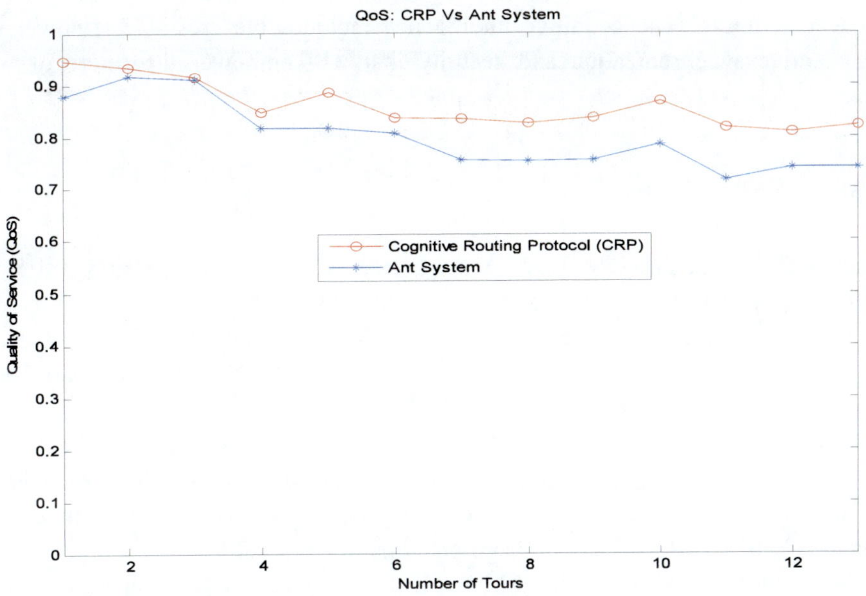

Figure 11. QoS Comparison for CRP Vs Ant System.

The QoS for an ITS system is primarily based on response time and traffic throughput. In Figure 11, the QoS under CRP and AS algorithms is compared for traffic conditions with low, medium and high message prioritization. The AS algorithm performs marginally with 79% QoS whereas, the CRP outperforms by 6%. The adaptive features of the CRP are based on the flexible cost function, and can be modified to any application requirement thus making it a generic solution to any new technology in the future.

6. Conclusion

There are tremendous advantageous in deploying sensor based ITS technologies. The advancement in this technology should lead to 4G network in a few years [34]. Therefore, the key solution should be self-adaptive, self-organizing algorithms, such as swarm intelligence for critical time sensitive applications. The robustness and de-centralized features of CRP attains a feasible solution even during node degradation. The MAC layer implemented here is a handshake protocol, which requires wait-time and is not feasible. Due to the application's time-sensitivity the MAC protocol will be further explored using heuristic approaches. POSets applied in CRP can be made adaptive using Bayesian network (BN), which requires real-time performance monitoring and initialization. Security is of great concern in modern applications. The proposed cognitive cross-layer protocol will be tested against Sybil attack [29], other cross layer denial-of-service (DoS) attacks [4].

Adaptivity of the CRP allows it to adapt to many wireless protocols including those involving Ultra-wideband (UWB). UWB provides better location estimates due to its short duration, and also it uses less energy for longer distances.[32]. The UWB pulses can be tailored to transmit both data and images simultaneously if two nearly orthogonal bands are designed different gaussian derivatives for the impulse UWB signals. We will be analyzing this in the near future. The cognitive intelligence includes the cross-layer, power control scheme, adaptive synchronization and security that will empower the future of wireless technology.

References

[1] Muraleedharan R., Osadciw L.A.,"A Predictive Sensor Network Using Ant System", *SPIE Aerosense*, Orlando, Florida, April 12-17, 2004,

[2] W. Chen, L. Chen, Z. Chen, S. Tu, "WITS: A Wireless Sensor Network for Intelligent Transportation System", In *Proc. of first International Multi-Symposiums on Computer and Computational Sciences (IMCCCS),* 2006

[3] Johnson J.C., Reed-Woodard M.A, "The future belongs to the unwired; Ultra-Wideband technology will lead the next wireless revolution", Article: Black Enterprise, Feb 2007.

[4] Muraleedharan R., Osadciw L.A., "Security: Cross Layer Protocols in Wireless Sensor Networks", IEEE Infocom 2006, Barcelona, Spain, April, 2006

[5] Udaya Shankar, C. Alaettinoglu, I. Matta, K. DussaZieger, "Performance comparison of routing protocols using MaRS: distance-vector versus link-state", In Proc. of the 1992 ACM SIGMETRICS and PERFORMANCE _92 Int_l. *Conf. on Measurement and Modeling of Computer Systems*, Newport, RI, USA, 1–5 June 1992, p. 181.

[6] Perkins C.E., Watson T.J., " Highly dynamic destination sequenced distance vector routing (DSDV) for mobile computers", In ACM SIGCOMM 94, *Conference on Communications Architectures*, London. UK, 1994.

[7] S. Murthy, J.J. Garcia-Luna-Aceves, " A routing protocol for packet radio networks", In *proc. of the first annual ACM International Conference on Mobile Computing and Networking*, Berkeley, CA, 1995, pp. 86-95.

[8] J. Broth, D.B.Johnson, D.A.Maltz, "They dynamic source routing protocol for mobile ad-hoc networks", Internet Draft, 1998.

[9] S. Das, C. Perkins, E. Royer, "Ad-hoc on demand distance vector (AODV) routing ", Internet Draft, Work in progress, 2002.

[10] V.D. Park and M.S. Corson. A highly adaptive distributed routing algorithm for mobile wireless networks. *In Proc. of IEEE INFOCOM*, Kobe, Japan, Apr. 1997.

[11] K. Xu, X. Hong, M. Gerla,, "Landmark Routing in Ad Hoc Networks with Mobile Backbones", *Journal of Parallel and Distributed Computing (JPDC), Special Issues on Ad Hoc Networks*, Volume 63 , Issue 2, 2002, pages: 110 . 122.

[12] M. Jiang, J. Li, Y.C. Tay, "Cluster based routing protocol (CBRP) functional specification", Internet 1998.

[13] E. M. Belding-Royer. "Multi-level Hierarchies for Scalable Ad hoc Routing.", Wireless Networking (WINET), Vol. 9, No. 5, pages 461-478, Sept. 2003

[14] S. Basagni, I. Chlamtac, V.R. Syrotivk, B.A. Woodward, "A distance effect algorithm for mobility (DREAM)", In *Proc of the Fourth Annual ACM/IEEE International Conference on Mobile Computing and Networking (Mobicom_98),* Dallas, TX, 1998.

[15] Lochert et al., "A routing strategy for vehicular ad hoc network in the city environments," Proc. of IEEE Intelligent Vehicles Symposium, June 2003

[16] J. Tian, L. Han, K. Rothermel, C. Cseh, "Spatially aware packet routing for mobile ad hoc inter-vehicle radio networks," *IEEE ITSC*,Oct 2003

[17] H. Wu, R. Fujimoto, R. Guensler, and M. Hunter, " MDDV: a mobility centric data dissemination algorithm for vehicular networks", In *Proc. Of the 1st ACM Workshop on Vehicular Ad Hoc Networks (VANET 2004),* pages 47–56, 2004.

[18] E. Royer, C. Toh, "A Review of Current Routing Protocols for Ad Hoc Mobile Wireless Networks*", IEEE Personal Communications*, pages 46--55, Apr. 1999.

[19] Wolfson, Xu , "Data on the road intelligent transportation system", *Proc. of the IEEE International Conference on Networking, Sensing, and Control (ICNSC 2004),* Taipei, Taiwan, Mar. 2004.

[20] S. Biswa, R. Morris, "Opportunistic Routing in Multi-Hop Wireless Networks", *Proceedings of the Second Workshop on Hot Topics in Networks (HotNets-II),* Cambridge, MA, USA. November 2003

[21] U. C. Kozat, I. Koutsopoulos and L. Tassiulas, "A Framework for Cross-layer Design of Energy-efficient Communication with QoS Provisioning in Multi-hop Wireless Networks", Technical Research Report, The Institute for System Research, The University of Maryland, 2004.

[22] P. Sinha, R. Sivakumar, V. Bharghavan, "CEDAR: a Core-Extraction Distributed Ad-Hoc Routing algorithm", In *Proceeding of IEEE Infocom*, Mar 1999.

[23] Martinoli A., Zhang Y., Prakash P., Antonsson E.K., Olney R.D., "Towards Evolutionary Design of Intelligent Transportation System ", *Proc. of the Eleventh Int.*

Symp. of the Associazione Tecnica dell'Automobile on New Technologies for Advanced Driver Assistance Systems (ADAS), 2002

[24] W. Sun, H. Yamaguchi, K. Yukimasa, S. Kusumoto, "GVGrid: A QoS Routing Protocol for Vehicular Ad-hoc networks ", 2006

[25] Zheng-Yu Wu, Han-Tao Song, "Ant-based Energy-aware Disjoint Multipath Routing Algorithm for MANETs", *The Computer Journal Advance Access*, Oxford University, Feb 2008.

[26] Barrett, A. Marathe, M. V. Marathe, M. Drozda, "Characterizing the interaction between routing and MAC protocols in ad-hoc networks", International Symposium on Mobile Ad-Hoc Networking and Computing, *Proceedings of the 3rd ACM international symposium on mobile ad hoc networking and computing*, Lausanne, Switzerland, 2002, Pages 92.103

[27] L. Qin, T. Kunz, "Survey on Mobile Ad-hoc network Routing Protocols and Cross-layer design", Carleton University, *System and Computer Engineering*, Technical report SCE-04-14, August 2004.

[28] M. Cordeiro, D.P. Agarwal, "Ad-Hoc and Sensor Networks Theory and Applications", World Scientific Publishing, ISBN: 981-256-681-3, 2006.

[29] Aickelin U and Li J, "Bayesian Optimisation for Nurse Scheduling", *Journal Annals of Operational Research,* 2005.

[30] V. Kawadia and P. R. Kumar, "A Cautionary Perspective on Cross Layer Design", Submitted to *IEEE Wireless Communication Magazine*. July 9, 2003.

[31] Woodward, P. M., *Probability and Information Theory, with Applications to Radar*, Pergamon Press, Inc., NY 1953.

[32] Chen Y., Zhao Q., "On the lifetime of Wireless Sensor Networks", *IEEE Communications letters*, Vol. 9, No. 11, Nov 2005, pg 976-978.

[33] Muraleedharan R., Osadciw L.A., "Balancing The Performance of a Sensor Network Using an Ant System", *37th Annual Conference on Information Sciences and Systems,* John Hopkins Unversity, 2003.

[34] Muraleedharan R., Osadciw L.A.,"Increasing QoS and Security in 4G Networks Using Cognitive Intelligence", *GLOBECOM 2007*, Washington D.C., Nov 2007.

[35] Marco Doringo, "The Ant System: Optimization by a Colony of Cooperating Agents", *IEEE Transactions on Systems, Man and Cybernetics-Part B*, Vol-26, No. 1, Sept1996,pp 1-13.

[36] Kennedy J, Shi Y, and Eberhart R.C., "Swarm Intelligence", Morgan Kaufmann Publishers, San Francisco, 2001.

In: Wireless Technologies in Intelligent Transportation… ISBN: 978-1-60741-588-6
Editors: Ming-Tuo Zhou et al, pp. 111-132

Chapter 5

TDMA MAC PROTOCOLS FOR DSRC-BASED INTELLIGENT TRANSPORTATION SYSTEMS

Fan Yu and Subir Biswas[1]
Electrical and Computer Engineering Department
Michigan State University, East Lansing, Michigan, USA

Abstract

This chapter presents an impact study of wireless Medium Access Control (MAC) Protocols on Intelligent Transportation System (ITS) applications using the Dedicated Short Range Communication (DSRC) standard. After a detailed review of the existing contention-based and TDMA-based MAC protocols for vehicular communication, a novel Medium Access Control protocol for inter-vehicle communication applications is introduced. The primary contributions of this chapter are the design of a self-configuring TDMA protocol VeSOMAC, and its evaluation for inter-vehicle data transfer applications. A novel feature of the proposed VeSOMAC is its in-band control mechanism for exchanging TDMA slot information during distributed MAC scheduling. It is shown that the in-band control mechanism can be used for fast protocol convergence during topology changes due to vehicle movements. Simulated performance comparison between VeSOMAC and the DSRC-recommended 802.11 MAC protocol demonstrates that the proposed protocol can accomplish better inter-vehicle data transfer performance through enhanced UDP and TCP throughput and fewer MAC layer packet drops compared to 802.11. Experiments in transient scenarios representing VeSOMAC slot reorganization due to network topology changes demonstrate that while the application level performance with VeSOMAC degrade during topology changes, it can still outperform the contention based 802.11 for data-intensive vehicular applications.

---

[1]E-mail address: sbiswas@egr.msu.edu. (Corresponding author)
This work was supported by research grants from Michigan Department of Transportation (M-DoT), National Science Foundation (SCI-0438271), and Michigan State University.

1. Introduction

1.A. Background and Motivation

The emerging Intelligent Transportation System (ITS) [1] architecture is currently being developed for enhancing vehicle safety, driving experience, fuel economy, and economic stimulation through safer and efficient transportation networks. A key component of the ITS architecture is vehicle-to-vehicle (V2V) and vehicle-to-roadside (V2R) communication for application data exchange between vehicles and road-side infrastructure.

ITS applications are broadly categorized into two kinds: safety oriented and non safety oriented data intensive applications. Examples of the safety oriented applications [2] include Cooperative Collision Avoidance (CCA), Cooperative Adaptive Cruise Control (CACC), Emergency Vehicle Preemption (EVP), and Intersection Collision Warning (ICW) systems. Notable examples of data intensive applications are vehicle based probe data collection for traffic congestion management, enhanced route planning and guidance, and entertainment applications such as inter-vehicle gaming and music downloads [2]. From the networking standpoint, the safety oriented applications have performance requirements in terms of lower and bounded delay [3], and better fairness. On the other hand, higher achievable end-to-end throughput is desirable for the data intensive applications.

Dedicated Short Range Communication (DSRC) [2] is an emerging V2V and V2R communication standard, which is being developed for an FCC-allocated 75 Mhz wide spectrum segment at the 5.9 Ghz band. Although IEEE 802.11p is currently recommended as the Medium Access Control (MAC) protocol within DSRC, like other random access protocols it has performance limitations at high traffic load conditions. At high data loads, the packet delivery latency increases due to increased collisions, retransmissions, self-competition within a single data flow, and cross-competition among multiple data flows [4]. Also, because of the underlying random access, at higher loads the delivery latency becomes unpredictable, and therefore not bounded [4, 5]. These effects could also bring down the achievable end-to-end throughput for inter-vehicle data transfer through connections spanning across multiple vehicles. This implies that for data intensive applications (and also for safety oriented applications in certain scenarios), the 802.11 family of protocols may not be able to provide the required end-to-end throughput and delivery latency.

To address those issues, we develop a Vehicular Self-Organizing MAC (*VeSOMAC*) which is self-configurable and fully distributed, and capable of providing satisfactory data throughput by dynamic MAC slot scheduling based on the relative location, speed and other context parameters of a vehicle platoon on highways and urban traffic intersections.

1.b. Related Work

The vehicular MAC protocols in the literature are categorized as contention-based and schedule-based. The contention based approaches have the advantage of not being sensitive to underlying node mobility and the subsequent topology changes. As a result, unlike for the schedule-based protocols, vehicle movements do not impose any reconfiguration overhead due to the network topology changes. This is a significant advantage. The unbounded delay, however, applies to all protocols in this category because of their underlying random access nature. Although a number of variations of CSMA/CA and 802.11 have been implemented [6-8] for mitigating the delay to some extents, the fundamental issue of delay unpredictability due to collisions at high loads, and cross-flow capacity contentions still remains for the protocols in this category.

These contention based approaches, however, are completely agnostic about the underlying mobility and topology changes. As a result, unlike the schedule-based protocols as explained later, the vehicle movements do not impose any MAC reconfiguration overhead due to the topology changes. This is a major advantage of the contention based approaches in vehicular networks in which the rate of topology changes can be very high.

The protocols in [9, 10], and [11],propose schedule-based TDMA mechanisms, in which TDMA slots are self-selected by the nodes in a distributed manner. While providing bounded latency and alleviating cross-flow capacity contentions, these protocols rely on contention based slot allocation process which requires stochastic resolution of collisions during the slot allocation. As a result, the slot reallocation due to topology changes may often incur a large upfront delay before the application data can resume flowing. The protocol in [12] proposes to configure a token ring protocol so that the maximum delivery delay is always bounded by the round-trip token time. Although bounded, the delay can be very large for large rings due to high vehicle crowding. This can be addressed by having multiple rings, but the ring management with frequently changing topology can cause huge upfront reconfiguration delays. OTRP [13] is another token ring protocol that can operate either in normal mode or in an emergency mode. By having a higher priority over normal token, the warning message latency in emergency mode is minimized. But still the management overheads of multiple rings remain an issue. MCS/CDMA [14] is another schedule based protocol that uses CDMA code scheduling. Each vehicle has to have parallel receiver matched filters corresponding to all the codes used. If the search for free code is done sequentially, it could take long time when the number of vehicles involved is large. This will add substantial reconfiguration latency in the event of a topology change. Also, when there are more vehicles than the number of codes, there could be a contention for free codes resulting in large delays in channel access. The protocol LCA [15] proposes a scheduled MAC, in which TDMA slots are allocated based on a vehicle's instantaneous geographical location, which is pre-allocated a TDMA slot. This mechanism offers bounded delay, and also there is no reconfiguration latency due to network topology changes. However, the system requires complete pre-mapping of geographical locations to TDMA slots, which may have practical limitations especially when the huge geographic coverage of a transportation system is considered. Also, dimensioning optimal cell size for varying application requirements and vehicle density is a non-trivial problem.

The following distributed TDMA protocols, which were originally proposed for wireless sensor networks, can also be considered in the context of vehicular networks. The protocols TDMA-W [16] and DRAND [17] both use an out of band handshake signaling mechanism for distributed TDMA slot allocation. Nodes send control packets during scheduled data slots, and that is how the neighbors' allocation information is disseminated. Based on its neighbors' allocation information, a node is able to select a collision-free transmission slot, which is used for subsequent data as well as control packet transmissions. Both these protocols have been designed for networks with limited node mobility, since frequent topology changes may require substantial reconfiguration latency caused by the two-way handshake signaling needed after every topology change event. This will cause the time out at upper level protocols, TCP for an instance, to degrade the data flow performance.

The protocol ZMAC [18] offers a hybrid solution, in which although each node is allocated a dedicated TDMA slot, the nodes are allowed to contend for bandwidth using CSMA with a goal to improve the bandwidth usage. This protocol too, due to its out-of-band allocation signaling, may suffer from high slot reallocation latency after a network topology change. The protocol D-MAC [19] allocates node and route aware TDMA slots with a goal of minimizing the end-to-end application delay. The primary difficulty of applying D-MAC to the targeted vehicular applications is that the MAC protocol assumes preset routes, which are not feasible in most vehicular applications. Additionally, the protocol does not provide syntax for completely avoiding hidden-collisions, which can prove detrimental. The protocol GWMAC [20] relies on a base station, which is not usually available in the V2V communication, to schedule the TDMA slot allocation. And it does not target on real-time application like the vehicular networks. The MLMAC [21] protocol achieves the TDMA allocation in a distributed way. But the authors do not address the reconfiguration penalty and the message delivery latency. It might be long for the ITS application requirements.

From the existing literature we conclude that while the schedule based protocols are more desirable for inter vehicle applications, it still remains an issue as to how to cope with frequent topology changes by fast TDMA slot reconfiguration. To address this, we propose a Vehicular Self-Organizing MAC (*VeSOMAC*) protocol that can achieve fast TDMA slot reconfiguration without relying on roadside infrastructure or virtual schedulers such as leader vehicles. Fast slot reconfiguration translates into low upfront latency (during topology changes) which was noted to be an issue for the deterministic protocols such as in [6] and [9].

2. Vesomac Protocol Details

2.a. Frame and Slot Structures

As shown in Figure 1:b, the transmission slots (and packets) in *VeSOMAC* are of constant duration τ, which includes necessary guard times for modem preambles and Tx-Rx switchover latencies. A frame is of duration T_{frame}, which defines the minimum periodicity of transmission from any vehicle. Therefore, the allocated rate to a vehicle is $\lambda_{alloc} = 1/T_{frame}$ packets per unit time. In *VeSOMAC* since a bitmap in the packet header is used for exchanging slot timing information, it is mandatory for each vehicle to send a packet every frame, even if no application data is available.

2.B. Synchronous and Asynchronous Operation

VeSOMAC can operate in both synchronous and asynchronous modes. In the synchronous mode all vehicles are assumed to be time synchronized, and therefore, they share the same frame and slot boundaries. In the asynchronous mode (see Figure 1:b), vehicles maintain their own frame boundaries and therefore the slots are also asynchronous.

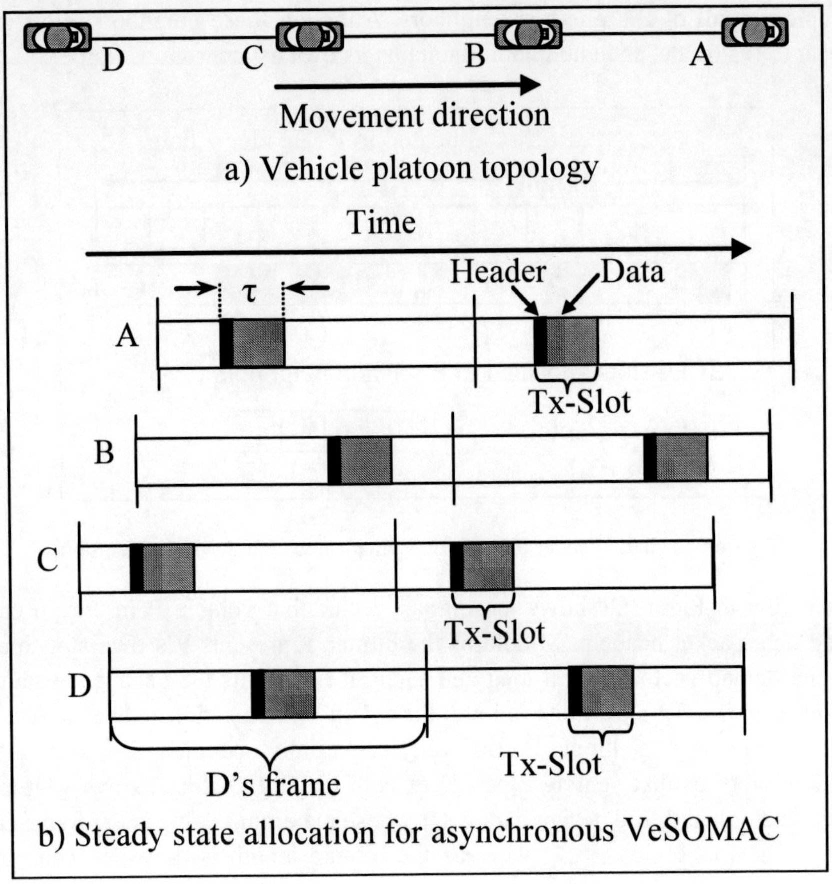

Figure 1. Example slot allocation in asynchronous VeSOMAC.

2.c. Protocol Logic

2.c.1. Slot Allocation

Slot allocation in VeSOMAC needs to satisfy the following **Timing Constraint**: No two one-hop or two-hop neighbors' slots can overlap. Overlaps between one-hop and two-hop neighbors cause direct and hidden collisions respectively.

A valid asynchronous VeSOMAC schedule for vehicles A-D is shown in Figure 1:b. Except the pair (A,D) all other vehicles are within up to two-hop transmission distance of each other. That is why all vehicles, except A and D, are allocated completely non-overlapping Tx-slots.

2.c.2. In-band Header Bitmap

Information about allocated slots is exchanged among the vehicles using a Bitmap Vector in each packet header. The concept is explained in Figure 2. The top segment illustrates a vehicle P's allocated Tx-slot within its own TDMA frame. The middle row depicts the Tx-slots occupied by all of P's one-hop neighbors. Although these neighbors' slots are shown with respect to P's frame, each neighbor maintains its own asynchronous frame.

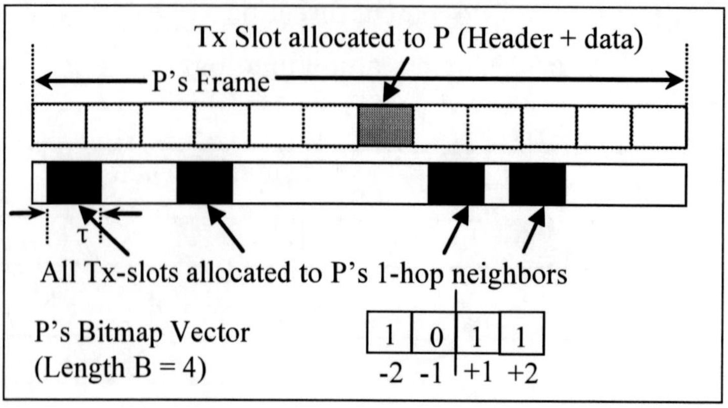

Figure 2. In-band header bitmap for asynchronous VeSOMAC operation.

The bottom row in Figure 2 shows the bitmap vector that vehicle P inserts in each of its transmitted data packet headers. Middle of the bitmap represents P's own slot time. In this example, the bitmap vector is 4-bit long and each bit represents the occupancy status of two slots around P's own Tx-slot. For example, the '1' in "+1" location indicates that two slots immediately following P's slot are already fully or partially occupied. Similarly, a '0' in the "-1" location indicates that vehicle P perceives both the slots before its own slot to be free. The bitmap vector length is a design parameter whose maximum value is the frame slot count. In Figure 2, the frame size is 12, whereas the bitmap length is 4, which can convey the occupancy information about only 8 slots. With a bitmap size 4, P is unable to represent the occupancy information about one of its neighbors' slots - the one in extreme left. To avoid this situation, the bitmap size should be chosen to be same as the frame size.

Using this header bitmap, a vehicle continuously informs its 1-hop neighbors about the slots occupied by its 1-hop neighbors. By listening to the bitmaps in all received packets, a vehicle can detect the slot locations of its 1-hop and 2-hop neighbors. This information can then be used for choosing a Tx-slot that satisfies the timing constraint in Section 2.c.1.

Since all timing information is relative, this bitmap based approach allows *VeSOMAC* to be implemented with or without network time synchronization. Although in the asynchronous version the bitmap efficiency is somewhat reduced by the overhead that one bit here represents the occupancy of two slots instead of one (see Figure 2), in the synchronous case this overhead does not exist. In the asynchronous *VeSOMAC*, the capacity efficiency is traded for more distributed deployments without the need for time synchronization across the network.

2.c.3. Transmission Slot Feasibility

A feasible transmission slot for a vehicle is one that satisfies the *timing constraints* as defined in Section 2.c.1. A *feasible time region* for a vehicle is defined by the region that is represented by shared '0's in the bitmaps transmitted by all its neighbor vehicles. A slot chosen from this feasible region is guaranteed to satisfy the *timing constraint*.

Consider the *VeSOMAC* example in a highway platoon scenario in Figure 3. A new vehicle *R* joins in between two unconnected vehicles *P* and *Q*. Bitmaps (with length 4) from *P* and *Q*, as received by the new vehicle *R*, are shown in Figure 3:a.

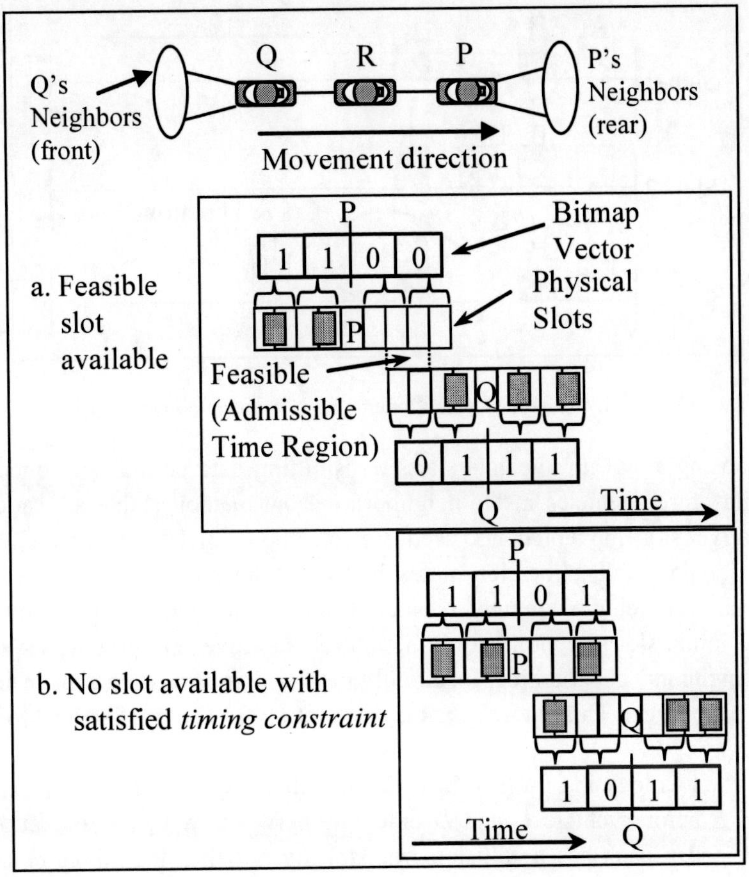

Figure 3. Slot feasibility scenarios in asynchronous VeSOMAC.

The shared '0's in the bitmaps of *P* and *Q* indicate a feasible time region for vehicle *R*. Since a shared '0' indicates that the corresponding feasible region is not used by any of *P's* and *Q's* 1-hop neighbors, a slot chosen in that region is guaranteed to be hidden collision free from all *R's* 2-hop neighbors. And, since the slot within the feasible region is within the bitmap of all *R's* 1-hop neighbors (*P* and *Q*), it is guaranteed not to be used by any of those 1-hop neighbors.

However, in Figure 3:b, since there are no shared '0' s, no feasible region is there for *R*. If *R* chooses a slot from the time region indicated by a '0' in *P's* bitmap, then it would collide with a 1-hop neighbor of *Q*. Therefore, because of the violated *timing constraint* a hidden collision may not be avoided.

2.c.4. Protocol Overview

A newly joined vehicle attempts to choose a collision free slot right after the slot of the vehicle immediately ahead.

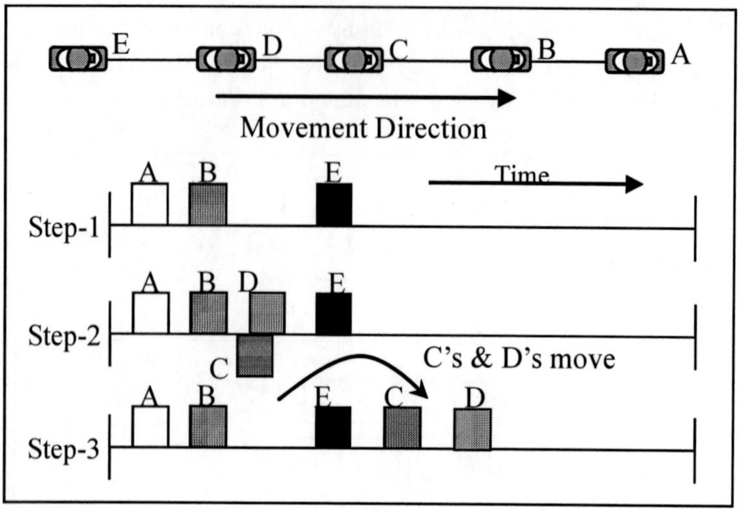

Figure 4. Iterative slot movements for allocation convergence.

Upon choosing a slot, the vehicle starts transmitting data periodically once per frame[2]. This action may force vehicles in the neighborhood out of their stable allocation state. But then an iterative slot movement is used for all the vehicles in the neighborhood to incrementally attain stable allocation states including the perturbing new vehicle. During these iterations, each vehicle attempts to place its slot behind the slot of its immediate front neighbor. The initial slot selection and the iterative slot movement process require a vehicle to know its neighbors' geographical location, which is assumed to be sent in the header of each transmitted packets. This is a reasonable assumption for ITS applications [6] with access to GPS systems.

Consider the topology in Figure 4, in which the allocation step-1 depicts Tx-slots chosen by vehicles *A-E* before vehicle *C* and *D* enter the network. With the allocation in Step-1, vehicles *A*, *B*, and *E* each has a collision free slot that satisfies the *timing constraint*. As a result, at Step-1 all vehicles are at a steady allocation state. Upon entering the network, *C* and *D* learns about the slot locations of their 1-hop neighbors *B* and *E* from their periodic transmissions, and the 2-hop neighbors (*A*) through the bitmaps in *B's* and *E's* packet headers. Then in Step-2, *C* and *D* attempt to independently select *timing constraint* compliant collision free slots. But in Step-2 in this example, nodes *C* and *D* happen to choose overlapping slots which make them unstable. Using the collision resolution mechanism, as described in Section 2.c.5, both *C* and *D* move their slots as shown in Step-3. All vehicles at Step-3 become stable. At this stage, if another vehicle joins causing a collision, all the affected vehicles will use the logic in Section 2.c.5 to choose slots again.

To summarize, the core idea of *VeSOMAC* is to let each vehicle iteratively move its slot following its immediate front neighbor's slot until a combined allocation pattern which is stable for all vehicles in the neighborhood emerges.

[2] For the asynchronous *VeSOMAC*, upon entering the network, a node arbitrarily chooses its frame starting point. For the synchronous version, it is chosen based on the frames of the existing vehicles or the timing information from onboard GPS.

```
/* VESOMAC Protocol Logic for Each Vehicle */
Arbitrarily choose the MAC frame boundary;
Listen:
   Listen to the existing network for W frames;
   // observe received transmissions to know the slot
   // location of all 1-hop neighbors; also observe the
   // bitmaps in the 1-hop neighbors' transmissions to
   // know the slot location of all 2-hop neighbors
   choose a slot using the logic described
   in Section 2.c.4.;
Evaluate:
   Periodically transmit packets using the chosen slot
   and listen from neighbors for W frames;
   if (W is over && bitmap/timing/ordering constraints
      are all satisfied)
        goto Stable;
   else{
        Move the slot using the logic described in
        Section 2.c.4.;
        goto Evaluate;
}
Stable:
    Periodically transmit packets using the chosen slot;
    If (violation of at least one constraint persists
       for W frames){
         Move the slot using the logic described in
         Section 2.c.4.;
         goto Evaluate;
    }
    If (new neighbor vehicle arrives or topology
       changes or packets get corrupted due to channel
       errors){/* do not change slot */
         goto Evaluate;
}
```

Figure 5. Pseudo code for the VeSOMAC protocol logic.

2.c.5. Collision Detection and Resolution

Packet collisions in *VeSOMAC* are detected using implicit acknowledgements through the bitmaps. From the bitmaps transmitted by all its neighbors, a vehicle can infer if all those neighbors have successfully received its own transmission. If not, the vehicle concludes that its transmission got either corrupted, or lost due to a collision. If the situation persists for a preset number of frames, a collision is declared. In Figure 3, if vehicles P and Q choose overlapping slots, a hidden collision will take place at vehicle R. Since R is not able to listen to *P's* and *Q's* transmissions due to the collision, it will simply indicate those two overlapping slots to be empty ('0') in its own bitmap. Upon receiving *R's* bitmap, *P* looks for its own slot location in that bitmap to see if *P's* transmission was successfully heard by *R*. A '0' corresponding to *P's* Tx-slot will indicate that there was a collision or packet corruption. If the situation persists for a preset number of frames then as explained in Figure 5, *P* will move its Tx-slot iteratively for resolving the collision. *Q* will also behave similarly. To summarize, a vehicle is required to be able to see a '1' for its own slot in the bitmaps of all its 1-hop neighbors. Otherwise, a collision is detected. This way, the bitmaps in *VeSOMAC* are used as implicit acknowledgements for detecting collisions.

2.c.6. Protocol Logic Pseudo Code

VeSOMAC protocol logic pseudo code is presented in Figure 5. The *Stable* state for a vehicle indicates the allocation steady state, and *Listen* and *Evaluate* are transient states. After a vehicle chooses a slot through the *Listen* state, it spends a preset (*W*) number of slots in the *Evaluate* state before getting into the *Stable* state. Any subsequent perturbations will force the vehicle to switch from the *Stable* state to the *Evaluate* state. Whenever a new slot is chosen by a vehicle, either as its first choice, or as a choice during the iterative allocation convergence, the vehicle enters in the *Evaluate* state. Subsequently, the slot is evaluated for *W* frames to make sure that the vehicle monitors its neighborhood activities to decide if its own allocation became stable. In our implementation, typical value of *W* was chosen to be 3 slot durations. When the protocol logic for all vehicles in a neighborhood reaches the *Stable* state, the protocol is said to have converged.

In *VeSOMAC*, a rare type of collision can happen when all vehicles in the network happened to have entered the network exactly at the same time and they have chosen the same slot. In such situations, the implicit acknowledgement described in Section 2.c.2 fails. To handle this, each vehicle is programmed to randomly skip one of its slots once in a large number of frames, so that during the skipped slot it can listen if any of its neighbors is transmitting during the same Tx-slot. If such a situation is detected, the vehicle simply chooses a new feasible slot, and transits to the Evaluate state. This type of collision is more frequent for the synchronous *VeSOMAC* because of its common frame timing across the vehicles.

3. Performance Evaluation

A hybrid simulation system for joint evaluation of different wireless network technologies, including their MAC and routing protocols, ITS applications, and vehicle following logic with drivers' behavior has been developed for evaluating different applications under different MAC protocols. Figure 6 depicts the architectural components of our VeNTSim system which is designed to be open for incorporating the evolving DSRC and other radio technologies, ITS applications [2], and their required network protocols. Using VeNTSim, Both DSRC and non-DSRC radio technologies, including WiMax [2] and Cellular mechanisms [2] can be architecturally evaluated for characterizing their impacts on heterogeneous ITS applications and use cases outlined by the research community and various standardization consortiums.

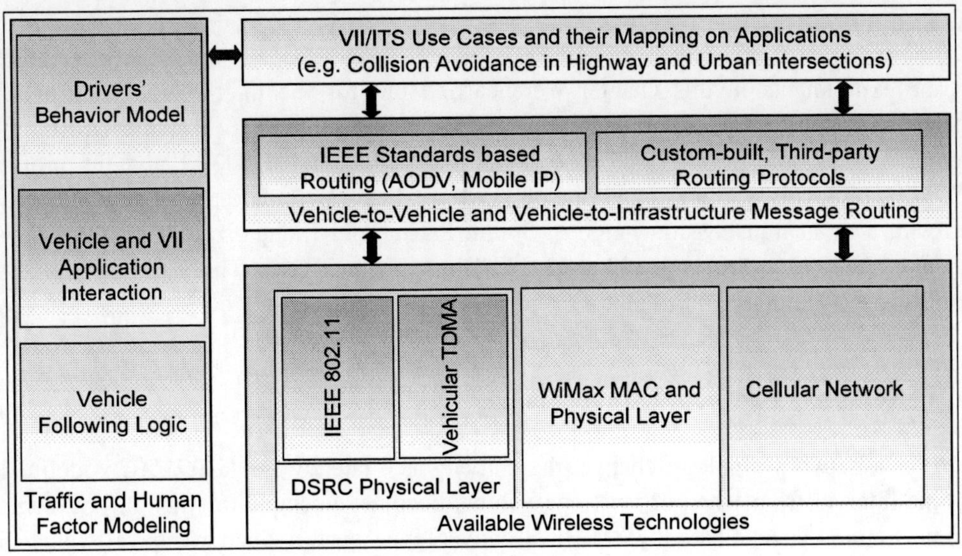

Figure 6. VeNTSim: ITS application and network evaluation/planning tool.

The networking functions in VeNTSim has been developed on top of ns-2 network simulator [22] by adding a vehicle mobility module that can react to the received wireless messages according to the modeled vehicle following logic with various drivers' reaction models. An ITS application modeling module capable of simulating various ITS applications such as cooperative collision control, cooperative cruise control and emergency vehicle preemption [2] has been also added. The synchronous version of *VeSOMAC* has been implemented at the ns-2 MAC layer, so that they can be compared with the 802.11 protocol running in the same radio environment. 802.11 is chosen for comparison because it is the current DSRC-recommended protocol. The architecture of VeNTSim is designed to be open for incorporating evolving other DSRC and ITS applications [2], and their required network protocols.

Table 1. Baseline experimental parameters

Vehicle and Scenario Related	
Platoon Size	50 vehicles
Vehicle Speed	68 mph (30 m/sec)
Inter-vehicle Spacing	45m
Network Related	
Channel	DSRC 5.9 GHz band, 24Mbps
Radio Model	Two ray ground
Radio Range	300m
MAC Protocols	IEEE 802.11 and *Synchronous VeSOMAC*
Packet Size	300 bytes (0.1 ms)
VeSOMAC Frame Size	100 packets (10 ms)

3.a. Experimental Parameters

The experiments in this chapter were carried out for the inter-vehicle data and file transfer applications on a one-lane highway platoon scenario [23] with DSRC radio communications. We have evaluated both DSRC-recommended 802.11 and the proposed *VeSOMAC* protocols with a channel rate of 24 Mbps. Two-ray ground propagation model was used with a radio transmission range of 300m. Packet and frame size for *VeSOMAC* were chosen to be 300 bytes and 100 packets (10ms) respectively. The baseline simulation parameters are summarized in Table 1.

3.b. VeSOMAC Protocol Convergence

After a network topology change, the convergence latency for *VeSOMAC* is defined as the time interval from when at least one vehicle becomes unstable to when all the vehicles become Stable. To evaluate *VeSOMAC's* convergence performance we have evaluated it during topology changes involving vehicle passing in a practical highway scenario. Convergence after a passing is needed because a passing vehicle's slot can violate the timing constraint of the vehicles that it is passing and the ones in the neighborhood of its new location in the platoon.

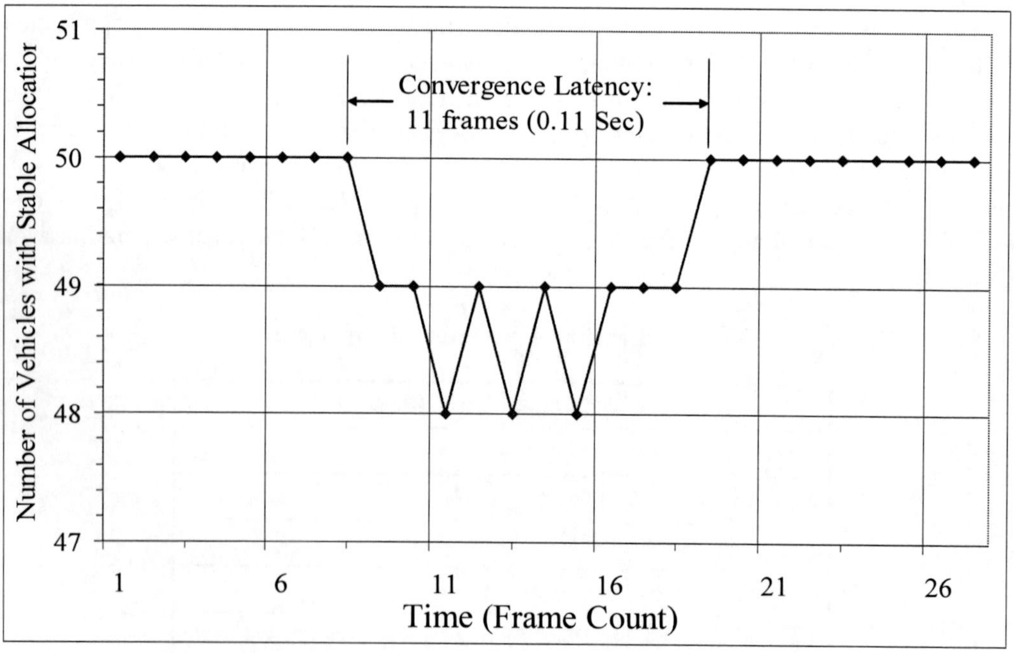

Figure 7. VeSOMAC convergence dynamics after a topology change.

The graph in Figure 7 corresponds to an instability triggered by a specific vehicle passing 40 vehicles in a highway platoon. From the *VeSOMAC* standpoint, the platoon is stable till the 8th frame, when the passing takes place. The vehicle-passing causes temporary instability by forcing at least two vehicles (not necessarily always the passing and the last-passed vehicles) out of their *VeSOMAC* stable states. But eventually, the protocol converges after 11 frames, which is 0.11 second for 10ms long frames.

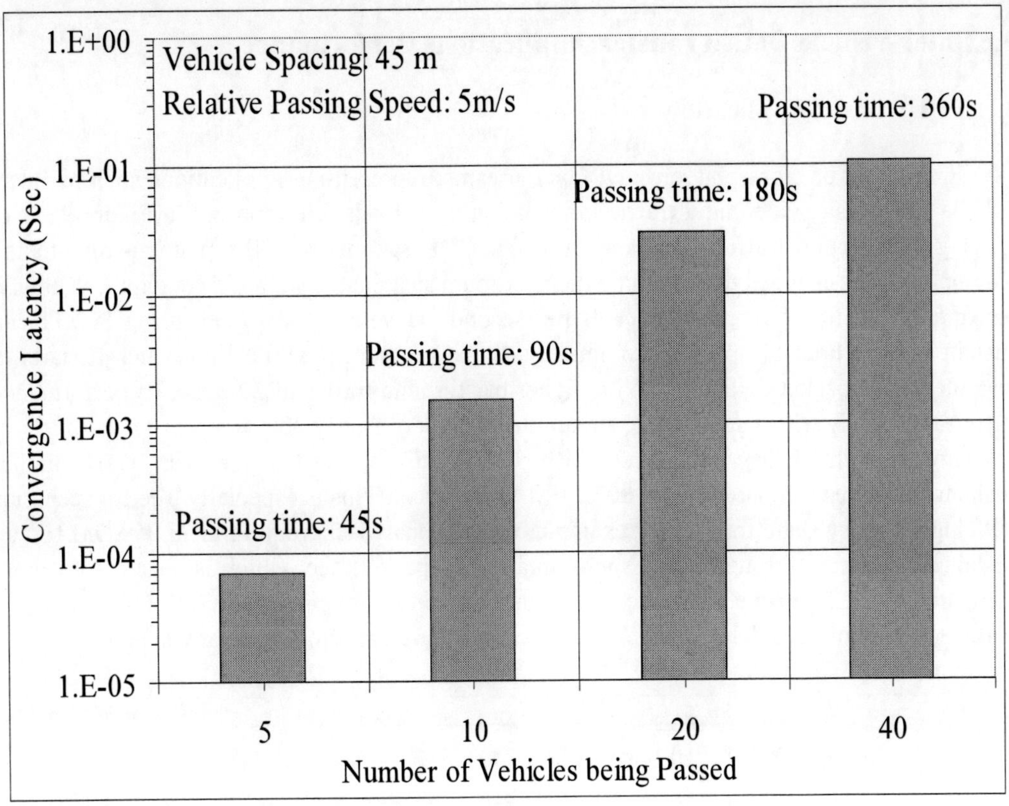

Figure 8. VeSOMAC convergence latency for intra-platoon vehicle passing.

Figure 8 reports *VeSOMAC's* convergence performance when a vehicle passes few vehicles in front to get ahead in the platoon. Convergence after a passing is needed because a passing vehicle's slot can violate the timing of the vehicles that it is passing and the ones in the neighborhood of its new location in the platoon. As expected, the convergence latency increases with longer passing events because more vehicles' slots are prone to be violated in these cases. For all the experimented scenarios within a 50-vehicle platoon, the post-passing allocations have always converged within 100 ms, when the ordering constraint was not used. Notice that the convergence time of 40 vehicle-passing is the largest. The reason is explained as follows. The average slot distance between two immediately neighboring vehicles is 2.5 slots, and the frame size is 100 slots in our experiments. This means that on an average every 40 vehicles, the slot position will be repeated. As a result, passing 40 vehicles has the largest probability to have a *VeSOMAC* slot collision. This explains why the convergence time of passing 40 vehicles is the largest.

VeSOMAC has a speed-aware component which prevents a passing vehicle from perturbing the allocation stability of the vehicles that it passes at a higher speed. The relative speed is inferred from the vehicle-location and timing information in the packet headers. A vehicle's *VeSOMAC* protocol logic does not react to the transmissions from neighbors with relative velocity larger than a pre-defined threshold. This way, the effects of passing shows up only after the passing vehicle settle down at a platoon location.

3.c. Inter-vehicle Data Transfer Applications Performance

3.c.1. UDP based Applications

Throughput of cross-platoon UDP data streams, representing applications such as inter-vehicle file, music, video, and traffic information downloads, are reported in Figure 9. With varying background traffic, the data rate of a UDP stream from the platoon-front to the platoon-end was changed to determine the maximum sustainable stream throughput. With low background traffic of 10 ppsv (packets per second per vehicle), a stream using 802.11 can sustain a throughput of up to approximately 47 packets/sec (pps), beyond which it starts to degrade due to packet collisions. With higher background traffic of 20 ppsv, its performance suffers drastically with a sustainable stream throughput of nearly 20 pps.

However, the UDP throughput with *VeSOMAC* is steadier and can reach higher sustainable values compared to the 802.11 MAC protocol. This is especially true for scenarios with higher background traffic. For example, with 20 ppsv background traffic, *VeSOMAC* can sustain a maximum throughput of approximately 45 packets/sec, which is nearly 80% more compared to the 25 packets/sec maximum sustainable throughput for the 802.11 MAC. Zero collisions and bounded delay of *VeSOMAC* account for this performance advantage.

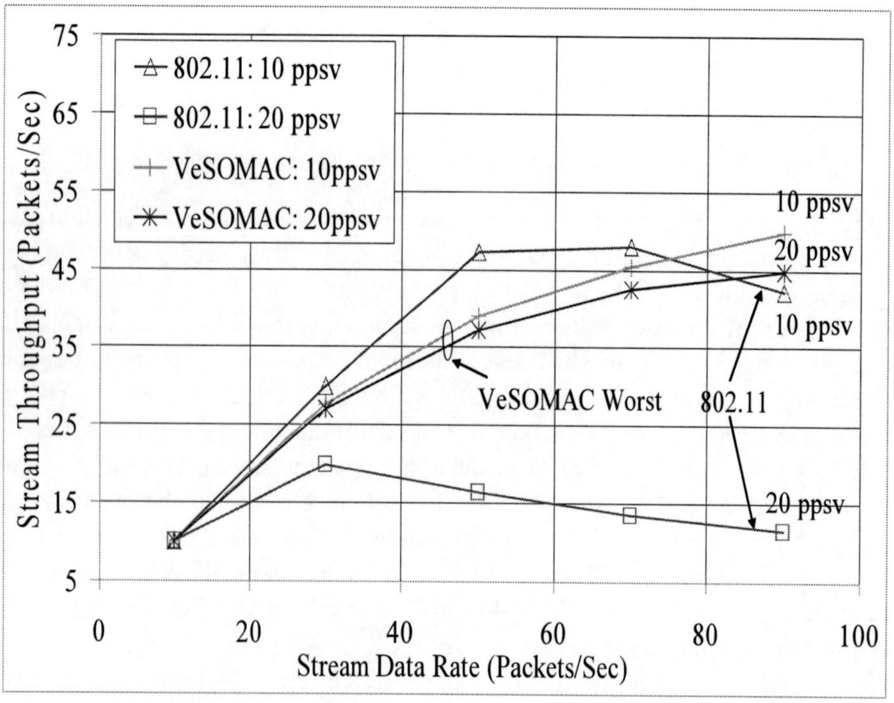

Figure 9. Throughput of cross-platoon UDP data streams with 802.11 and *VeSOMAC*.

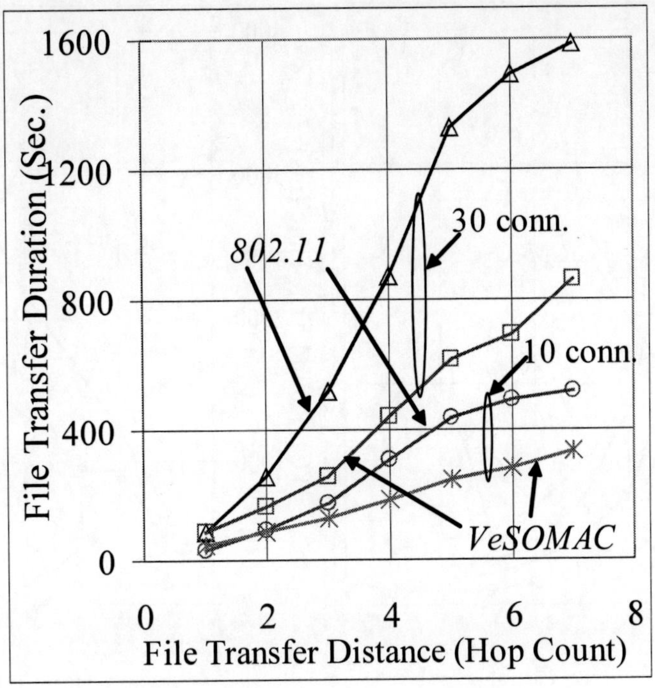

Figure 10. FTP duration with varying hop counts.

3.c.2. TCP based Applications

Figure 10 reports the FTP (running over TCP) completion time for an 1 Mbyte data file transferred across vehicles which are multi-hop apart within a highway platoon of 50 vehicles. The reported results indicate the average FTP completion time computed over 30 and 10 simultaneous file transfer sessions active across different vehicle pairs within the platoon. In each experiment, a given number of connections with a specified connection length have been established with arbitrarily chosen source and destination vehicles within the 50-vehicle platoon. Each point on the graph represents the average duration computed from 200 such independent experimental runs.

FTP has been run over TCP over 802.11 or *VeSOMAC*, and the TCP connections were set up using DSDV MANET [24] routing protocol. Note that such FTP based data transfer corresponds to a large number of data intensive applications such as maps, navigation and location based service, and music download [2].

The graphs in Figure 10 demonstrate that in majority of the situations, the file transfers are faster with the *VeSOMAC* compared to with 802.11, especially for scenarios when the involved vehicles are located farther apart within the platoon. The reason for this is better TCP performance with *VeSOMAC* compared to with 802.11 as reported in Figure 11. Note that the only scenario in which 802.11 performs better is when files are transferred across only 1-hop, and with lesser number (10 in this case) of simultaneously active connections within the platoon.

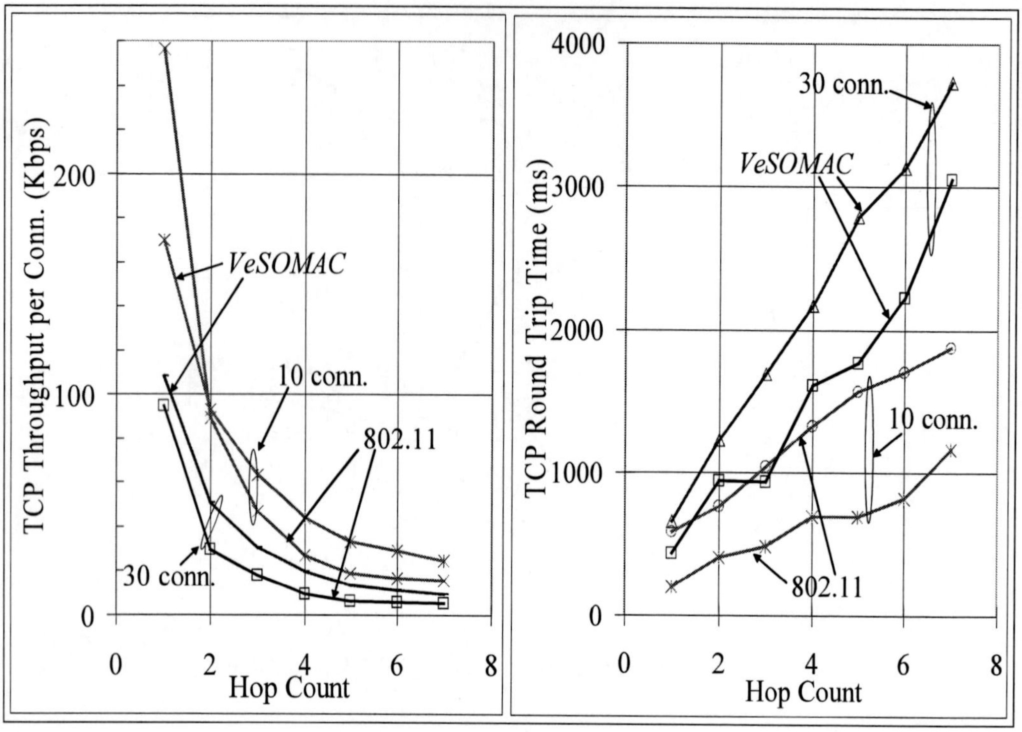

Figure 11. TCP Performance: a) throughput, b) round trip time.

Figure 11:a reports the corresponding TCP throughputs using 802.11 and *VeSOMAC*. Observe that with longer connections, the TCP throughput per connection reduces for both the MAC protocols due to increased user data traffic in the platoon. However, the proposed *VeSOMAC* enables higher TCP throughput compared to 802.11 for all scenarios except for 1-hop connections under small connection counts (10 in this case). For example, when the connections are 2-hop long and there are 10 connections present in the platoon, *VeSOMAC* can enable a per connection TCP throughput of 51 Kbps, which is a 66% improvement over the 802.11 enabled throughput of 30 Kbps. As explained later, higher MAC layer packet drops for 802.11, in all scenarios except for the 1-hop connections under small connection counts, turn out to be the reason [4] for the low TCP throughput.

Generally higher TCP throughput using *VeSOMAC* can be explained through the TCP congestion window evolution as reported in Figure 12:a and 12:b (for 30 simultaneously active 5-hop connection). Observe that due to frequent MAC layer packet drops, the congestion window for the 802.11 case is reset much more frequently compared to the *VeSOMAC* case, in which there are no MAC drops. TCP segment drops due to 802.11 drops in this experiment are reported in Figure 13:a. The occasional window resets for *VeSOMAC* are due to out-of-sequence packet delivery in the events of rerouting caused by the routing protocol DSDV's periodic route refresh operations. Zero packet drops for the *VeSOMAC* allows the congestion window to grow higher, which in turn yields better TCP throughput.

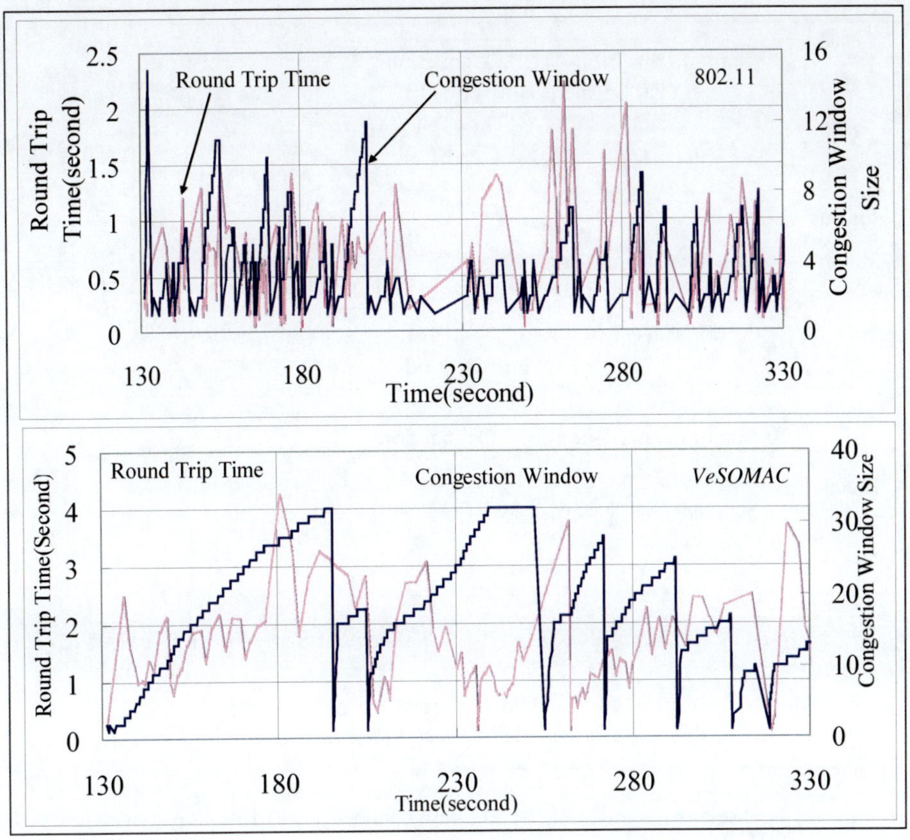

Figure 12. Dynamics of TCP : a) TCP-over-802.11, b) TCP-over-VeSOMAC.

Note, however, from Figure 11:b and 12 that in spite of its better throughput, the TCP round trip delay for the *VeSOMAC* case is actually higher than that with 802.11. This is due to the higher load for *VeSOMAC* caused by larger TCP congestion windows. Larger TCP load in the *VeSOMAC* case is evident from the higher sent segment sequence numbers compared to the 802.11 case as shown in Figure 13:a. Higher TCP load forces *VeSOMAC* to operate at a higher MAC load point, which results into larger MAC queuing and subsequent TCP round trip time.

TCP loading for 1-hop connections under small connection counts (10 in this case) are reported in Figure 13:b. Unlike in Figure 13:a, 802.11 in this case enables higher TCP loading compared to the *VeSOMAC*. Lack of self-competition [4] for the 1-hop connections and mild inter-connection competition due to fewer connections help significantly reduce the MAC layer drops for the 802.11. This explains 802.11's better TCP and file transfer performance for 1-hop connections under small connection counts as shown in Figure 10 and 11.

3.c.3. Application Level Impacts of VeSOMAC Slot Reorganization

The results in this section report the impacts of *VeSOMAC* slot reorganizations on inter-vehicle file transfer applications. While a certain number of TCP-based FTP sessions (10 concurrent 7-hop connections in this case) are active on a 50-vehicle highway platoon, MAC layer slot reorganizations are triggered by a vehicle passing event.

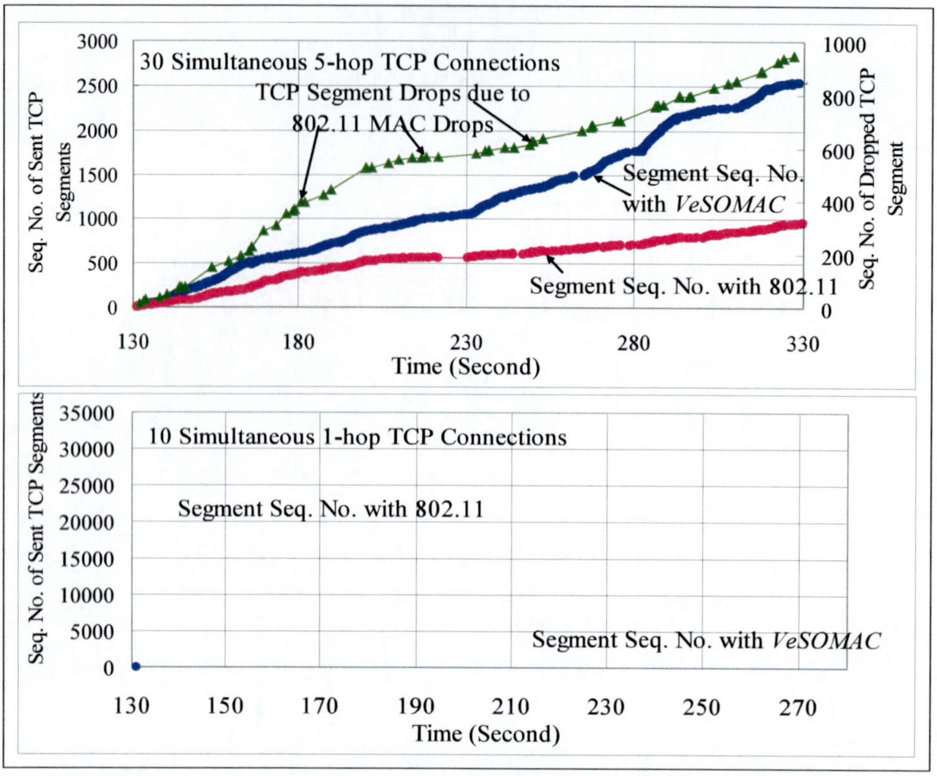

Figure 13. TCP segment generation: a) 30 5-hop connections b) 10 1-hop connections.

A vehicle passing event here involves a single vehicle passing a certain number of other vehicles in the platoon. This process is controlled in such a way that the original routes for all the TCP connections do not get altered. However, the passing vehicle causes temporary slot collisions with vehicles near which the passing vehicle settles down after the passing event. These affected nodes are always on the original TCP routes as intermediate nodes. This controlled vehicle passing process ensures that the impacts of MAC layer slot reorganization can be studied without perturbing the routing layer operations, which is not the subject of this study.

Figure 14 reports the average FTP transfer duration of an 1 Mbyte file with different vehicle passing rates. The FTP transfer duration of 802.11 is not prone to any changes with changing vehicle passing rate because of its inherent topology unawareness property. On an average, it takes around 550 seconds for transferring the 1 Mbyte file with 802.11 as the underlying MAC. In contrast, the *VeSOMAC* MAC can enable a much faster file transfer at the smaller switching rates, at which the *VeSOMAC* protocol is at near steady state with very little differences to no slot reorganization. Although the FTP transfer duration with *VeSOMAC* has a clear rising trend with increasing vehicle passing rate, it saturates around 550 seconds when the passing rate reaches a reasonably high value [1] of around 5.3 passings per minute. Here the point to be observed that the FTP performance with *VeSOMAC* is better than that with 802.11 not only at the protocol steady state at lower vehicle passing rates, but this comparative trend is maintained for situations with fairly high slot reorganizations caused by high vehicle passing rates.

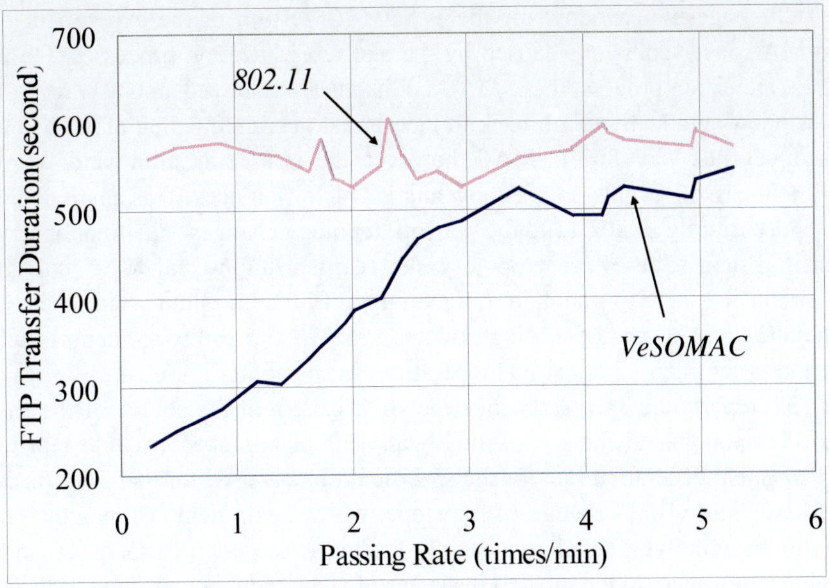

Figure 14. FTP duration with varying vehicle passing rate.

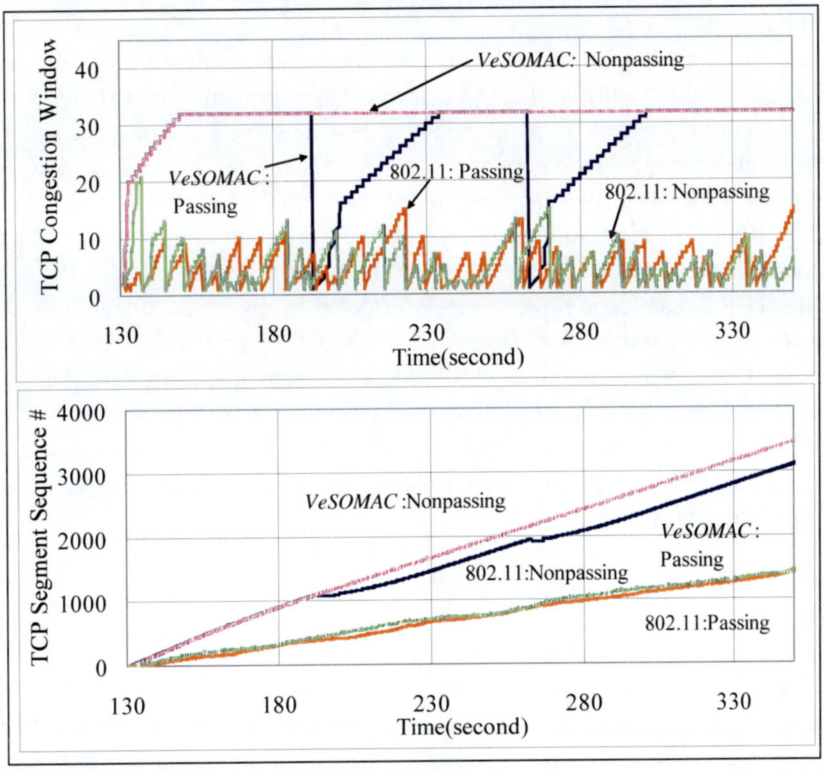

Figure 15. TCP Dynamics with passing events: a) congestion window, b) segments generation.

To investigate the results in Figure 14 further, we have collected the TCP congestion window evolution (for a specific 7-hop connection) with two separate vehicle passing events as shown in Figure 15:a. The passing events happen at time instants 190s and 260s, at which points the TCP congestion window for the *VeSOMAC* version of the experiment evidently falls due to MAC layer collisions caused by the slot reorganization processes. Observe that after the slot reallocation process, these MAC collisions are removed, and as a result the TCP congestion windows gradually climb back to its original maximum value of 32, as before the vehicle passing events. With 802.11 MAC, however, the TCP congestion window evolutions with and without passing events do not show any visible differences. This again demonstrates the desirable insensitivity of 802.11 to the network topology changes. The final observation in Figure 15:a is that in spite of its temporary slot reorganizations and MAC collisions, the *VeSOMAC* protocol is able to maintain a much larger TCP congestion window compared to that with the 802.11 MAC protocol. This explains *VeSOMAC's* ability to support higher TCP throughput and faster inter-vehicular FTP operations as shown in Figure 14.

Figure 15:b reports the TCP segment generation pattern at the source of the same TCP connection corresponding to the results in Figure 15:a. The slope of a line represents the overall TCP segment generation rate for the specific protocols used for that line. As expected, for 802.11 there is no visible change of the slope, which again demonstrates the protocol's insensitivity to the underlying topology changes. For the scenario with *VeSOMAC* and vehicle passing enabled, however, notice the two instances of slope reduction at times 190s and 260s, which are the instances of the vehicle passing events. The segment generation rate recovers soon after the vehicle passing events indicating a convergence of the MAC layer slot reorganization process.

Note that for all the experiments in this section, it was made sure that the effects of topology change manifests only at the MAC layer, while the end-to-end routes for all TCP connections remain unchanged. Experimentally it was observed that in practical scenarios when the routes also change due to network topology changes, the impacts of MAC layer slot reorganization in *VeSOMAC* becomes negligible compared to those caused by the route changes. This is because the rerouting delay overhead is usually much larger [20] (usually hundred of ms) compared to that for the MAC slot reorganization [8] (usually 10s of ms), and as a result the TCP dynamics is more severely affected by any route disruptions caused by events such as vehicle passing. Also, the routing disruption affects the vehicular networks with 802.11 MAC as severely as those with *VeSOMAC* protocols. Considering these facts, it can be safely concluded that for data intensive ITS applications such as inter-vehicular file transfer, the slot reorganization overhead of *VeSOMAC* does not cause any substantial performance degradation on top of the performance loss due to routing disruptions caused by network topology changes.

4. Conclusion

In this paper we have evaluated the performance of 802.11 and a new distributed TDMA MAC protocol for inter-vehicle data intensive applications. Simulation results demonstrate that unlike the 802.11 style contention-based protocols, the proposed *VeSOMAC's* TDMA mechanism can offer better application level file transfer performance due to its higher UDP and TCP throughput, which is enabled by zero MAC layer packet drops. It was also shown that during topology changes, *VeSOMAC* can reallocate TDMA slots with a fast protocol convergence. Experiments in transient scenarios representing TDMA slot reorganization due to network topology changes have demonstrated that while the application performance with

TDMA degrade during severe topology changes, the *VeSOMAC* can still outperform the traditional 802.11 family of protocols. Future work on this topic will include extending the results in urban intersection scenarios, evaluating other transport protocols such as RTP, and application of *VeSOMAC* for safety applications such as cooperative collision avoidance in highway and urban traffic intersections.

References

[1] "Intelligent transportation systems," United States Department of Transportation, www.its.dot.gov/index.htm.

[2] L. Armstrong, "Dedicated Short Range Communications (DSRC)," www.leearmstrong.com/DSRC/DSRCHomeset.htm.

[3] Y. Liu, F. Dion, and S. Biswas, "Dedicated Short-range Wireless Communications for Intelligent Transportation System Applications: State of the Art," *ITS and vehicle-highway automation,* vol. 1910, 2005.

[4] Y. Wang and B. Bensaou, "Achieving Fairness in IEEE 802.11 DFWMAC with Variable Packet Size," in *IEEE Global Telecommunication Conference*, 2001.

[5] T. Pagtzis, P. Kirstein, and S. Hailers, "Operational and Fairness Issues with Connection-Less Traffic Over IEEE802.11b," in *IEEE International Conference on Communication (ICC)*, 2001.

[6] N. H. Vaidya, "Mobile Ad-hoc Networks: Routing, MAC and Transport Issues," http://www.crhc.uiuc.edu/~nhv.

[7] Q. Xu, T. Mak, J. Ko, and R. Sengupta, "Vehicle-to-Vehicle Safety Messaging in DSRC," in *the 1st ACM International Workshop on Vehicular ad hoc networks*, October 01-01, 2004.

[8] R. M. Yadumurthy, A. Chimalakonda, m. Sandashivaiah, and R. Makanaboyina, "Reliable MAC Broadcast in Directional and Omni-directional Transmissions for Vehicular Ad Hoc Networks," in *the 2nd ACM International workshop on Vehicular ad hoc networks*, 2005.

[9] T. Liu, J. A. Sylvester, and A. Polydoros, "Performance Evaluation of R-ALOHA in Distributed Packet Radio Networks with Hard Real-time Communication," in *IEEE Vehicle Technology Conference* 1995.

[10] R. Verdone, "Multihop R-ALOHA for Intervehicle Communications at Millimeter Waves," *IEEE Transaction on Vehicular Technology,* vol. 4, 1997.

[11] V. Rajendran, K. Obraczka, and J. J. Garcia-Luna-Aceves, "Energy-Efficient, Collision-Free Medium Access Control for Wireless Sensor Networks," in *International Conference on Embedded Networked Sensor Systems*, Nov. 2003, pp. 181-192.

[12] M. Ergen, D. Lee, R. Sengupta, and P. Varaiya, "WTRP-Wireless Token Ring Protocol," *IEEE Transaction on Vehicular Technology,* vol. 53, 2004.

[13] J. Zhang, K.-H. Liu, and X. Shen, "A Novel Overlay Token Ring Protocol for Inter-Vehicle Communication," in *IEEE ICC'08* Beijing, China, May 2008.

[14] T. Nagaosa and T. Hasegawa, "Code Assignment and the Multicode Sense Scheme in an Inter-Vehicle CDMA Communication Network," *IEICE TRANSACTIONS on Fundamentals of Electronics, Communications and Computer,* vol. E81-A, 1998.

[15] S. Katragadda, G. M. C.N.S., R. R. M.S., K. S. M, and R. Sachin, "A Decentralized

Location-based Channel Access Protocol for Inter-vehicle Communication," in *Vehicular Technology Conference (VTC 2003-Spring)*, 2003.

[16] Z. Chen and A. Khokhar, "Self organization and Energy Efficient TDMA MAC Protocol by Wake up for Wireless Sensor Networks," in *1st Annual IEEE Communications Society Conference on Sensor and Ad Hoc Communications and Networks 2004 (IEEE SECON' 04)*, Oct. 2004, pp. 335-341.

[17] I. Rhee, A. Warrier, J. Min, and L. Xu, "DRAND: Distributed Randomized TDMA Scheduling for Wireless Ad-hoc Networks," in *the 7th ACM international symposium on Mobile ad hoc networking and computing*, Florence, Italy, May 22-25, 2006,.

[18] I. Rhee, A. Warrier, M. Aia, and J. Min, "Z-MAC: A Hybrid MAC for Wireless Sensor Networks," in *ACM SenSys 2005*, 2005.

[19] G. Lu, B. Krishnamachari, and C. S. Raghavendra, "An Adaptive Energy-efficient and Low-Latency MAC for Data Gathering in Wireless Sensor Networks," in *18th International Parallel and Distributed Processing Symposium*, Apr. 2004, p. 224.

[20] A. Elsaify, P. Padhy, K. Martinez, and G. Zou, "GWMAC- A TDMA Based MAC Protocol for a Glacial Sensor Network," in *4th ACM PE-WASUN 2007* Chania, Crete Island, Greece, October 2007.

[21] S. Mank, R. Karnapke, and J. Nolte, " An Adaptive TDMA based MAC Protocol for Mobile Wireless Sensor Networks," in *International Conference on Sensor Technologies and Applications, 2007. SensorComm 2007*, October 2007.

[22] "The network simulator: NS-2," www.isi.edu/nsnam/ns.

[23] X. Yang, J. Liu, F. Zhao, and N. Vaidya, "A Vehicle-to-Vehicle Communication Protocol for Cooperative Collision Warning," in *the First International Conference on Mobile and Ubiquitous Systems: Networking and Services*, 2004.

[24] C. Perkins and P. Bhagwat, "Highly Dynamic Destination-Sequenced Distance-Vector Routing (DSDV) for Mobile Computers," in *the ACM SIGCOMM*, October 1994.

In: Wireless Technologies in Intelligent Transportation... ISBN: 978-1-60741-588-6
Editors: Ming-Tuo Zhou et al, pp. 133-174

Chapter 6

SECURITY OF VEHICULAR AD HOC NETWORKS

Zhengming Li, Zhou Wang and Chunxiao Chigan
ECE Department, Michigan Technological University

Abstract

As emerging civilian instantiates of Mobile Ad Hoc Networks (MANETs), Vehicular Ad Hoc Networks (VANETs) are critical components in Intelligent Transportation Systems (ITS). VANETs, though special cases of MANETs, have distinct characteristics: high mobility with limitation of road topology, rapidly changing network topology, unbounded network size, and time-sensitive data exchange, to name only a few. These unique characteristics, together with the fact that many VANET applications are safety-related, make security provisioning in VANETs a challenging while indispensable task. In this chapter, we systematically investigate the security provisioning issues of VANETs, to give a rather complete understanding on the requirements, challenges, potential solutions, open issues, and the future research directions from different perspectives. Based on the analysis of threat models and different applications requirements, we categorize the VANET security requirements into three major classes (the information security, the data security, and the trust management), along with considerations about network performance constraints. In-depth investigations of these security provisioning issues and future research directions are presented. A security provisioning framework that enables interactions of different security provisioning modules is also proposed.

Keywords: VANETs, ITS, Security Provisioning, Network Performance

1. Introduction

Vehicular Ad Hoc Networks (VANETs) are emerging civilian instantiates of Mobile Ad Hoc Networks (MANETs), which are attracting the attention of industry and academia. VANETs are systems of communication technologies enabling the vehicles to communicate with one another (inter-vehicle communication or V2V) and to communicate with roadside units (V2R), as shown in Figure 1. Behind the roadside units (RSUs) there are infrastructure components providing specific services to the vehicles. RSUs are also connected to other networks, especially the Internet. In the near future vehicles will become *smart*, equipped with advanced features such as Event Data Recorder (EDR), (D)GPS Receiver, communication facilities (Mobile, Bluetooth, DSRC [1], etc.) [2]. The communication features generally are integrated into the On-Board Unit (OBU), with which the *smart* vehicles are able to communicate with one another or RSUs. In VANETs the network nodes mainly are vehicles, so hereafter the terms vehicle and node will be used interchangeably.

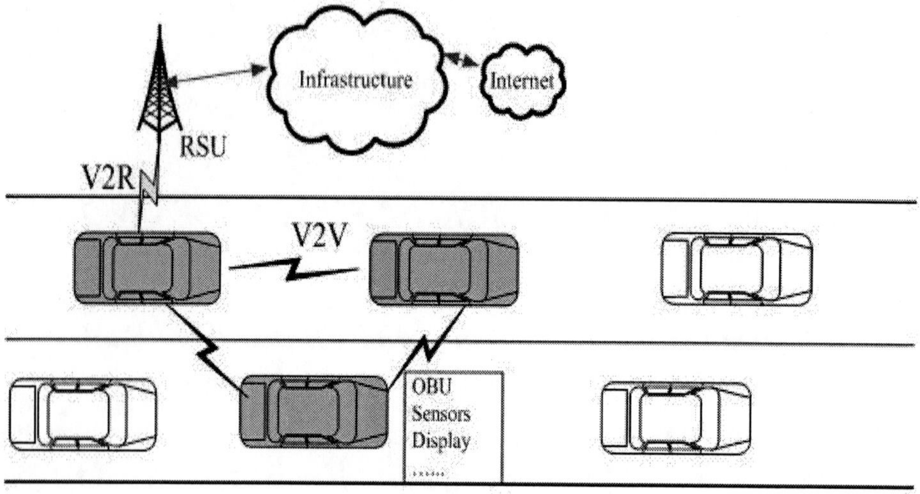

Figure 1. VANET Model: V2V and V2R Communications. The dark vehicles are equipped with VANET technology. The equipped vehicles can communication with each other and with the RSU. The application-specific servers for VANETs reside in the Infrastructure.

The roadside infrastructures consist of RSUs as network and service access points, and the infrastructures which provide advanced management functions (certification, registration, membership management) and value-added services (Location-based Services). Although roadside infrastructures may not prevail initially due to the heavy investment involved, in the long run they surely will be rolled out on a massive scale, probably in the form of Intelligent Transportation System (ITS).

VANETs open the door for a plethora of applications and services, ranging from automated highway systems to distributed passengers teleconference [3]. In the context of ITS, VANETs are becoming an enabling factor in many ITS applications by providing additional communication capabilities.

In VANETs most safety-related applications, such as Traffic Signal Violation Warning, Left Turn Assistant specified in [1], Collision Avoidance Warning (CAW), etc., are delay- and error- sensitive, since any error or delay of these applications may endanger the vehicles and the drivers. Therefore, VANETs have strict security requirements on authenticity, non-repudiation, integrity, as well as performance requirements including real-time and reliability. Meanwhile, VANETs also have many application-specific security requirements, namely node privacy and message confidentiality. All these security and performance requirements will be more formally analyzed later in this chapter. The stringent security requirements raised by the safety-related applications alone qualify VANETs as a distinct and promising research field. To better understand the security requirements of VANETs, the differences between VANETs and MANETs will be identified.

1.1. MANETs vs. VANETs

Under close scrutiny, VANET security differs from MANETs security as follow:

1) *Security Requirements*: In VANETs the security requirements stem not only from network functions, but also from specific application and services that differentiate VANETs from MANETs. Thus VANET security has more application-specific requirements than that of MANETs. For example, node privacy, while not stressed in MANETs, is indispensible to VANETs.
2) *Security Solutions:* The unique characteristics of VANETs provide new implications to security solutions, which makes many existing security solutions for MANETs not directly applicable to VANETs.

According to our understanding and several existing literatures [4] [5] [6], the characteristics of VANETs with security significance can be summarized as: *high mobility with the constraint of road topology*, *impact of road layout*, *rapidly changing network topology*, *initially low market penetration ratio (MPR)*, *unbounded network size*, *anonymous addressee*, *time-sensitive data exchange*, *support from infrastructure*, *abundant resources*, and *better physical protection*. All these characteristics are relevant to the design and implementation of security solutions in VANETs.

High mobility with the constraint of road topology: The constraint of road topology increases the predictability of vehicle motion, and makes it difficult to protect the location privacy of vehicles. And the high mobility of nodes poses challenges to the reliability of communication connections. However, it may be helpful to messages authentication by node location verifications.

Impact of road layout: The road layout has direct impacts on the network connectivity in VANETs. Within downtown areas the roads have multiple lanes and the road density is high, so VANET nodes tend to have good connectivity. However, in a country road the nodes will have few neighbors, resulting in poor connectivity. Thus, the road layout has direct impact on the security schemes relying on cooperation among the neighbor nodes.

Rapidly changing network topology: The rapidly changing network topology caused by high node mobility makes it extremely challenging to provide reliable inter-vehicle communications. This in turn makes it difficult to design and implement security schemes in general. Specifically, it will be more difficult to maintain node accountability and historical records of trust and reputation which are helpful to punish the selfish and malicious nodes.

Initially Low MPR: This adds difficulty to the reliable communication in VANETs. It also will be difficult to detect and evict the misbehaving nodes if the node density is low.

Potentially unbounded network size: This characteristic poses scalability requirement to all security solutions in VANETs, especially key management and trust management.

Anonymous addressee: In VANETs sometimes the receiver nodes are addressed anonymously. For example, one accident warning is relevant to the nodes within one specific region, so it is enough to send this warning to all the nodes within that region, without explicitly addressing any specific node. This paradigm helps the protection of node privacy.

Time-sensitive data exchange: For many VANET applications time-sensitive data exchange is a must, and it poses strict network performance requirements. Adopting strong security mechanisms generally will degrade network performance especially by introducing considerable message transmission delay, which is contradictive to the performance requirements of many applications.

Support from infrastructure: VANETs will enjoy support from the roadside infrastructure with the help of RSUs. Although RSU is unlikely to be pervasive in the beginning, it will be installed at some hot spots, e.g. downtown areas and toll gates. This provides opportunities of designing and implementing better security solutions which would be otherwise unfeasible.

Abundant resources: In VANETs the nodes have practically unlimited energy supply and high computation capacity. This provides an opportunity to security provisioning, for example the application of asymmetric cryptographic algorithms in VANETs. At the same time this also brings risks of more powerful attacks launched by the compromised nodes.

Better physical protection: The physical protection of nodes in VANETs is better than that in MANETs, so VANET nodes are relatively more difficult to compromise.

Based on the above analysis, security provisioning in VANETs is an interesting and challenging research field.

1.2. Organization of This Chapter

In this chapter we aim to provide a systematical and comprehensive analysis of security provisioning in VANETs, addressing security from the perspective of various application requirements. The rest of this chapter is organized as follows. We identify and analyze the common security requirements and the threat model of VANETs in section 2. In section 3, a security provisioning framework for VANETs is proposed, with all relevant factors identified and considered. Four modules are selected as the main topics of this chapter: *Information Security*, *Data Security*, *Network Performance* and *Trust*. Then the state of the art of the Information Security is investigated in section 4 with emphasis on node privacy and node accountability. The state of the art of both Data Security and Network Performance is investigated in section 5, due to their close relations. In section 6 the existing schemes for trust model and trust management are studied. Section 7 discusses the open issues in the highlighted topics, which are also the promising future research directions in VANET security provisioning. In the end, section 8 concludes this chapter.

2. Security Requirements and Threat Model

In VANETs various applications have various security requirements, which define the final (or ideal) states in which these applications should be performed and corresponding services should be provided. Meanwhile, VANETs have to put up with numerous possible attacks from the potential attackers. In order to meet the security requirements of different applications, security schemes should be systematically designed and deployed in VANETs. This involves two prerequisites for VANET security provisioning: security requirements and threat model. Security requirements define the requirements the security provisioning schemes have to meet; threat model defines what potential attacks and attackers (threats to VANET security) the security provisioning schemes have to countermeasure. This section will thoroughly investigate these prerequisites.

2.1. Security Requirements of VANETs

As previously discussed, numerous VANET applications have been identified and it is highly possible that many more will be identified in the future. Thus it will be rigid and cumbersome to analyze every specific application to get a holistic view of security requirements of VANETs. Besides, the security requirements are mainly about the messages, which have smaller cardinality than the applications, so existing papers [5] [7] mainly analyze the security requirements of different VANET messages. [7] classifies VANET messages into three categories: data unchanged (fixed spatial data, for example the position of traffic lights), data changing over time (spatio-temporal data) and data changing with positions and time (mobile data). [5] categorizes the safety-related VANET messages into traffic information messages, general safety messages and liability-related messages. Based on these categorizations the security requirements of each message group are studied in [5] [7].

This approach, though intuitive, has several inherent drawbacks. Firstly, the message categories are coarse-grained, and usually have overlaps and omissions. Secondly, the same message may have different security requirements in various application scenarios. Last but not the least, it omits the security requirements on other important aspects in VANETs, for example the relations among the nodes.

A novel approach for grouping security requirements is proposed here, which focuses on the security requirements of the important assets in VANETs. Information, resource and relation are identified as the most important assets in VANETs, because they are most crucial to the users and most appealing to the attackers. They are the ultimate goals of any security schemes and potential attacks. Thus, in this approach the nature of security requirements and the intentions of potential attacks can be effectively stressed.

Here *resource* includes network bandwidth, radio access time, and services provided by the specific applications. It focuses on the available services or benefits provided by the networks. The resources of individual nodes are not considered, since VANET nodes have abundant energy and computation resources. The asset *information* consists of the ID of the nodes, personal communication of the drivers, user accounts, passwords, etc.

For each asset we group the security requirements and potential attacks based on the attacker intentions: rational and malicious. The attackers with rational intention try to launch attacks to get access to unauthorized resources or information, or to get an unfairly large share of them. The malicious intention means that the attackers try to deprive other nodes of their due share of information or resource. A preliminary categorization of the common security requirements and potential attacks is shown in Table I. For convenience, hereafter the security requirements of resource are regarded as *data security*, and those of *information security*. These concepts will be further investigated in the following parts of this chapter.

Just like the human society, VANETs also consist of *relations* among their participants (nodes). Thus it is necessary to securely set up and manage the node relations. Roughly speaking, the relations among nodes can be depicted by the following examples:

1) *Behavior evaluation*: The selfish nodes may try to get advantage over other nodes. The malicious nodes may intentionally bring harms to other nodes and the network. Such nodes should be detected and punished in order to guarantee a cooperative environment in VANETs.

2) *Authorization:* Node A gives the access right to its resource to its friends. One of its friends, say node B, may want to transfer its access right to its friend node C. There should be a systematic way to decide whether C can get access to A in this case.
3) *Certification:* Suppose node A fully trusts node B. In the future A may run into node C, which hold a certificate signed by node B. Thus node A need decide whether to trust node C or not, based on the certificate signed by B. This case can become more complicate if between these two nodes there are numerous intermediate nodes signing certificates for one another.

Table I. Attack and Security Properties Grouped by Intention

Item	Attack	Security Requirements	Attacks
Information	Rational	Message Confidentiality, Node Privacy, Access Control	Eavesdropping, Offline Password Guessing, Traffic Analysis; ID revealing, Position Tracking, User Profiling
	Malicious	Message Authenticity, Information Availability, Node Accountability Message Integrity,	Impersonation, False Message Injection, Replay, Sybil Attack; DoS, Packet Dropping, Refusal to Generate Public Packets, Bad-Mouthing; packet tampering;
Resource	Rational	Access Control; Fairness	Unauthorized Access; MAC Protocol Violation; Physical Compromise;
	Malicious	Availability; Reliability (On the three lower layers)	Packet Dropping, Route Tampering, Packet Tampering, DoS; DoS, Physical Interference, Physical Compromise,

Currently the concept *Trust* is used to model and manage relations among participants in computer networks. There exist many reported efforts on trust modeling in distributed systems or wireless networks [8] [9] [10] [11], but the research on trust management in VANETs [12] [13] has just begun and needs serious investigations.

2.2. Threat Model

As an ancient Chinese adage goes, *in order to win you should know both yourself and your adversaries.* In order to meet the various security requirements, it is necessary to study the potential threats to VANET security: attacks and attackers (adversaries). VANETs can be better secured only based on a thorough understanding of the potential threats.

Various definitions of attack and attacker exist in the context of security, but the main ideas remain similar. In [14] attack is defined as a sequence of attack actions associated with the system trying to get advantage over the security schemes, and attacker is the entity launching attacks. In [15] an attacker is a device deviating from the definition of protocols. In this chapter an attacker is defined as a node which chooses not to comply with the protocols due to either selfish or malicious reasons.

The potential attacks and attackers in VANETs have been classified in several existing papers. In [5] attackers are checked in three dimensions: (a) *Insider* (nodes with valid authentication key and participating in the network) vs. *Outsider* (attackers who are not VANET nodes and try to launch attacks from outside), (b) *Malicious* (attackers who only aim to harm the VANETs and the nodes) vs. *Rational* (attackers who only want to take advantage of VANETs, without harming themselves), and (c) *Active* (attackers who can actively send invalid or harming messages) vs. *Passive* (attackers who can only passively eardrop the communication in VANETs). Several specific attacks are also mentioned in [5] and the associated attackers are checked against these three dimensions. In [16] five attacker types are identified: *Greedy Drivers, Snoops, Pranksters, Industrial Insiders, and Malicious Attackers.* These attackers are grouped based on their intentions and potential attack capabilities, which is different from [5] in that it does not distinguish the different dimensions, identifying the attackers based on their overall characteristics instead.

Our approach explained in subsection 2.1 aims to categorize the security requirements in the light of the assets of VANETs. Thus it is sufficient to address only rational and malicious intentions of the attackers. However, the static categorization approaches in [5] [16] are obviously insufficient to this end. [5] [16] group the attackers by their intentions and capabilities only in a coarse way: the attackers are characterized with static and unique properties. In reality attackers may change their intentions according to the results of their previous attacks, and their future attack strategies and capabilities will follow suit. Thus the identification and categorization of attackers and attacks are only the first step in research of threat model. The intent, objectives and strategies of the attackers, as well as the interactions between VANETs and the attackers need to be systematically modeled. Based on this, the most possible attacks and attack strategies could be formally inferred to provide useful information to the design and implementation of VANET security schemes.

In [14], the incentive-based modeling and inference of attacker intent, objectives and strategies for network security are studied with the help of game theory. The interplay between attackers and the network is described as games, and the potential attack strategies are inferred accordingly. The proposed approach provides useful hints to the research on the threat model in VANETs.

The cost and potential revenue of the security provisioning schemes also deserve further investigation, since they are directly related to the implementation of VANETs. In [17] the behaviors of attackers based on economic considerations are studied, and its principles could be adopted in the security research of VANETs.

In summary, several issues are still open in the research of threat model for VANETs.

Firstly, concrete work should be done to formally model the intent, incentive and attack strategies of attackers, so that they can be conveniently used to design security schemes. The existing work on threat model [5] [16] only identifies (and enumerates) the potential attackers and attacks, where the intent of attackers is only implicitly addressed. The objectives and strategies of attackers are yet to be studied.

Secondly, the interplays between attackers and VANETs still need further investigation. It is meaningful to study these issues such as how the behaviors of attackers will change according to the security strength of VANETs, and how the defense strategies of VANETs should be adjusted in face of different types of attackers, etc. [14] provides helpful hints, but further work should be done to adapt this approach to VANETs.

Thirdly, the economic issues in the security provisioning schemes should be investigated to maximize the ratio of revenue to cost of these schemes. This is especially meaningful when the security provisioning schemes are to be selected for implementation in VANETs. Reference [17] provides some help to this end, but more work should be done to thoroughly investigate this topic.

These issues are both appealing and promising for further research efforts.

3. Vanet Security Provisioning Framework

As previously discussed, many important issues have been identified for VANET security provisioning: information security, data security, trust management, and threat model. Due to the unique characteristics of VANETs, network performance also is an indispensible issue in security provisioning. Application requirements and the underlying technologies should also be considered due to their implications to security provisioning. A security provisioning framework is proposed here to incorporate all these relevant issues in VANET security provisioning. In this section this framework will be presented, based on which the relevant issues and their interplays are investigated. In the end, the most important security modules are identified for further investigation in this chapter.

3.1. Framework Overview

The proposed **V**ANET **S**ecurity **P**rovisioning **F**ramework (VSPF) is shown in Figure 2. VSPF consists of five indispensible models: *Application Model, Security Model, Network Model, Threat Model, and Trust Model.*

3.1.1. Security Model

The Security Model defines the security requirements and security solutions for VANETs. As shown in Figure 2, its upper part depicts the security requirements on information and data: information security and data security. The lower part shows the potential solutions to meet these security requirements. The security requirements shown in Figure 2 mainly follow Table I, with the security requirements grouped into information security and data security. Network performance is also included in this model because it is directly related to data security and it has significant implications to information security.

This model is influenced by all the other models. With the specific requirements and constraints from the other models, security requirements can be identified and proper security solutions can be designed and implemented. The implications of the other models will be investigated in details when they are introduced as follow.

3.1.2. Application Model

The Application Model contains the existing and potential applications supported by

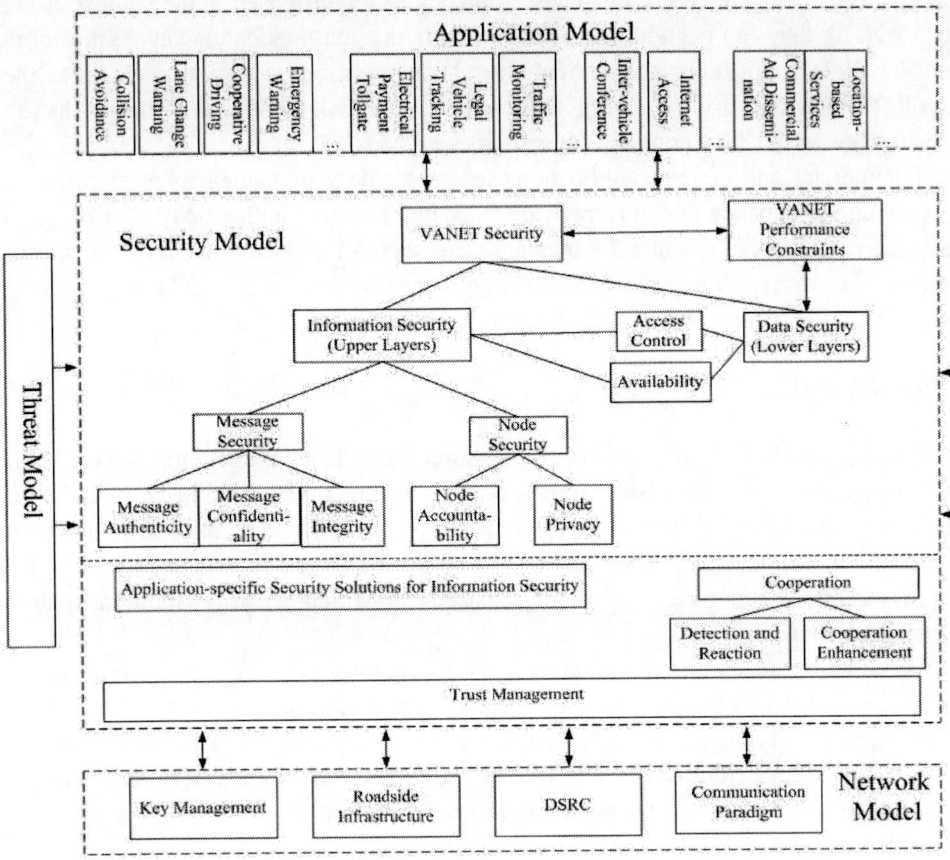

Figure 2. VSPF: A Comprehensive Framework for VANET Security Provisioning.

VANETs, and determines the concrete requirements on security and network performance. Different applications will pose different requirements to the Security Model, so a concrete and realistic Application Model will be the prerequisite to the Security Model.

The detailed characteristics and requirements of each possible VANET application are related to the problem domain, which is out of the scope of this chapter. Here to facilitate our investigation on security requirements we only briefly investigate the potential applications in VANETs. According to our understanding, four groups of applications exist in VANETs: safety-related applications, traffic management applications, commercial applications and general applications, as shown in the application model in Figure 2.

The safety-related applications are designed to improve the driving safety with the messages exchanged among the vehicles in VANETs. Typical applications include Forward Collision Avoidance, Lane Change Warning, Emergency Warning, etc. In these applications one node generally make decision based on the messages from other nodes. Thus, the authenticity of the messages is critical. These applications also pose stringent requirements on the network performance, since any long delay in message transmission may result in disaster.

Traffic Management applications rely on the messages collected from the VANET nodes to perform management and optimization of the traffics. This group includes traffic monitoring, electric tollgate and so on. These applications are proposed in the context of ITS, but with VANETs they can get additional support from the communication capabilities of the vehicles. For these applications the authenticity of the messages is also critical. Since they have no direct impact on driving safety, these applications will have lower priority than the safety-related ones when they compete for resources in VANETs.

The commercial and general applications have no safety or management significance. Instead, they aim to enhance the convenience of driving. Here the authenticity of messages is not the concern of VANETs, since the messages are should be verified only by the specific applications. However, these applications still require message integrity, message confidentiality, node accountability and node privacy.

3.1.3. Network Model

In the Network Model all the underlying/supporting technologies for VANETs are defined: key management, roadside infrastructure (potentially available in the future), communication technologies (DSRC [1] and other technologies), and communication paradigms (group communication or general communication), as shown in Figure 2. This model provides the supporting environment based on which security solutions can be designed and implemented.

In this model the roadside infrastructure provides opportunities of security provisioning schemes which will not be feasible otherwise. Key management is a crucial module, defining the capabilities of cryptographic key management for VANETs. The communication technologies provide a certain level of service quality in terms of reliability and real-time, which form the constraints for the security solutions. The communication paradigm indicates how the nodes perform their functions: individually or in group.

The modules other than key management and communication paradigm are orthogonal to the focus of this chapter, and will not be discussed in details. Key management in itself is a separate topic worthy of thorough investigation, and it is our future work. Here we will only identify and clarify the requirements on key management of the existing security schemes. Communication paradigm lies in the upper layers of VANETs, and has direct impacts on information security. Thus it will be investigated in the context of information security.

3.1.4. Threat Model

The Threat Model defines the potential attackers and attacks which have to be considered in the context of a certain application. In other words, it defines what VANETs should handle in order to meet the security requirements. This is an indispensible factor to any security provisioning scheme, defining the security issues to be solved. In subsection 2.2 we have briefly discussed several principles for the threat model of VANETs. These principles are useful to the construction of threat model for any concrete application scenario.

3.1.5. Trust Model

The Trust Model consists of the policies and rules of trust establishment and management for the specific applications, based on which Trust Management in the Security Model can be done, as shown in Figure 2. The policies and rules of trust evaluation, trust maintenance and trust propagation should be predefined in the Trust Model. In order to establish these policies both technical issues and application issues (for example the need of authorization and the goals of trust management) should be considered. Here we will only focus on the technical issues, such as trust metrics, trust modeling and trust evaluation.

3.2. Highlighted Topics

Based on the previous discussions and our understanding, we identify four important topics (modules) within VSPF: information security, data security, network performance and trust. Information security, data security and trust define the vital security requirements on the most important assets in VANETs: information, resource and relation. At the same time, network performance poses constraints to any security schemes in VANETs. As we will show later, these modules are of vital significance to VANETs and have many open issues, calling for thorough investigations.

3.2.1. Information Security

Information dissemination is the principal function of VANETs, so information security is critical to VANETs. As shown in Figure 2, information security consists of several components: message security (Message Authenticity, Message Confidentiality, Message Integrity), node security (Node Accountability and Node Privacy), and holistic security (Access Control and Information Availability).

Message Authenticity means that the messages propagated in VANETs should be authentic and valid. The messages reporting false events should not be propagated in VANETs. Message Confidentiality requires that the contents of messages will be accessed only by the intended receivers. Message Integrity means that messages should not be tampered during transmission. And once the messages are illegally changed it will be detected by the receivers.

Node Accountability (Node Non-repudiation and Node Authentication) requires that the nodes can be authenticated by other parties in VANETs, and that no one can deny what it has done in case of liability determination. Node Privacy means that the identity and location information of any node will be securely concealed from the unauthorized parties in VANETs.

Access Control provides a method of controlling the access right to the information available in VANETs. Only the nodes with proper authorization should be able to access the sensitive or protected information. Information Availability means that the information should be available to the targeted nodes at any time. Under closer scrutiny we observe Information Availability and Access Control can be satisfied if all the other security requirements are met at the same time. Thus, they will not be investigated in details here.

Information security can be directly experienced by the users of VANET applications, and its absence will make VANET applications less appealing or even unacceptable to the users. Thus it is meaningful to investigate the existing solutions for information security and identify open issues.

3.2.2. Data Security

Data security is the security regarding the resources in VANETs, and it is mainly relevant to the lower layers: Physical, MAC and Network layers. Data Security in essence can be represented as Availability and Access Control. Availability means that VANET applications and services should be available to the intended users at any time. Access Control requires that the resources in VANETs should be accessed by the authorized nodes as predefined.

Data security aims to guarantee the availability and fair sharing of resources in VANETs. Thus it will directly influence the network performance in terms of real-time and reliability.

The resources (network bandwidth, radio access time, etc.) are collaboratively provided by all the VANET nodes, and at the same time competitively shared by them. Thus, data security can be essentially reduced to the cooperation issues in VANETs, and the cooperation enhancement schemes can be regarded as the main approach to guarantee data security in VANETs. This chapter will mainly focus on the cooperation issues in the Network layer.

3.2.3. Network Performance

As stated above, data security can help guarantee the network performance in VANETs. At the same time security schemes are likely to consume the network and computation resources, negatively impactint network performance. Most safety-related applications in VANETs are highly delay-sensitive and error-sensitive [1], so it is critical to consider the tradeoffs between security provisioning and network performance. In the following parts the existing security provisioning schemes will be investigated with proper consideration of their impacts on network performance.

3.2.4. Trust Model and Management

Trust model is important, since it defines rules and policies for trust management in practice. Trust management somehow overlaps service security and data security, but it is a separate research topic in and of itself. Trust management incorporates the separate concepts in service security and data security, and establishes cooperative environment in VANETs by monitoring and evaluating the nodes' behaviors. Actually, one important applications of trust management is cooperation enhancement of VANET nodes. Thus, this topic is both meaningful and important to VANETs. The fact that the nodes' historical records are not always available in VANETs makes this topic simulatenously more difficult and appealing.

4. Information Security

Information security mainly is concerned with the upper layers of VANETs, so communication paradigm as one upper layer factor will directly impact information security. Thus, in this section two communication paradigms and their impacts on information security will be studied before the specific information security requirements and schemes are investigated in details.

4.1. General Communication and Group Communication

Two communication paradigms exist in VANETs including *general communication* and *group communication*, which define how the VANET nodes communicate with one another. In general communication each node behaves as an individual and communicates with other nodes on behalf of its own interests. In group communication several nodes form a group according to a set of predefined rules, and the group will communicate with other groups or other nodes as a single entity.

An exemplary network with both group communication and general communication is shown in Figure 3. The groups are formed by the spatially adjacent nodes according to predefined rules. There also exist individual nodes which do not join any groups. Any group communicates with other groups and other individual nodes as a whole.

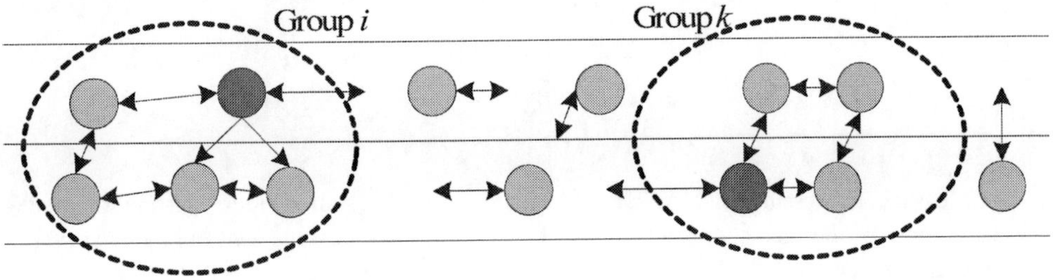

Figure 3. Group Communication and General Communication in VANETs. Two groups exist in this figure within the circles. Within the groups the dark nodes are the group leaders, which manage the groups and communicate with outside world in behalf of their groups. The nodes outside the groups can communicate with others as individual nodes.

Group communication can be used to boost specific applications, such as spaced platoons [18] [19] [20] [21] and automated highways co-operative driving [22] [23], or to enhance the security of the group members. In the light of information security, group communication has important implications.

First of all, the main challenge in group communication lies in the secure and efficient group formation and management. The nodes should be authenticated first in order to join a certain group. The misbehaviors of the group members, especially the group leaders, should be detected and promptly stopped. Otherwise the group will become an attack tool of the misbehaving nodes. The mobility of the nodes and the scarcity of network resources make it necessary to design and implement lightweight group communication schemes in VANETs. Existing schemes [7] [24] [25], to our best knowledge, need to be further refined to meet all these requirements. These issues described here can be handled with the authentication and trust schemes, which will be investigated in details in this section.

Secondly, group communication can enhance the node privacy and message authenticity. The whole group interfaces with the outside as a single node, so the individual node identities are concealed from the outside. As long as proper methods are adopted in group management to prevent the group members from disclosing the privacy of its peer members, group communication can enhance the privacy of nodes. The messages sent out by the group will be verified by the majority of the group members, and this reduces the possibility that any invalid message can be initiated by the group. Thus group communication is commonly used to authenticate messages in VANETs [25].

With the important security challenges and benefits mentioned above, group communication has been extensively investigated in security provisioning for VANETs. This will be shown in our investigation of the existing schemes for the specific information security requirements.

4.2. Message Security

As shown in Figure 2, message security contains the security requirements about the information contained in the messages: message authenticity, message confidentiality and message integrity. Message is the main carrier of information in VANETs, so message security is critical to information security in VANETs.

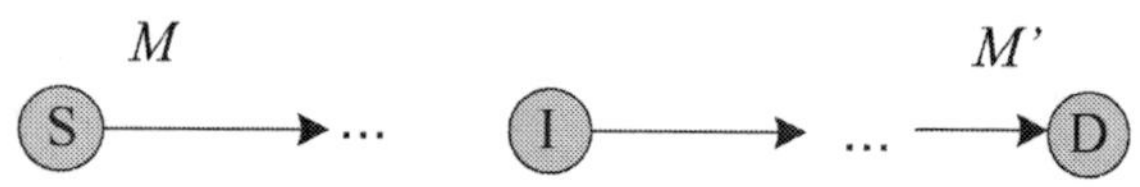

Figure 4. Message Processing in VANET. S, D and I are the source node, destination node and intermediate nodes of the message, respectively. *M* is the original message sent out by S, and *M'* is the message received by D.

Here we consider the generic message processing procedures in VANETs, to clarify the aforementioned message security requirements. The security requirements reside in each step of message processing in VANET, as shown in Figure 4. Message confidentiality requires that except for S and D no node can get access to the content in the message. Message integrity means that if any node changes *M* during the message transmission, D can detect this change. Message authenticity has two-fold meanings: S should only send out messages with valid contents; D can verify both the origins and the contents of the received messages and discard the invalid messages.

There are many cryptographic tools to ensure message confidentiality and integrity, such as public key algorithms and secret key algorithms. It is quite straightforward to apply these tools in VANETs. In [26] [27] [28], one-time session keys are set up between the source and destination nodes, and messages confidentiality is achieved with the session keys. Since message confidentiality and message integrity can be guaranteed through the cryptographic tools, their provision in VANETs can be realized in the context of key management. Message confidentiality requires message encryption, and message integrity requires message signature. All these in turn require proper setup of keys among the originators and destination nodes. Due to the characteristics of VANETs, it is challenging to implement Public Key Infrastructure (PKI) or Key Distribution Center (KDC). Meanwhile, the emerging cryptographic techniques such as IBS (Identity-based Signature) and IBC (Identity-based Cryptography) [29] [30] [31] could also be adopted in VANETs for key management. However, this issue need to be further investigated in the future.

Comparatively message authenticity is more complicate and difficult in that it cannot be ensured by the implementation of cryptographic tools alone. This is where the characteristics of VANETs become more relevant to security provisioning, and it is our main focus in this subsection.

4.2.1. Scope of Message Authenticity

In VANETs the origin and content of the messages should be verified to guarantee their authenticity. With techniques such as digital signature it is easy to verify the source node of a certain message in VANETs, which will be investigated in the context of Node Accountability.

Comparatively it is more difficult to check the authenticity of the message contents in VANETs. False information attack results from the false and invalid messages. And this is a challenging issue since it is difficult to distinguish whether the false message has been generated by a malicious node or caused by some failures/changes in the network topology. Without message authenticity, the VANET application can become a security pitfall where the malicious nodes launch false information attacks and cause panic, traffic jamming or even car accidents [32].

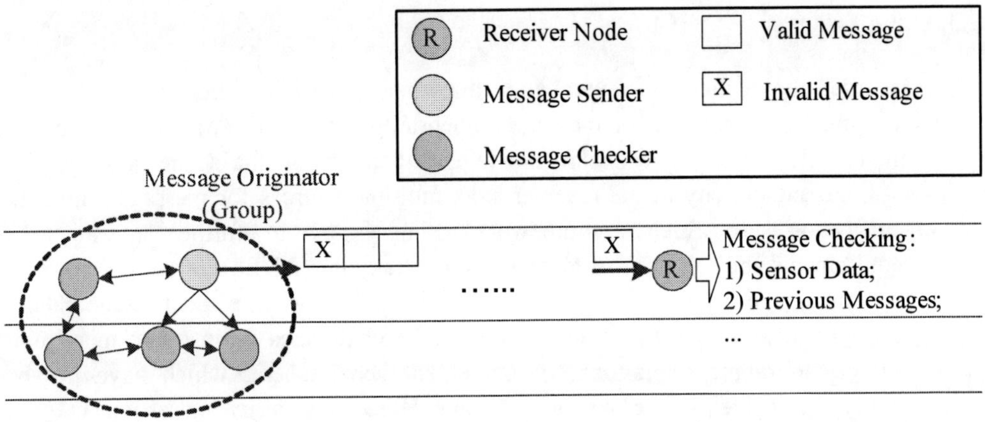

Figure 5. Message Authentication Approaches. At the message originator side, the Message Checkers try to authenticate the messages sent out by the Message Sender, to make sure that only valid messages can be sent out. At the receiver side, the received messages are checked against 1) sensor data of the receiver, or/and 2) the previously received messages. Most existing message authentication schemes only apply one of these two approaches.

4.2.2. Existing Schemes

The schemes for message authenticity in VANETs follow two approaches: to ensure message authenticity at the source nodes and at the receiver side.

In the first approach, as shown in Figure 5, the messages from an originating node will be checked by the neighboring nodes for validity before they are further propagated in VANETs, and the invalid messages will be stopped. Group communication and threshold cryptography could be used to implement this localized message checking in VANETs. A secure aggregation scheme is proposed in [25] to efficiently aggregate information in VANETs with the help of group communication. The nodes form groups based on their locations, and the messages will first be checked by all the group members before they can be propagated by the group header. In this way, the group as a whole is likely to generate authentic messages, given that the majority in any group is honest. The location-based group management can reduce the overhead in the group formation and maintenance. However, the frequent changes in group members, especially the group headers, will entail additional communication overheads. In fact, in this approach it still remains a challenge to securely and efficiently organize the neighboring nodes into groups. Thus the second approach (message authenticity checking at the receiver side) is still indispensible.

In the second approach, the authenticity of the received messages is verified by the receiver nodes, as shown in Figure 5. The verification may be realized by checking the message content against the relevant data obtained by the sensors in the receiver nodes [33] [34], or by checking the message content against the model maintained by the receiver nodes [35]. The schemes in [33] [34] focus only on the position data. The model proposed in [35] promises more application significance, assuming the approach of message aggregation. However, this scheme relies heavily on the completeness and validity of the preset model of VANETs. If the pattern of one possible attack is not recorded in the model, the resulted messages will possibly be accepted by the receiving nodes as valid.

4.2.3. Open Issues

As discussed above, there are several appealing open issues in this field.

1) A complete framework for message authenticity should be proposed to incorporate all the relevant facts: to ensure message authenticity at the sender side, to verify message authenticity at the receiver side, multiple sources for message authenticity verification, etc. Such a framework is necessary to further investigate the requirements and solutions of message authenticity in VANETs.
2) In the message originator side, a secure and efficient scheme for message checking and agreement still needs to be designed. The challenges lie in group management and cryptographic operations for agreement achievement, which have not been satisfyingly solved in the existing schemes. Here the security and efficiency should be deliberately balanced, in order to get a realistic security scheme.
3) In the receiver side the most appealing method of message verification is to check the received message against the model maintained by the receiver node. However, the existing schemes have not provided complete solutions including formal presentation of the model, evolution of the model, and efficient usage of the model. A static model as in [35] is not sufficient to adapt to the messages reflecting the changing environment and events in VANETs. More efforts should be put into this direction to enable secure judgment and aggregation of messages in VANETs.

4.3. Node Accountability and Node Privacy

4.3.1. Scopes of the Concepts

Based on our understanding of the characteristics of VANETs, the scopes of node accountability and node privacy are grouped regarding node identity and node location, as shown in Table II. Node authentication, node non-repudiation, and identity privacy are the requirements on node identity. Node localization and location privacy are the security requirements on node location, which becomes significant in VANET applications.

These concepts are closely related to one another, and even contradict one another, so they all should be considered. In general, the contradictory requirements from node accountability and node privacy have to be reconciled in order to design a feasible security solution to VANETs. Either one is too important to be ignored in any solution. These counterparts and the existing solutions will be investigated in details in this subsection.

Table II. Scope of Node Accuntaility and Node privacy

Concept		Significance	Counterpart
Node Accountability	Node Authentication	To verify that the node is a valid/authorized VANET user.	Identity Privacy
	Node Non-repudiation	No node could deny what it has done or transmitted in case of liability dispute.	Identity Privacy
	Node Localization	The node can determine the true locations of its neighbors.	Location Privacy
Node Privacy	Location Privacy	The location information of any node should be kept safe from the unauthorized users.	Node Localization
	Identity Privacy	The identity information of any node should be kept safe from the unauthorized users.	Node Non-repudiation; Node Authentication

4.3.2. Node Localization and Location Privacy

Besides the liability dispute scenarios, the precise and valid location information of vehicles is useful for routing, data dissemination, platooning, etc [38]. So it is critical to enable vehicles to obtain location information of their neighbors. Meanwhile, location privacy, the ability to prevent other parties from linking one's current or past location [37], is also critical for any VANET node. If location privacy cannot be properly protected no one would like to participate in VANETs.

Node Localization

In node localization, the nodes periodically broadcast packets containing their positions. These packets are used by other nodes or RSUs to determine their locations. The invalid position information from the nodes should be detected and handled. Several schemes in this light are proposed in [33] [34] [35], which have already been reviewed in the context of message authentication.

In the future, one potential research direction is to adopt the secure localization schemes proposed for sensor networks, which have become rather mature and could provide useful hints to localization schemes for VANETs. In this direction, [2] proposes two solutions for vehicle tracking with the triangulation algorithms. In [38] a data fusion model is proposed to incorporate multiple information sources to get a better estimation of the node and its neighbors.

Location Privacy

Pseudonym is the most common way of achieving location privacy in VANETs. Each node is equipped with a set of pseudonyms and uses them, instead of their unique identities, in communications. The nodes need to periodically change their pseudonyms to prevent the attackers from obtaining their location histories by linking the pseudonyms they use at different times.

Due to the constraints of road topology and layout, it is easy to predict and track the node's location. So simple pseudonym changing cannot sufficiently protect one node's location privacy. As pointed out in [37], to simply change pseudonyms will result in waste of pseudonyms, since in this case it is easy for the attackers to link the old and new pseudonyms of any node. The fact that the nodes may change their pseudonyms at different time further decreases the effect of pseudonym changing. For example, if at one time in a certain area only one node changes its pseudonym, the attackers could easily link the new pseudonym to the old one. These issues are studied in [39] [40].

Thus, existing schemes try to optimize the pseudonym changing strategies in different scenarios. Several pseudonym changing strategies have been proposed.

Random silent period: In [39] a random silent period is inserted between pseudonym changes in order to mix the location of the node changing its pseudonym to other nodes during this period. This has become a common practice in the following schemes.

Context Selection: Several schemes focus on the optimal selection of context for pseudonym changing in VANETs.

In [41] pseudonym changes are optimized with the concept of mix-context, where enough anonymity can be achieved by changing pseudonyms. One mix-context where one node has enough neighbors is presented in the paper. However, in this paper the example of mix-context is simplistic and its strength is not theoretically analyzed.

In [5] a simple scenario is considered where attackers use RSUs to track vehicles. The proposed scheme takes into consideration the speed, mobility pattern of the vehicles, and the distances between the base stations. Again, this scenario and the attack model are simple and this algorithm has no great security strength.

Synchronization in pseudonym change: With synchronization in nodes for pseudonym changing, the anonymity set can be increased, and the location privacy can be better protected. The anonymity set is the set of nodes where one node is indistinguishable from another node.

In [40] two facts about anonymity set are observed: 1) to increase privacy the probability distribution of the nodes in the anonymity set should be equal; and 2) to increase privacy the cardinality of the anonymity set should be maximized.

A user-centric scheme Swing is proposed in [40] to loosely synchronize pseudonym changes in the nodes in order to increase anonymity. As shown in Figure 6(a), one node will change its pseudonym only when it is changing its mobility states, such as its direction and velocity. In this way, the predictability of mobility is reduced, conforming to the first fact. At the same time, the node will send message to its neighbors to inform them of its pseudonym change. Other nodes may choose to change their pseudonyms too, increasing the size of the anonymity set.

However, in Swing the anonymity set only includes the nodes changing pseudonyms simultaneously. As shown in Figure 6(a), the node k is not in the anonymity set because it does not change its pseudonym along with other nodes. To further increase the anonymity set, Swap, an extension to Swing, is proposed in [40]. Swap enables the nodes to exchange their pseudonyms, as shown in Figure 6(b). Thus, node k is also included in the anonymity set, even though it hasn't changed its pseudonym. This is because the pseudonym change has happened, and all the nodes within the neighborhood have been mixed up.

Swing and Swap are interesting schemes in that they loosely synchronize the pseudonym changing in the neighbor nodes. More work has to be done in order to put these schemes into practice in VANETs, especially in the context of applications requiring consistent communication connections.

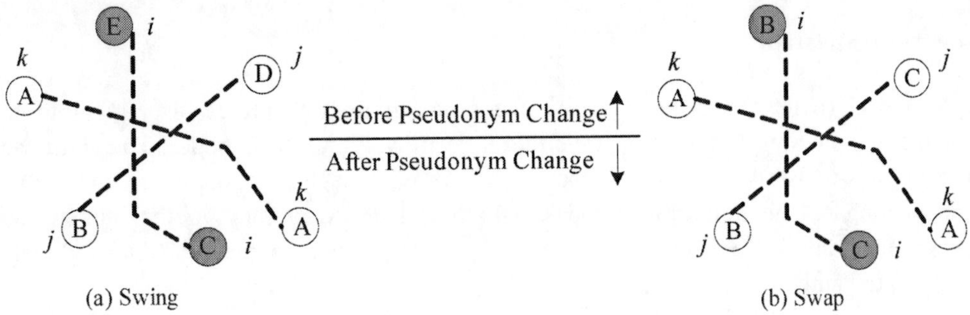

Figure 6. Illustration of Swing (a) and Swap (b). Circles represent nodes with the letters inside being their associated identifiers. Dashed line indicates the period when node is silent.

Group communication: Reference [42] proposes a location privacy scheme called CARAVAN which combining the group navigation with a random silent period enhancement technique to mitigate tracking of a vehicle. Group signature [51] is used to provide anonymity and to prevent spoofed identities while still preserving the driver's privacy expectations. It enables one group member to sign a message so that other group members are able to verify the message origin without identifying the actual sender. In this way the location privacy and usage profile privacy are preserved. An expansion to [42], AMOEBA, is proposed in [43], in which location privacy in presence of location-based services is additionally considered.

Reconciling Node Localization and Location Privacy

The existing schemes either focus on either node localization or location privacy, which has contradictory requirements regarding broadcast of position information. The localization schemes rely on periodical position information broadcast to localize the neighboring nodes. However, the pseudonym changing and random silent period adopted in the location privacy schemes surely will impact the broadcast of position information. Further investigation should be done to investigate the interaction and tradeoff among these two concepts.

4.3.3. Node Authenticity, Node Non-repudiation and Identity Privacy

As shown in Table II, node authentication and node non-repudiation seemingly contradict identity privacy. The core problem here is to determine how and to whom the identity of a node should be revealed. Once this problem is solved the balance among these three concepts can be achieved. Currently the common solution for this problem is *conditional privacy* [44] or *conditional anonymity* [5]. These concepts hint that node privacy should be protected in the common application scenarios and against the common nodes, but the identity information should be revealed to the authorized participants in critical application scenarios, such as liability dispute. In the existing schemes this concept is widely adopted. This is consistent with the application requirements of VANETs.

Here the existing schemes for node authentication, specifically anonymous authentication and schemes against Sybil attack will be investigated. Then the schemes to protect identity privacy in VANETs will be reviewed, and their considerations on node non-repudiation will also be investigated.

Node Authentication

In VANETs the nodes need to use the proof of their identities, usually in form of credentials, in communications to prove that they are valid VANET nodes. The certification authority and registration authority usually will provide such a proof to the nodes. While authenticating to the authorities, nodes risk their identity privacy in the authentication process. Thus several existing schemes pay special attention to the protection of identity privacy in the authentication process.

The schemes in [45] [46] try to perform node authentication without revealing the real node identity to the authentication authority. In [45] privacy is regarded as a user-specific concept and the user can select to which degree his privacy should be protected. An adaptive privacy-preserving protocol is proposed to enable the nodes to authenticate themselves while they are requesting data or services from the RSU. The nodes are grouped, and they only need to prove that they are in one group during the authentication, by disclosing different parts of the group (public keys of the group members) and verifying its knowledge of the public keys. Privacy is preserved since the RSU or other authorities cannot figure out the identity of the node if it chooses to authenticate using a large part of the group. However, this indicates that node non-repudiation cannot be ensured in this scheme. And the protocol involves high communication and computation overheads, mainly caused by the asymmetric operations.

In [46] a symmetric random key-set based scheme is proposed to protect node privacy. Each node is equipped with α symmetric keys randomly chosen from the key pool with all available symmetric keys. Whenever one node authenticates itself to a RSU, it will use a random key-set selected from its full key set. In this way, the RSU cannot identify the specific node, since each key is shared by multiple nodes. The full key-sets in all the vehicles are only stored in one key distribution center (KDC), which can help to identify the vehicles with a probability when it is necessary, in order to provide node non-repudiation. However, in this scheme node non-repudiation can only be achieved with a probability. And if several dishonest nodes exchange their full key sets they can create new full key sets, which cannot be identified by the KDC and still will be accepted by the RSUs.

In light of node authentication, Sybil attacks [47] deserve special attention. In Sybil attacks one malicious node tries to impersonate multiple nodes by participating in network activities with different identities. With pseudonyms available in VANETs, it becomes easy to launch Sybil attacks. Sybil attacks can bring great harm to the local voting schemes or routing schemes. To countermeasure Sybil attacks the presence of the claimed nodes should be evaluated by their neighbor nodes. For example, in [48] the positions of the nodes are checked against the RSSI measurement at the receiver node. In [49] the network topology of the VANET is visualized and analyzed. The unreasonable topology characteristics are used to detect the Sybil nodes.

Node Privacy and Node Non-repudiation

According to the concept of conditional privacy, node non-repudiation and privacy should be achieved simultaneously, but regarding different participants. Reference [50] investigates various application scenarios entailing node privacy and node non-repudiation from the perspective of a vehicle manufacturer.

The existing schemes for node privacy all allow for tracing the node identity in case of liability dispute scenario. They fall into two categories: schemes with privacy-preserving primitives and schemes with common cryptographic primitives.

Schemes with privacy-preserving primitives: Several cryptographic primitives, namely group signature [51] and ring signature [36], allow for the verifier to verify the signatures without knowing the identities of the signer, so they are privacy-preserving themselves.

In group signature one group consists of group members and one Group Manager (GM). A group member can sign the message with its private key, which can be verified using the group public key. The message receiver can verify whether the signature is produced by a valid group member or not, without knowledge of the sender's identity. Once necessary, the group manager can *open* the identity of the node. Formal descriptions of the properties of group signature can be found in [51]. Comparatively, ring signature [36] can achieve total anonymity where no one can reveal the identity of the signer of a message. In a ring signature scheme anyone can choose a set of possible signers, and sign a message with his own private key and others' public keys. The verifier can verify that the signer is a valid user without knowing his identity. Since traceability of the node's identity is not built in ring signature, it is rarely used in VANETs.

Due to the salient characteristics of group signature, it is extensively adopted as the cryptographic build block to provide node privacy and non-repudiation in VANETs.

In [52] group signature and pseudonym authentication are combined to enable the vehicles to generate and certify pseudonyms for themselves periodically. Equipped with a group private key gsk_v, a node can generate a pair of public/private keys K_v^i and k_v^i, and certify K_v^i with its group signature $gsk_v(K_v^i)$. During communication this node attaches its signature on the message *m* using its generated private key. The format of the packet is

$$m, \sigma_{k_v^i}(m), K_v^i, gsk_v(K_v^i) \tag{1}$$

The receiver of the packet can verify the sender's public key using the group public key, and then verify the digital signature of the sender with the sender's public key. Afterwards the sender will only attach K_v^i to the data packet to reduce communication overhead, since K_v^i has already been verified by the receiver. The nodes use these pseudonyms and self-generated certifications to authenticate themselves to one another. In this way, the nodes need not disclose their real identity during authentication. Once necessary a CA can disclose the real identity of a node by perform *open* operation on the signature $gsk_v(K_v^i)$. Here the nodes can generate and certify their own pseudonyms, and the overhead of pseudonym distribution by a centralized authority is removed.

A protocol of **G**roup **S**ignature and **I**dentity-based **S**ignature (GSIS) is proposed in [44], which protect node privacy while providing secure authentication and identity traceability at the same time. In GSIS, the OBUs use group signature keys to sign each generated message. In this way, the identities of vehicle nodes are protected from unauthorized parts. At the same time, RSUs sign their messages with private keys obtained from an IBS scheme. In this way, the load of key distribution and management is removed.

The main procedure of GSIS is shown in Figure 7. Here the role of GM is assumed by two parts: Membership Manager (MM) and Tracing Manager (TM). In the beginning, TM generates a set of system parameters, including the private key for itself. Other parameters are securely sent to MM to perform vehicle registration. When a vehicle registers itself at MM, MM will generates a secret key for the vehicle based on the system parameters, and keep a record of the mapping between the private key and the identity of the vehicle.

When it is necessary to disclose the identity of one vehicle, only TM is able to retrieve the private key of the vehicle based on its signed messages. And only MM is able to retrieve the real identity of the vehicle based on its private key. As shown in Figure 7, it requires the cooperation of TM and MM to disclose the identity of any vehicle, which can prevent the abuse of the power of GM.

One of the major drawbacks of GSIS is the fact that message confidentiality is not considered. Another drawback lies in the heavy communication overhead caused by the long signatures generated using group signature and IBS. According to [44], the signature part of the messages from OBU is 192 bytes, much larger than the payload of the messages (100 bytes).

In summary, one of the major drawbacks of these schemes is the computation overhead and communication overhead, both incurred by group signature and group management. Another drawback is the lack of an efficient and scalable group management scheme. Without this the group signature schemes can only remain theoretic. Reference [53] proposes a security framework for VANETs based on group signature, where vehicle grouping is preliminarily discussed. More efforts should be put in this direction.

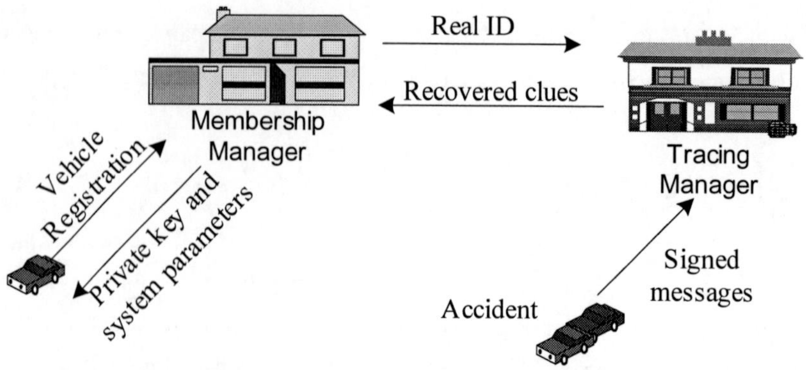

Figure 7. Communication between OBUs using Group Signature. The nodes get their private keys and system parameters from MM. During communication they sign the messages with their private keys, which could be used by TM and MM to recover their real identities in liability dispute.

Schemes with common primitives: With delicately designed procedures, identity privacy can also be protected with common cryptographic primitives. This always involves complicated message formation and pseudonym usage, as shown in [54]. In [54] a scheme built on symmetric cryptographic primitives is proposed, which uses short-term pseudonyms to preserve node privacy against peer nodes, and uses long-term pseudonyms to preserve privacy and accountability against base stations (front-end authorities). It assumes that nodes can trust front-end authorities and back-end authorities in VANETs, which can cooperate to disclose the identities of nodes if necessary.

4.3.4. Putting Pseudonyms into Practice

Due to the popularity of pseudonym in the node privacy schemes, it is meaningful to investigate its implementation issues: 1) what can be used as pseudonyms for the nodes; 2) how to put pseudonym changing into practice; 3) the impact of pseudonym changing to VANETs applications.

In [55], the cross-layer addresses (MAC address, Position Based Routing address, and IP address) of a node are regarded as its pseudonym. In order to effectively change its pseudonym, the node has to change these addresses simultaneously. This paper studies the feasibility of changing MAC addresses and IP addresses during communication and its impact on data forwarding. Several solutions for pseudonym changing are proposed in this paper. These solutions should be further optimized in the future due to the limitations and performance inefficiency.

In [56] two parameters in pseudonym changing in VANETs are examined through simulation: node re-interaction (how often and in what time intervals do nodes re-interact), and node quiet-time (how long should a node stay quiet before selecting another pseudonym). An analytical equation is proposed to determine the node quiet-time based on the characteristics of node mobility. The simulation results and analytical equation are helpful to designing protocols of pseudonym changing in VANETs.

Since pseudonym is an important method of privacy protection in VANETs, the research on the implementation of pseudonym changing is becoming more and more important, especially in the face of applications requiring consistent communication connections from the nodes to the infrastructure. Extensive research should be performed in this direction to make pseudonym changing a reality in VANETs.

4.3.5. Open Issues

In summary, many techniques such as pseudonym, group communication, group signature and ring signature can be used to preserve the location privacy and identity privacy of VANET nodes. However, the existing schemes all consider node privacy in specific applications, but the interplays and balance among node authentication, node non-repudiation and identity privacy still need to be thoroughly investigated in the context of different applications. The open issues of node privacy and node accountability will be discussed in detail in section 7.

5. Data Security and Network Performance

5.1. Data Security

In VANETs, basic network layer functions such as routing and packet relaying are performed by all nodes instead of any dedicated ones. If individual nodes refuse to cooperate with or serve for others, the whole network will suffer from Denial-of-Service (DoS) attack and degradation of basic network services, which threatens unavailability of critical services. For example, as stated in [57] 10%-40% misbehaving nodes will result in 16%-32% average throughput decrease.

At the Network layer, the reliance on unknown nodes for network functions presents opportunities for the potential attackers which can easily launch numerous attacks such as route tampering, message alteration and fabrication, data dropping, etc. Thus, cooperation among nodes is the core issue for Data Security. Only when all nodes comply with the protocols and cooperate with one another to perform network functions will VANETs satisfy the security requirements of Data Security.

There exist two kinds of approaches for ensuring cooperation among nodes in Ad Hoc networks: detection-and-reaction approaches and cooperation-promotion approaches. The cooperation-promotion approaches aim to persuade the nodes to cooperate with incentives. In this way the selfish nodes can be attracted from un-cooperation to cooperation, since their aim is to maximize their personal gains. The cooperation-promotion approaches combine economic incentives with basic network functions, typically packet forwarding, and urge the selfish nodes to cooperate by providing them with incentives for forwarding packets for other nodes, as in [58] [59] [60] [61]. The detection-and-reaction approaches try to detect and tackle the "stubborn" nodes, especially malicious nodes, which still misbehave despite the incentives for cooperation. The misbehaviors of the nodes, once detected, will influence their reputation or trust records. Each node reacts to the requests from others according to their reputation or trust records. Thus many schemes in this category [57] [62] [63] [64] [65] [66] have to use nodes' reputation or historic information to identify and isolate malicious nodes.

Both kinds of approaches are necessary since they are complementary to each other as discussed above. The existing schemes of both approaches will be investigated and special attention will be paid to their applications to VANETs in this section.

5.2. Cooperation-Promotion Approach

In this category, the pay-for-service model usually is adopted for cooperation promotion. As shown in Figure 8, in this model a node has to pay for the network service it uses, and it get payment for the services it provides. The payment usually is in the form of incentives or virtual currency.

5.2.1. Existing Schemes

As shown in Figure 8, the relaying node needs to get payment for its services. Thus two questions follow: 1) *how to pay* for the service: which node will pay, and in what form; and 2) *how much to pay*: what will be the amount of the payment for any specific data packet relaying. These two problems are the core problems tackled by the existing schemes.

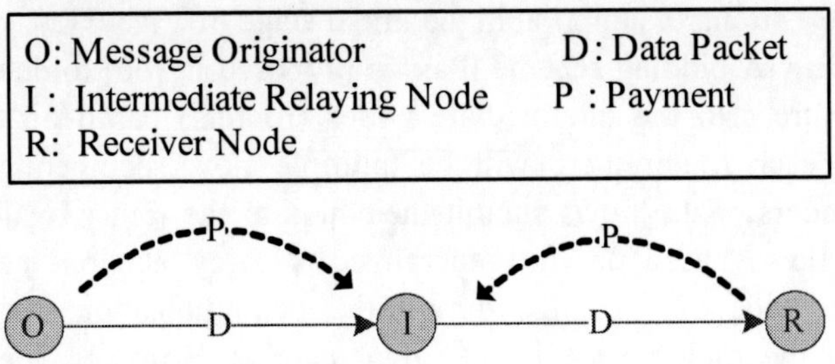

Figure 8. Pay-for-service Model. For simplicity only one intermediate relaying node is shown here. The relaying node relays the data packet from the message originator to the receiver. To do this it costs its own resources, so the message originator or/and the receiver will have to pay for its service.

How to pay: This problem means how to pay for the services provided by the intermediate nodes. Specifically, this includes how to represent the virtual currency in the network, which node to pay for the service (the sender or the receiver), and how to judge the service quality of the service provider.

In [58] the nodes are rewarded with chips, a kind of virtual currency, if they relay data packets for their neighbors. Each node maintains a record of the chip counts of all its neighbors. If one neighbor relay a data packet for it, it will increase the chip count for this neighbor. The node will decide whether to relay the data packet for its neighbor, based on this neighbor's recorded chip count. However, this is only a preliminary effort in this direction, since the secure representation of chip in the nodes is not considered.

A centralized entity, Credit Clearance Service (CCS), is proposed in SPRITE [61] to verify the transmission and relaying actions of a node and to charge and reward it in terms of credit accordingly. In SPRITE [61] each node will get a receipt for forwarding a data packet, and will get an amount of credit from CCS for this receipt. At the same time each node needs to pay a number of credits for sending out data packets. The pricing policy is designed based on game theory so that any node only can get maximum gains by abiding the rules and cooperating with others. Although SPRITE is a centralized solution, it might be applicable to VANETs after the regular road infrastructure, such as tolling stations and RSUs, are implemented. However, SPRITE treats safety-related and other messages equally, charging the message originators for packet transmission. In reality the safety-related messages will benefit both the receivers and the intermediate nodes, so it is unfair to charge the message originator in this case. Another drawback of SPRITE is that the pricing policy for the sender and the intermediate nodes is centralized, determined by the CCS. Thus, the resource constraints in the intermediate nodes are not considered, which may make the incentive unappealing to some nodes with resource constraints.

A virtual currency called Nuglet is proposed in [59] as an incentive to enforce cooperation in MANETs. Nuglet is stored in the tamper-proof hardware module in the nodes, and represents the wealth of the nodes. Three payment models are proposed for packet transmission: 1) Packet Purse Model (PPM) which asks the sender of packets to pay Nuglet for the transmission; 2) Packet Trade Model (PTM) which requires the receiver to pay for the packet transmission, and 3) hybrid model by combining PPM and PTM. However, the dynamics and potential attacks are not thoroughly investigated in this scheme, so several pitfalls remain. For example, in PPM the impact of packet loss and overpay of packet transmission is not properly handled. Besides, Nuglet requires a tamper-proof device in the node, which may be a strong assumption in the initial stage of VANETs.

How much to pay: A bidding scheme iPass is proposed in [60] to determine the price of the bandwidth resources in the intermediate nodes (routers) based on the competition of multiple flows. In each router there will be multiple flows competing for the available bandwidth. The senders of the flows submit their bids to the router, which will allocate its bandwidth to the flows based on the generalized Vickrey auction model. The required bandwidth of each node may change during the communication, so that the resource allocation result in the router also changes dynamically. However, the characteristics of generalized Vickrey auction guarantee that the bandwidth can be efficiently allocated, since the rational senders will place their bids according to their true value of the bandwidth.

Although iPass provides a feasible scheme for pricing in MANETs, several drawbacks make it unfeasible to VANETs. The bidding procedures will bring heavy communication overhead and latency to VANETs. The bidding per packet scheme makes the available bandwidth for each flow fluctuate rapidly, as shown in the evaluation part of [60]. This makes the QoS of each flow unacceptable, especially for the flows with lower bids. In the extreme cases, one flow may fail the bidding and get no bandwidth at all. Indeed, new pricing schemes should be proposed for VANETs, which can balance efficiency and fairness in resource allocation at the same time, with proper QoS guarantee for the flows in VANETs.

5.2.2. Open Issues

The existing schemes [58] [59] [60] [61] are designed for MANETs, and all messages are treated equally. However, in VANETs the messages related to safety applications should be differentiated. The pay-for-service model is no longer fit for such messages, since their significance justifies their being forwarded for free. So existing schemes can be only applied to the messages which are not related to safety applications and will benefit the receivers and senders at the cost of other relaying nodes. Meanwhile, these schemes also should be modified to adapt to the unique characteristics of VANETs.

5.3. Detection-and-Reaction Approach

In this category, the behaviors of the nodes are monitored and evaluated, and corresponding reputation or trust values are assigned to them. Based on the reputation record, other nodes can decide whether to serve one node or not, or how to serve. Sometimes a node with bad reputation is then denied service by other nodes, and hence isolated from the network. Even more often is the case that other nodes will not ask the nodes with bad reputation to relay packets for them again.

5.3.1. Existing Schemes

There have been many reputation-based schemes in MANETs. [65] develops a reputation system in mobile networks, and [66] proposes a pair wise-evaluating buddy system. Several schemes [62] [57] [64] use the watchdog mechanism to monitor the packet forwarding behaviors of the downstream neighboring nodes within the transmission range to detect non-cooperating behavior. Different policies are adopted to punish the misbehaving nodes.

However, the information maintenance in [62] [57] [64] and evaluation information accumulation in [63] will be unfeasible in the highly dynamic VANETs. The system proposed in [65] requires human interaction while the solution in [66] requires good connectivity to establish buddies. In a word, further improvements are required before applying those schemes to VANETs.

In [67] a set of indexes are introduced to implicitly promote cooperative data transmission and non-cooperative nodes are excluded from the VANET communications. However this scheme assumes that each node can get enough knowledge regarding another node's behavior within its operational duration, which seems hard to achieve in extremely dynamic VANETs.

In a word, the applicability to VANETs of these schemes mentioned above is questionable.

To our best knowledge, at present there are only two published schemes specifically designed for VANETs: VARS (Vehicle Ad-Hoc Reputation System) [63], and DTT (Dynamic Trust Token) [26]. VARS [63] makes use of direct and indirect trust as well as opinion piggybacking to enable confidence decisions on event messages. Opinions are appended during message forwarding.

DTT [26] uses Watchdog to guarantee cooperative data packet forwarding in VANETs as shown in Figure 9. In Figure 9, one node after relaying a packet to the next hop node becomes a Predecessor (Watchdog), and monitors the packet forwarding behavior of the next-hop node (Relayer). The Relayer relays the packet further to its own next-hop node (Successor). The data packet relayed by the Relayer will be checked by the Predecessor by comparing it to the data packet previously sent out by the Predecessor itself. A Trust Token packet will be generated by the Predecessor and sent to the Successor to indicate the validity of the data packet. The Successor will decide whether to accept the data packet or not based on the Trust Token. If the Trust Token indicates that the data packet is valid, the Successor will start a new Transmission Session. In the new Transmission Session, the Relayer becomes a Predecessor, the Successor becomes a Relayer, and the further next hop node becomes a new Successor. Thus, in each hop only those data packets indicated to be valid by the corresponding Predecessors will be further relayed.

Although DTT assigns instant trust to the Relayer based on its packet relaying behaviors and is especially fit for VANETs, it does have several security flaws. Firstly, it assumes that the Predecessor will always give correct Trust Token, which is not necessarily true in realistic VANETs and subjects it to the bad-mouthing attacks launched by the dishonest Watchdogs. Secondly, the potential collusion between Predecessors and Relayers is not considered in this scheme. These flaws severely weaken the security strength of DTT scheme, and allow the propagation of tampered data packets in VANETs.

Based on DTT [26] we further propose a novel detection and reaction scheme based on Watchdog: Autonomous Watchdog Formation with 2-hop Neighborhood Awareness (AWF-NA) [28]. AWF-NA is a distributed scheme which focuses on guaranteeing the secure data packet relaying in VANETs. Its security strength is provably complete in various attack scenarios. The main contribution of AWF-NA is a scheme for watchdog selection and activation during the transmission of the data packets in VANETs. In each hop, the Relayer will have multiple watchdogs monitoring its packet relaying behaviors. Thus the misbehaviors of the Relayer can be detected and tackled, as long as the majority of its watchdogs are honest and cooperative. Several measures are taken to enhance the possibility of more cooperative watchdogs than uncooperative ones.

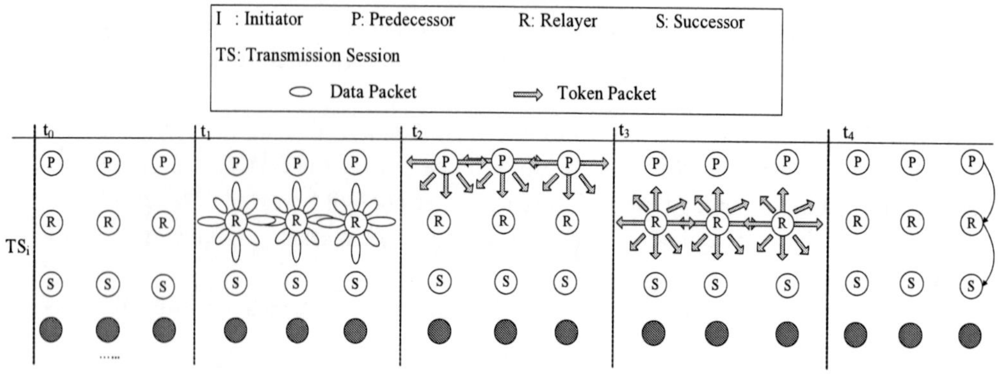

Figure 9. Message Flows in DTT Scheme.

5.3.2. Open Issues

It is difficult to maintain and manage the reputation record of VANET nodes, so the concept of instant trust in [26] [28] may be a useful direction for future research work. The main idea of instant trust is to judge the behaviors of VANET nodes based on the instant observations and take actions accordingly. This concept is especially fit for the scenarios where tampered data packets are harmful to safety-related applications.

On the other hand, powerful approaches should be developed to efficiently and effectively identify nodes' misbehaviors in the Network layer, in the presence of uncooperative watchdogs. Taking advantage of the fact that the misbehaving nodes are only the minority in VANETs, we could devise schemes to enable as many neighbor nodes as possible to function as watchdogs. The communication and computation overhead should be minimized, but not at the sacrifice of the security of the schemes. AWF-NA is a promising scheme in this direction, which will be further refined in our future work.

5.4. Security Provisioning vs. Network Performance

Once deployed, the core network services (safety-related services) of VANETs must be always available to the nodes and service users, which results in the requirement of Availability. For VANET messages, especially the safety-related messages, usually both security requirements and performance requirement are necessary [68].

However, security provisioning schemes usually consume network resources of VANETs, which makes it necessary to achieve tradeoff between security and network performance requirements.

In [69] a resource-aware self-adaptive network security provisioning scheme for MANETs is proposed, which tries to achieve tradeoff between security and performance based on the available security protocols and potential attacks. Though this scheme is helpful to VANETs, it has to be further refined in order to become applicable, specifically in terms of performance metrics of the security protocols.

By studying the implementation issues about the certificate lifetime and key set size, [5] finds that one of the most time consuming and critical overhead of a given public key cryptography algorithm is the signature verification time. With certain modification of cryptography algorithms (e.g., to use smaller signature sizes), the time constrain requirement of VANETs could be satisfied.

In [25] the group aggregation scheme is designed to balance the security requirements and performance requirements of VANETs. In this scheme the groups are formed based on the locations of the vehicles, which supposedly could reduce communication overhead and latency caused by group management. This is a good example of incorporating performance considerations into security provisioning.

Reliability is another network performance requirement in transmission of safety-related messages. For example, in [16] it is stated that an application functioning correctly 99.99999% of the time may probably malfunction on one vehicle, given so many vehicles in the world. And without reliable network service, it is hard, if not impossible, for an application in VANETs to function correctly all the time.

At present very few reported research efforts are dedicated to achieving tradeoff between security and network performance in VANETs, and this topic remains underdeveloped. In the future, extensive research should be done to investigate the interplay between security provisioning and network performance, in order to get insights into this topic.

6. Trust

As discussed in subsection 2.1, trust is a critical concept in VANET security which deserves thorough investigation. Here we first investigate the concept and significance of trust in the context of VANETs. After that trust modeling and trust management are surveyed in details. In the end, the emerging research efforts for trust management specifically for VANETs are investigated. Open issues are identified in all these issues.

6.1. Trust: Concept and Scope

There are multiple definitions for trust, which explain trust from various perspectives: sociology, psychology, technology, etc. Currently there is no clear consensus on the definition of trust. However, it is a trend to define trust as the belief in another party fulfilling a certain commitment, or as the probability of one party performing a certain function. For instance, [70] defines trust in distributed networks as a tuple {*subject*: *agent*, *action*}, which means the *subject* believes that *agent* will perform a specific *action*.

Based on our understanding of trust in all related fields, we identify trust's three-fold meanings: 1) authorization to others for access to its own resources or the right of performing certain tasks in one's own name, 2) general belief in others obtained from their properties such as valid ID and keys, or friendship, and 3) confidence in others' ability and intent to perform a certain task based on behavior evaluation and monitoring. They are all relevant to security provision in VANETs. The first two meanings are related to the ways the nodes treat one another in terms of resource access, friendship, etc. They can be preset based on the management policies. The third meaning gives the nodes the opportunity of evaluating their peers on-the-fly based on their interactions. The previously trustworthy nodes may become compromised, and the nodes may display different level of capability in terms of different applications. Thus it is critical to evaluate the nodes and obtain the valid opinions regarding their capabilities and aptitudes in context of different applications.

The first two meanings of trust belong to the Trust Model, and the third meaning of trust falls in the category of Trust Management, as shown in Figure 2.

6.2. Trust Modeling and Metrics

Trust has been regarded as a powerful tool in distributed systems for misbehavior prevention and cooperation enhancement [8]. The nodes with valid certificates and authorizations could be compromised, and be used as the tools to launch attacks against others. Some nodes might choose to disobey the protocols for their own interests. All these potential attackers and their attacks should be detected, and trust is one of the tools for this aim. This may be the reason that currently most of the research efforts [8] [9] [11] [70] [71] [72] [73] on trust in wireless networks focus on the trust modeling and management based on behavior evaluation in networks.

In [8] [9] [70] [71] by Yan Lindsay Sun etc., trust modeling and management in distributed networks and Ad Hoc networks are investigated based on information theory.

In [9] [70], an information theoretic framework for quantitative trust modeling and trust propagation is proposed, in which trust is measured with entropy about other nodes in regard of a certain action. The characteristics and properties of trust and trust propagation are summarized in four axioms. The first axiom states that uncertainty is the measurement of trust, indicating the nature of trust. The other axioms are about trust propagation and aggregation in the networks, as shown in Figure 10.

The second axiom states that concatenation propagation of trust does not increase trust, as shown in Figure 10 (a). T_{BC} is the trust of B on C's performing a certain task, and R_{BC} is the trust of A on B's recommendation. No matter how A trusts B on its recommendation, that is, how big R_{AB} is, T_{AC} should not be greater than T_{BC}. The underlying philosophy of this axiom is that every node trusts its own observations the most.

The third axiom states that the multipath propagation of trust does not decrease the value of trust, as shown in Figure 10 (b). In this scenario, the trust of A2 on C2 is calculated based on two paths with the same values as the single path of A1 on C1. Whatever T_2 is, the absolute value of $T_{A_2B_2}$ is greater than the absolute value of $T_{A_1B_1}$. This axiom indicates that the recommendations from multiple nodes should be taken more seriously.

The forth axiom indicates that trust based on multiple recommendations from a single node should not be larger than the trust based on the recommendations from multiple independent sources, as shown in Figure 10 (c). Since the recommendation from B1 to A1 is highly correlated, A1 should not trust it more than it does the recommendations from multiple sources.

Two models (entropy-based model and probability-based model) for trust evaluation and trust propagation are proposed, which conform to these four axioms. After that the scheme for establishment of trust among nodes based on observations is proposed, which allows nodes to establish trust relations among themselves based on observations of the actions performed by other nodes.

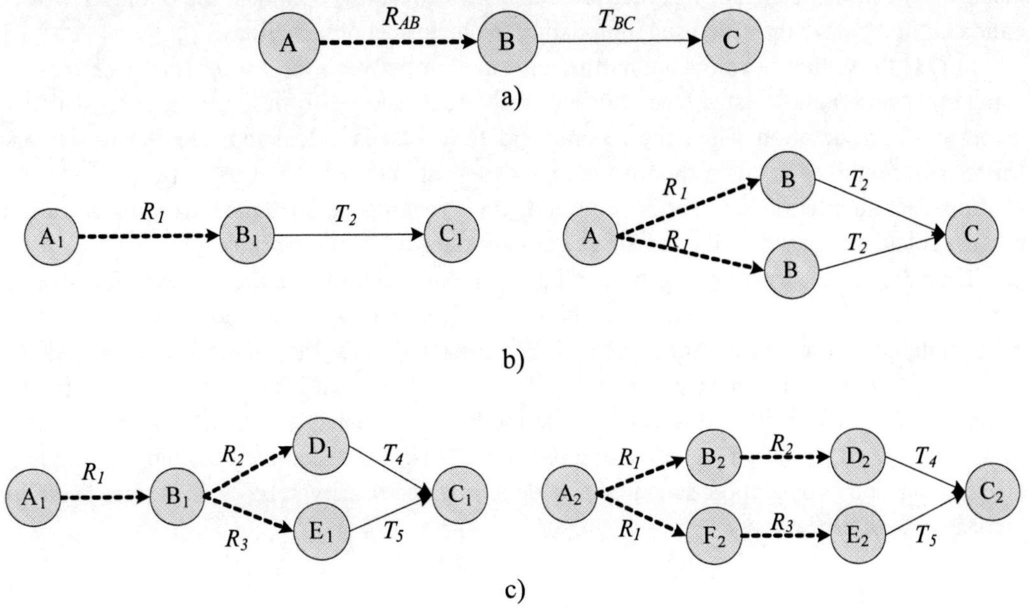

Figure 10. Trust Propagation and Calculation.

In [8] [71] the attacks on trust evaluation in Ad Hoc networks are analyzed and several algorithms are provided to countermeasure them. Four typical attacks to trust evaluation are handled: Bad Mouthing Attack, On-off Attack, Conflicting Behavior Attack and other attacks. The effects of the algorithms on these attacks are evaluated through simulations.

A comprehensive concept of trustworthiness based on both trust and confidence is proposed in [11]. Trust is calculated based observations of the packet delivery behaviors of other nodes, using beta distribution. The concept of confidence measures the reliability of trust based on the historical observations. Trust and confidence are integrated into trustworthiness, which indicates how trustworthy a node is in light of packet delivery. The trustworthiness of the nodes is taken into consideration while a route from source node to the destination node is selected. The idea of measuring confidence on the trust evaluation in this scheme is both interesting and meaningful.

A Bayesian network based trust modeling scheme for MANETs is proposed in [72]. Here a node's trust on a server for a specific service is calculated based on direct trust and indirect trust, which represent the trusts obtained from direct experience and the recommendation from other nodes respectively. The aggregation of trust based on direct trust and indirect trust is performed with the help of Bayesian network. This scheme is fit for evaluation of service quality of some centralized servers in MANETs and VANETs, but it is not suitable for the behavior evaluation of VANET nodes. What's more important, in this scheme the potential attacks on trust evaluation are not considered, and all the recommendations from other nodes are regarded as valid.

An Objective Trust Management Framework (OTMF) which takes into consideration direct experience and opinions from other nodes is described in [73]. This scheme also uses beta distribution and Bayesian theory to evaluate trust and confidence based on the historical observations on a certain node. However, here other nodes' opinions are treated equally as the experience by the node itself, which supposedly justifies the claim of objective trust management in this scheme. In order to countermeasure false opinions, the opinions will be validated first before they are used in updating the historical observations.

In [74] the authors exploit the mutual information between the truster (the node trusting some other node) and trustee (the node which is trusted to perform a certain task) which is taken into consideration when the recommendation trust is calculated. For the truster and trustee adjacent to each other, the trust is regarded as direct trust. The trust between the truster and trustee with intermediate nodes between them is regarded as recommendation trust. These two are calculated based on the transitive graph of the Ad Hoc network.

The schemes for trust aggregation and propagation are useful in the context of VANETs. However, the scheme for updating trust based on observations of the nodes' behaviors could be unsuitable, or at least hard, for VANETs. In VANETs the chances that two nodes frequently interact with each other are quite small, due to the inherent characteristics of VANETs. This makes it hard to accumulate the historical behavior records for trust update. Thus, a new trust paradigm has to be proposed for VANETs, which includes the existing trust aggregation and propagation schemes and does not necessarily rely on historical behavior records for trust update.

6.3. Authorization in Trust Model

We will only consider the distributed trust model here, since the centralized trust model where a centralized authority assigns trust to every nodes is not closely related to VANETs.

PGP [75] tries to establish trust relationship among nodes based on the recommendation and ownership of valid keys, proposing the concept of Web of Trust. This is a classic scheme for establishment of trust relations based on recommendations, but its quantification of trust is simplistic, which cannot represent all the possible trust values in realistic applications.

In [76] it is pointed out that traditional authorization schemes are not sufficient for Internet, and trust-management approach should be used to handle this problem. The scheme proposed in [10] is a typical one which can be used to handle the delegation and authorization problems in wireless networks. Also the scheme in [77] gives a framework for trust calculation in Ad Hoc networks, which can also be used in the field of authorization.

In [77] the properties of trust calculation along a path and convergence from several components are studied, and it is shown that they are similar to the properties of semiring operators (s, $\oplus$, $\otimes$). The operator $\oplus$ can be used to calculate trust based on the recommendations from multiple nodes, and the operator $\otimes$ can be used to calculation trust propagation along a path. Within this framework several trust evaluation models are proposed in [77], which conform to the semiring model. The semiring operators represent the properties of trust in networks, as do the four axioms in [70].

In [10] a practical approach for compute trust in wireless networks is proposed, which is based on a transitive signature scheme. Here the node in wireless networks is regarded as a node in a delegation graph G, and a path from a source node to a destination node is mapped to an edge in the corresponding transitive closure graph G^*. Based on this and the transitive signature scheme, the trust is computed. This approach can be used to manage authorization in wireless networks.

6.4. Trust in VANETs

These existing schemes cannot be directly applied to VANETs due to the characteristics of VANETs. Several reported research efforts [12] [13] begin to study trust modeling and management in VANETs.

A data intensive reputation management scheme is proposed in [12], which is used in the broadcast data dissemination model where some anchored sources periodically broadcast data packets about local conditions to nearby vehicles. The main idea here is to verify the data packets received from other nodes by comparing them against those received from the anchored sources, or by achieving agreement among all the packets received from all the nodes. The reputation for one node will be increased if it propagates valid data packets. Otherwise its reputation will be decreased. In this scheme it is questionable whether the vehicles will run into each other frequently enough to make the reputation record available for future use. Meanwhile, no recommendation from other nodes is considered when the reputation of one node is estimated.

[13] proposes a security architecture for VANETs with trust and privacy emphatically considered. The messages or data from any entity are regarded as context and stored in a context database. The items in the context database are tagged with trust and confidence, which are obtained from proper estimations (by self checking, context comparison or certification verification etc.). Though the idea of trust tagged context in [13] is useful, the discussions are basically in the conceptual level. Further work should be done in order to put this architecture into application.

According to the current status of research of trust in both common wireless networks and VANETs, more work should be done to design realistic and applicable trust model and trust management schemes for VANETs. The open issues in this field will be analyzed in detail in section 7.

7. Open Issues and Future Work

Based on above investigations and discussions, it can be concluded that VANET security provisioning is challenging due to the distinct VANET characteristic. Several research topics, namely node privacy vs. accountability, data security, trust management, and threat model, are currently open, and deserve extensive future research efforts. Here we focus on these major topics, identifying potential research problems and presenting our initial proposals.

7.1. Privacy vs. Accountability

The protection of node privacy is difficult due to the pervasive network communication in VANETs. It becomes even more difficult because in security provisioning sometimes node's identity and location are necessary. For example, in legal vehicle tracking the identity and location of the tracked node are requested by VANETs, which risks breaching privacy such as the speed privacy, direction privacy and location privacy of users [3] [42]. Thus, the major challenge is to balance node privacy and node accountability, so that both requirements can be met in the realistic VANET applications. To tackle this challenge, several open issues should be thoroughly investigated.

Firstly, clear and consistent definitions of node's privacy and accountability should be made, which should cover the characteristics of VANET applications. These definitions should be abstract enough to present the essences of privacy and accountability. Meanwhile, they also need to provide concrete metrics for privacy and accountability, and to reflect against who the privacy or accountability should be achieved. These definitions should be flexible enough to incorporate different requirements posed by different application scenarios. These definitions, missing in the current literature, are necessary for the design of schemes to protect node privacy and accountability.

Secondly, a hierarchical framework for protecting node privacy, providing accountability, and for balancing them should be proposed. The basic concepts of privacy (identity privacy and location privacy) and accountability (node localization, node authentication and non-repudiation) should be considered together within this framework. One basic idea is to give different privileges to different users or applications, with which they can access node's identity and location information accordingly. Conditional anonymity [5] can be regarded as one preliminary instantiate of this framework, where privacy preservation and anonymity are conditional on the scenarios if the safety and liability requirements matters more. The concept of adaptive privacy in [45] is also helpful to the setup of this framework.

Thirdly, within this framework concrete schemes should be proposed to handle the existing issues, such as legal tracking of vehicles, tracing back the messages to the source node, anonymous authentication, etc. Improvement on existing schemes can also be made with the help of this framework. In a word, with concrete understanding of privacy and accountability, and with the help of the framework, much more work can be done to better secure VANETs in terms of privacy and accountability.

In the end, it is necessary to analyze the feasibility and performance impact of the presented schemes in order to put them into practice. For example, the feasibility and efficiency of changing pseudonyms (in terms of IP and MAC address), as well as their impact on network performance, in VANETs should be considered. [55] is a preliminary step in this direction, and much more work should be performed in the future.

7.2. Data Security

Data Security in VANETs is directly related to network performance, and it in essence is the counterpart of Availability and Access Control of resource in VANETs. However, the current state of the art of this field is not satisfying, and there are still many open issues.

7.2.1. Cooperation Enhancement

Different from the conventional MANETs where no infrastructures exist, VANETs may utilize the road infrastructure to promote cooperative communications. This provides new opportunities to the design and implementation of cooperation-promotion approaches.

First of all, it is necessary to determine reasonable and attractive prices for the services provided by the vehicular nodes, which is the foundation of the existing and future incentive-based cooperation promotion schemes. Based on the resource demand of different services, reference prices should be provided to the VANET nodes, for them to decide whether to accept them or to give their own prices based on other considerations. Here the message types should be taken into consideration: for example the safety-related messages should be relayed by all the nodes for free, while the general messages can be reasonably charged. At least the message initiator node should not pay anything for the safety-related messages; otherwise there will be no incentive for the nodes to generate safety-related messages.

Thus speaking, the pricing policy should take into consideration all relevant factors, such as resource requirements, message types, the beneficiary and benefactor of the messages, etc. Game theory may be a useful tool here to design pricing policies for VANETs.

Secondly, the impact of cooperation-promotion schemes should be thoroughly analyzed. New schemes which are both effective and efficient should be proposed, which should be able to provide attractive and robust incentives to the nodes and incur low requirements on network resources.

7.2.2. Detection and Reaction

Network layer, (even MAC layer and physical layer) will all be involved in the detection and reaction of node misbehaviors. The major challenges include how to diagnose nodes misbehaviors, how to detect misbehaving nodes and how to punish or discourage misbehaviors in a highly mobile and fully distributed VANET.

Due to the high mobility, fast changing topology, short connection time, and constant changing neighbors, it is difficult for the nodes to maintain long-term history information about their neighbors. This makes the current reputation based schemes difficult to be applied in VANETs. So, novel schemes for detecting uncooperative nodes that do not necessarily rely on historical records should be designed. The schemes in [26] [28] belong to these schemes, but they focus on misbehaviors in packet relaying. Thus further research should be done to design similar schemes for other types of misbehaviors in VANETs.

7.2.3. Comprehensive Cooperation Schemes

In reality the cooperation-promotion schemes and detection and reaction schemes are complimentary, and both are indispensible in VANETs. Thus, it is meaningful to study the integration of these two types of schemes. Their interplay should be studied to enable them to provide friendly interfaces to each other. For example, the reputation in detection-and-reaction schemes can be used to differentiate the nodes in pricing policy in the cooperation promotion schemes. Additionally, the efficiency of the integration should be thoroughly studied to obtain an overall gain larger than the addition of the separate gains.

7.3. Trust Management

Based on the previous discussions several open issues can be identified in this field.

First of all, novel paradigms for authorization and certification should be proposed to exploit the unique characteristics of VANETs. New applicable policies/rules in the trust model should be designed, in order to properly handle authorization and resource access problems. For example, the potential support from infrastructure may be exploited to store the authorization actions performed by one node, so that other nodes can easily access the proof of these authorization actions. Meanwhile, limited network redundancy in VANETs should be considered in this procedure.

Secondly, since in VANETs historical records of nodes may not be available, the original trust evaluation schemes based on observation become not applicable in VANETs. In this sense the concept of trust should be updated to adapt to this situation. The concept of instant trust in [26] [28] provides a new direction to trust management in VANETs. Besides, the trust modeling and evaluation schemes also should be adapted to accommodate the possible unavailability of historic records in VANETs. Multiple sensors exist in any VANET node which can provide sensing data about the environment. Meanwhile, potential support from roadside infrastructure may also provide a new source for trust calculation. These two facts, as mentioned in [13], provide a good starting point for trust in VANETs.

In summary, more work should be done in this field to establish proper trust model and trust management system in VANETs.

7.4. Threat Model

As mentioned in section 2.2, new threat model for VANET security, as well as new approach to analyze the relations between attackers and VANETs, should be proposed. The intention, objectives and strategies of the attackers of VANETs should be thoroughly investigated to lay the foundation for cost-efficient and effective security provisioning in VANETs. Game theory and other social engineering theories are good candidates for research in this direction.

8. Conclusion

As potentially the largest instantiates of MANETs, Vehicular Ad hoc Networks (VANETs) open the door for a plethora of applications and services, as well as provide additional supports to the ITS applications. In VANETs, most applications, especially the safety-related applications, are delay- and error- sensitive. Therefore, VANETs pose strict security and performance requirements: Message Authentication, Message Integrity, Message Confidentiality, Node Accountability, Node Privacy, Availability, Real-time and Reliability, all of which have been thoroughly investigated in this chapter. The interplays among these requirements, together with the distinct characteristics of VANETs, make security provisioning in VANETs an indispensible, challenging and appealing issue.

In this chapter the security issues in VANETs are systematically investigated. First the common security requirements of VANETs are thoroughly investigated, and the necessity of a new threat model with dynamic relations to VANETs considered is pointed out based on review of the existing threat model for VANETs. Based on this a VANET security provisioning framework VSPF is proposed, within which four major security modules are identified and highlighted: Service Security, Data Security, Network Performance and Trust Management. In-depth investigation of all these security provisioning modules is presented in details. Among these modules, major topics are identified which are currently underdeveloped and still need extensive research in the future.

Finally, we present our insights on open issues and future research directions. In general, different application scenarios have their special requirements and it is impossible to satisfy all the requirements with one mechanism or two. Therefore VANET security mechanisms should be application-oriented within VSPF.

VSPF and the open issues identified in this chapter could serve as helpful guidelines for security provisioning in VANETs. With them thorough and complete understanding of VANET security provisioning could be obtained, and concrete research work to solve the specific security issues in this field could be better performed.

References

[1] CAMP Vehicle Safety Communications Consortium, "Vehicle safety communications project: Task 3 final report: identify intelligent vehicle safety applications enabled by DSRC," National Highway Traffic Safety Administration, U.S. Department of Transportation, Washington DC 2005.

[2] J. P. Hubaux, S. Capkun, and Jun Luo, "The security and privacy of smart vehicles", *IEEE Security and Privacy Magazine*, vol. 02, issue 3, pp. 49-55, May-June 2004.

[3] S. Roy, D. Saha, S. Bandyopadhyay, T. Ueda and S. Tanaka, "Service differentiation in multi-hop inter-vehicular communication using directional antenna," in *Proc. IEEE Semiannual Vehicular Technology Conference*, May 17-19, 2004, Milan, Italy, pp. 2176-2180.

[4] J. Blum, A. Eskandarian and L. Hoffman, "Challenges of intervehicle ad hoc networks," *IEEE Transportation Systems*, vol. 5, pp.347 – 351, Dec. 2004.

[5] M. Raya and J. Hubaus, "The security of vehicular ad hoc networks", *EPFL Tech. Rep.* IC/2005/009, March 2005.

[6] Mario Gerla, Biao Zhou, Yeng-Zhong Lee, Fabio Soldo, Uichin Lee, and Gustavo Marfia, "Vehicular grid communications: the role of the internet infrastructure", in *Proc. 2nd Annual International Workshop on Wireless Internet (WICON 2006)* , Boston, NY, August 2006. (Invited Paper)

[7] M. Zarki, S. Mehrotra, G. Tsudik and N. Venkatasubramanian, "Security issues in a future vehicular network," In *Proc. European Wireless 2002 Conference*, Florence, Italy, Feb. 2002.

[8] Y. L. Sun, Z. Han, W. Yu, and K. J. R. Liu,"A trust evaluation framework in distributed networks: vulnerability analysis and defense against attacks", In *Proc. IEEE INFOCOM 2006*, pp. 1-13.

[9] Yan Sun, Wei Yu, Zhu Han, and K.J.R. Liu, "Trust modeling and evaluation in ad hoc networks", In *Proc. IEEE GLOBECOM '05*, vol. 3, 28 Nov.-2 Dec. 2005.

[10] Huafei Zhu, Feng Bao, and R.H. Deng, "Computing of trust in wireless networks", In *Proc. IEEE 60th Vehicular Technology Conference*, 26-29 Sept. 2004, vol. 4, pp. 2621 – 2624.

[11] C. Zouridaki, B. L. Mark, M. Hejmo, and R. K. Thomas, "A quantitative trust establishment framework for reliable data packet delivery in MANETs," In *Proc. 3rd ACM Workshop on Security of Ad Hoc and Sensor Networks (SASN 2005)*, Alexandria, VA, USA, November 7, 2005.

[12] Anand Patwardhan, Anupam Joshi, Tim Finin, and Yelena Yesha, "A data intensive reputation management scheme for vehicular ad hoc networks", In *Proc. 3rd Annual International Conference on Mobile and Ubiquitous Systems: Networking and Services*, July 2006, pp. 1-8.

[13] Matthias Gerlach, "Trust for vehicular applications," In *Proc. 8th International Symposium on Autonomous Decentralized Systems (ISADS '07)*, 21-23 Mar. 2007, pp. 295 – 304

[14] Peng Liu, and Wanyu Zang, "Incentive-based modeling and inference of attacker intent, objectives, and strategies," In *Proc. 10th ACM Conference on Computer and Communications Security (CCS '03)*, Washington D.C., USA, October 2003, pp. 179 – 189.

[15] T. Leinmüller, L. Buttyan, J.-P. Hubaux, F. Kargl, R. Kroh, P. Papadimitratos, M. Raya, and E. Schoch, "SEVECOM - secure vehicle communication," in *Proc. IST Mobile and Wireless Communication Summit,* June 2006.

[16] B. Parno and A. Perrig, "Challenges in securing vehicular networks," In *Proc. 4th Workshop on Hot Topics in Networks (HotNets-IV)*, November 14-15, 2005.

[17] Cremonini M. and D. Nizovtsev. "Understanding and influencing attackers' decisions: implications for security investment strategies," In *Proc. Fifth Workshop on the Economics of Information Security (WEIS 2006)*, Cambridge, England, 2006.

[18] O. Gehring and H. Fritz, "Lateral control concepts for truck platooning in the CHAUFFEUR project," In *Proc. 4th World Conngress on Intelligent Transport Systems*, Oct. 1997.

[19] W. Li, "Design and implementation of digital radio communications link for platoon control experiments," *PATH Research Report* UCB-ITS-PRR-95-2, University of California, Berkeley, Jan. 1995.

[20] M. Ohomo, R. Kimura, S. Fukushima and N. Fujii, "Automatic following system utilizing vehicle-to-vehicle communication," In *Proc. IEEE International Conference on Intelligent Vehicle*, Oct. 1998, pp. 381-384.

[21] S. Sachs and P. Varaiya, "A communication system for the control of automated vehicle," *PATH Tech. Memorandum 93-5*, University of California, Berkeley, Sept. 1993.

[22] M. Aoki and H. Fujii, "Inter-vehicle communication: Technical issues on vehicle control application," *IEEE Communications Magazine*, vol. 34, no. 10, pp. 90-93, Oct. 1996.

[23] W. Kremer and W. Kremer, "Vehicle density and communication load estimation in mobile radio local area networks (MR-LANs)", In *Proc. 42nd Vehicular Technology Society Conference*, May 1992, vol. 2, pp. 698-704.

[24] L.Briesemeister, "Group Membership and Communication in Highly Mobile Ad Hoc Networks," PhD Thesis, School of Electrical Engineering and Computer Science, Technical University of Berlin, Germany, 2001

[25] Maxim Raya, Adel Aziz, and Jean-Pierre Hubaux, "Efficient secure aggregation in VANETs", In *Proc. 3rd International Workshop on Vehicular Ad Hoc Networks (VANET '06)*, September 2006, Los Angeles, CA, USA, pp. 67 - 75.

[26] Z. Wang and C. Chigan, "Countermeasure Uncooperative Behaviors with Dynamic Trust-Token in VANETs," In *Proc. IEEE International Conference on Communications (ICC 2007)*, June 2007, pp.3959 – 3964.

[27] K. Plossl, T. Nowey and C. Mletzko, "Towards a security architecture for vehicular ad hoc networks", In *Proc. 1st International Conference on Availability, Reliability and Security*, 20-22 April 2006, pp. 374 – 381.

[28] Z. Li, C. Chigan and D. Wang, "AWF-NA: A Complete Solution for Tampered Packet Detection in VANETs," In *Proc. IEEE Global Communications Conference (Globecom'08)*, Nov. 2008..

[29] X. Boyan, "Multipurpose Identity-based Signcryption: A Swiss Army Knife for Identity-based Cryptography," *In Proc. Of CRYPTO 2003*, LNCS 2729, Springer-Verlag, Berlin. pp. 383-399.

[30] J. Malone-Lee, "Identity-based Signcryption," *Cryptology ePrint Archive*, Report 2002/098, 2002, [Online]: http://eprint.iacr.org/2002/098.

[31] T. H. Yuen and V. K. Wei, "Fast and proven secure blind identity-based signcrption from pairings, " in *Proc. Of CT-RSA 2005*, Springer-Verlag, 2005, *LNCS* 3376, pp 305-322.

[32] J. Blum and A. Eskabdaruab, "The threat of intelligent collision," *IT Professional*, vol. 6, Jan.-Feb. 2004, pp. 24-29.

[33] Gongjun Yan, Gyanesh Choudhary, Michele C. Weigle and Stephan Olariu, "Providing VANET Security Through Active Position Detection," *Poster of the fourth ACM International Workshop on Vehicular Ad Hoc Networks (VANET '07)*, 2007.

[34] Tim Leinmüller, Christian Maihöfer, Elmar Schoch and Frank Kargl, "Improved Security in Geographic Ad Hoc Routing Through Autonomous Position Verification," In *Proc. 3rd International Workshop on Vehicular Ad Hoc Networks (VANET '06)*, September 2006

[35] P. Golle, D. Greene, and J. Staddon, "Detecting and correcting malicious data in VANETs," in *Proc. 1st ACM Workshop on Vehicular Ad hoc Networks (VANET '04)*, 2004, pp. 29-37.

[36] R. L. Rivest, A. Shamir, and Y. Tauman, "How to leak a secret," In *Proc. 7th International Conference on the Theory and Application of Cryptology and Information Security (ASIACRYPT 2001)*, *LNCS* 2248, pp. 552-565.

[37] R. Beresford and F. Stajano, "Location privacy in pervasive computing," *IEEE Pervasive Computing*, vol. 2, Jan. 2003, pp. 46–55.

[38] Boukerch et al., "Vehicular Ad Hoc Networks: A New Challenge for Localization-Based Systems," *Computer Communication (2008)*, vol. 31, July 2008, pp. 2838 – 2849.

[39] L. Huang, K. Matsuura, H. Yamane and K. Sezaki, "Enhancing wireless location privacy using silent period," in *Proc. IEEE Wireless Communications and Networking Conference*, Mar. 2005, vol. 2, pp. 1187-1192.

[40] Mingyan Li, Krishna Sampigethaya, Leping Huang, Radha Poovendran, "Swing and swap: user-centric approaches towards maximizing location privacy," In *Proc. 5th ACM Workshop on Privacy in Electronic Society (WPES '06)*, Oct. 2006, Alexandria, Virginia, USA, pp. 19 - 28..

[41] M. Gerlach and F. Guttler, "Privacy in VANETs using Changing Pseudonyms - Ideal and Real," In *Proc. IEEE 65th Vehicular Technology Conference*, 22-25 April 2007, pp. 2521 – 2525.

[42] R. Sampigethaya, Leping Huang, Mingyan Li, Radha Poovendran, K. Matsuura and K. Sezaki, "CARAVAN: providing location privacy for VANET," in *Proc. 3rd workshop on Embedded Security in Cars (ESCAR)*, Nov. 2005, Cologne, Germany.

[43] K. Sampigethaya, M. Li, L. Huang and R. Poovendran, "AMOEBA: robust location privacy scheme for VANET", *IEEE JSAC*, vol.25, pp. 1569 – 1589, Oct. 2007.

[44] Xiaodong Lin, Xiaoting Sun, Pin-Han Ho and Xuemin Shen, "GSIS: a secure and privacy-preserving protocol for vehicular communications", *IEEE Tran. on Vehicular Technology*, vol. 56, pp. 3422 – 3456, Nov. 2007.

[45] K. Sha, Y. Xi, W. Shi, L. Schwiebert and T. Zhang, "Adaptive privacy-preserving authentication in vehicular networks," In *Proc. International Workshop on Vehicle Communication and Applications*, Oct. 2006.

[46] Yong Xi, Kewei Sha, Weisong Shi, Loren Schwiebert and Tao Zhang, "Enforcing privacy using symmetric random key-set in vehicular networks", In *Proc. 8th International Symposium on Autonomous Decentralized Systems*, 21-23 Mar. 2007, pp. 344 – 351.

[47] J. R. Douceur, "The Sybil Attack," In *Proc. 1st International Workshop on Peer-to-Peer Systems (IPTPS '02)*, Mar. 2002, pp. 251–260.

[48] Bin Xiao, Bo Yu and Chuanshan Gao, "Detection and localization of sybil nodes in VANETs,", In *Proc. Workshop on Dependability Issues in Wireless Ad Hoc Networks and Sensor Networks*, Sep. 2006, pp. 1-8.

[49] Weichao Wang and Aidong Lu, "Visualization assisted detection of sybil attacks in wireless networks," In *Proc. 3rd international workshop on Visualization for computer security (VizSEC '06)*, Nov. 2006, pp. 51 – 60.

[50] F. D¨otzer, "Privacy issues in vehicular ad hoc networks," in *Proc. Workshop on Privacy Enhancing Technologies*, Cavtat, Croatia, May 2005.

[51] D. Chaum and E. van Heyst, "Group signatures," *Advances in Cryptology — EUROCRYPT '91,* vol. 547 of Lecture Notes in Computer Science, pp. 257-265.

[52] Giorgio Calandriello, Panos Papadimitratos, Jean-Pierre Hubaux and Antonio Lioy, "Efficient and robust pseudonymous authentication in VANET," In *Proc. 4th ACM International Workshop on Vehicular Ad Hoc Networks (VANET '07)*, Sep. 2007, pp. 19 – 28.

[53] J. Guo, J.P. Baugh and S. Wang, "A group signature based secure and privacy-preserving vehicular communication framework," In *Proc. of MOVE workshop in IEEE INFOCOM' 07*, May 2007, pp. 103 - 108.

[54] Jong Youl Choi, Markus Jakobsson and Susanne Wetzel, "Balancing auditability and privacy in vehicular networks," In *Proc. 1st ACM International Workshop on Quality of Service and Security in Wireless and Mobile Networks*, Oct. 2005, pp. 79 - 87.

[55] E. Fonseca, A. Festag, R. Baldessari and R.L. Aguiar, "Support of anonymity in VANETs - putting pseudonymity into practice," In *Proc. IEEE Wireless Communications and Networking Conference*, Mar. 2007, pp. 3400 – 3405.

[56] Stephan Eichler, "Strategies for pseudonym changes in vehicular ad hoc networks depending on node mobility," In *Proc. IEEE Intelligent Vehicles Symposium*, June 2007, pp. 541 – 546.

[57] S. Marti, T. Giuli, K. Lai and M. Baker, "Mitigating routing misbehavior in mobile ad hoc networks," in *Proc. 6th Annual International Conference on Mobile Computing and Networking*, 2000, pp. 255 - 265.

[58] S. Bansal and M. Baker, "Observation-based Cooperation Enforcement in Ad Hoc Networks," *Tech. Rep.*, Stanford University, July 2003.

[59] L.Buttyan and J. P. Hubaux, "Nuglets: A virtual currency to stimulate cooperation in self-organized mobile ad hoc networks," *Tech. Rep.* DSC/2001/001, Department of Communication Systems, Swiss Federal Institute of Technology, 2001.

[60] K. Chen and K. Nahrstedt, "iPass: an incentive compatible auction scheme to enable packet forwarding service in MANET," in *Proc. 24th IEEE International Conference on Distributed Computing Systems (ICDCS 2004)*, Mar. 2004, pp. 534 - 542.

[61] S.Zhong, J. Chen and Y. R. Yang, "Sprite: a simple cheat-proof credit-based system for mobile ad-hoc networks," In *proc. 22nd Annual Joint Conference of the IEEE Computer and Communications Societies (INFOCOM 2003)*, March-April 2003, vol. 3, pp. 1987-1997.

[62] S. Buchegger and J. Boudec, "Performance analysis of the CONFIDANT protocol (cooperation of nodes-fairness in dynamic ad-hoc networks)," in *Proc. of IEEE/ACM Workshop on Mobile Ad Hoc Networking and Computing (MoBiHOC)*, June 2002, pp. 226 - 236.

[63] F. Dotzer, L. Fischer and P. Magiera, "VARS: a vehilce ad-hoc network reputation system," In *Proc. 6th IEEE International Symposium on a World of Wireless Mobile and Multimedia Networks (WoWMoM'05)*, 2005, pp. 454-456.

[64] P. Michiardi and R. Molva, "Core: a collaborative reputation mechanism to enforce node cooperation in mobile ad hoc networks," In *Proc. IFIP-Communication and Multimedia Security Conference*, Sept. 2002, pp. 107 - 121.

[65] S. Moloney and P. Ginzboorg, "Security for interactions in pervasive networks," in *Proc. of 1st European Workshop on Security in Ad-Hoc and Sensor networks*, Aug. 2004.

[66] S. Tahnrich and P.Oreiter, "The buddy system: a distributed reputation system based on social structure," *Tech. Rep. 2004-1*, Universität Karlsruhe, Faculty of Informatics. March 2004.

[67] V. Oberoi and C. Chigan, "A generic packet-dropping detection mechanism augmented with efficient power saving in ad hoc networks," In *Proc. SPIE Symposium on Defense and Security*, Mar. 2006, pp.1-10.

[68] T. Leinmuller, E. Schoch and C. Maihofer, "Security requirements and solution concepts in vehicular ad hoc networks," In *Proc. Fourth Annual Conference on Wireless on Demand Network Systems and Services*, 24-26 Jan. 2007, pp. 84 – 91.

[69] Chunxiao Chigan, Leiyuan Li and Yinghua Ye, "Resource-aware self-adaptive security provisioning in mobile ad hoc networks," In *Proc. IEEE Wireless Communications and Networking Conference*, Mar. 2005, vol. 4, pp. 2118 – 2124.

[70] Yan Lindsay Sun, Wei Yu, Zhu Han and K.J.R. Liu, "Information theoretic framework of trust modeling and evaluation for ad hoc networks," *IEEE JSAC* vol. 24, Feb. 2006, pp. 305 – 317.

[71] Y. L. Sun, Z. Han, W. Yu and K.J.R. Liu, "Attacks on trust evaluation in distributed networks," In *Proc. 40th Annual Conference on Information Sciences and Systems*, Mar. 2006, pp. 1461 – 1466.

[72] Chung Tien Nguyen, Olivier Camp and Stephane Loiseau, "A bayesian network based trust model for improving collaboration in mobile ad hoc networks," In *Proc. IEEE*

International Conference on Research, Innovation and Vision for the Future, Mar. 2007, pp. 144 – 151.

[73] Ruidong Li, Jie Li, Peng Liu and Hsiao-Hwa Chen, "An objective trust management framework for mobile ad hoc networks," In *Proc. 65th IEEE Vehicular Technology Conference*, 22-25 Apr. 2007, pp. 56 - 60.

[74] Huafei Zhu and Feng Bao, "Quantifying trust metrics of recommendation systems in ad-hoc networks," In *Proc. IEEE Wireless Communications and Networking Conference (WCNC '07),* Mar. 2007, pp. 2904 – 2908.

[75] Phil Zimmermann. Pretty Good Privacy (PGP), *PGP User's Guide*, MIT, October, 1994.

[76] M. Blaze, J. Feigenbaum and J. Ioannidis, "The role of trust management in distributed systems security," Chapter in *Secure Internet Programming: Security Issues for Mobile and Distributed Objects*, (Vitek and Jensen, eds.) Springer-Verlag, 1999.

[77] G. Theodorakopoulos and J.S. Baras, "On trust models and trust evaluation metrics for ad hoc networks," *IEEE JASC,* Vol. 24, Feb. 2006, pp. 318 – 328.

In: Wireless Technologies in Intelligent Transportation…
Editors: Ming-Tuo Zhou et al, pp. 175-204
ISBN: 978-1-60741-588-6

Chapter 7

HANDOFF MECHANISMS IN IEEE 802.16 NETWORKS SUPPORTING INTELLIGENT TRANSPORTATION SYSTEMS

Melody Moh, Bhuvaneswari Chellappan, Teng-Sheng Moh and Swathi Venugopal
Department of Computer Science, San Jose State University
San Jose, CA, USA

Abstract

With the recent rapid revolution in wireless network technology, wireless metropolitan area networks (WMAN) and vehicular networks (VN) have both gained significant research focuses, in academia as well as industry. VN are vital for the realization of intelligent transportation systems (ITS). The short-range communications for ITS are most likely supported by IEEE 802.11p WAVE (Wireless Access for Vehicular Environment), whereas the long-range communications would be realized by infrastructure-based WMAN such as 3G technology or IEEE 802.16 WiMAX (Worldwide Interoperability for Microwave Access). It is thus timely to study how WiMAX may be utilized in the emerging VN for ITS. This chapter surveys the recent developments in mobility management of IEEE 802.16 networks, and presents a comprehensive feasibility study of 802.16 handoff schemes for the support of ITS. With the strong worldwide deployment of WiMAX networks, we believe that the study is a solid contribution in the practical realization of ITS

1. Introduction

Intelligent Transport System (ITS) refers to the endeavor that attempts to equip transport infrastructures and vehicles with intelligent information and communication technologies. The main goal is to improve safety and transportation efficiency, including reduced transportation time and fuel consumption. This chapter focuses on the wireless communication technology behind ITS, i.e., Intelligent Vehicular Ad-hoc Networks (InVANET), or simply Vehicular Ad-hoc Networks (VANET),

VANET is a specialty of Mobile Ad-hoc Networks (MANET). Such a system uses every moving vehicle that participates in the network as a router or a mobile node. The network provides communications among vehicles (Vehicle-to-Vehicle or V2V), communication between vehicles and nearby fixed equipments (Vehicle-to-Roadside, or V2R), and Internet access through infrastructure networks to and from vehicles (Vehicle-to-Infrastructure, or V2I).

Many safety and non-safety applications have been proposed for ITS. Examples of safety applications are collision warning, congestion notification, communication among fire, ambulance and police vehicles, emergency message dissemination, etc. Non-safety applications include broadband access and mobile commerce while on the road, tracking automobiles, inter-vehicle communication for entertainment and information exchange, high-speed automatic toll payment, and automatic parking fee payment.

Current vehicular communications mainly rely on passengers using cellular devices and on roadside traffic monitoring sensors. They are still quite restrictive and are not well suited for many foreseeable vehicular applications including those mentioned above. Hence, the need for wireless communication technologies supporting ITS has rapidly increased. VANET has thus become a highly focused research area.

Several wireless communication technologies have been proposed for ITS. Short-range communication technologies are mainly based on IEEE 802.11. Specifically, IEEE 802.11p Wireless Access in the Vehicular Environment WAVE and Dedicated Short Range Communications Standard [1]. Many research works in VANET are based on these short-range technologies [2].

On the other hand, many believe that long-range communication technologies for ITS will be based on infrastructure wireless technologies, such as IEEE 802.16 and Global System Mobile Communications (GSM) or 3G. As these technologies either have or will soon have wide deployments in many countries, they are therefore considered as highly competitive candidates for wireless components of ITS. The focus of this chapter is on IEEE 802.16, specifically, how it can support long-range V2I and V2R communications in ITS.

The IEEE 802.16 was originally developed as a Broadband Wireless Access (BWA) standard for the global deployment of broadband Wireless Metropolitan Area Networks (WMAN) [3]. Although the 802.16 family of standards is officially called WirelessMAN, it is popularly known as "WiMAX" (Worldwide Interoperability for Microwave Access). A flavor of 802.16 – IEEE 802.16e – offers many useful services supporting wireless mobility [3, 7].

Due to the high mobility of most vehicles, mobility management is evidently an important aspect in the wireless communications of ITS. There have been many existing works on mobility management of IEEE 802.16 and 802.16e. Some issues that are vital for a typical 802.16 application scenario, however, may not be relevant to ITS. This chapter thus aims to study the feasibility of IEEE 802.16e mobility management for ITS. Specifically, we survey a number of major handoff protocols proposed for IEEE 802.16e, and evaluate their feasibility in the context of supporting ITS for V2R and V2I communications.

The chapter is organized as follows. Section 2 presents the background information, covering IEEE 802.16, 802.16e, and its basic handoff scheme. Section 3 studies the handoff problem in ITS, including major issues, examples, and a simple classification. Sections 4 and 5 present layer-2 and layer-3 802.16 handoff schemes, respectively. Each section consists of several handoff schemes; for every scheme, the protocol description is first presented, followed by a study of its feasibility in supporting ITS. Finally, Section 6 concludes the chapter.

2. Background: IEEE 802.16 and Its Basic Handoff Scheme

This section presents the WiMAX background. Section 1 briefly introduces the IEEE 802.16 technology. Section 2 introduces IEEE 802.16e, and finally Section 3 illustrates the basic 802.16e handoff mechanism.

2.1. IEEE 802.16 (the WiMAX Standard)

The recent rapid advances in wireless networking have motivated the evolution of next-generation wireless technologies that aims to achieve higher range, higher bandwidth, and better quality of service (QoS). Among them, The IEEE 802.16 standard specifies the "Air Interface for Fixed Broadband Wireless Access Systems" [4]. It has gained immense popularity and research focus, and has been rapidly deployed in many regions of the world. Table 1 depicts its potential usage scenarios and the associated technical feature requirements [5].

Table 1. WiMAX usage scenarios

	Flexible Architecture	**High Security**	**WiMAX QoS**	**Quick Deployment**	**Multi-level Service**	**Interoperability**	**Portability**	**Mobility**	**Cost Effective**	**Wider Coverage**	**NLOS**	**High Capacity**
Cellular Backhaul				x					x			x
WSP Backhaul				x					x			x
Banking Networks	x	x	x						x		x	
Education Networks	x		x						x	x		
Public Safety	x	x	x	x			x	x			x	
Offshore Communications	x		x				x	x		x	x	
Campus Connectivity	x	x	x									x
Temporary Construction			x	x			x				x	
Theme Parks	x		x				x	x			x	
WSP Access Network		x	x		x	x			x		x	x
Rural Connectivity			x			x			x	x		
Military Battlefield	x	x		x			x	x				

As mentioned in Section 1, IEEE 802.16 was developed as a BWA standard for WMAN. As a wireless standard, there should be no stringent constraints on physical movement of a device. Thus, one of the most critical issues to be addressed is that of mobility. A strong solution supporting mobility would allow service providers to reach nomadic and mobile customers equipped with portable devices, and allow outdoor roaming across geographical areas of coverage. In another word, WiMAX mobility features will allow service providers to offer new high-value multimedia services to their subscribers. As a result of this vision, the IEEE 802.16 standard has gone through several major steps of evolution, as can be seen in Table 2 [6].

Table 2. Evolution of IEEE 802.16 standard

	802.16	**802.16a**	**802.16e**
Completed	December 2001	January 2003	Mid 2004
Spectrum	10-66 GHz	< 11GHz	< 6 GHz
Channel Conditions	Only line-of-sight	Non line-of-sight	Non line-of-sight
Bit Rate			Up to 15 Mbps in 5 MHz channel bandwidth
Modulation			Same as 802.16a
Mobility			Nomadic portable
32 – 134 Mbps in 28 MHz channel bandwidth	Up to 75 Mbps in 20 MHz channel bandwidth		Same as 802.16a with uplink subchannels
QPSK, 16QAM, and 64QAM	OFDM 256 sub-carriers, QPSK, 16QAM, and 64QAM		2-5 km
Fixed	Fixed as well as portable		
20, 25, and 28 MHz	Scalable from 1.5 to 20 MHz		
2-5 km	7-10 km max; range 50 km		

2.2. IEEE 802.16e (the WiMAX Mobility)

The IEEE 802.16e amendment [7] has been introduced to the standard to support *mobility*, apart from providing other extensions. More specifically, it aims to provide seamless mobility to end users in their homes, offices, and during transit, thus spurring the demand for innovative mobile services. It would allow users to take advantage of BWA while traveling at vehicular speeds. The IEEE 802.16e can provide BWA up to 30 miles (approximately 50 km) for fixed stations, and 3 - 10 miles (approximately 5 - 15 km) for mobile stations. It therefore provides a strong mobility platform for vehicular communications.

To provide mobility in a WiMAX network, IEEE 802.16e implements a mobility node called 'Access Service Network' (ASN) gateway in the WiMAX architecture. This gateway connects to multiple WiMAX base stations (BS) and interfaces with other network elements, such as Authentication, Authorization, and Accounting (AAA) servers, Home Agents and Domain Host Configuration Protocol (DHCP) servers to provide seamless mobility to end users. Its capabilities include high-speed data processing, determination of optimal path, compatibility with WiMAX resource management protocols such as routing protocols, address allocation protocols, Internet Group Management Protocol (IGMP) for multicast, Public Key Management (PKM) protocol and Internet Key Exchange (IKE) for security, etc. [6].

2.3. The Basic IEEE 802.16e Handoff Scheme

A base station (BS) provides support to mobile node or mobile station (MS) to remain connected to the network; i.e., to remain reachable within a given geographic area (cell). As the MS moves from one cell to another, the BS of the source cell (the *serving* BS) should *handover* the support to the BS that covers the target cell (the *target* BS); the handover should be done with minimum delay and disruption. A handover or handoff (HO) mechanism thus defines the steps in this process of information and resource transfer, involving the serving BS, the target BS, and the MS, so that the MS remains connected to the wireless network infrastructure [4].

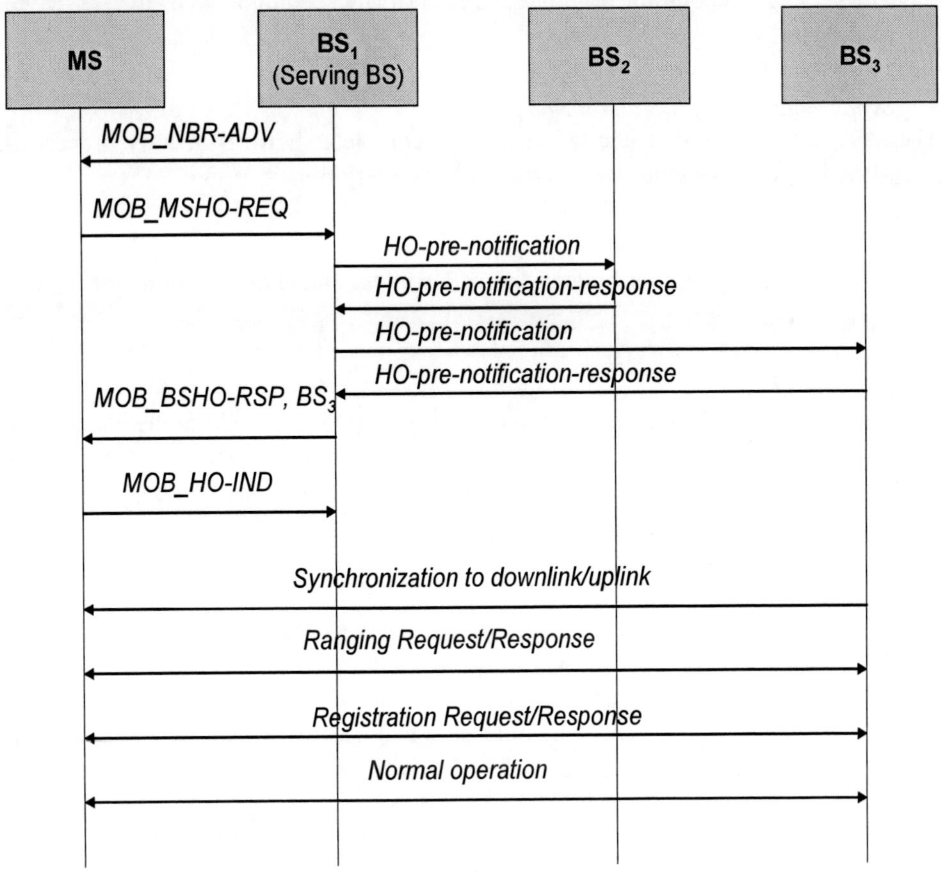

Figure 1. 802.16e Handoff (Example, per MSS request).

The operation of IEEE 802.16e handoff process is depicted in Figure 1. The handoff protocol involves the following stages [5, 8]:

- *Determination of Potential target BSs:*

The MS may acquire information about neighboring BSs through the MOB_NBR-ADV (mobile neighbor advertisement message) broadcast by the serving BS. Also, the serving BS may allocate scanning intervals periodically or upon-request to the MS, during which the MS can scan and assess potential target BSs. During this stage, the MS may or may not have decided to perform a handoff.

- *HO Decision and Initiation:*
    - When the MS decides to perform a handoff, it sends the MOB_MSHO-REQ (MS HO request message) to its serving BS. The serving BS sends HO pre-notification messages to potential target BSs. Based on the HO pre-notification responses received from potential target BSs, the serving BS determines the target BS.
    - The serving BS then responds with a MOB_BSHO-RSP (BS HO response message), in which it includes the recommended target BS.
    - The MS finally sends the MOB_HO-IND (HO indication message) to the target BS. The serving BS could also initiate the handoff by sending a MOB_BSHO-REQ (BS HO request message).

- *Synchronization of MS and target BS:*

The MS synchronizes with the target BS and completes network reentry process. After this, the MS and its new serving BS engage in normal operation.

- *Termination of MS Context:*

The original serving BS terminates the MS context (including information in queues, counters and timers).

- *HO Cancellation:*

After the handoff decision, the MS may cancel its decision to perform the handoff, before expiration of a timer called the resource retain timer. In such a case, the MS transmits a MOB_HO-IND with HO-cancel option. The handoff is then cancelled and the MS continues its normal operation with the serving BS.

3. Handoff in ITS: Issues, Examples, and Classification

Handoff has been a long-focused research problem in wireless cellular communication systems. When applying to ITS, there are old issues that need to be considered, as well as new challenges arose that need to be addressed. In this section, first, the key issues in providing high quality handoff in VANET, specifically to V2I (Vehicles to Infrastructure) VANET, are discussed. Next, a brief survey of a few examples of VANET handoff mechanisms is presented. Finally, we illustrate a broad classification of IEEE 802.16 handoff schemes that support V2I VANET.

3.1. ITS Handoff Issues

In order to support ITS, the underlying VANET needs efficient handoff mechanisms to provide desired QoS for various applications. For examples, high-availability is needed for safety applications while guaranteed delay is vital for real-time applications. In the following, we discuss the main issues of building an efficient handoff mechanism for V2I VANET. (Note that below we refer a vehicle as Mobile Station or MS; and refer BS, AP, or roadside assistant to which vehicle is connected as simply BS.)

- Accuracy and Timeliness

ITS handoffs that are consistently both accurate and timely can result in higher capacity and better overall link quality [10]. This is especially important, as cells have been getting smaller in order to achieve higher capacity. For a given average vehicular speed, the number of handoffs increases as the cell area decreases. Thus, accuracy and timeliness in handoff becomes even more critical.

- Ping-Pong Handoff

This phenomenon refers to the situation when a MS is unnecessarily handed over to neighboring target BS and the target BS then returns the MS to the original serving BS within a very short period of time. These unnecessary handoffs would waste network resources and decrease communication quality.

- Handoff Latency and Service Disruption

The time duration between the time a MS leaves the serving BS's coverage to the time it completes the entire handoff is called handoff latency. A well designed ITS handoff mechanism should aim to reduce this latency and thereby increase data throughput and reduce packet losses. (Note that while IEEE 802.16 covers only the physical and MAC layer, layer 3 handoff also needs to be completed for the MS to resume normal operations with the target BS.)

During a handoff period, if the MS does not receive data during handoff, the service may be disrupted for the MS. A good ITS handoff scheme would try to reduce service disruption by enabling the MS to communicate with its target BS even before the entire handoff is complete, thus resulting in lesser service disruption time compared to handoff latency [11]. The ultimate goal is to achieve seamless connectivity and data exchange during handoff.

- Frequent Link Disconnection and Packet Loss

Due to high vehicular speed and small cell area, a MS may frequently be disconnected from its serving BS. Unless an ITS handoff scheme is designed to handle packets delivered during the handoff period, packet losses will be severe. Though a minimal loss will not affect real-time applications, this has significant effect on non-real time applications like FTP and HTTP. These protocols rely on TCP in which lost packets have to be retransmitted. To avoid this problem, good handoff schemes employ techniques such as buffering, tunneling, and bicasting [12, 18].

- Signaling Overhead

A handoff process involves the transmission of control signals among MS, the serving BS, and potential target BS's. As a VANET protocol serving the ITS becomes more complex, the handoff process would involve the exchange of QoS and security information; this causes higher signaling overhead. This in turn will result in increased handoff latency. Hence, a well-designed protocol should try to keep signaling overhead and signaling delay to a minimum.

- Layer-2 and Layer-3 handoff

A complete ITS handoff process must cover both data-link and network layers. This depends on the number of subnets or domains present in the network, number of BS's present in each domain, and the location of MS in the domain. An efficient ITS handoff scheme must be intelligent to make the handoff decision at the accurate time, position, and at proper layers.

- Energy Consumption

While most mobile devices are battery-powered and thus severely power-constrained, VANET of ITS can rely on the battery that is powered by vehicle engine. Thus, energy consumption is not a cruel issue for ITS handoff schemes.

3.2. ITS VANET Handoff Examples

In this section, we describe some existing handoff schemes proposed for VANET. As accuracy and timeliness are important issues in VANET handoff, the first two deal with handoff decisions – one is based on pattern recognition of signal strengths and the next on vehicle movement estimation. The third example is a proposal based on the IEEE 802.11p WAVE technology.

3.2.1. Handoff Decisions Based on Pattern Recognition

One significant technique used in handoff algorithms for wireless cellular systems is pattern recognition. In the handoff decision process, a MS makes use of the "relative signal strength," which is the difference in signal power level associated with a new BS from that associated with the original serving BS. The decision is based on a predefined threshold in dB. Due to factors such as fading effects, however, the relative signal strength may be positive for a brief period of time, and the undesirable ping-pong handoff effect may occur. Using hysteresis in the decision process may in turn solve this issue, yet it may also increase decision delay [13].

Wong and Cox proposed to use statistical pattern recognition techniques to overcome the shortcomings discussed above. Their algorithms exploit previous knowledge about propagation characteristics of the environment. In other words, the algorithms are trained on the signal strength measurements at locations where handoffs may be desired, thus gain specific knowledge about the propagation characteristics at these places [9]. They also applied this technique to corner-turning situations. These situations are particularly challenging for handoff decisions, considering many roads crossing one another and numerous obstacles such as tall buildings exist in busy downtown and urban areas [10]. The authors have successfully improved accuracy and timeliness of handoff decisions over the use of traditional pattern recognition techniques. As a result, both the number of unnecessary handoffs and the handoff decision delay have been reduced.

3.2.2. Handoff Decisions Based on Mobility Estimation

Several other factors have been taken into consideration for handoff decisions including cell traffic load and location information. Kim et al proposed to use signal strength measurements to estimate vehicle mobility, which is then used adaptively to control handoff decisions [20]. Specifically, the algorithm measures 'handover mobility factor' (HMF), which is the average value of the difference between the changing rate of the measured signal strength of the serving BS and that of a potential target BS. The authors found that HMF provides accurate information on both direction and speed of a MS movement, especially at the border of a serving cell. The authors further adaptively adjust the handoff execution time to avoid ping-pong handoffs [20].

3.2.3. Handoff Scheme Based on IEEE 802.11p MAC Protocol

The 802.11p Wireless Access in the Vehicular Environment (WAVE) defines amendments to IEEE 802.11 to support ITS applications [1]. It is therefore expected to address handover issues such as frequent disconnection, high mobility, and time varying channel condition. In addition, there are new challenges specific to the 802.11p protocol, including stateless channel access, opportunistic frame scheduling, and absence of data caching during handoff [2]. Towards this effort, Choi et al proposed an extension to IEEE 802.11p to support V2R communications [2]. The authors defined a new operation mode that is solicitation-based, such that only users initiate data frame transmissions. Using throughput analysis, they showed that the proposed scheme is able to achieve a high, stable throughput independent of the number of contending stations or the number of leaving nodes.

3.3. Classification of Handoff Schemes

There have been quite a few handoff schemes proposed for IEEE 802.16 in the literature. Many of them may be adopted to be used to support V2I VANET for ITS. Handoff schemes may be classified in different ways such as soft and hard handoff, micro- and macro-domain handoff, etc. In this chapter, we adopt a grouping that follows the network hierarchy, as described below:

1. *Layer-2 or Data-Link Layer Schemes*

These mechanisms propose enhancements to the IEEE 802.16 standard at the data-link layer.

2. *Layer-3 or Network Layer Schemes*

These mechanisms aim to achieve handoff at the network layer in an IEEE 802.16 network.

In addition, there can be a third, and much smaller category that is "domain-perspective". It considers an IEEE 802.16 network as a domain, and may serve as a fit for 802.16 networks between macro and micro domains. Due to length limitation, we focus on the first two main categories.

These handoff schemes improvise, enhance or substitute some of the features of the basic handoff mechanism in IEEE 802.16 (Section 2.3). Figure 2 depicts the three major categories, each with a list of 802.16 handoff schemes that will be studied in a subsequent section.

Following this section, the next two sections provide a brief description for these schemes in each of the two categories, respectively. In addition, they also analyze the feasibility of these schemes supporting long-range V2I and V2R communications in ITS.

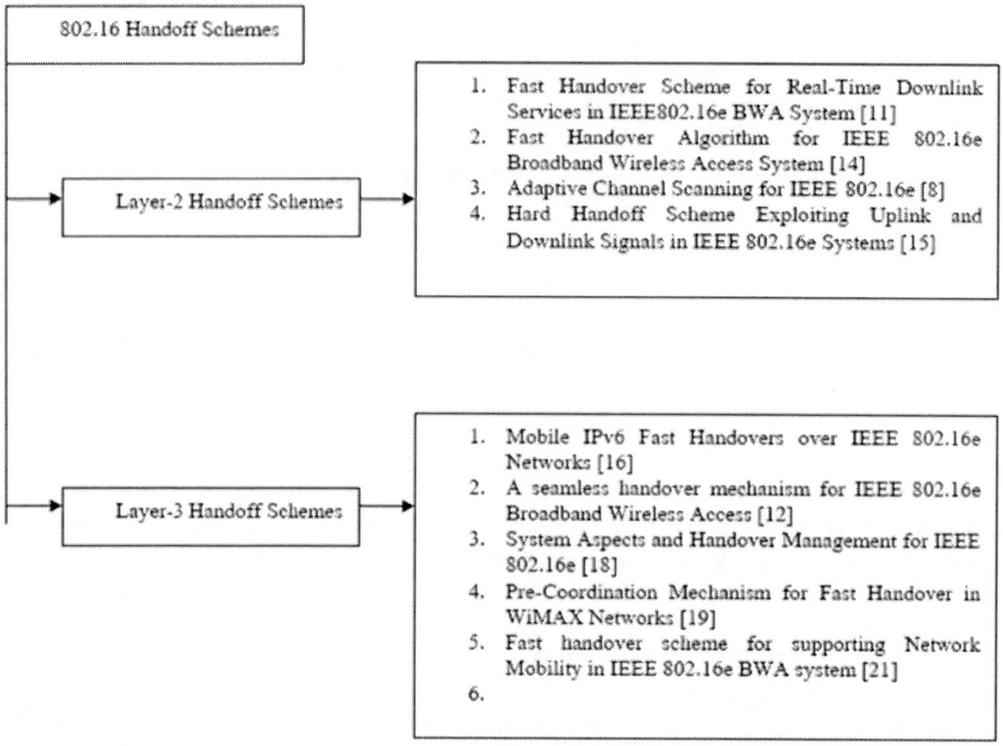

Figure 2. Classification of 802.16 Handoff Schemes for ITS Support.

4. IEEE 802.16 Layer-2 Handoff Schemes

In this section, we describe four layer-2 handoff schemes proposed for IEEE 802.16. For each scheme, a feasibility study of the scheme for supporting V2I and V2R communications in ITS is also included. Finally, at the end of this section, we compare these four schemes in the form of tables, in terms of basic scheme handoff, assumptions and/or requirements specific in VANET, advantages, disadvantages, and suggestions. For simplicity, we use "IVN" (Intelligent Vehicular Networks) to represent "VANET for supporting ITS".

4.1. Fast Handover Scheme for Real-Time Downlink Services in IEEE 802.16e BWA System [11]

4.1.1. Scheme Description

This scheme, proposed by Choi, et al, suggests improvements so that real-time data reception is not affected by handoff latency and packet loss [11]. The main idea of the scheme revolves around the idea of synchronization downlink before the entire handoff is complete. Therefore, real-time data reception on the downlink could resume before the entire handoff is complete. Since real-time data beyond a certain delay (playout delay) will be discarded, the scheme claims to reduce the packet loss probability. This scheme is depicted in Figure 3.

This scheme considers the hard handoff mode. When the MS sends the handoff request (MOB_MSSHO_REQ), the serving BS communicates with neighboring BSs and determines a suitable target BS for MS. The serving BS then sends a handoff response (MOB_HO_RSP) to MS containing information about the suitable target BS.

Then, the MS sends handoff indication. The serving BS transfers the security related information of the MS to the target BS and starts forwarding the real-time data destined for the MS to the target BS (MSS-Data-Forwarding). Upon completion of downlink synchronization, the MS will be able to receive the real-time data in the downlink through the reception of a new message, Fast_DL_MAP_IE. The remaining steps in 802.16e handoff are performed (uplink synchronization, ranging, registration etc.) simultaneously. After handoff is complete, normal operation proceeds, and the old serving BS stops forwarding data.

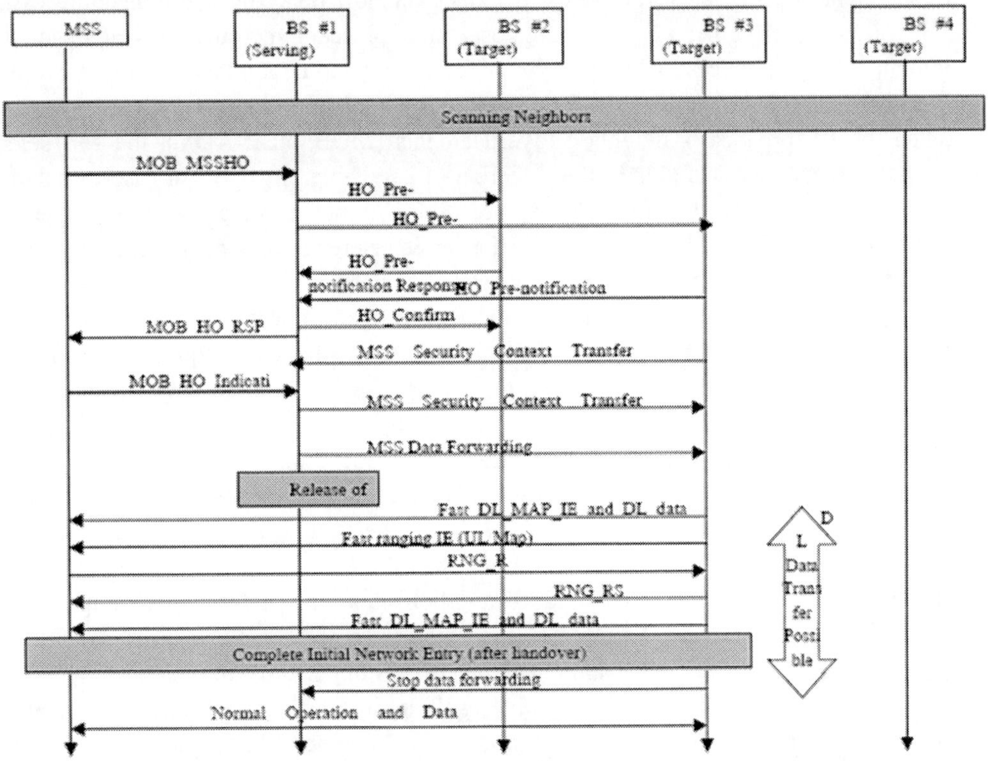

Figure 3. Message Sequence Chart of handover process (Choi, et al).

Since this scheme emphasizes on synchronizing only downlink communication before the completion of handoff, this scheme is suitable only when the MS is the receiver. It does not cover the scenario where the MN is sending real-time data. In addition, the scheme suggests that, after handoff is complete, data forwarding from old BS should stop. This would be fine in a micro-mobility scenario. In the case of macro-mobility, the layer-3 handoff should be initiated and completed to avoid greater delay for real-time packets.

4.1.2. Feasibility of Supporting IVN

This scheme expects the serving BS to be capable of scanning neighboring BSs of vehicles. It also expects the serving BS to choose the best suitable target BS for vehicle. Furthermore, this scheme makes downlink communication to vehicle faster by allowing it to occur even before handoff is complete. So, it helps receiver vehicles. We therefore suggest that the scheme can be implemented in IVN by allowing both uplink and downlink communications before the handoff is complete, or by requiring scanning operations to be performed by the MS.

4.2. Fast Handover Algorithm for IEEE 802.16e Broadband Wireless Access System [14]

4.2.1. Scheme Description

This scheme, by Lee et al, proposes to reduce handoff delay by avoiding unnecessary scanning of neighboring BSs [14]. It suggests that by scanning only the most suitable potential target BSs, unnecessary interleaving of scanning and data intervals can be avoided. Hence, throughput would not degrade. This scheme is depicted in Figure 4.

When the BS broadcasts neighbor advertisements (MOB_NBR-ADV), the MS selects some BSs for scanning. This aids in the selection of a suitable target BS in case of handoff. This scheme suggests that MS should "prove preambles" of neighboring BSs and choose only the most suitable BS based on CINR (carrier to interference-plus-noise-ratio) and ATD (arrival time difference), for synchronization and association. In the left half of the above figure, the MS “proves preambles” of BS1 and BS2 instead of scanning both BSs. Then, it selects neighboring BS, BS2 only to scan. Hence, it sends the scan request to serving BS (MOB_SCN-REQ), receives scan response (MOB_SCN-RSP) from serving BS and finally synchronizes only with BS2.

The scheme proposes "Fast Synchronization and Association" (depicted in right half of figure) to allow the MS to simultaneously engage in data transmission as well as synchronization and association. They propose that if the serving BS could avoid downlink and uplink transmissions at the appropriate time, then the MS could engage in simultaneous operations. Finally, they suggest that handoff initiation should be performed at an appropriate time.

Thus, this scheme tries to use a single antenna of MS to perform simultaneous operations with two BS. However, this scheme would result in increased processing burden at the serving BS in order to decide the appropriate slots due to "Fast Synchronization and Association"

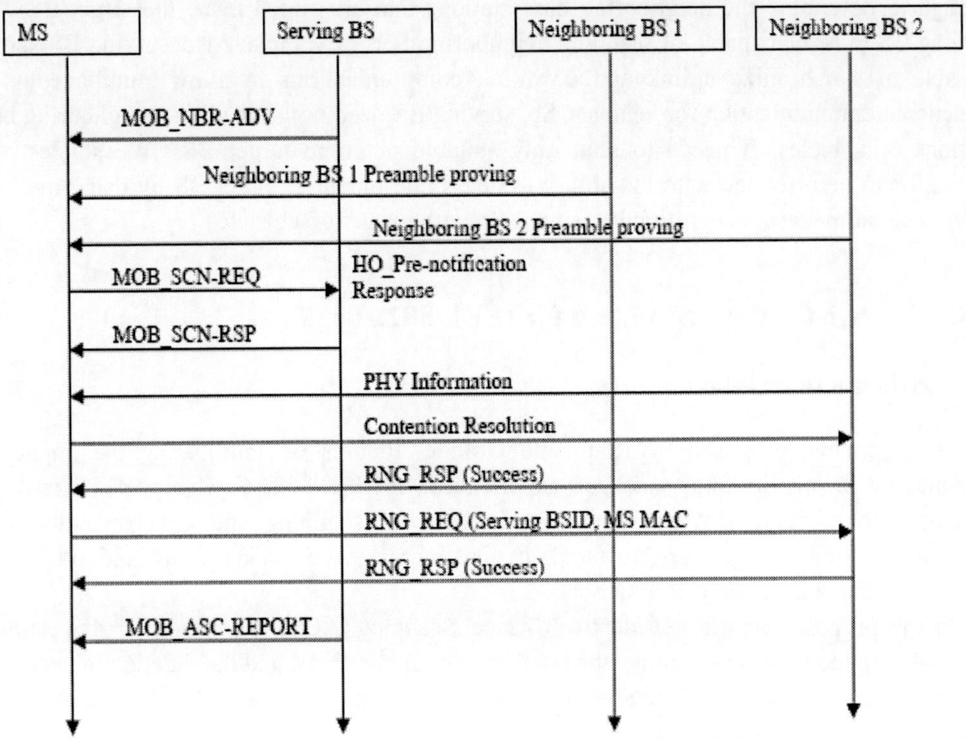

Figure 4a. Proposed Network Topology Acquisition (Lee et al).

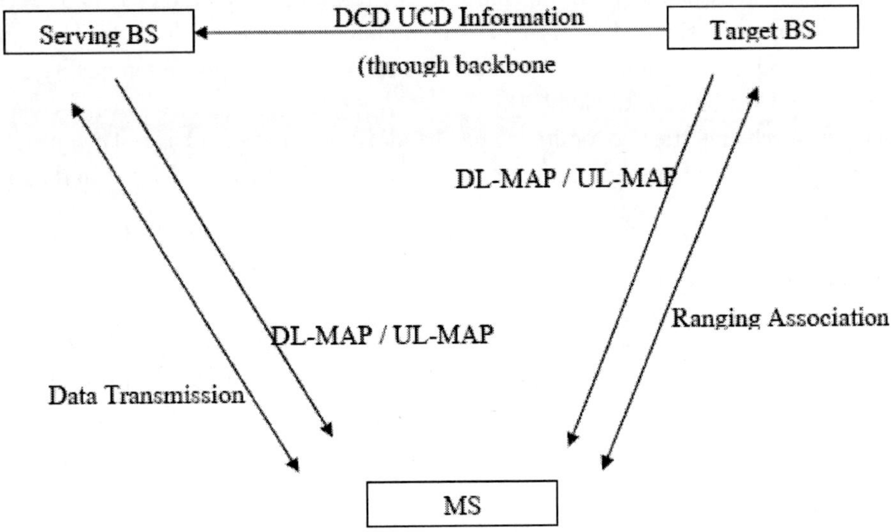

Figure 4b. Proposed Fast Association Model (Lee, et al).

4.2.2. Feasibility of Supporting IVN

In order to implement this scheme in IVN, vehicles would need to be equipped with devices to select neighboring base stations to be scanned. Vehicles would be intelligent enough to determine the neighboring base stations that are suited to be the target BS. The serving BS must be capable of scanning neighboring BSs of vehicle. Also, serving BS should be able to synchronize uplink and downlink communications to allow simultaneous bi-directional communication for vehicles. So, serving BS need not scan all the neighboring base stations of vehicles. It needs to scan only suitable potential target BSs. We suggest that vehicles can be provided with the ability to determine potential target BS by using methods such as signal pattern recognition [9], signal strength measurements etc.

4.3. Adaptive Channel Scanning for IEEE 802.16e [8]

4.3.1. Scheme Description

This scheme, proposed by Rouil and Golmie, focuses on minimizing the impact of scanning on the quality of service QoS experienced. In IEEE 802.16, when the MS performs scanning, the serving BS buffers incoming data, as scanning and data operations are interleaved. Since packets are buffered, it may increase the packet delay and jitter, thus affecting the quality of service experienced.

In this proposed algorithm, Adaptive Channel Scanning (ACS), the authors tried to estimate the total time required for scanning, the number of scan iterations, and the duration of each scan interval. The time duration required for scanning depends on the total time required for synchronization and association. The time required for association, in turn, depends on the level of association, namely whether ranging will be contention-based or without contention, etc. The authors have suggested a formula to compute the total scanning time based on time required for synchronization and association. For this purpose, the relevant information has to be exchanged over the backbone, to facilitate the serving BS to compute the scanning duration. The number of scan iterations and the duration of each are decided based on the "application quality of service requirements and available bandwidth". [8]

Thus, this scheme tries to optimize the handoff preparation phase. This phase has a significant impact on the throughput as data operations are interleaved with scanning. Further, packet delay and jitter will be reduced.

4.3.2. Feasibility of Supporting IVN

This scheme expects vehicles to be capable of scanning neighboring BSs, and be equipped with devices that can run Adaptive Channel Scanning (ACS) Algorithm. According to this scheme, if a vehicle is scanning its neighboring base stations, the serving BS buffers the data packets destined to vehicle. ACS estimates total time required for scanning, number of scan iterations and duration of each scan interval. This information helps a vehicle to avoid buffering of data packets at serving BS thereby reducing packet delay. However, implementing this scheme in IVN has an overhead of calculating total time required for scanning, number of scan iterations, and duration of each scan interval.

We suggest that the overhead of ACS may be reduced by periodically allocating a pre-determined period for scanning. Also, allocate a scanning period when data traffic is low so that less data packets are buffered at the serving BS.

4.4. Hard Handoff Scheme Exploiting Uplink and Downlink Signals in IEEE 802.16e Systems[15]

4.4.1. Scheme Description

This scheme, by Cho et al, aims to base the handoff decision on both downlink and uplink signals. Specifically, based on threshold values shown in Table 3, the serving BS makes the decisions of when to perform handoff and the direction of handoff.

Table 3. Threshold values (Cho et al)

Threshold	Description
TH_1	Threshold to determine handoff initiation using uplink signal level
TH_2	Threshold to determine handoff initiation for the AMC (Adaptive Modulation and Coding) users
TH_3	Threshold to determine handoff direction using uplink signal level (UL hysteresis)
TH_4	Minimum downlink signal level to prevent call drop
TH_5	Threshold to determine handoff direction using downlink signal level (DL hysteresis)
TH_6	Minimum uplink signal level to prevent call drop

The serving BS monitors the strength of uplink signals. When this drops below TH_1 (as shown in the first decision box in Figure 9), it sends the NBR_MON_REQ to neighboring BSs. This is a request to the neighboring BSs to monitor and report the uplink signal strength of the MS. The serving BS sends the relevant information for this purpose. To measure the downlink signal, the serving BS sends DL_MEASUREMENT_REQ to the MS. On receipt of this message, the MS monitors downlink signal from serving BS as well as neighboring BSs, and reports the result to the serving BS. Based on the results received and the thresholds maintained, the serving BS decides the direction of handoff. As the MS assists the BS in the handoff decision, this scheme comes under the category of "mobile assisted handoff" [15].

As the entire handoff making decision is at the BS, the complexity of the BS increases. Furthermore, as the number of MSs under a BS increases, the amount of bookkeeping at the BS also increases. Further, the applicability of this scheme provides maximum benefit only to those MSs with traffic in both uplink and downlink directions. Hence, this scheme may not cover the scenario when the MS is engaged in real-time video streaming, where there is mostly data in the downlink.

However, this scheme reduces the scanning overhead at the MS and hence there is more time for data operations. The authors' simulation shows that this scheme performs well in terms of outage probability and average number of handoffs performed.

4.4.2. Feasibility of Supporting IVN

In order to implement this scheme in IVN, vehicles need to be equipped with devices to measure downlink signal strength from BS. BSs need to be capable of measuring uplink signal strength from vehicles. This scheme considers both uplink and downlink network loads for a handoff decision.

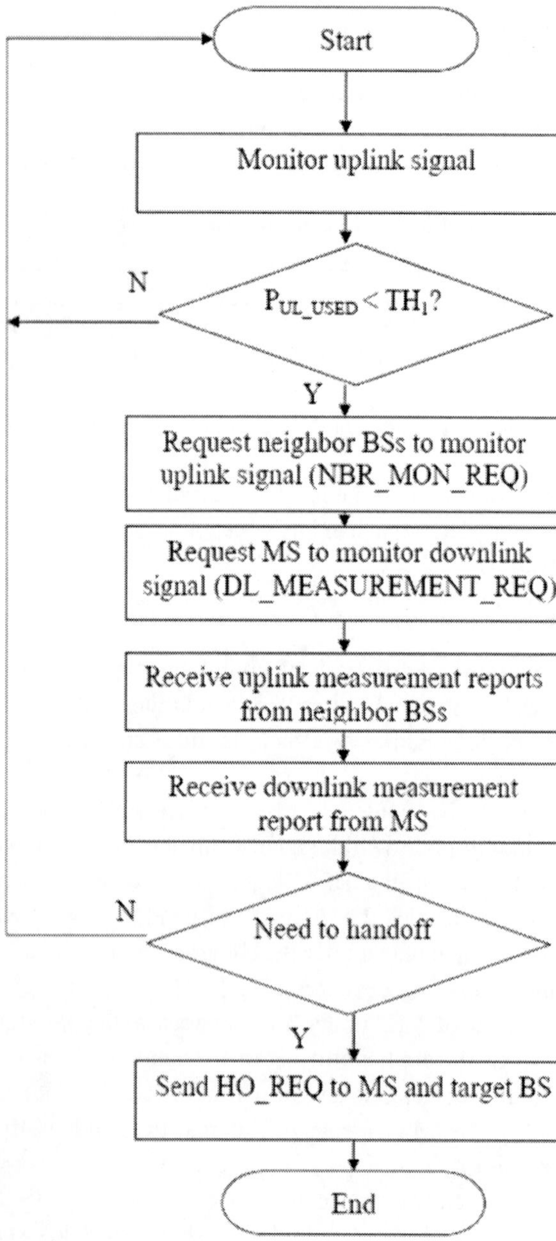

Figure 5. Handoff process in serving BS (Cho et al).

So, it is suitable for vehicular communications that need bi-directional communication. It may not be suited to vehicles using only uni-directional communication. Note also that BS will have higher responsibility of making handoff decision.

4.5. Summary of 802.16 Layer-2 Handoff Mechanisms

This section summarizes and compares the four layer-2 HO schemes presented in this section with a table. For better readability, we separate the table into two parts. Table 4(a) provides a comparative overview of each basic scheme and its assumptions and/or requirements for IVN. Table 4(b) summarizes their advantages, disadvantages, and provides some suggestions and comments.

Table 4(a). Comparison of Layer-2 HO schemes with respect to IVN

Scheme	Scheme Overview	Assumptions and/or Requirements for IVN
Fast Handover Scheme for Real-Time Downlink Services in IEEE 802.16e BWA System [11]	Downlink data received before completion of handoff Fast DL_MAP_IE used for sending real-time downlink data from target BS to MS	Serving BS capable of scanning neighboring BSs of vehicle Serving BS capable of choosing the best suitable target BS for vehicle
Fast Handover Algorithm for IEEE 802.16e Broadband Wireless Access System [14]	Reduced scanning overhead "Fast Synchronization and Association" helps in simultaneous synchronization and data operations	Serving BS capable of scanning neighboring BSs Serving BS can synchronize uplink and downlink communication to allow simultaneous bi-directional communication Vehicles capable of selecting neighboring BS to be scanned
Adaptive Channel Scanning For IEEE 802.16e [8]	Scanning phase improved considering QoS "Adaptive algorithm" suggested for calculation of scanning related parameters	Scanning and data transmission are mutually exclusive operations If vehicle is scanning its neighboring BS, serving BS buffers the data packets destined to vehicle Vehicle capable of running ACS algorithm
Hard Handoff Scheme Exploiting Uplink and Downlink Signals in IEEE 802.16e Systems [15]	Serving BS decides handoff based on downlink and uplink signal measurements Predefined thresholds are used to make handoff decision Neighboring BSs help in uplink signal measurements and MS helps in downlink signal measurements	BS capable of measuring uplink signal strength from vehicles Vehicles capable of measuring downlink signal strength from BS

Table 4(b). Comparison of Layer-2 HO schemes with respect to IVN (Cont'd)

Layer-2 Schemes	Advantages	Disadvantages	Observations, Suggestions or Comments
Fast Handover Scheme for Real-Time Downlink Services in IEEE802.16e BWA System [11]	Real-time data will experience less latency and hence better QoS Packet loss reduced Advantageous to receiver vehicles.	Not suited for vehicles that have heavy traffic on both up and down links	Allow both uplink and downlink communication between vehicles before handoff is complete
Fast Handover Algorithm for IEEE 802.16e Broadband Wireless Access System [14]	Serving BS scan only suitable potential target BSs. Throughput will not drop during HO; less HO operation time	Vehicle needs to be intelligent enough to determine the neighboring BS's that are suited to be target BS.	Vehicle can determine potential target BS using signal pattern recognition, signal strength etc
Adaptive Channel Scanning (ACS) for IEEE 802.16e [8]	ACS estimates total time required for scanning, number of scan iterations and duration of each scan interval based on QoS This information helps vehicle to avoid buffering of data packets at serving BS thereby reducing packet delay.	Overhead of calculating total time required for scanning, number of scan iterations, and duration of each scan interval, which are dependent on the level of association.	Allocate scanning period when data traffic is low so that not many data packets are buffered at the serving BS.
Hard Handoff Scheme Exploiting Uplink and Downlink Signals [15]	HO decisions depend on both uplink and downlink network load Suited to vehicles using bi-directional communication	BS's have high responsibility of making HO decision	Suited for VANET where communications are mostly bi-directional

5. IEEE 802.16 Layer-3 Handoff Schemes

In this section, we describe five layer-3 handoff schemes proposed for IEEE 802.16. For each scheme, a feasibility study of the scheme for supporting long-range V2I and V2R communications in ITS is also included. Finally, at the end of this section, similarly to the previous section, we compare these five schemes in the form of tables, in terms of basic scheme handoff, assumptions and/or requirements specific in VANET, advantages, disadvantages, and suggestions. As in the last section, we also use "IVN" (Intelligent Vehicular Networks) to represent "VANET for supporting ITS".

5.1. Mobile Ipv6 Fast Handovers Over IEEE 802.16e Networks [16]

5.1.1. Scheme Description

This scheme, proposed by Jang, et al, aims to reduce handoff latency by initiating steps for an impending handoff in advance. It is based on Mobile IPv6 Fast Handover protocol (FMIPv6), an IETF draft [17]. Under this scheme, the handoff procedure of FMIPv6 has been reused to suit the 802.16 link layer technology. It uses the primitives proposed by IEEE 802.21 for performing a handoff. In this scheme, there are two modes of handoff, called predictive mode and reactive mode. The distinction between these two modes will be described later .We will initially describe the predictive mode of handoff illustrated in Figure 6. Any reference to step numbers is with reference to Figure 6.

After the 802.16e handoff mechanism (steps 1-2), the L2 (layer 2) of MN notifies its L3 (layer 3) that a new link is detected through the LD (Link_Detected) primitive. In order to find those ARs to which potential target BSs are attached, the MN and PAR (previous access router) exchange RtSolPr (router solicitation for proxy advertisement) and PrRtAdv (proxy router advertisement) messages (step 4).

When L2 decides to perform a handoff (through the exchange of handoff request-response mechanism), it notifies L3 through the LHI (Link_Handover_Imminent) primitive (step 6). Then, the L3 of MN sends FBU (fast binding update) to the PAR (step 7). PAR sets up a tunnel with NAR (new access router) through the exchange of HI (Handover Initiate) and HACK (Handover Acknowledge) messages. Then the PAR sends an FBACK (fast binding acknowledgement) to the MN.

Note that if the MN receives FBACK before sending MOB_HO-IND, it runs in Predictive mode.

To enable the MN to postpone sending handoff indication until FBACK is received, the HC (Handover_Commit) primitive is used. Once FBACK is received, L3 issues this signal to L2, so that it can send the handoff indication. After the MN sends MOB_HO-IND, it stops monitoring traffic from the serving BS. As the MN performs 802.16e network entry with the target BS (step 9), the PAR forwards packets to the NAR through the tunnel established in step 7. Hence, when the handoff process of L2, namely, 802.16e is completed, the IP (internet protocol) packets are already in transit to the NAR with which the target BS is connected. After the MN completes 802.16e handoff, L2 signals L3 to send FNA (Fast Neighbor Advertisement) through the Link_Up (LUP) primitive. L3 of the MN sends FNA to the NAR, which verifies that the care of address is acceptable and then forwards the packets to the MN.

If the MN does not receive FBACK before sending MOB_HO-IND, it runs in Reactive mode. In this case, after the MN completes 802.16e handoff, it sends FNA with an FBU inside (step 10). The NAR forwards this FBU to the PAR and establishes the tunnel (step 11). The PAR then forwards the data, which is forwarded to the MN by the NAR.

Thus, this mechanism tries to make handoff fast by coinciding L3 and L2 procedures. In the absence of this, after the MN's L2 completes 802.16e link layer handoff, the L3 handoff would be initiated. This undoubtedly would cause some delay. By initiating L3 handoff simultaneously with L2 handoff, a seamless connectivity can be achieved.

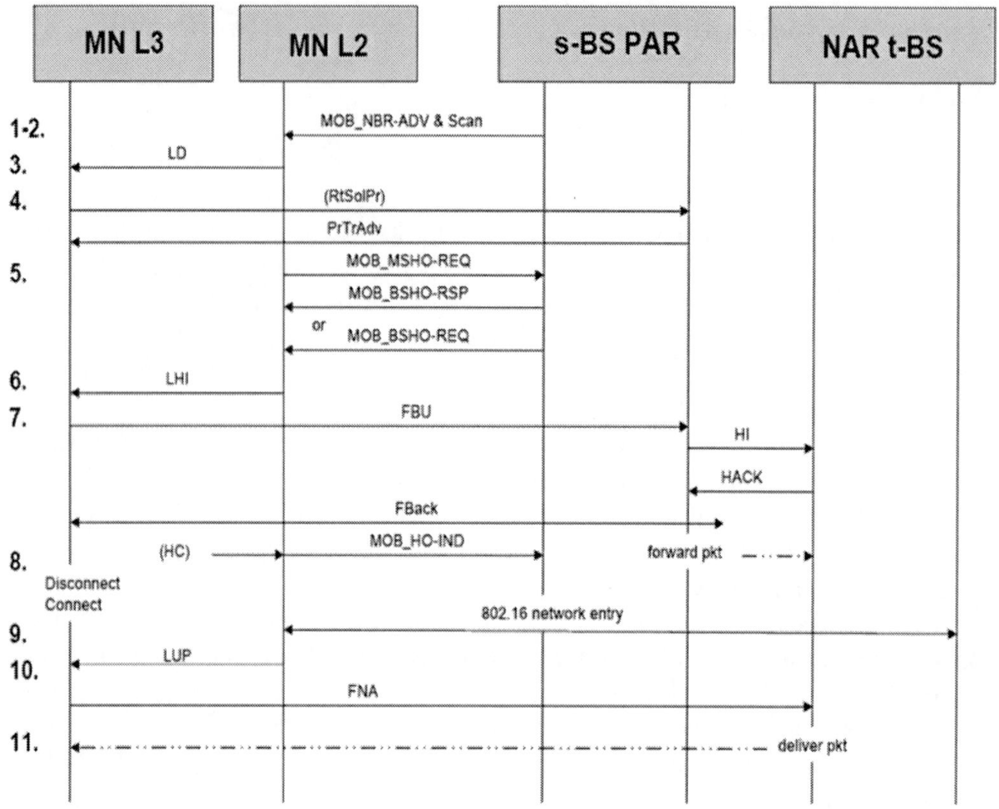

Figure 6. Predictive Mode (Jang et al).

5.1.2. Feasibility of Supporting IVN

This scheme may reduce handoff latency as it triggers HO at both layers 2 and 3 simultaneously. The scheme expects access routers to be installed with roadside units. BS should be capable of scanning neighboring BS. Also, BS must be compatible with Mobile IPv6. In order to implement this scheme in IVN, the capability of BS can be increased to act as Access Routers. Since IVN is in its initial stage, making IVN Mobile IPv6 compliant is feasible.

5.2. A Seamless Handover Mechanism for IEEE 802.16e Broadband Wireless Access [12]

5.2.1. Scheme Description

This scheme, proposed by Kim, et al, aims to achieve loss-free handoff for intra-domain mobility [12]. This in turn would improve TCP (Transmission Control Protocol) performance.

Re-transmission due to lost packets can be avoided. This scheme, illustrated in Figure 7, is called LPM or Last Packet Marking.

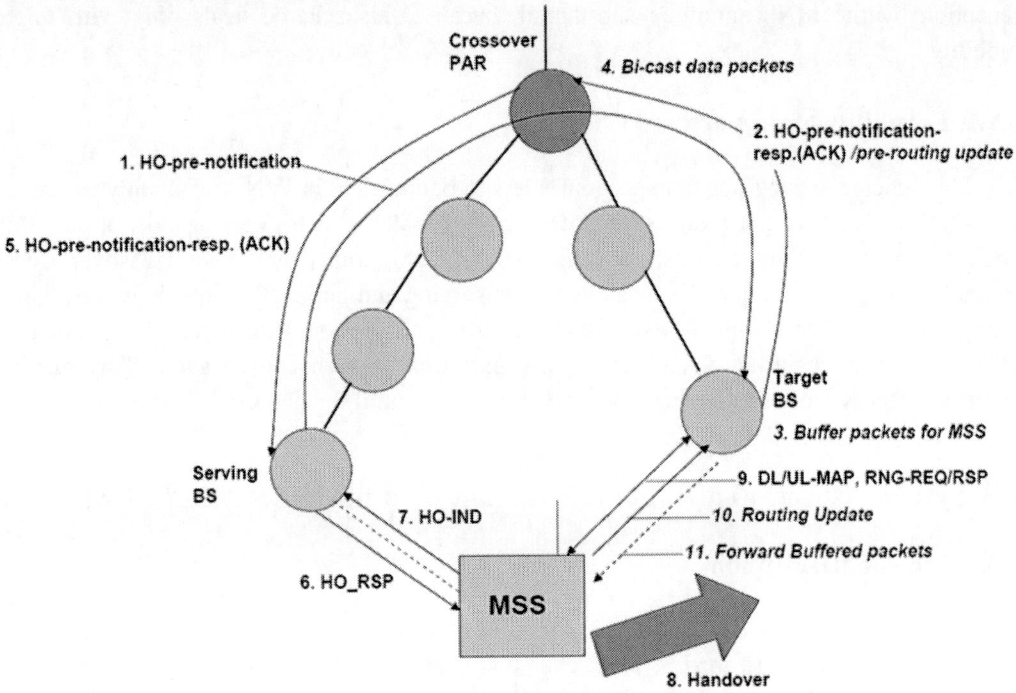

Figure 7. Last Packet Marking (Kim et al).

When the MN sends the handoff request to the serving BS, the serving BS sends a handoff pre-notification (step 1) to potential target BSs, which may respond positively or negatively. This is similar to the 802.16e standard. When a target BS responds with acknowledgement (positive), shown in step 2, a pre-routing update is sent and all PARs (Packet Access Router) in the path from the target BS to the crossover PAR update their cache. The crossover PAR is the root of the tree between the serving BS and the target BS.

From then, the crossover PAR bicasts data (step 4) destined to the MN. This means that the crossover PAR sends data (with MS as destination) to both the serving BS and the target BS. When the MN completes handoff with the target BS (steps 8 and 9), it sends a routing update, which is an indicator to stop bicasting to the crossover PAR (step 10). The target BS forwards the buffered data to the MS (step 11).

Thus, this scheme tries to proceed with L2 and L3 handoff simultaneously and thus reduce handoff time. It also employs bicasting, so that when handoff is complete the MS will have data destined for it at the target BS. The advantage of this approach is that the MN achieves seamless mobility. By the time L2 handoff is complete, it has packets destined for it already in the target BS. As the authors pointed out, in the worst case, however, many target BSs may respond with a positive acknowledgement and so the same data may be bicasted to several base stations until the routing update timeout expires.

To avoid unnecessary data transmission, one approach could be to maintain a buffer at the crossover PAR (attached to the backbone) once a handoff pre-notification response is sent. The data has to be buffered at the crossover PAR from the time the pre-routing update is received to the time a routing update is received from MN after handoff. After this, the crossover PAR could forward the buffered data. Note that the extra delay incurred in packet reception by the MN cannot be substantial, because this scheme deals only with micro mobility.

5.2.2. Feasibility of supporting IVN

This scheme expects separate access routers to be installed in IVN. It is mainly useful for intra-domain handover. It reduces handoff latency in IVN as it triggers handoff at data-link and network layers simultaneously. It achieves seamless mobility by making crossover access router bicasting data destined for vehicle to both serving and target BS. Note, however, that if many BS respond to Handoff-pre-notification, crossover Access Router may bi-cast data to all of them unnecessarily. Such unnecessary data transmission can be avoided by making crossover access routers buffer data destined for vehicle until the handoff is complete.

5.3. System Aspects and Handover Management for IEEE 802.16e [18]

5.3.1. Scheme Description

This scheme, proposed by Das et al, describes a complete mobility solution covering L2 and L3. It uses Mobile IP for L3. This scheme is depicted in Figure 8. Any reference to step numbers is with reference to Figure 8.

After BS sends handoff response with list of potential target BSs to MS, the source ASN-GW (access service network gateway) establishes tunnels to the ASN-GWs associated with potential target BSs (step 3). Once the target BS is decided, all the other tunnels are torn down (step 6) and the MAC state is migrated (step 7). When the MN completes handoff (steps 8 and 9), it receives data seamlessly as the source ASN-GW has transmitted data to the target ASN-GW through an established tunnel (step 10). Next, the foreign agent should be migrated from source ASN-GW to target ASN-GW through MIP (mobile IP) registration.

The basic method used in this mechanism is bicasting and tunneling data. This approach and the scheme - "A Seamless Handover Mechanism for IEEE 802.16e Broadband Wireless Access" [11] are comparable, in the sense that both try to bicast the data to the potential target BS, anticipating the impending handoff. By this, they try to ensure that the mobile node has data flow seamlessly when it completes handoff. This scheme provides a complete solution as it covers macro-mobility also and includes foreign agent migration. Thus, their scheme is a comprehensive handoff solution.

In the worst case, the serving ASN-GW may set up tunnels to a large number of target ASN-GWs, unnecessarily bicasting the same information on multiple tunnels. Thus, a lot of bandwidth is wasted and cost of extra tunnel establishments is an overhead to the serving ASN-GW. The 802.16e standard says that it is recommended but not required that the MS conducts a handoff to a target BS recommended by the serving BS in handoff response. Thus, if the MS decides to handoff to another BS not recommended in handoff response, the tunnels established by this mechanism represent an overhead.

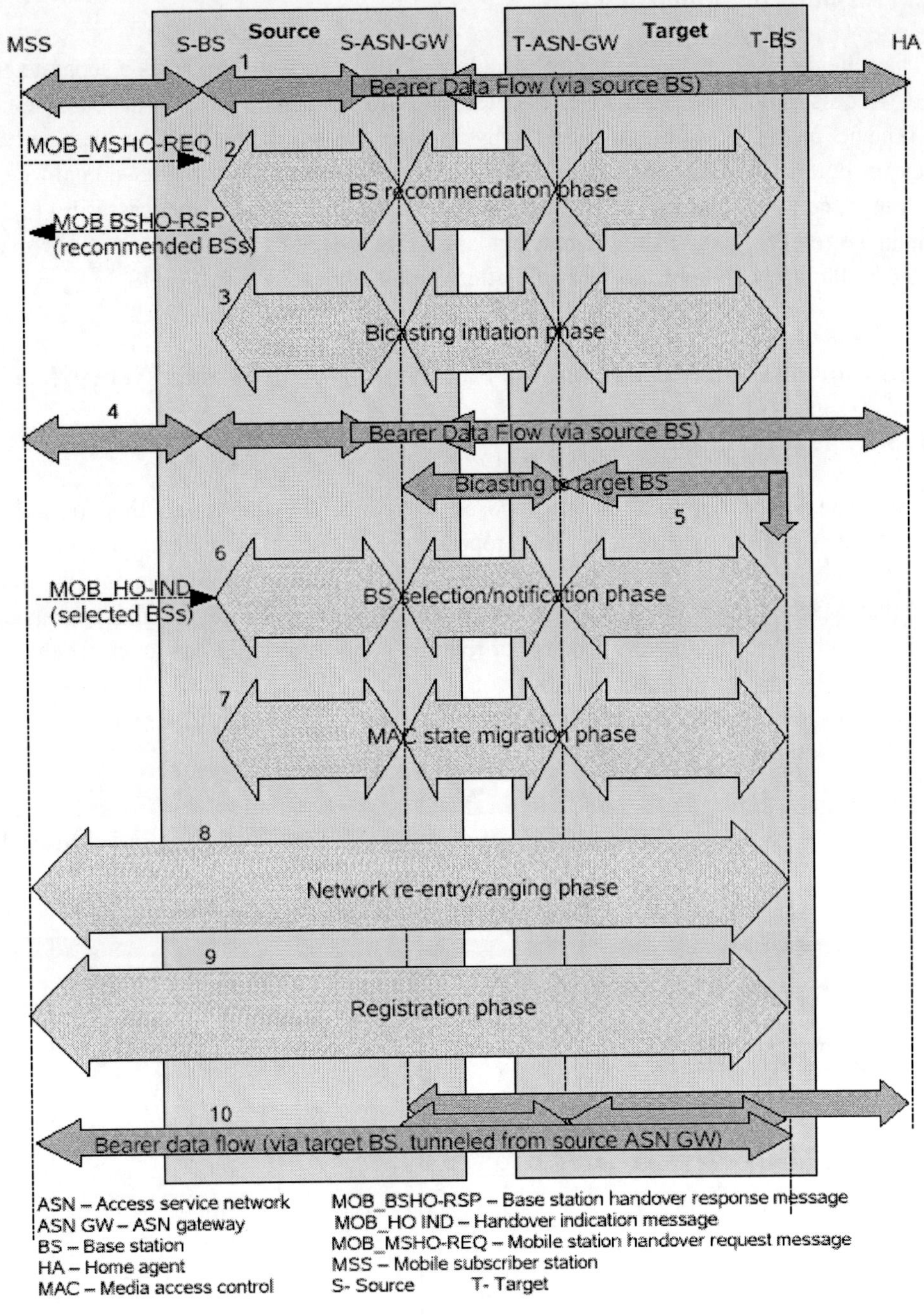

Figure 8. MAC migration and data flow (Das et al).

One approach to avoid some of the overhead would be to require the source ASN-GW to establish tunnels with target ASN-GWs, but not bicast data until MS sends MOB_HO-IND to BS. The 802.16e standard says that when MOB_HO-IND is sent, the MS is not responsible for monitoring downlink traffic from serving BS (after expiration of resource retain timer) [7]. The advantage of this approach is that unnecessary transmission of data on multiple links is avoided until a handoff indication is sent by MS (which has the target BS ID).

5.3.2. Feasibility of Supporting IVN

The scheme assumes that Mobile IP is adopted for IVN. It also expects a separate entity, called "Access Service Network Gateway" (ASNG), to be installed at BS (or roadside units). The scheme provides seamless mobility by making ASNG bicasting data destined for a vehicle to both its Serving and target BS. It is therefore suitable for inter-domain handoff (long-range communications). However, ASNG may bicast data unnecessarily to many potential target BS. Establishing tunnels to non-target BSs is an overhead. This may be avoided by having ASNG bicast data only after vehicle chooses its Target BS.

5.4 Pre-coordination Mechanism for Fast Handover in Wimax Networks [19]

5.4.1. Scheme Description

This scheme, by Chen et al, is proposed to reduce the disruption time of handover especially for real-time applications. The proposed pre-coordination scheme predicts the time of handoff by measuring the distance between Mobile Subscriber Station (MSS) and BS. (Note that MSS is used in this paper for MS.) Then the scheme pre-allocates available resources required for handover. It does not require, as in traditional handover schemes, BSs to periodically transmit characteristics of neighbor BSs to MSS through neighbor advertisement messages during potential handover. The scheme therefore reduces the wastes of network resources.

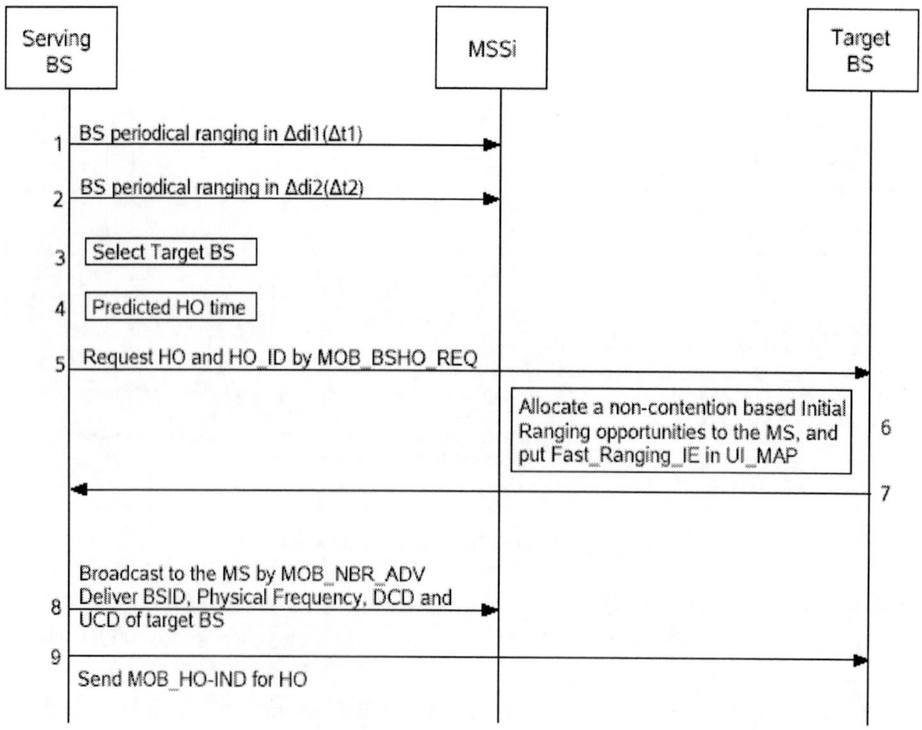

Figure 9. Pre-coordination Handover Mechanism (Chen et al).

The pre-coordination scheme is depicted in Figure 9. In this scheme, in the presence of MSS, only Serving and target BS participate in the handover scheme. During a potential handover, serving BS determines the target BS and predicts handover time (T_{HO}). Then, the serving BS requests the pre-coordination from the target BS (MOB_BSHO-REQ). It periodically updates T_{HO} based on the response (MOB_BSHO-RSP) from target BS. This response contains information such as target BS ID, Physical frequency etc. serving BS also updates T_{HO} based on the position, distance and direction of MSS. When MSS requests handoff, serving BS broadcasts information received from target BS to MSS. Then serving BS sends handoff indication message (MOB_HO-IND) to target BS.

5.4.2. Feasibility of Supporting IVN

In order to implement this scheme in IVN, we should be able to measure vehicle position and distance between vehicle and base station. We should also be able to predict direction of vehicle movement. This scheme therefore may make use of some previous works that assume that vehicle movement, speed, and direction may be predicted [13, 20]. Note that, in urban areas or highways, the direction of vehicle movements can be easily predicted.

5.5. Fast Handover Scheme for Supporting Network Mobility in IEEE 802.16e BWA System [21]

5.5.1. Scheme Description

This scheme, proposed by Zhong et al, aims to reduce handoff latency in the Network Mobility (NEMO) 802.16 network. Mainly, it consists of three stages: neighbor discovery, advance preparation, and handover execution. It uses two types of triggers, Link Going-Down (LGD) and Link Switch Complete (LSC), to co-ordinate Layer-2 and Layer-3 handover processes. The handover scheme is depicted in Figure 10.

In the first stage, the serving BS periodically broadcasts the link-layer neighbor advertisement message (MOB_NBR-ADV). This message carries channel information of the neighbor BS, the number of neighboring BSs, their BSIDs, their corresponding Access Router (AR) IP addresses, network prefix, etc. The MS collects this information for future handovers. At the same time, MS monitors the serving BS radio signal strength and if the signal strength crosses a pre-determined threshold, MS requests serving BS to scan neighboring BSs and send their information. Based on the scanning result, the MS may trigger a handover.

In the advance preparation stage, the MS carries out handover process in both network and link layers simultaneously to save time. MS sends handoff request (MOB_MSHO-REQ) to the serving BS and Fast Binding Update (FBU) to the Home Agent (HA). Upon receiving the FBU, the HA sets up a tunnel with the new access router (NAR) by sending the handover initiate (HI) message. NAR checks the validity of the NCoA with a duplicate address detection (DAD) procedure and then responds to MS with handover acknowledgment (Hack) message.

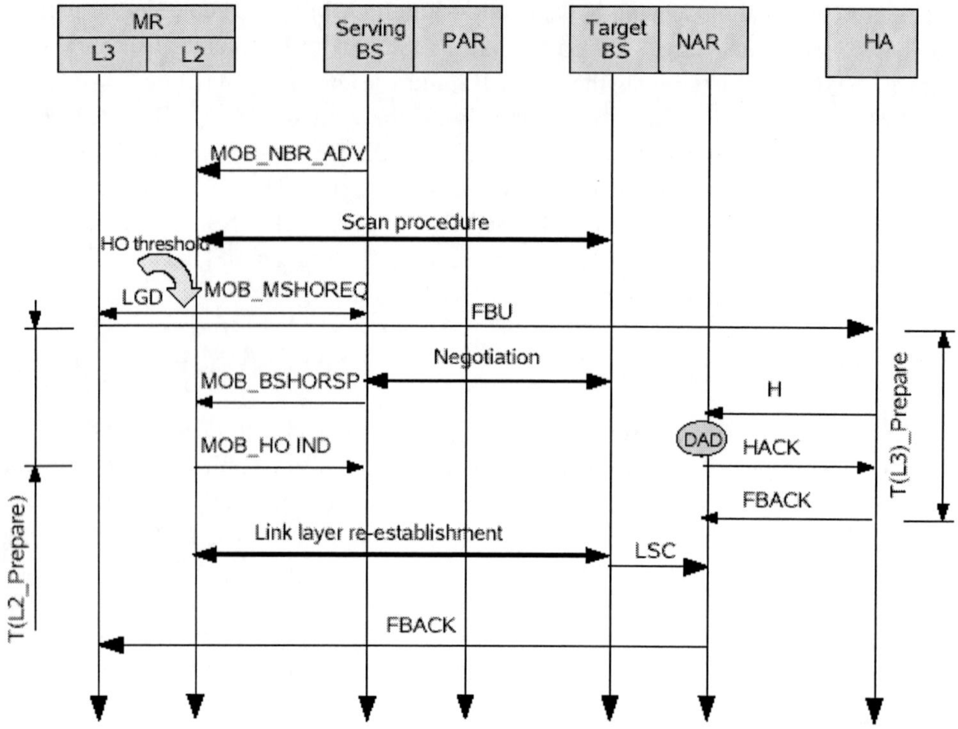

Figure 10. NEMO basic support protocol [21].

In the handover execution stage, the MS sends handoff indication (MOB_HO-IND) message to serving BS. After the IEEE 802.16e network re-entry process, target BS sends LSC trigger to the New AR (NAR) informing NAR to send Fast Binding Acknowledgment (FBACK) message to MS. Then MS continues with normal operation.

In this scheme, HA directly tunnels the data to MS via NAR without using PAR. This eliminates a round trip and encapsulation by the PAR, which is defined in 802.16e Fast Handoff (FH80216e). Also, HA even handle the authorization as it responds with FBACK message. This scheme utilizes FH80216e message format. When the MS is just a node, FH80216e sends its FBU to Previous AR, whereas when it is a mobile router, FNEMO sends its FBU to the corresponding HA directly.

5.5.2. Feasibility to Support IVN

To apply to IVN, this scheme assumes both Mobile IP and Network Mobility (NEMO) structure in IVN. The scheme does not require BS or vehicles to take the responsibility of triggering handover process. Instead, mobile router takes handoff decision at accurate time. The function of mobile routers may also be implemented in the roadside unit or BS within the IVN structure.

5.6. Summary of 802.16 Layer-3 Handoff Mechanisms

This section summarizes and compares the five layer-3 HO schemes presented in this section with a table. For better readability, as in Section 4, we separate the table into two parts. Table 5(a) provides a comparative overview of each basic scheme and its assumptions and/or requirements for IVN. Table 5(b) summarizes their advantages, disadvantages, and provides some suggestions and comments.

Table 5(a). Comparison of Layer-3 802.16 HO Schemes with respect to IVN

Scheme	Scheme Overview	Assumptions and/or Requirements for IVN
Mobile IPv6 Fast Handovers over IEEE 802.16e Networks [16]	Scheme based on FMIPv6 (Koodli, 2005) for 802.16e IEEE 802.21 primitives used Defines two modes: Predictive and Reactive Tunneling used to transmit packets during HO	BS are capable of scanning neighboring BS BS are compatible with Mobile IPv6 Access routers needed (may be in BS or roadside units)
A Seamless Handover Mechanism for IEEE 802.16e Broadband Wireless Access [12]	Provides seamless data exchange through bicasting Minimizes packet loss by making packets available at target BS during HO	Used for intra-domain handover Access routers needed (may be in BS or roadside units)
System Aspects and Handover Management for IEEE 802.16e [18]	Tunneling and Bicasting used to make data available at target BS during HO Migration of Foreign Agent subsequent to HO provides a complete solution to mobility	VANET uses Mobile IP Access Service Network Gateway (ASNG) needed
Pre-Coordination Mechanism for Fast Handover in WiMAX Networks [19]	Predicts the time of handover by measuring the distance between Mobile Subscriber Station (MSS) and BS. Pre-allocates available resources required	Distance between vehicle and BS can be calculated Direction of vehicle movement can be predicted
Fast handover scheme for supporting network mobility in IEEE 802.16e BWA system [21]	Neighbor Discovery: MS learns about neighboring BSs from Serving BS. Advance Preparation: MS initiates handover in both layers 2 and 3 simultaneously. HO Execution: MS moves from Serving BS to Target BS.	IVN uses Mobile IP and NEMO with mobile routers

Table 5(b). Comparison of Layer-3 802.16 HO Schemes with respect to IVN

Layer-3 Schemes	Advantages	Disadvantages	Observations, Suggestions, and Comments
Mobile IPv6 Fast Handovers over IEEE 802.16e Networks [16]	Complete solution (both L2 L3) thereby reducing HO latency and packet loss Uses 802.21 triggers and hence for heterogeneous HO Takes care of both predictive and reactive mode	Need for installing access routers (or integrated with roadside units or BS)	Enhance BS so that they can also act as access routers Need to be Mobile IPv6-compliant
A seamless handover mechanism for IEEE 802.16e Broadband Wireless Access [12]	Reduce HO latency by triggering HO at both layers 2 and 3 Achieves seamless mobility by having crossover access router bicasting data to both serving and target BS.	If many BS respond to HO-pre-notification, crossover access router may bi-cast data to all of them unnecessarily. Need for installing access routers (or integrated with roadside units or BS)	To avoid unnecessary data transmission: crossover access router may buffer data until HO is complete. Enhance BS to function as access routers.
System Aspects and Handover Management for IEEE 802.16e [18]	Provides inter-domain HO Complete solution including migration of foreign agent Provides seamless mobility by making ASNG bicast data to both Serving and target BS.	ASNG could unnecessarily bicast data to many potential target BS. Overhead of tunnel establishments to non-target BSs.	Enhance BS to function as ASNG gateway Make ASNG bicast data only after vehicle chose its target BS
Pre-Coordination Mechanism for Fast Handover in WiMAX Networks [19]	HO disruption reduced by predicting HO time Reduces message complexity by avoiding neighbor advertisements Pre-allocation ensures resources available during HO	Distance measured between vehicle and BS may not be accurate as vehicles moves with high speed	Direction of vehicle movement and distance between vehicle and potential target BSs may be estimated using signal strength In urban areas or highways direction of motion of vehicles can be easily predicted
Fast handover scheme for supporting network mobility in IEEE 802.16e BWA system [21]	BS and vehicles need not be responsible for triggering HO Fast, accurate HO by mobile router	Implementing mobile routers adds to complexity of VANET	BS may be enhanced by mobile router functionality

6. Conclusion

The IEEE 802.16 WiMAX technology is of paramount importance as it provides long-range broadband wireless access. Many believe that it would also provide long-range communication for ITS. In this chapter, we focus on mobility management of vehicular networks for supporting ITS. A comprehensive study of IEEE 802.16 handoff schemes is presented, analyzing their feasibility of supporting V2R and V2I in ITS. On-going and future works include enhancement of some of these schemes for supporting QoS and security in ITS [22 - 24], and applying some of them in 802.11p for short-range communication support of V2V in ITS.

References

[1] Jiang, D. and Delgrossi, L. (2008, May). Towards an International Standard for Wireless Access in Vehicular Environments. *Proceedings of IEEE Vehicular Technology Conference*, 2036-2040.

[2] Choi, N., Choi, S., Seokt, Y., Kwon, T., and Choi, Y. (2007, May). A Solicitation-based IEEE 802. ip MAC Protocol for Roadside to Vehicular Networks. *Proceedings of IEEE - Mobile Networking for Vehicular Environments*, 91 – 96.

[3] *The IEEE 802.16 Working Group on Broadband Wireless Access Standards*. Retrieved June 28, 2008, from http://wirelessman.org/

[4] IEEE 802.16 Working Group on Broadband Wireless Access. (2004). *Part 16: Air Interface for Fixed Broadband Wireless Access Systems*. Retrieved January 22, 2007, from http://standards.ieee.org/getieee802/download/802.16-2004.pdf

[5] WiMAX Forum (2005). *Can WiMAX Address Your Applications?* Retrieved April 29, 2007, from http://www.wimaxforum.org/technology/downloads/ Can_WiMAX_Address_Your_Applications_final.pdf

[6] *Enabling Mobility in WiMAX Networks*, Retrieved June 04, 2008, from http://www.wimax.com/research/whitepapers/WiMAX_07Nov06.pdf

[7] IEEE 802.16 Working Group on Broadband Wireless Access. (2005). *Part 16: Air Interface for Fixed and Mobile Broadband Wireless Access Systems Amendment* **2**: *Physical and Medium Access Control Layers for Combined Fixed and Mobile Operation in Licensed Bands and Corrigendum 1*. Retrieved April 9, 2007 from http://standards.ieee.org/getieee802/download/802.16e-2005.pdf

[8] Rouil, R. and Golmie, N. (2006). Adaptive Channel Scanning for IEEE 802.16e. Military Communications Conference, 1- 6.

[9] Wong, D. and Cox, D. (2000, July). A pattern recognition system for handoff algorithms. *IEEE JSAC*, **18**, 1301-1312.

[10] Wong, D. and Cox, D. (2001, March). Two-state pattern-recognition handoffs for corner-turning situations. *IEEE Transactions on Vehicular Technology*, 354 – 363.

[11] Choi, S., Hwang, G., Kwon, T., Lim, A., and Cho, D. (2005). Fast Handover Scheme for Real-Time Downlink Services in IEEE 802.16e BWA System. *IEEE* **61***st Vehicular Technology Conference,* **3**, 2028 - 2032.

[12] Kim, K., Kim, C., and Kim, T. (2005). A Seamless Handover Mechanism for IEEE 802.16e Broadband Wireless Access. *Proceedings of 5th International Conference on*

Computational Science - ICCS 2005 Atlanta, GA, Lecture Notes in Computer Science, **3515**(II), 527 – 534, Springer Berlin/Heidelberg.

[13] Vijayan, R. and Holtzman, J., M. (1993, August). A model for analyzing handoff algorithms. *IEEE Trans. Vehicular Technology,* **43**, 351 – 356.

[14] Lee, D., H., Kyamakya, K., and Umondi, J., P. (2006). Fast Handover Algorithm for IEEE 802.16e Broadband Wireless Access System. *2006 1st International Symposium on Wireless Pervasive Computing*, 1 – 6, IEEE Press.

[15] Cho, S., Kwun, J., Park, C., Cheon, J., Lee, O., and Kim., K. (2006). Hard Handoff Scheme Exploiting Uplink and Downlink Signals in IEEE 802.16e Systems. *Vehicular Technology Conference*, 2006, 3, 1236 – 1240.

[16] Jang, H., Jee, J., Han, Y., Park, S., D., and Cha, J. (2007). *Mobile IPv6 Fast Handovers over IEEE 802.16e Networks*. IETF Internet Draft. Retrieved September 28, 2007, from http://tools.ietf.org/html/draft-ietf-mipshop-fh80216e-03

[17] Koodli, R. (2005). *Fast Handovers for Mobile IPv6.* IETF Internet Draft. Retrieved September 28, 2007, from http://tools.ietf.org/html/rfc4068

[18] Das, S., Klein, T., Rajkumar, A., Rangarajan, S., Turner, M., and Viswanathan, H. (2006). System Aspects and Handover Management for IEEE 802.16e. *Bell Labs Technical Journal*, **11**(1), 123 - 142.

[19] Chen, J., Wang, C., and Lee, J. (2007, August). Pre-Coordination Mechanism for Fast Handover in WiMAX Networks Pre-Coordination Mechanism for Fast Handover in WiMAX Networks. *The 2nd International Conference on Wireless Broadband and Ultra Wideband Communications* (AusWireless 2007), IEEE Computer Society, pp. 15.

[20] Kim, S., K., Kang, C., G., and Kim, K., S. (2004, Fall). An adaptive handover decision algorithm based on the estimating mobility from signal strength measurements. *Proceedings of Vehicular Technology Conference - IEEE 60th, 2, 1004* – 1008.

[21] Zhong, L., Liu, F., Wang, X., and Ji, Y. (2007, September). Fast Handover Scheme for Supporting Network Mobility in IEEE 802.16e BWA System. *International Conference on Wireless Communications, Networking and Mobile Computing, WiCom 2007*, 1757 – 1760, IEEE Press.

[22] Chellappan, B., Moh, T., and Moh, M (2009, April). Supporting Multiple Quality-of-Service Classes in IEEE 802.16e Handoff. *Proceedings of the International Conference of Computing, in Engineering, Science and Information,*" Fullerton, CA, Apr. 2009.

[23] Kim, H. J, & Moh, M. (2009, Aug). Performance Study of FMIPv6-based Cross-layer WiMAX Handover Scheme For Supporting VoIP Service. *IEEE Pacific Rim Conf. on Communications, Computers and Signal Proc. (PACRIM),* Victoria, B.C., Aug 2009.

[24] Huynh, P.Q., Jangyodsuk, P. and Moh, M (2010, March). Supporting Video Streaming and VoIP over Mobile WiMAX Networks by enhanced FMIPv6-based Handover. *Accepted to be presented at the Fourth International Conference on Information Systems, Technology and Management (ICISTM-10), to be held in Bangkok, March* 2010.

In: Wireless Technologies in Intelligent Transportation... ISBN 978-1-60741-588-6
Editors: Ming-Tuo Zhou et al, pp. 205-222

Chapter 8

BROADCAST TECHNIQUES FOR VEHICULAR AD HOC NETWORKS

***Anis Laouiti*[1,*]*, Paul Muhlethaler*[2,†] *and Yasser Toor*[2,‡]**
[1]Telecom SudParis
Every France
[2]INRIA Paris Rocquencourt
Le Chesnay France

Abstract

VANET applications can be classified into two categories: user applications and safety applications. Both of them must rely on efficient vehicular ad hoc routing protocols in order to deliver the information to the corresponding destination(s). Many possible VANET applications, e.g., accident avoidance or distributed traffic management applications, do not need explicit unicast data exchange, but locally aggregate and process the data broadcast by other vehicles. One-to-many communications is of great importance for VANETs since vehicles are likely to exchange data by diffusion rather than in a unicast manner. Many broadcast techniques have already been proposed for MANETs, but classical MANET flooding techniques may not be suitable for VANETs and do not take advantage of the additional information available in vehicles. In this chapter we investigate various broadcast techniques that may be used in VANETs.

1. Introduction

Intelligent Transportation Systems are a subject of intense interest nowadays because they can enhance road safety and provide in-vehicle entertainment for passengers. This has become important as an increasing number of people are able to afford cars and travel. Many government organizations and car manufacturers are carrying out extensive research geared towards developing such transportation systems. Examples include Network On Wheels (NOW) [3], Car2Car Consortium [1], European Telecommunications Standards

*E-mail address: Anis.Laouiti@it-sudparis.eu
†E-mail address: Paul.Muhlethaler@inria.fr
‡E-mail address: Yasser.Toor@inria.fr

Institute (ETSI) [2], Partners for Advanced Transit and Highways (PATH) [4], etc. This work has be greatly aided by the recent progress in wireless telecommunications, especially Mobile Ad hoc Networks (MANETs) since they can provide the necessary wireless communications capability.

Mobile ad hoc networks enable mobile nodes to communicate with each other without the need for a preexisting infrastructure in a dynamic, random and multihop topology. Since MANETs are autonomous networks, the mobile nodes can be located on anything that moves, e.g., airplanes, ships, trucks or cars. Intelligent transportation systems need a special kind of MANET called **V**ehicular **A**d Hoc **NET**work (VANETs) which are formed between vehicles using **V**ehicle-to-**V**ehicle (V2V) communication or between vehicles and an infrastructure using **V**ehicle-to-**I**nfrastructure (V2I) communication (Figure 1).

VANETs have several properties [13] which distinguish them from other MANETs. Although VANETs do have dynamic topologies, they are not completely random. The movement of nodes in a VANET is relatively predictable because it is restricted by the roads on which the vehicles travel. This has both advantages and disadvantages for applications and routing protocols. The lack of available bandwidth is also exacerbated due to congestion and contention problems at intersections and traffic jams. The presence of buildings beside the roads, especially in an urban environment, increases fading effects. VANETs also have the potential to grow to a very large scale. This means that the probability of network partitions is higher and end-to-end connectivity is lower [15]. The good news is that, unlike other MANET nodes, VANET nodes are not constrained by computing power or energy since they are located on vehicles and normally have access to both.

These VANET characteristics, especially their high mobility, dynamic topologies and limited bandwidth, make reliable information exchange between vehicles very difficult. Since the radio range of each node is limited, multi-hop dissemination protocols are required to allow information from one node to reach nodes that are beyond its range. Unfortunately, current conventional MANET protocols [5], [24], [18], etc. do not take these characteristics into account. Furthermore, some VANET applications e.g., collision avoidance, co-operative driving, etc., have very strict requirements and can fail with possibly fatal results if these requirements are not met. Therefore, it is necessary either to modify existing protocols or to develop completely new ones that take advantage of the characterstics of VANETs and fulfill the requirements of VANET applications.

In the rest of this chapter, we explore the ideas presented above, particularly focusing on broadcast schemes in VANETs, their classification and explanation with the aim of giving the reader a thorough grounding in this important topic. In section 2. we give an overview of VANET applications, their characteristics and the main difficulties in designing broadcast protocols. In section 3. we categorize different broadcast protocols and describe some example protocols in sections 4.,5. and 6.. We present other approaches to data dissemination in section 7. before concluding in section 8..

2. Vehicular Ad Hoc Networks

We believe that VANETs are likely to revolutionize our traveling habits. Several applications have been proposed that will make traveling on the roads safer as well as more

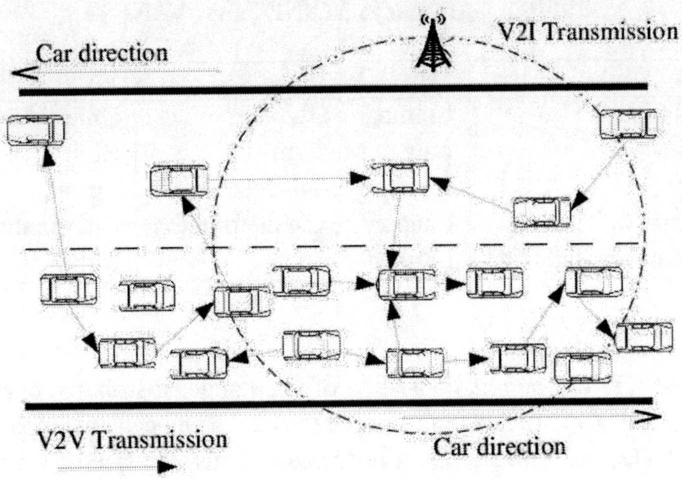

Figure 1. A simple VANET showing inter-vehicle communication .

enjoyable. In this section we focus on these VANET applications and how VANET characteristics affect them.

2.1. VANET Applications

VANET applications can be divided into two major categories: User and Safety applications. User applications provide value-added services, e.g., entertainment, news, advertisements, Internet access, etc. These applications can be unicast or multicast-oriented. Additionally, they can be delay-dependent as well as delay-independent.

Safety applications, on the other hand, increase overall safety on the roads by increasing the situational awareness of drivers and reducing their reaction time. These applications are likely to be deployed first and will have the greatest impact for normal road users e.g., post-crash warning, emergency braking, fog/rain updates, etc. In such applications the information must be diffused to the subset of vehicles concerned with as little delay as possible. Therefore, they are considered as delay-dependent and broadcast-oriented, which is why developing well-designed, efficient and fast broadcast protocols for VANETs has become a priority research topic and the focus of this chapter.

In order to better understand the challenges faced in designing broadcast protocols for VANET applications, in the next section we will look at the characteristics of VANETs that make them different from MANETs.

2.2. VANET Characteristics

A VANET is a special type of MANET but there are several important differences between them. We have to consider these differences because they affect any application or broadcast protocol that is used in a VANET.

Table 1. Summary: MANETs vs. VANETs

	MANETs	VANETs
Area and Size	Limited	Unlimited
Mobility	Low to Medium	High
Topology	Random, 2-dimensional	Linear
Node Density	Random, Even distribution	Uneven distribution
Available power and CPU	Limited	Unlimited

1. **Size:** MANETs can be big, consisting of 1000 nodes distributed over a square kilometer but VANETs are not at all limited by area. They can encompass large areas in which vehicles can move without restriction. Thus, the size of a VANET can vary from a small number of cars forming a platoon up to a city-wide VANET under the local traffic authority. Since VANETs are not bounded by area, they can consist of up to millions of vehicles.

2. **Mobility:** MANETs usually have low to medium mobility e.g., sensor networks are practically stationary. When MANETs do have mobility, it is often a random waypoint pattern. This means that the nodes in a MANET are usually spread out whereas in VANETs, they are concentrated on the roads and are highly mobile. They might be traveling at relative speeds of up to 260 kph. This means that nodes in a VANET usually have very little time to communicate with other nodes before they are out range. Furthermore, the network topology changes quickly and frequently.

3. **Network Topology:** VANET network topology is usually linear, i.e., restricted to roads, which is quite different from MANET networks, which are distributed evenly over the network and more often than not, have multiple one-hop neighbors.

4. **Node Density:** Node density in MANETs is generally randomly but evenly distributed over the area, but in VANETs node distribution can vary from sparse to very dense and is usually uneven. This kind of node density leads to frequent network partitioning/fragmentation.

5. **Other Constraints:** Nodes in VANETs, unlike in MANETs, do not usually have power or CPU constraints. However, due to the large size of VANETs, they might run out of storage space.

Based on these characteristics (summarized in Table 1), several difficulties in designing a VANET broadcast protocol can be foreseen. In the next section we describe these issues in detail.

2.3. Issues for VANET Broadcast Protocols

The great size of VANETs means that using current MANET protocols in VANETs would result in huge routing states at each node. They would require a lot of storage space and CPU cycles to go through them for each packet. Thus, VANET broadcast protocols have to be *scalable*.

VANETs are highly mobile networks: topology changes can happen frequently and quickly. This means that the broadcast protocol has be to very quick and reactive to these changes. Current MANET broadcast protocols are suited to MANET topologies where mobility is low to medium and hence they usually take some time to converge to the right route. In a VANET, the topology might have already changed before the route has been calculated using such MANET protocols. Therefore, VANET broadcast protocols have to be *fast*.

Furthermore, VANETs are much more likely than MANETs to suffer from network partitioning because of the uneven distribution of the nodes in the network. MANET protocols usually assume an even, albeit random distribution of nodes. Therefore, it is assumed that a relay will usually be found. VANET broadcast protocols cannot assume this and consequently should be *robust* against network partitioning.

Reliability is also very important for VANET safety applications, and this could be achieved, for example, by multiple retransmissions or by using some kind of acknowledgement technique. But it is important to avoid having too much redundancy because it would consume excessive bandwidth and may lead to unnecessary transmissions. This could be avoided by *duplicate packet detection*. *Redundancy*, *duplicate packet detection* and *reliability* are linked. Another point is that, for safety applications, the delivery of the first packet plays an important part in crash avoidance. Several protocols place greater importance on this delivery delay of the first warning packet than on the overall average delay. This *delay* should be as low as possible, even in multi-hop situations.

In the light of the above analysis, we conclude that generic MANET broadcast protocols cannot be used in VANETs. These protocols would have to be modified, keeping in mind the following important requirements. Otherwise, VANET applications would not be able to function properly.

- The protocol should be scalable.
- The protocol should perform well in dense as well as sparse networks.
- The protocol should use the linear topology of VANETs to its benefit.
- The protocol should work in highly mobile scenarios.
- The protocol should be robust and reliable in response to network segmentation and topology changes.
- The end-to-end delay should be minimized.
- The packet delivery ratio should be maximized.
- Packet redundancy should be minimized (Duplicate packet detection).
- Congestion should be avoided and channel contention and packet collisions minimized.
- Overheads (Signaling, Packet, etc.) should be minimized.

Table 2. Summary: Important parameters for VANET broadcast protocols

Parameters	Preferred Values
End-to-End Delay	Small
Packet Delivery Ratio	High
Reliability, Robustness	High
Redundancy	Low
Signaling,Network Overheads	Low
Bandwidth Usage	Low

- The protocol should make use of location information.

It should also be noted that most of the protocols proposed in the literature are CSMA-based although it has limitations e.g., channel access unfairness, hidden node problems and collisions. On the other hand, TDMA schemes suffer from several problems due to the high mobility of VANETs and the lack of a centralized scheduling entity to synchronize the channel access; clearly, further research is needed in this area. All the protocols presented in this chapter are based on CSMA.

In the next section, we present a classification of the protocols that have been developed for broadcast traffic in VANETs.

3. Classification of Broadcast Protocols

As stated before, VANETs are an important research topic and as a result several broadcast protocols have been proposed. Starting from the simple pure flooding protocols, they have evolved into much more sophisticated protocols. This evolution took place due to the specific nature and demands of VANETs, as described in the previous section. Already existing techniques e.g., MANET routing protocols and geographic protocols have been adapted for VANETs. These broadcast protocols can be broadly classified as below, but it should be borne in mind that due to on-going research, new protocols are emerging that combine the properties of previous protocols.

1. **Flooding-based:** Flooding is a simple method of disseminating information, where each node within the network relays the broadcast packet once. Its main advantage is the good delivery ratio obtained, even in high mobility scenarios. However, in dense networks, flooding can lead to network overload due to broadcast storms.

2. **Table-based:** Table-based protocols maintain a list of neighbor nodes that is periodically updated with the help of control messages e.g., the HELLO messages used in proactive protocols. The neighbor information can also be gathered on the fly whenever an originator has a packet to broadcast. The main idea behind such protocols is that the sender selects the next relay node with the help of this local table. These protocols perform well in low mobility scenarios but have performance issues in high mobility scenarios.

3. **Position-based:** In position-based protocols, the node decides to rebroadcast the packet according to its geographic position as well as the geographic position of the source node and sometimes the destination area. For example, a node might rebroadcast the packet depending on the distance to the last relayer. Position-based broadcast protocols are usually scalable, as well as robust to mobility and partitions.

These different classes are discussed in detail in the following sections.

4. Flooding-Based Broadcast Protocols

Data dissemination in wireless networks like VANETs can be done by simply broadcasting the data. Every node in the transmission range will receive the data and rebroadcast it to its neighborhood once, and so on. This technique is also called pure, classical or simple flooding. Pure flooding is the simplest broadcast technique to implement since it does not require any sophisticated neighbor discovery procedure or additional control overhead. However, there have been several modifications proposed to optimize its performance. For example, a simple way to control the propagation is to use a packet time-to-live counter. Geographic position information, obtained from GPS receivers, could also be used to limit the flooding area. Information contained in the packet is used to decide where to stop packet retransmissions. Since some safety applications require the dissemination of the information in a specific direction, e.g., a crashed vehicle may start broadcasting emergency messages in order to inform the vehicles coming behind it (Figure 2). In this case, the geographical information could be used to direct the propagation of the emergency broadcast message behind the crashed vehicle. Another possible technique in this situation is to use directional antennas. The restricted, one-dimensional topology of VANETs means that the use of directional flooding can be very effective.

One of the major advantages of flooding-based protocols is their robustness and reliability. For example, in an error free channel, simple flooding ensures that each node in the network will eventually receive the broadcast message, as well as redundant copies (as many as the number of its neighbors). Even in real conditions, with mobility, network fragmentation, transmission collisions and errors, the mechanism of flooding ensures high reliability and robustness.

However, pure flooding has some drawbacks because once a broadcast packet has been transmitted and received by the neighbors, they all start to compete to access the channel and eventually relay this message. This process continues until all the nodes in the network have received and retransmitted the broadcast message once. Consequently, in a network of n hosts, there will be n transmissions of the message, even if all the hosts are within radio range of the initial transmitter. This behavior leads to excessive message redundancy, collisions, channel contention and bandwidth consumption problems. The worst case scenario is a broadcast storm [23] in the network.

Message redundancy can be reduced if each node detects duplicate packets, e.g., by using duplicate packet tables, and stopping their retransmission. To cope with the channel contention problem, and prevent simultaneous transmissions by neighboring nodes, the relaying is usually delayed by a random duration. This reduces the risk of simultaneous transmissions and consequently the number of collisions is reduced within the network. On

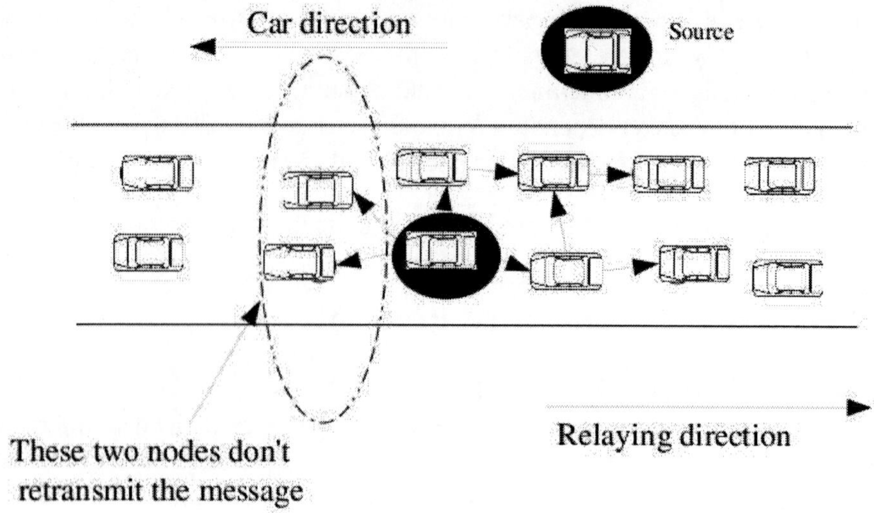

Figure 2. Directional flooding .

the other hand, this added jitter will increase the global propagation delay (the amount of time needed for a packet to reach all the nodes in the network). For post-crash warning and collision avoidance warning applications, this delay is one of the most important parameters to measure and minimize.

4.1. Example Protocols

Dolphin [26] is an example of pure flooding designed at the MAC level for VANETs. In this protocol, the authors assume that the vehicles know their respective locations. When an emergency message is first generated, it must include the position of the originator. Hence, the receivers will forward the message only if it comes from the front. Messages generated or relayed by any vehicle behind it are simply ignored. Each new, non-duplicate received message, is periodically repeated n times in order to cope with possible packet collisions at the receiver nodes.

In [9], authors present two broadcast schemes based on the flooding technique. The first one is called *naive broadcast* in which the originator of the warning message repeats the message n times. All the vehicles behind it that receive this message will execute a procedure similar to Dolphin. Eventually, every vehicle will receive the broadcast message. The propagation of the warning message is stopped only when it reaches the last vehicle in the platoon.

The second scheme in [9] is called *I-BIA (Intelligent Broadcast with Implicit Acknowledgment)*. It uses implicit acknowledgements to cope with packet loss due to collisions. This decreases the delivery latency of the message, and reduces the number of packets transmitted in the network. The broadcast is initiated by a vehicle after detecting an emergency situation. It starts transmitting a message periodically to warn the cars coming behind it. Subsequently, if it overhears the same packet being relayed by another car behind it, then this means that the latter vehicle is now responsible for relaying the message and the vehi-

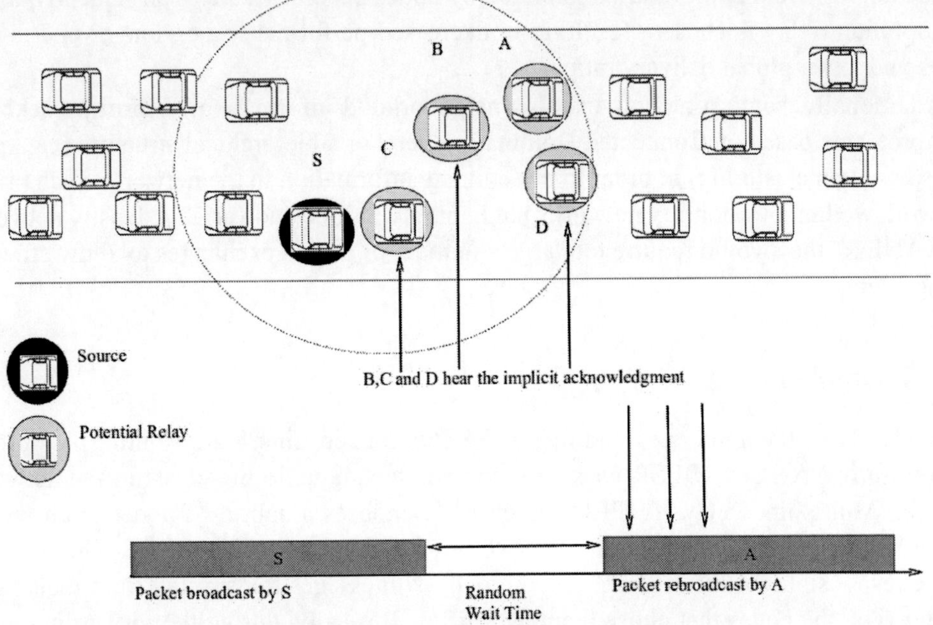

Figure 3. Intelligent Broadcast with Implicit Acknowledgment.

cle stops retransmitting the message. This is referred to as an implicit acknowledgement. Intermediate vehicles use the same principle but they wait for a random time period during which they listen for the same message from behind. If they receive such a message, they know that the message has been propagated further in the network, and they stop the procedure. Otherwise, they start their periodic transmissions of the message. Although the average delay is increased by this scheme, it reduces the total number of transmissions compared with *naive broadcast*. Another disadvantage is that the choice of the relay is not always optimum due to the randomness of the wait period. This can be seen in Figure 3 where although *D* is the best node to act as a relay, *A* transmits first.

5. Table-Based Broadcast Protocols

In a table-based broadcast scheme, nodes maintain information about their neighborhood so that the sender can select the best relay to forward the broadcast packet. This neighborhood information is either exchanged in a proactive manner or when the originator has a message to broadcast within the network, hence the relayer vehicle can be easily chosen by the sender. It implies that only a subset of the nodes within the network will repeat the broadcast messages in order to reach all of them. This family of broadcast protocols is known to have better performances than classical flooding in terms of packet end-to-end delay and delivery ratio. It also dramatically reduces bandwidth consumption in the network, especially when the network is relatively stable with a low rate of topology changes. The majority of these protocols were first proposed for MANET networks. However, in VANETs, the topology may change rapidly and frequently in an unpredictable way, leading to situations where the information gathered by nodes becomes stale. Consequently, a less than optimum relay is chosen, resulting in decreased performance i.e., longer end-to-end delays and lower global delivery ratios.

Additionally, some protocols use this neighborhood information to form a backbone (e.g., protocols based on Connected Dominating Sets) or a hierarchical structure (e.g., protocols based on clustering) in order to disseminate information in the network. Such protocols work well in low mobility networks but in highly mobile and variable density networks like VANETs, they would require too large a number of packet exchanges to allow efficient management.

5.1. Example Protocols

In [10] [21], the authors extend the OLSR (Optimized Link State Routing) [5] protocol for use in VANETs. OLSR uses an optimized technique to broadcast in the network, based on Multipoint Relays (MPRs). Each node chooses a subset of nodes from within its one-hop neighbors, which covers the two-hop neighbors (see Figure 4) . These MPRs will be responsible for forwarding all the broadcast messages transmitted from their MPR selector (i.e., the nodes that chose them as MPRs). Basically, during the forwarding, each MPR repeats the first copy it receives once if it is an MPR of the sender, otherwise it discards the packet. OLSR uses a periodic exchange of Hello messages in order to obtain the neighborhood knowledge and enable the nodes to select their own relays. The frequency of control message exchange has an important role in the efficiency of this selection. It should be high enough to reflect the correct topology of the network. In the case of high mobility it means that the participating nodes should increase the frequency of the control messages, which implies an additional overhead. Moreover, when choosing MPRs, the nodes tend to select the furthest nodes (in terms of radio range) among its one-hop neighbors. When congestion and simultaneous transmissions occur, these distant MPRs are more vulnerable to collisions. In VANETs, this becomes worse since the network topology is linear in most cases. This implies that each vehicle will choose only two MPR nodes (one in front, and another one behind). A collision on one MPR causes the propagation of the message to be stopped. This behavior is not acceptable in emergency cases where a warning message must be transmitted to all the following cars. In [21], the authors propose using an implicit acknowledgment system to cope with this weakness. In this configuration, MPRs will retransmit the received message periodically until they conclude that their respective MPRs have received their transmissions and have started their own transmissions. These transmissions are considered as implicit acknowledgements.

The second example broadcast protocol [25] is a kind of reactive protocol, in the sense that the originator and the forwarders select the relays when they have a packet to broadcast, (in contrast with the previous protocol where the relays are chosen in a proactive manner). This protocol uses the location information of the cars, and the global objective remains the same as the previous one, i.e., optimizing bandwidth utilization. This protocol is based on two steps. The reactiveness of this protocol comes from the first step, during which the originator vehicle sends a request to its neighborhood asking for position information. Then, it waits for the reply messages for a predetermined duration. According to the replies, the originator vehicle classifies its neighbor vehicles into

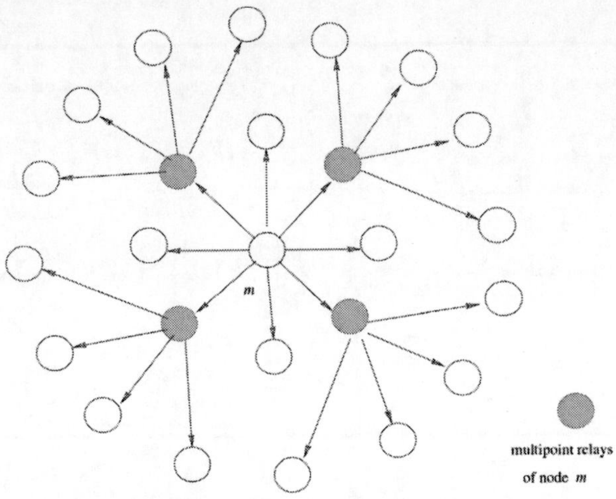

Figure 4. Multipoint relays selection.

five classes (Same_Road_Same_Direction_Ahead, Same_Road_Same_Direction_Behind, Same_Road_Opposite_Direction_Ahead, Same_Road_Opposite_Direction_Behind, and Different_Road), see Figure 5. Depending on the strategy used, the originator selects the relays (called *Border Vehicles*) and puts their IDs in the message and transmits it. The neighbors that receive the message can deduce whether they are responsible for rebroadcasting it by checking whether their ID is included in the message or not. Each relayer then executes the same relay selection procedure described above until the destination area has been reached. Different selection strategies could be used, depending on the message requirements, e.g., the message could be intended for all the cars in the area of the originator, or only for vehicles travelling in the same lane. If the second strategy is used, the *Border Vehicles* selected will be the farthest vehicles in Same_Road_Same_Direction_Ahead and Same_Road_Same_Direction_Behind sets. The authors also propose two algorithms for the classification of the neighbor vehicles based on their positions (see [25] for more details on this). It should be noted here that the classes are used only to control the direction of the flooding and not to decide on a particular relay. Other table-based protocols are proposed in [6] and [20].

6. Position-Based Broadcast Protocols

Position-based broadcast protocols propagate packets towards their destination using relays that are self-selected dynamically per packet optimizing some criterion. This criterion is usually based on the knowledge of nodes' positions e.g., the distance between the sender and the potential relay. The complexity of the protocol is reduced since for each hop of the packet, potential relays are self-selected from among all the *receivers* of the packet. Furthermore, the routing state of each node is also significantly reduced compared to the routing states in topology-based broadcast protocols. As a result, these protocols are much more scalable than flooding-based and table-based protocols. These protocols are also able to handle high vehicle mobility because unlike table-based protocols, they can quickly adapt

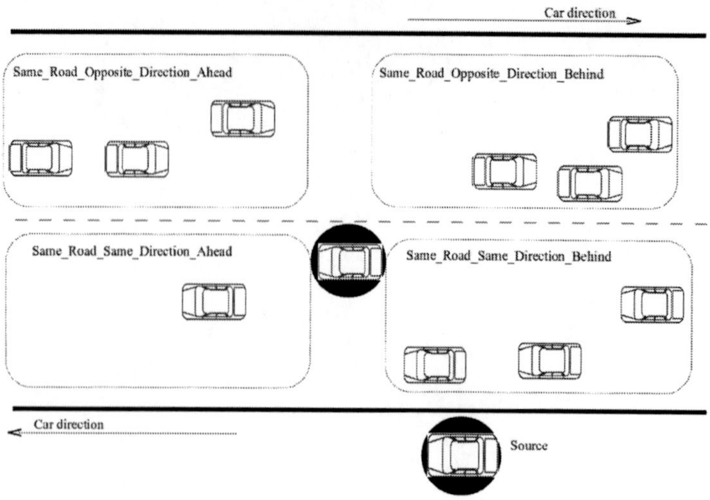

Figure 5. Five classes of *Border Vehicles*.

to changes in the network topology since they do not have to be aware of the entire network configuration to route the packets correctly.

Moreover, the effects of packet collisions and fading effects are nullified because the relay is self-selected among the neighboring nodes which have received the packet *correctly*. In contrast, in table-based protocols, relays are pre-selected by the sender; and if there is a packet loss (due to fading or collisions) at the pre-selected relays, the propagation is stopped until the sender retransmits it again. Position-based protocols also work well both in dense and sparse networks while maintaining low duplicate transmissions. The process of self-selection allows a sufficient number of relays to be selected while keeping the number of duplicate transmissions to the minimum both in sparse and dense networks.

The nodes can get their position using GPS receivers, a location service or some other mechanism. On the other hand, one of the drawbacks of position-based protocols is that the nodes have to know their position. While this information can be obtained through several means, as pointed out above, current technology does not provide position information that is accurate enough or is updated fast enough to meet the requirements of VANET safety applications. However, it is likely that within a short timespan, newer technology and techniques will overcome this issue.

6.1. Example Protocols

The first position-based protocol that we will discuss is "Role-based Multicast" [11]. It is called role-based multicast because the set of receivers is defined by a set of constraints that include e.g., a possible danger to the node and the node's ability to avoid that danger. The danger to a node is decided by the node's direction and its position on the digital map. Its ability to avoid the danger is decided by its braking distance. The braking distance of a node is calculated by taking into consideration the velocity of the node, its maximum deceleration value and the reaction time of the driver.

The proposed protocol works on top of the data link layer that keeps track of the neighborhood and informs the protocol of changes in topology. The protocol maintains a set

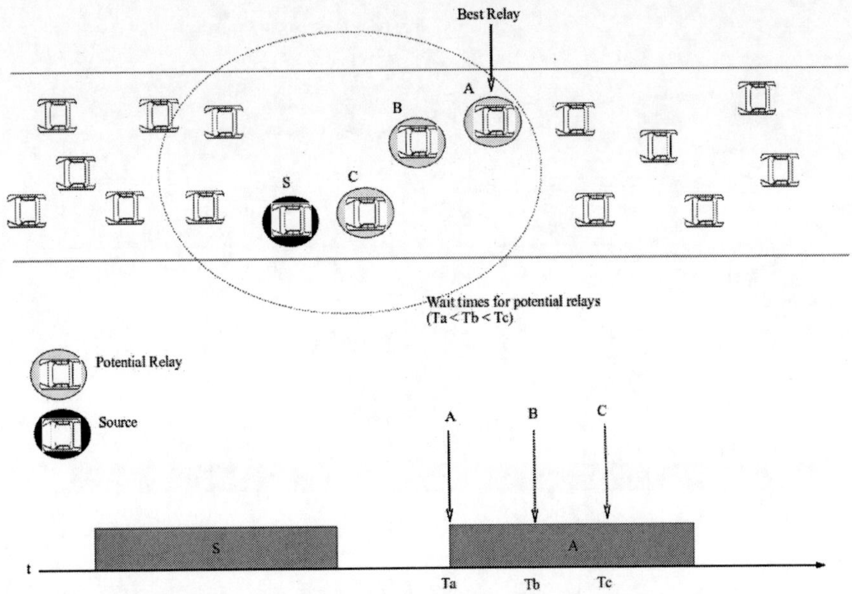

Figure 6. A simple position-based protocol .

of neighbors according to these notifications. It should be noted that this information is not used for relay selection, but only to avoid unnecessary transmissions. When a node receives a warning message, it can enter into one of two possible states. It will enter the *WaitToResend* state if it has neighbors other than the sender. The "wait time" is calculated such that it is shorter for the more distant nodes and longer for the closer nodes (see Figure 6). The nodes do not transmit immediately, because as mentioned above, CSMA does not perform well in an overloaded channel. Hence, the protocol tries to avoid putting too much load on the network by forcing the nodes to wait. Due to this wait time, it is mainly the hosts near the limit of the transmission range that will take part in the retransmissions. On the other hand, if the node has no neighbors other than the sender, then it will enter the *WaitForNeighbor* state. In this state, it will wait until a new neighbor arrives in the neighborhood. It should also be noted that if during the *WaitToResend* state, the neighborhood becomes empty, the node enters the *WaitForNeighbor* state. The results of the simulation show that this protocol is suitable for sparse networks or networks where the market penetration is low. The protocol performs well when the market penetration is below 50%; after that, however, the advantage of the wait time is lost and it performs almost like flooding.

A similar idea is presented in [8], [12] and [22] but in [19] the authors take into account the possibility that there are no nodes near the border of the transmission range. Consequently, nodes closer to the source will wait unnecessarily before retransmitting the message, thus increasing the end-to-end delay. In their protocol, the intermediate nodes choose a "time window" rather than a specific wait time. The size of this time window is inversely proportional to the node's distance from the previous relay. Thus distant nodes will have a smaller time window and closer nodes will have larger time windows. The nodes will then choose a wait time at random from this time window. As a result, distant nodes will have a higher proability of being selected as relays because they will have a higher probability

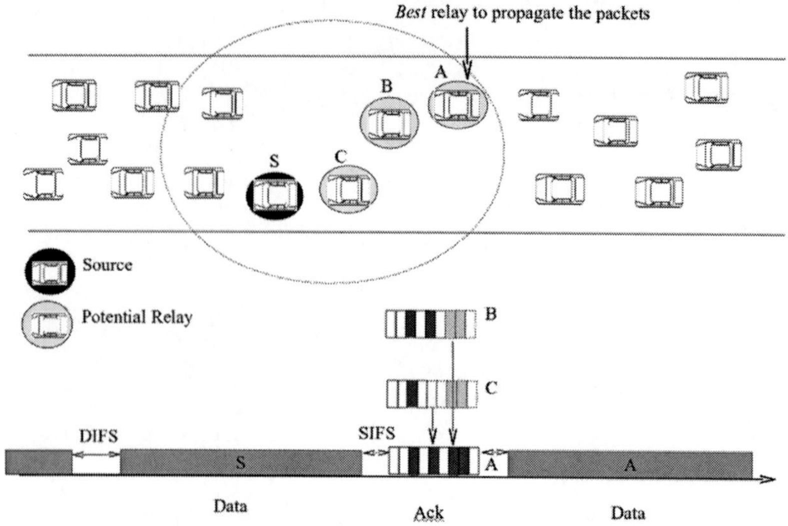

Figure 7. A MAC-level position-based protocol.

of choosing a shorter wait time. Simulations show that in low density situations, where it is likely that there are no distant nodes, this technique considerably reduces the end-to-end delay.

The second position-based protocol [7] we describe uses a MAC-level smart acknowledgement to select the best relays. The potential relays are selected from among the nodes that have captured the packet correctly. These nodes then send a burst to acknowledge the received packet. This burst consists of a sequence of intervals of the same length in which a given node can either transmit or listen (see Figure 7). During this active signaling phase, each receiver applies the following rule: if it detects a signal during any of its listening intervals, it quits the selection process (i.e., it stops transmitting during the entire remaining part of the active signaling phase). It does this because the detection of a transmission during a listening interval implies that a better relay has also captured the broadcast data sent in the first part of the slot. This scheme can be interpreted as a modified CSMA; the back-off technique usually used to compete is replaced here by the active signaling scheme. These signaling bursts can best be represented by a binary sequence, where 0 denotes a listening interval and 1 denotes a transmission interval. This binary sequence is computed by the receiver as a function of the criterion that we wish to optimize for our 'best' relay. These bits may simply code this criterion in base 2. The criterion need only satisfy the following rule: the 'better' the relay is, the larger the criterion becomes. The last bit is always set to 1, to give evidence of the reception of the packet by the best selected relay.

7. Other Approaches

The classification of protocols presented above is based on classical relaying techniques. In these techniques the packet routing decision is taken as fast as possible after

Table 3. Advantages and disadvantages of different broadcast techniques

Type	Advantages	Disadvantages
Flooding-based	simple implementation low overhead reliable robust	high bandwidth consumption broadcast storm problem not scalable
Table-based	moderate control overhead	not scalable suitable for low mobility
Position-based	scalable suitable for high mobility	location information needed

packet reception. This decision is based on the information contained in the packet header (usually using the destination IP address to find out the next hop router). However in this section we present other approaches for data dissemination and relaying i.e., content-based, gossip-based and epidemic protocols. Content-based schemes [17] use information filtering to relay or drop a packet. In these protocols, each node looks and analyzes the information in the packet before making a decision to relay the packet. This implies that the delivery ratio of a packet will depend on the information filters of the intermediate nodes. In gossip-based protocols, each node that receives a message for the first time retransmits it to a given subset of nodes in the network until it is disseminated to all or to a pre-defined number of nodes. Gossip-based protocols can provide reliable and efficient communication in VANETs due to their inherently decentralized and robust nature. In [14], the authors present research challenges related to the application of gossip-based protocols for VANETs. In epidemic routing [16], the intermediate nodes retransmit the received messages with some probability. It works well in networks with an unstable configuration like VANETs, giving a higher delivery ratio without the need to maintain stable routes. It should be noted that gossip-based and epidemic-based protocols are more suitable for delay-tolerant applications. Message delivery can be ensured by exploiting store-and-forward techniques. Additionally these schemes can be coupled with other information to help a node to process the packet, such as GPS information, digital map or data carried within the packet.

8. Conclusion

We have presented various applications that can be used in VANETs and shown that the broadcast protocols will be mainly used by safety applications. We have also presented the main characteristics of VANETs and especially how they differ from the standard MANETs. We have enumerated the different issues, characteristics and the main metrics that should be taken into consideration when designing or evaluating a VANET broadcast protocol for safety applications. We have classified the existing broadcast protocols into three main types: flooding-based, table-based and position-based. We have discussed the main advantages and disadvantages of these protocols and described two example protocols for each class (see Table 3 for the summary).

Flooding-based broadcast protocols are the simplest to implement and are robust and very reliable in highly mobile VANETs. However they consume a lot of network bandwidth and generate too many duplicate packets. In the second category table-based broadcast protocols try to minimize their bandwidth utilization. However they are unsuitable for VANETs with even moderate mobility because they require a large number of control packets to keep their tables synchronized. Position-based protocols are the most suitable broadcast protocol for VANETs because they are well adapted to the high mobility of VANETs. The main reason for this advantage is that for each hop of the packet, potential relays are self-selected dynamically from among all the receivers. This mechanism counteracts the effects of packet collisions and fading effects. It works well in dense and sparse networks while maintaining low duplicate transmissions.

References

[1] Car-2-Car Consortium, www.car-to-car.org.

[2] European telecommunications standards institute ETSI, www.etsi.org.

[3] NOW, www.network-on-wheels.de.

[4] PATH, www.path.berkeley.edu.

[5] C. Adjih, T. Clausen, P. Jacquet, A. Laouiti, P. Minet, P. Mühlethaler, A. Qayyum, and L.Viennot. Optimized Link State Routing protocol (OLSR). *RFC* **3626**, October 2003. Network Working Group.

[6] H. Alshaer and E. Horlait. An optimized adaptive broadcast scheme for inter-vehicle communication. In *Vehicular Technology Conference, 2005. VTC 2005-Spring. 2005 IEEE 61st*, volume 5, pages 2840–2844 Vol. 5, 2005.

[7] Paul Mühlethaler Anis Laouiti and Yasser Toor. Opportunistic broadcast in vanets (OB-VAN) using active signalling for relay selection. Technical report, INRIA, January 2008.

[8] Abderrahim Benslimane. Optimized dissemination of alarm messages in vehicular ad-hoc networks (VANETs). In *High Speed Networks and Multimedia Communications Journal*, pages 655–666, 2004.

[9] S. Biswas, R. Tatchikou, and F. Dion. Vehicle-to-vehicle communication protocols for enhancing highway traffic safety. *Communications Magazine, IEEE*, **44**, Issue1(6):74–82, 2006.

[10] L. Bouraoui, A. de la Fortelle, and A. Laouiti. OLSR improvement for distributed traffic applications. In *Fourth Annual Mediterranean Ad Hoc Networking Workshop, MedHocNet'2005*, Porquerolles, France, 2005.

[11] Linda Briesemeister and Gunter Hommel. Role-based multicast in highly mobile but sparsely connected ad hoc networks. In *MobiHoc '00: Proceedings of the 1st ACM*

international symposium on Mobile ad hoc networking & computing, pages 45–50, Piscataway, NJ, USA, 2000. IEEE Press.

[12] C.-F. Chiasserini, R. Gaeta, M. Garetto, M. Gribaudo, and M. Sereno. Efficient broadcasting of safety messages in multihop vehicular networks. *International Parallel and Distributed Processing Symposium*, 2006.

[13] S. Corson and J. Macker. Mobile ad hoc networking (MANET): Routing protocol performance issues and evaluation considerations, 1999.

[14] Paolo Costa, Daniela Gavidia, Boris Koldehofe, Hugo Miranda, Mirco Musolesi, and Oriana Riva. When cars start gossiping. In *ACM Middleware for Network Eccentric and Mobile Applications Workshop (MiNEMA'08)*. ACM Press, 2008.

[15] Olivier Dousse, Patrick Thiran, and Martin Hasler. Connectivity in ad-hoc and hybrid networks. *INFOCOM 2002. Twenty-First Annual Joint Conference of the IEEE Computer and Communications Societies. Proceedings. IEEE*, **2**, 2002.

[16] Patrick T. Eugster, Rachid Guerraoui, Anne-Marie Kermarrec, and Laurent Massoulie. Epidemic information dissemination in distributed systems. *Computer*, **37**(5):60–67, 2004.

[17] Holger Füßler, Jörg Widmer, Michael Kusemann, Martin Mauve, and Hannes Hartenstein. Contention-based forwarding for mobile ad hoc networks. *Elseviers's Ad Hoc Networks*, **1**(4):351–369, 2003.

[18] David B. Johnson, David A. Maltz, and Yin-Chun Hu. The Dynamic Source Routing protocol for mobile ad hoc networks (DSR), 2004.

[19] Tae-Hwan Kim, Won-Kee Hong, and Hie-Cheol Kim. An effective multi-hop broadcast in vehicular ad-hoc network. In *Architecture of Computing Systems - ARCS 2007*, pages 112–125.

[20] Gökhan Korkmaz, Eylem Ekici, Füsun Özgüner, and Ümit Özgüner. Urban multi-hop broadcast protocol for inter-vehicle communication systems. In *VANET '04: Proceedings of the 1st ACM international workshop on Vehicular ad hoc networks*, pages 76–85, New York, NY, USA, 2004. ACM.

[21] A. Laouiti, P. Mühlethaler, and Yasser Toor. Comparison of flooding techniques for safety applications in VANETs. In *Proc of the 7th International conference on Intelligent Transport Systems*, pages 444–449, 2007.

[22] M. N. Mariyasagayam, T. Osafune, and M. Lenardi. Enhanced Multi-Hop Vehicular Broadcast (MHVB) for Active Safety Applications. In *ITST '07. 7th International Conference on ITS.*

[23] Sze-Yao Ni, Yu-Chee Tseng, Yuh-Shyan Chen, and Jang-Ping Sheu. The broadcast storm problem in a mobile ad hoc network. In *MobiCom '99: Proceedings of the 5th annual ACM/IEEE international conference on Mobile computing and networking*, pages 151–162, New York, NY, USA, 1999. ACM.

[24] C. Perkins, E. Belding-Royer, and S. Das. Ad hoc On-Demand distance Vector (AODV) routing. *RFC* **3561**, 2003. Network Working Group.

[25] Min te Sun, Wu chi Feng, Ten-Hwang Lai, Kentaro Yamada, Hiromi Okada, and Kikuo Fujimura. GPS-Based message broadcasting for inter-vehicle communication. In *ICPP '00: Proceedings of the 2000 International Conference on Parallel Processing*, page 279, Washington, DC, USA, 2000. IEEE Computer Society.

[26] K. Tokuda, M. Akiyama, and H. Fujii. Dolphin for inter-vehicle communications system. In *Proc. of IEEE ITS*, pages 504–509, 2000.

PART 3.
SYSTEMS AND APPLICATIONS

In: Wireless Technologies in Intelligent Transportation… ISBN: 978-1-60741-588-6
Editors: Ming-Tuo Zhou et al, pp. 225-265

Chapter 9

WIRELESS AND INFORMATION TECHNOLOGIES SUPPORTING INTELLIGENT LOCATION-BASED SERVICES

C. Ray[1,*], F. Comblet[2], J.-M. Bonnin[3] and Y.-M. Le Roux[3]

[1] Department of computer science, Naval Academy Research Institute
29240, Brest, France

[2] Ecole Nationale des Ingénieurs des Etudes et Techniques d'Armement
2 rue François Verny, 29806, Brest, France

[3] Institut Telecom-Telecom Bretagne
Technopole Brest Iroise, 29238, Brest, France

Abstract

The continuing and emerging advances in the development of wireless technologies significantly accelerate the design and operation of new intelligent transportation systems. These advances have to be combined with other fields of research and technologies in order to derive end-users' services. The research presented in this chapter proposes to combine three fundamental ITS technologies, i.e. the wireless communications, geolocalisation technologies and the geographic information systems in order to design transportation services able to monitor in real-time mobile objects (e.g. pedestrians, vehicles). This chapter is illustrated by disaster response management and coordination that aims at monitoring terrestrial (indoor and outdoor) and maritime rescue operations.

1. Introduction

Intelligent transportation science has long considered Intelligent Transportation Systems (ITS) as services related to vehicle and road management. Emergent and continuing researches oriented to specific branch called location-based services is more and more

[*] E-mail address: cyril.ray@ecole-navale.fr. Tel +33 298233611, Fax +33 298233857. Address correspondence to: Ray C., Ph.D., Institut de recherche de l'Ecole navale, Ecole navale, 29240, Brest Naval, France.

considering the design of transportation systems trough a global approach able to integrate technologies such as localisation, information systems and communication transparently for users. Considering wireless communication facilities as the key technology that bring real-time exchanges, information system as a core of that define services returned to users and localisation technologies as emerging devices to be integrated in all future users' appliances, the aim of this chapter is to demonstrate that Intelligent Transportation System can also be seen as Intelligent Transportation Services that requires such technologies together. This integrated approach is inherent to that next evolution of ITS that will have to cope with increased mobility and ubiquity of users. This leads to transportation services that tend to multi-modal transportation services where users would like a continuity of service whatever their transportation modes (e.g. individual car, then by foot, then by boat).

One of the main objectives of these location-based services is to integrate a large number of mobile objects (e.g. persons, vehicles) through a distributed architecture and wireless communications, and to offer different levels of visualisation facilities to mobile end-users. Several aspects have to be considered for the design of such a system:

- An information system able to manage positioning information and that supports delivery of real-time geo-referenced data as well as additional contextual (e.g., what's around a mobile user) and spatio-temporal data (e.g., speed, direction).
- Some medium and communication protocols adapted to the application context. Aerial, terrestrial and maritime transportation systems use a large spectrum of wireless telecommunications (e.g., WiFi, WiMax, GSM, GPRS, EDGE, UMTS, HSxPA and VHF) for the integration of location-based information.
- A sensor-based system that delivers positioning information. The choice of such a given location-based system is often driven by technologies available and the environment where the mobile object evolves (e.g., indoor or outdoor positioning). These location-based sensors can for instance rely on GPS, radio-based systems (GSM or WiFi), inertial systems or a combined approach [42].

The research introduced in this chapter aims at presenting (1) state of the art and improvements in these fields of ITS science and, (2) ways to combine wireless communications, localization techniques and information systems for the design of location-based services. Principles and methods have been linked and illustrated through an application that supports the real-time following of rescue services [41].

Giving to rescue brigades in terrestrial and maritime operations, accurate localisation, communication means and real-time information can improve their efficiency, while preserving their own security. Indeed, in numerous cases of accident or a disaster occurring either on the road, in a factory, in a city, in a harbour or on a boat cruising close to the coast, intelligent transportation services are of the highest importance to improve the rapidity (e.g. navigation to disaster) and security (e.g. tracking firemen) of the operations. It also favours exact, complete and up-to-date information exchanges between firemen, headquarters, vehicles and public authorities.

From a scientific and technological point of view, such application context has the advantage to combine most of the issues inherent to the conception of an integrated and multi-modal transportation system. The location-based information system being designed in this context combines NEMO mobile routers, wireless communications, integrated localisation system based on GPS and inertial system and a geographical information system that provide end-users geo-services.

The following deals with the fundamentals of these three aspects and describes the underlying key technologies. Section 2 introduces the principles of information systems for the design of transportation services. These elements are applied to the design of a location-based service able to combine maritime, vehicle and pedestrian real-time tracking. Section 3 presents issues related to the design of wireless communications infrastructure that supports this tracking system. Section 4 addresses accurate location issues that allow providing of a continuity of tracking and navigation services, through combined technologies.

2. Designing Information Systems for Transportation Services

Geographical Information Systems (GIS) relates to the science that addresses methods and tools for capturing, managing, analysing, and displaying all forms of geographically referenced information. Such geographic information provides first spatial features but might also contain temporal aspects and is often tied to metadata (i.e. data about data that describe content, quality, behaviour, and other characteristics). For example, a trajectory of a mobile object (e.g. a vehicle) is a set of geographic locations (spatial data) with moves and stops that evolve over time and at a precise moment (temporal data). Such geographic information might be accompanied by metadata providing either quantitative or qualitative information (departure location is home, arrival location is a shopping center).

Recent works showed that combination of transportation systems and Geographical Information System [13] [14] is of great interest for many terrestrial, maritime and aerial applications related to the monitoring and analysis of mobility and traffic phenomena in which represented people and vehicles locations are changing in a fast and almost continuous mode. Location-based services addressed in this chapter are at the crossroads of many engineering and scientific domains. Their development is largely driven by progress in wireless communications, interfaces and geo-positioning systems [15].

The development of location-based services is often closely related to research in transportation systems and the development of distributed services for public transportation [16] [33]. The search for normalised and standardised platforms, exploration of novel mobile-based services is active areas of interest. Integration of geographical information standards under the umbrella of the Open Geospatial Consortium[1] (OGC), Web specifications imposed by the W3C endowment have lead to the proposal of several distributed and location-based GIS architectures [29]. Additional services for mobile systems have been explored for neighbourhood retrieval systems [25], directional queries [23], diffusion of multi-resolution maps [22], personalisation of geographical maps in mobile environments [31], diffusion of three-dimensional geographical data [24] through mobile appliances and management of geographical data [27].

[1] http://www.opengeospatial.org

2.A. Distributed Information System

An information system has been designed to illustrate needs and benefits of a combination between accurate geo-positioning, wireless communications, data management and visualisation services to users. This information system acts as a located-based service able to integrate, store, model, analyze and visualize spatially related data in time. In the retained application domain (i.e. real-time monitoring of rescue services) these dynamic data to locate and display are firemen and vehicles and even ships (e.g. passenger ships) moving in coastal and port areas [17].

The platform, developed so far, has been designed with four tiers client-server architecture and organized through a distributed data and processing model. The objective of the platform is to provide an integrated and flexible system for the monitoring and tracking of different types of mobile objects in real-time. This information system illustrated in Figure 2.1 is a distributed platform based on four components corresponding to three functions: (1) integration of positioning information, (2) data management and processing and (3) visualisation.

Localisation on Ground

Tracking of mobile objects on ground such as pedestrians and their vehicles relies on a portable Java-based platform designed for mobile objects (Figure 2.1, component 1) embedded on lightweight mobile appliances (e.g. PDA, UMPC or enhanced mobile phones). This software component has three main functions that every similar transportation system should have. First, it acquires in real-time localisation data that comes from GPS, MEMs and other environmental sensors. Second, this component parses and interprets these raw data and finally it stores analysed (and correct) information in a local spatial database (PostGIS) for data manipulation. The use of such spatial database brings numerous benefits for the design transportation systems. The two main ones are an efficient storage and management of mobile objects and related indexes.

Fast moving objects that combine indoor and outdoor localisation systems embed MEMS and GPS (cf. Section 4). In order to apply a fusion algorithm, as described in Section 4.d, data stored locally have to be associated using unified timestamp. Intrinsically, GPS and MEMS sensors generate raw data at different granularity level (e.g. 1Hz for GPS and 100Hz for MEMS) and using different time representation (absolute vs. relative). One of the aims of the Java-based platform designed for mobile objects is to identify amongst GPS locations, received and stored, the most accurate ones that can be identified to initialize fusion process.

The second component (Figure 2.1, component 2) of the distributed architecture corresponds to a mobile router preferably based on NEMO and IPv6 standards (cf. Figure 3.2). This router brings mobile wireless communication facilities (cf. Section 3) to mobile objects participating to a rescue operation. This feature interconnects every mobile client on site (pedestrian, vehicles) and allows for wide-range connections (e.g. Internet) in order to collect maritime traffic information (cf. next section) and produce visualization facilities (cf. Section 2.c) to distant users (e.g. rescue services control center).

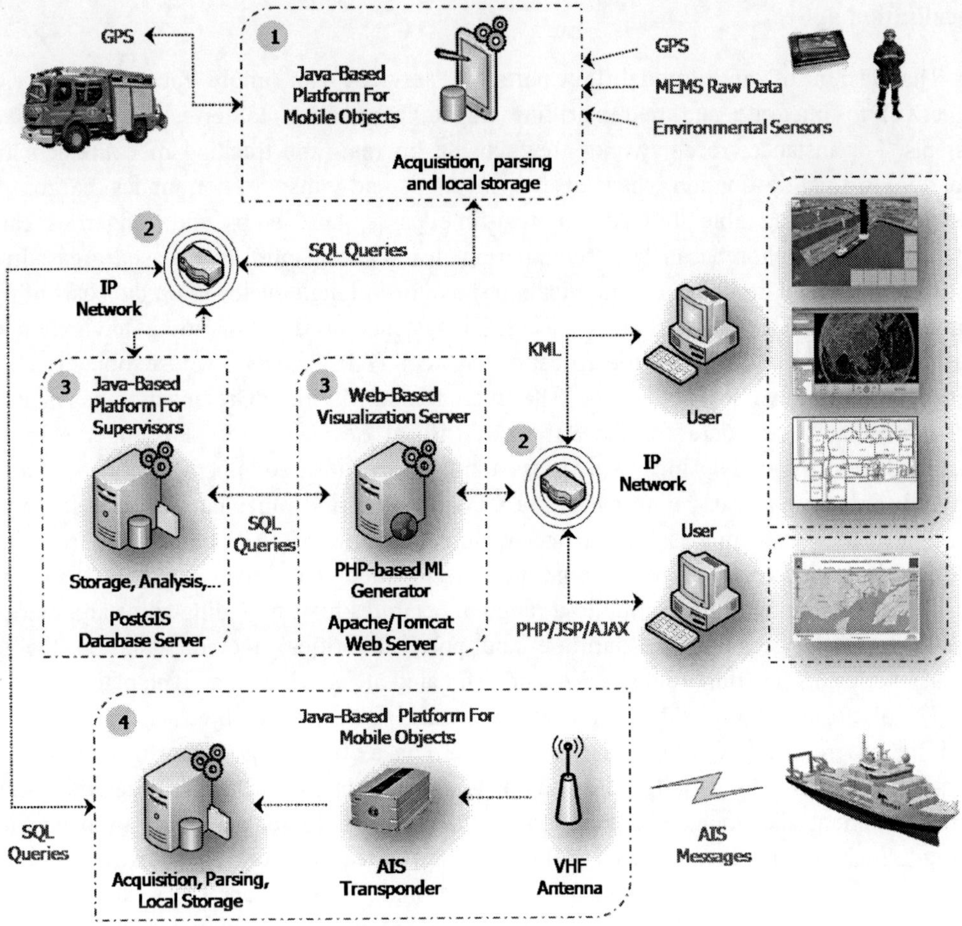

Figure 2.1. Architecture.

The central component (Figure 2.1, component 3) of the distributed architecture is the Java-based platform for supervisors. This component includes a server linked to the mobile router (for instance physically located on the firemen mobile control center). The software acts as the core of the information system and has two objectives. Based on a central spatio-temporal database (PostGIS database), a first objective of this component is to collect and store location-related information generated by all mobile objects (Figure 2.1, components 1). The gathering of these data is triggered on an almost continuous mode as far as wireless connectivity is ensured. A second objective is to offer a Web-based visualisation server to users (cf. Section 2.c).

Visualisation of location-based information on mobile appliances (Figure 2.1, component 1) might require the processing of a fusion algorithm and a map-matching step. These fusion algorithms and matching process can be applied either locally, on mobile appliance, or remotely by the Java-based platform for supervisors (Figure 2.1, component 3). The favoured method, as far as communication is possible, is to get back information to the central platform that have a complete view of all mobile objects. In this context, users can get (through relational queries) on their mobile appliances not only selected positioning information but also all related meta-information.

Localisation at Sea

The design of multi-modal transportation services can imply (according service's objective) to integrate and tied maritime traffic information to terrestrial transportation systems. For instance, recent works are focusing on real-time tracking of containers from departure to final destination whatever their locations and transportation modes (cargo, train, truck, plane). Being able to locate in real-time passengers' ships and dangerous cargos approaching or anchored can be also important for securing port and coastal areas. In the maritime and aerial domains, communications have been long developed on the basis of VHF communications. One of the most successful systems used in maritime navigation and positioning is the Automatic Identification System (AIS) system whose objective is to identify and locate vessels at distance. The International Maritime Organization has made the AIS a mandatory standard for the Safety of Life at Sea (SOLAS). Passenger ships and international commercial ships with gross tonnage are now equipped with AIS. An AIS generally integrate a transceiver system, a GPS receiver and other navigational sensors on board such as a gyrocompass and a rate of turn indicator. An AIS transponder runs in an autonomous and continuous mode, regardless its location (e.g., open sea, coastal or inland areas). The transmitted information to a range of 35 miles to surrounding ships and maritime Vessel Traffic Systems (i.e., maritime and port authorities) include data on the ship characteristics, its position, route, speed, and estimated arrival time to a given place.

The platform allows for the monitoring and tracking of mobile objects active in a coastal maritime environment. Component 4 on figure 2.1 is a computing system that includes the Java-based platform for mobile object that acts, in this case, as an AIS management server. This component also includes a PostGIS spatial database for data manipulation and local storage. Such a component can be installed in port area and rely on an AIS transponder (listening mode only), a VHF antenna to get positioning information. The system is scalable as it can be easily extended to additional sites. When Internet communication through the mobile router (Figure 2.1, component 2) is established, the Java-based platform for supervisor (Figure 2.1, component 3) collects maritime traffic information on a regular basis (e.g. every 30 seconds). This information is then stored in the database server and available for visualisation.

2.B. Data Integration and Filtering

A location-based system whose aim is to locate in real-time mobile objects has to integrate data coming from many different sensors including GPS and MEMS but also standard such as Automatic Identification System (AIS) for maritime traffic or Automatic Dependent Surveillance-Broadcast (ADS-B) for aerial traffic.

The way mobile data should be integrated and manipulated within a computing and spatio-temporal database systems have been areas of active research. This has been considered as data representation and storage issues where the objective is to design appropriate physical indexes, query languages and prediction mechanisms for moving objects [18] [26] [32] [35] [37] [38]. Recent work concerns the development of conceptual languages and data mining algorithms [36] for the manipulation of trajectory at the interface level, and the search for trajectory patterns in space and time.

As often in location-based systems, the data volumes generated are relatively large thus making data processing and analysis a computationally costly operation. The development of data filtering and compression algorithms for moving points has been the object of recent attention [19] [20] [30] [34]. Amongst these works, accuracy of locations, reliability of trajectories, management of errors underlying the filtering processes have generated a great interest, particularly in terms of legal issues the simplification algorithms might raise. For instance, [19] uses line-simplification to reduce the size of a trajectories database of location-aware mobile appliances and analyse the effectiveness of multi-distance line-simplification algorithms able to bound errors inherent to spatio-temporal queries on simplified trajectories. Reference [34] develops heuristic prediction, based on sampling that take advantage of the spatial locality and temporal timeliness inherent in trajectory streams, to distinguish which locations to preserve in the filtered trajectories. Such filtering algorithms are also useful to address another lack of naive location data integration. They allow understanding of trajectories patterns and semantics at the appropriate granularity level instead of considering a trajectory as a simple set of discrete locations.

In the context of rescue services illustrating this chapter, three types of mobile objects are of interest: firemen, their vehicles and boats in coastal and ports areas. In order to optimize, without loss of knowledge, the volumes of the trajectory data produced by such moving objects, two filtering processes have been studied, applied and implemented for vehicles and boats.

Ship Trajectory Analysis

Regarding boats trajectories, similarly to [30], the filtering process proposed is based on a line simplification algorithm initially introduced by [21], and that preserves directional trends in a line using a tolerance distance. In fact, it rapidly appears that this polyline-based approach is computationally efficient. It is particularly adapted to the main characteristics of maritime trajectories that are most often linear over long periods of time [17]. The trajectory of a mobile object at sea can be defined as an ordered set of positions (P_i) over time. A given position P_i is represented by a n-tuple [x, y, speed, direction, time]. A filtering process selects the key positions of a given trajectory. A position is considered as a key position when either the speed or the direction changes significantly. The other positions are removed. The principles of the original algorithm are as follows. A straight line segment connects the start and end points of a given polyline. Perpendicular offsets for all intervening end points of segments are then calculated from this segment, and the point with the highest offset is identified. If the offset of this point is less than the tolerance distance, then the straight line segment is considered adequate for representing the line in a simplified form. Otherwise, this point is selected, and the line is subdivided at this point of maximum offset. The selection procedure is then recursively applied to the two parts of the polyline until the tolerance criterion is satisfied. Selected points are finally chained to produce a simplified line. This simplification algorithm can be adapted for trajectory filtering. Conversely to [19] and [30] who used Euclidean distance between points at a same time, the choice of the great-circle distance might be better for maritime trajectories. This distance is the shortest distance between any two points on the surface of a sphere measured along a path on the surface of the sphere. The perpendicular distance is therefore derived as a spatio-temporal distance.

Vehicles Trajectory Analysis

Understand and manage trajectories of vehicles appropriately imply to consider a semantic-based trajectory modelling that cope with constrains raised by the underlying urban network. The first step of a filtering process for vehicles is to consider network modelling [51] [48] [39]. The second step is to consider a map-matching algorithm that correlate location of mobile objects and the designed network [40]. Finally the third step is to define the filtering algorithm adapted to network structures and trajectory properties on different network environments.

Algorithms for this filtering step can be classified in four types. The most straightforward trajectory data representation approach combines location point and coordinate-based spatial reference (absolute coordinates) is often used in applications such as fleet management and automatic vehicle location [54] [49] [50] [53] [47]. Despite its simplicity, the critical drawback relies on the large data volumes generated. The second type belongs to network-based models to explicitly represent location point of trajectories using a network coordinate system [46]. However, either points or segments, without explicit consideration of the temporal dimension, represent trajectories. Furthermore, as these models do not explicitly make the difference between basic trajectory points and trajectory points of importance in the network, the data volumes generated is still very large. When semantic representations are associated to coordinate-based approaches [52] [55] [56], this introduces a model where a trajectory is modelled as a dynamic attribute, using a temporal function to dynamically calculate moving object's location. Points of importance in the network are selected using velocity and direction criteria. However, trajectory data are referenced by a coordinate-based spatial reference. More recently, a semantic network-based has been introduced [45]. This fourth type combines key points of a trajectory and network-based modelling to extract relevant point of a trajectory. The approach studied for vehicles of firemen belongs to this last type. It is based on a key point selection method that integrates different network-based selection criteria [28]. These criteria include network, velocity, and direction parameters and allow apprehending semantic relationships between moving objects and the network. Figure 2.2 illustrates the benefit of the algorithm on a raw trajectory composed by 1608 positions. Filtered trajectory, while composed by only 373 positions, describes appropriately original vehicle's movements.

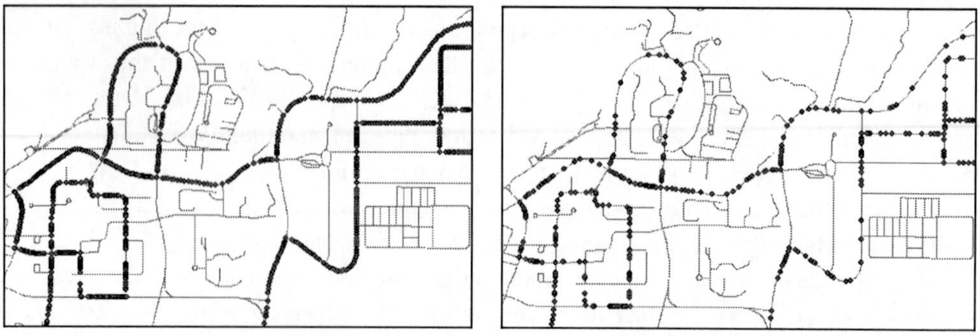

Raw trajectory data (1608 locations) Filtered data (373 locations)

Figure 2.2. Filtering vehicles' trajectory.

2.C. Visualisation

Advances in models, standards and developments of the Web, especially in the field of semantic, mobile and collaborative Web but also regarding Web services bring new interests and challenges in the diffusion, usage, and visualisation of geo-referenced data. Geovisualisation refers to techniques and methods that allow visualising, in an interactive manner, spatio-temporal phenomenon such as location-based applications [12]. The visualisation of data describing mobilities in space and time requires the use of displays based either on raster (e.g. pixels) or vector (e.g. point, polylines and polygons) 2D or 3D maps. When such a visualisation considers the presentation of localised geographic data on the Web, one speaks about "*Geospatial Web*" or "*Geoweb*".

Principles and technologies of "*Geospatial Web*" are supported with "*geobrowsers*" and "*geo-API*" that allow visualising location-based data on Web-based applications. Amongst Geoweb solutions, virtual globes such as NASA World Wind, Google Earth or OssimPlanet or Web-mapping site such as Google Maps, Windows Live Local, Yahoo Maps and Géoportail and API such as World Wind Java SDK or GeoAPI have the advantage to facilitate the diffusion of geographic information to final end-user without being expert in geographic information science.

Based on existing standards and recommendations of the Open Geospatial Consortium (OGC) and the W3C for the design of Web services, the approach proposed for the real-time following of multi-modal transportations has been designed as a Web-based visualisation tool that can report dynamic location-based information on maps. It has been designed with a Java Server Page (JSP) and Apache Tomcat approach, in order to provide external access to the spatial-temporal database (cf. Figure 2.1).

This information system has to cope with the diversity of, on one hand sensors, and on the other hand visualisation tools. Between sensors and computing devices (components 1 and 4 in Figure 2.1) data exchange is often either constrained by sensor design (GPS basically exports NMEA), or structured in a standard meta-language [43] such as SensorML or TransducerML. The platform integrates a large set of mechanisms that allow data integration in appropriate language. Between computing systems, SQL queries have been favoured for efficiency purposes (also allows spatio-temporal queries). Regarding visualisation, the Web-based visualisation server (cf. Figure 2.1, component 3) has been designed with a PHP-based generator of Keyhole Markup Language (KML) files (KML is an OGC standard) to apprehend the diversity of Geoweb applications. This allows end-users to express and personalize geographic contexts, annotations and visualizations features through an XML-based language schema. Instead of SDL and GML description languages (both are also OGC standards), the KML language combines data description and visualisation features within a same structure and has the advantage to specify a large set of features (placemarks, images, polygons, 3D models, textual descriptions, etc.). It also favours the integration of the location-based information on existing GIS software (e.g. ESRI tools, Google Earth) or GeoWeb-based applications (e.g. GeoServer, Google Maps and Mobile).

3. Mobile Wireless Networks

The design and deployment of intelligent transportation systems able to track pedestrians and vehicles movements and to broadcast traffic information (but not only) in real-time to mobile end-users relies on great progresses that have been made in wireless communication technologies in the last few years.

3.A. Motivation

These communication technologies have been long designed through dedicated wireless networks. The deployment of such networks for government related services such as police, army or civil security is however very expensive, long to set up and could not be economically profitable for generic purposes. Moreover, such networks offer limited services compared to what is available to anybody on wide spreading public wireless networks. These dedicated networks (e.g. ANTARES for French civil security) have anyway good properties in terms of reliability and availability and could not be replaced by public networks that do not have the same constraints and could be unavailable in case of grave crisis.

Researches as well as daylife uses are more and more oriented towards public networks to provide users with advanced services that require more bandwidth and appropriate quality of service. As it is common today to have, at the same place, several wireless access networks using various access technologies (WiFi, WiMAX, UMTS, HSxPA,…), it is possible to use these networks to provide advanced services using multimedia data such as image (3D map of a building), video, voice, etc. Unfortunately, the usage of these public networks is still expensive, each of them has a specific billing plan that could be flat, depends on the amount of data exchanged or on the duration of the connection. It is then necessary to manage carefully the different attachments and to take into account the way the billing is applied while routing the traffic through a specific access network.

Compared to a dedicated network, it remains difficult to guarantee a certain level of quality of service. Moreover, the availability of the network could not be guaranteed. But, all public networks have also availability constraints coming from their commercial nature. Operators have strong incentives to properly maintain their networks in operation. Even stronger if other networks are also available in the same area, as it is often the case.

Regarding the illustrative context of this chapter, if a dedicated network is guaranteed to be operational even during the worst crisis, the technology diversity (i.e. the ability to benefit from all available technologies) could provide a sufficient level of reliability and a larger. Persons in rescue operations could therefore use various media (at least image, video, data and voice have been identified) to ease their work and to reduce the amount of paper they have to fill up once the returned at the commandment center.

3.B. Requirements

Needs in terms of communication principles and technologies when considering location-based service for rescue brigades in operation are representative of underlying scientific issues raised by emerging transportation systems. A preliminary survey made with rescue services identified several communication scenarios (cf. Figure 3.1). In the first scenario devices and software embarked in vehicles (cf. Figure 2.1, Components 2 and 3) should be able to communicate with the SDIS (departmental rescue services) headquarter. To avoid too much unnecessary communications it would be preferable to allow the headquarter to be the initiator of the communication. If the remote device could only initiate the communication, a keep alive should be sent on a regular basis to allow the receiver to detect a dysfunction, even if there is nothing to send. When it is allowed to, the server can request for a keep alive only when it is necessary depending on the current situation.

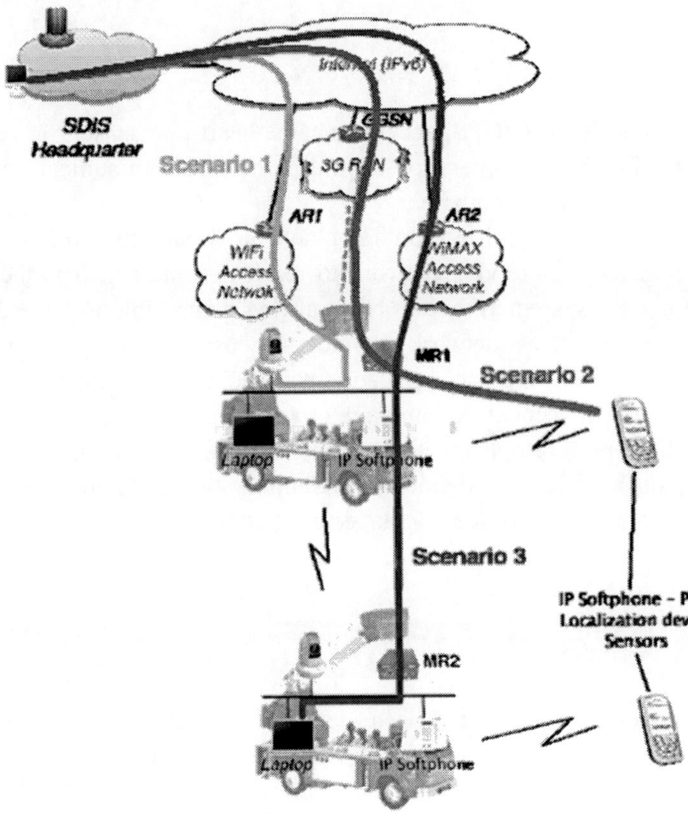

Figure 3.1. Envisaged communication scenarios.

Devices should be able to send information to the headquarter through the Internet with an appropriate level of privacy. Of course the appropriate level of privacy depends on the data and should be decided at the application level. In the second scenario, devices attached to person in operation could send information to the mobile commandment center (Figure 3.1, Mobile Router 1). The same devices could reach directly the commandment center for specific data flow (VoIP). In the third scenario, a second vehicle having its own devices with which it will link to the first one to reach the Internet. This would participate in provided an uncovered area with Internet connectivity temporarily.

Reliability/Robustness/Availability

It is obviously not possible to guaranty the availability of the network in any place at any time. But as we already stated, the probability to get a reasonable Internet connectivity is pretty good, even if it could be necessary to use simultaneously several wireless technologies to improve the Internet connectivity. Once the Internet connectivity is established, the robustness depends on the different computers involved for each application. Then the robustness is much better while the number of involved materiel is reduced. This militates for the use of end to end communications since they could be easily resumed if something goes wrong in between (e.g. a router failure).

Transparency

Going from simple sensors (GPS, accelerometers, heart rate sensors, ...) to full featured computers, all useful devices in operation should be able to communicate with the mobile commandment center and the headquarter.

Nevertheless, very simple devices would not have the capability to manage their mobility while the vehicle changes its serving network to reach the Internet. It is then necessary to provide them with a transparent IP access. It is always impossible to have all this devices featured with long range wireless network interfaces for cost and energy consumption raisons.

One could have imagined that the application will take in charge the mobility (the change of IP addresses) instead of specialized hardware or software in the network. This would have complexified application development without any benefit. Making application unaware of the mobility will allow easier development and simplify the deployment of the solution as they behave just as if they have been deployed in a static environment somewhere on the Internet.

Simple Quality of Service

One of the requirements that have been identified regarding the service differentiation concerns the fact that some flows are more critical than others. It is then necessary to insure that these flows will receive the better service as possible. Anyway when the resources become scarce this could lead to drop number of packets belonging to less important flows. Moreover, depending on the situation, it could be necessary to filter some flows:

- To avoid extraordinary cost due to the usage of an expensive wireless interface to send expendable high bandwidth flows;
- To avoid making unusable a wireless interface sending too much data on it. For example, sending a huge amount of data over GPRS network will fill up all the buffers in the Radio Access Network and leads to unexpected delay for the next data packet which could be urgent;
- To avoid sending critical flows over unsecured wireless networks.

Interface Management

For several raisons it is necessary to manage carefully wireless interface. Energy management and the ability to keep alive wireless devices is a strong incentive to reduce the number of active wireless interfaces and to choose the more frugal ones. This is even more critical for a vehicle when the engine has been powered off. Also, in certain critical situation, some of the wireless technologies should not be used. Everybody knows for example that it is forbidden to use cellular phones in a gas station. It is then easy to understand that some of the wireless interfaces could provoke dangerous spark that could make the disaster worst than before. The interface management piece of software should then take into account the available wireless networks and the running applications. In such contexts, geographical information systems are able to provide location-aware information regarding tracked mobile objects (e.g. vehicles) as well as facilities and dangers in the area of concerns (e.g. where are hydrants in a range of 250m) in order to provide decision-making support.

3.C. The Mobile Router

It is now commonly admitted that a mobile router will take place in vehicles in the near future [60]. Such mobile router (MR, Figure 3.1) has to take in charge the connection of the mobile network standing in the vehicle to the Internet. It could have multiple wireless interfaces and share a single antenna pod between multiple wireless technologies to take advantages from the technology diversity. It could play only the role of a router (at the network level) or deal with the application flows to optimize the bandwidth usage.

Of course, a mobile router will have to manage carefully its wireless interfaces to provide the best connectivity as possible while roaming over the Internet. Moreover the mobile router movement should remain transparent to these devices and to the applications.

Solutions and Products on the Market

Nowadays we could find some product on the shelves. For example, Cisco has product[2] running with IPv4 and with IPv6. The IPv6 solution uses the IETF standard will describe later for network mobility. Kerlink is a SME that has developed a mobile router using a Linux operating system[3]. They are selling a complete solution comprising the material (the mobile router and the software) and a service platform somewhere on the IPv4 Internet. This solution mount a tunnel between the MR that is attached to a public IPv4 network and an access server collocated with the service platform. Other products or papers use the same type of solution. In this solution devices attached in the mobile network will get a private address since it is really difficult to give an IPv4 prefix to all mobile networks. The Network Address Translator could stand in the MR or in the home network. This complexifies correspondent-initiated connections since it requires a complex configuration in the NAT (as for home gateway) [72,62].

---

[2] http://www.cisco.com/en/US/products/hw/routers/ps272/index.html

[3] http://www.kerlink.com/rubrique.php5?SiteID=1&LangueID=2&RubriqueID=141

Mobile Router Management Solution

Recently (in 2008), few commercial offers has appeared in Europe. For example travellers using the high speed train Thalys will be provided with Internet connectivity while the train cross over the countryside at 350 km/h. This solution uses commercial router that manage satellite and 3G/UMTS interfaces and the technology proposed to the customer is the WiFi. The router automatically moves the Internet traffic to the UMTS interfaces when the train pass through a tunnel or in the town premises. In this solution the interface management is quite simple; only two interfaces are considered and a simple priority mechanism could be enough since the model cost applied depends only on the amount of data exchanged. In such a situation the 3G connection is maintained during the entire trip and used only when the satellite link fails. This will not lead to an unsustainable amount of signalling in the 3G operator network since the network tracks the terminal at a cell-basis when traffic is actually exchanged. As we already stated, in our scenario the interface management and routing should be smarter than that and has to take into account higher level constraints.

Standardization in the Field

Numerous works and standards design OBE (On Board Unit) that embarked only one or two types of wireless interfaces. It is IEEE 802.11p [70] for the IEEE WAVE architecture [69] or the WiFi-based interfaces for the Car-to-Car Communications Consortium (C2C CC) architecture [60]. In both cases developments remain complex and do not fit with needs expressed for the illustrative case study. The interface management mechanisms are simplified by the homogeneity of the interfaces and often rely on a specific support in the RSU (Road Side Unit) or in the network. We could not make this assumption since we want to provide continuous communication using generic public networks.

Another body decided few years ago to design evolutive multi-technologies communications architecture for vehicles. ISO has then created the working group 16 in the technical committee 204. This group designs the CALM (Communications Access for Land Mobiles[4]) architecture that is “the first open way to combine public access network (e.g. GPRS) with vehicle-optimized WLAN technologies. It is not a complicated collection of new, unproven radio technologies[5]. As it has already been done for other architectures, the working group chose IPv6 and the IETF standards to manage network mobility and Internet connectivity [61].

3.D. Technical Design

In order to reduce as much as possible development costs and interoperability issues, it is reasonable to use current or ongoing standards currently being designed at the IETF. Development related to positioning and specific applications should not be impacted by the choices to be made for the network and could take advantages from a continuous Internet access. The choice of IPv6 as the network protocol has to be favoured to avoid complex

[4] Formerly, Continuous Air-interfaces Long and Medium range

addressing issues that would have been managed at the application level. Of course, it would be possible to deploy only one IPv4 addressing space comprising mobile networks, persons and fix locations (e.g. headquarter) but it would dramatically complexifies the interaction with external server.

Use of NEMO BS

Going with IPv6 and NEMO already standardized, the ongoing work called MCoA (Multiple Care-of Address) has been selected to ease the management of simultaneous Internet access. The supplementary open tunnels would rather be used in order to keep active communication interface to be able to reroute timely the traffic if needed, i.e. if the current interface could not continue to serve the traffic. Indeed, once a tunnel is established all link layer (attachment to an AP and security check performed) and network layer (configuration of a public IPv6 address) procedure are done and the traffic could use this interface without any delay but the exchange of a binding update with the HA.

Public wireless access networks often fail to offer a fully ubiquitous wireless access. This is why we chose to use technology diversity to reduce the number of places where no Internet access is possible. Anyway it could be not sufficient since any wireless access network does not cover some places.

To increase further the coverage and to deploy a complete infrastructure once the public ones have been damaged by a disaster, a solution is to extend existing infrastructure. A lot of works propose solutions to deploy this kind of network through the design of specific mesh router devices [58,67].

We have rather chose to use the ability to nest NEMO networks. Indeed, NEMO basic support allows a network to attach itself to another mobile network in order to get access to the Internet. This way, the nested mobile network will provide its embarked node with a full Internet connectivity through the use of the Internet connectivity provided by the other mobile network. Of course, this is not really efficient in terms of tunnelling overhead but it allows to easily deploying simple infrastructure without any specific development. When several vehicles come to the operation theatre one of them have to step back in order to be in the coverage of at least one public access network. Then, it offers a wireless access (WiFi or WiMAX) to other vehicles and persons in operation. This principle could be extended using satellite access when the location of the sinister is far from any public wireless access.

Interface Management Mechanism

In [73,74] an architecture dedicated to multiple interfaces management has been described. This architecture relies on a specific set of drivers and on kernel modification to be able to manage multiple interfaces simultaneously and to provide multiple interfaces support to higher layers. An implementation has been realized within the Cyberte [59] project that aimed at offering a seamless mobility management solution to IPv6 mobile terminals. The

[5] To paraphrase Knut Evensen

mobile device distributes dynamically flows over available interfaces taking into account policies given by the administrator and needs announced by applications.

Derived from the Cyberte project, the Ubique framework has been developed to implement a comprehensive multi-interfaces selection mechanism on mobile station without modifying in depth the protocol stack. Later, Ubique has been adapted to mobile router [75].

The purpose of Ubique framework (cf. Figure 3.2) was to bring smart dynamic interface selection capabilities to the communication stack. On one hand, Ubique gathers applications requirements and administrator's policies concerning interfaces management in a local database [57]. Together, application's requirements and administrator's policies or preferences could be considered as routing policies. The profile manager uses them to produce a simple set of rules the selection mechanism will consider choosing the interfaces to be activated and configured. In other words, the Ubique framework is able to deliver an ordered list of preferred and forbidden interfaces merging the policies given by the applications and the availability messages given by the interfaces.

This multi-interfaces management framework manages different interfaces taking into account both the availability of access networks and the administrator preferences (cost, energy consumption, security, priority, etc.). It can automatically manage interfaces activation, attachment, configuration) and to propose to NEMO the best available interface. It provides NEMO implementation with much more than the best tunnel to use: a full list of the available tunnels ordered from the best to the worst and also a list of forbidden tunnels.

Ubique associated with the support of multiple tunnels would improve significantly the performances of the NEMO implementation. In the following paragraphs, we study different alternatives to achieve the support of multiple tunnels and to distribute flows simultaneously on several active interfaces regarding administrator's policies (security, cost and QoS requirements).

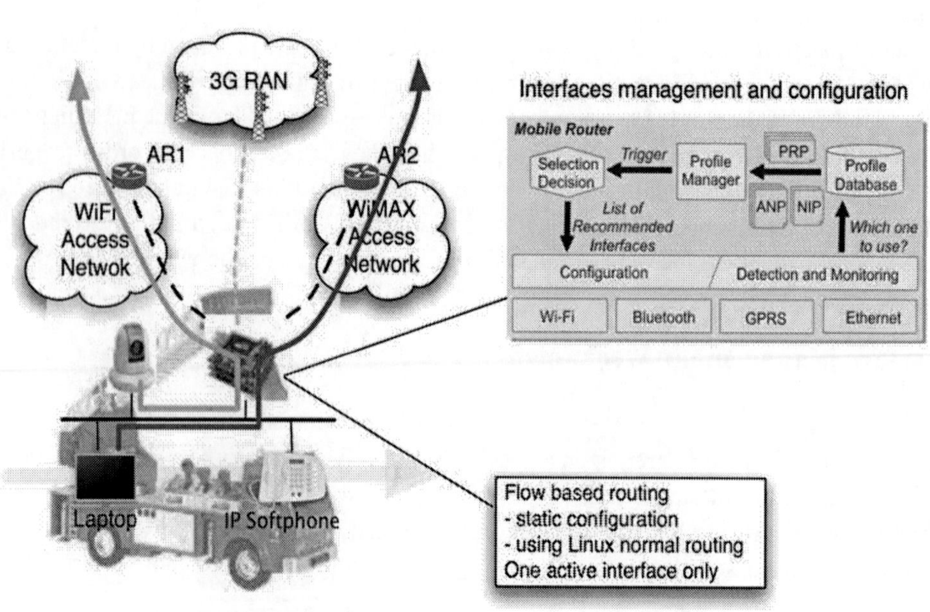

Figure 3.2. Interface management in Mobile Router.

To meet identified requirements a solution could be to integrate Ubique with a NEMO implementation including MCoA support. As for interfaces, when a new tunnel appears or an active one goes down, the tunnel monitoring module will detect the variation and trigger the internal modules to update the available tunnel list [64].

3.E. Interface Management

Various policy/preferences routing mechanisms can be imagined in order to let the mobile end-users and/or the applications to choose how a packet should be treated depending on the interfaces currently in use. A number of improvements are currently under discussions at the IETF. Unfortunately, even if their importance is recognized [66], the decision mechanisms for selecting "best" egress interface (also known as path selection) are not within the scope of the IETF work. Nevertheless, the optimum interface selection problem within the multi-interface mobile nodes communicating in heterogeneous radio environments has already leaded to research works [68,63].

An important part of the Mobile Router architecture is dedicated to the definition and the management of profiles. Generally, the profiles are files stored in profile databases and they summarize key information about the components of the system and its interactions with the environment that is the mobile users, the network operators, or the service providers. Therefore, the *Profile Manager* determines if the *Selection Decision* module needs to be informed or not about the changes occurring within the mobile router. Finally, based on the triggers received from the Profile Manager, the Selection Decision component queries various Profile Databases and suggests the "best" egress interface to be used for communication. The architecture comprises a set of three supporting modules (see Figure 3.3):

- The first one detects the available networks and provides real-time information about the interfaces and access networks capabilities;
- The second "lower-layer" module makes on-demand network interface configuration;
- The third one handles the selection execution process, i.e., it actually maps the network traffic on the preferred outgoing interfaces. The simplest and obvious way is to use a static configuration (i.e. rules) to control the way interfaces are used by the mobile router.

The middleware can control, through the use of a comprehensive set of scripts, all these lower layers modules by initiating network configurations and performing interface activation/configuration decisions.

Selection Decision

After receiving a trigger from the *Profile Manager*, the *Selection Decision* module uses various profiles for choosing the most suitable egress interface to be used for communication. Thus, in our approach the *Selection Decision* entity does not need to know how the parameters are collected within profiles or how the selection decisions are enforced.

The outcome of the *Selection Decision* module consists of a list of recommended interfaces. This list contains two parts: The first part comprises the preferred interfaces that are already configured and can be employed for communication immediately; the second part contains the interfaces that may be activated and employed only if the configured interfaces fail to offer the expected services.

In this kind of architecture the decision to use or not an interface is usually made for all existing traffic. In other works, the *Selection Decision* module proposes the best suited interface for each flow, and thus, several interfaces may be used simultaneously on the mobile routers.

3.F. QOS Management

The mobile router is in charge to route flows depending on their type and how they have been classified. A set of static rules is configured in the mobile router to associate specific behaviour to different type of flows. If this is enough to feed the interface management process or even to share the load between active interfaces, another mechanism is needed to be sure urgent traffics experience the best service as possible.

It is obvious that safety related applications have such kind of requirements since some packets warn for an immediate danger (a fireman pass out) while others just give supplementary description. There is also background traffic that could be postponed without damage.

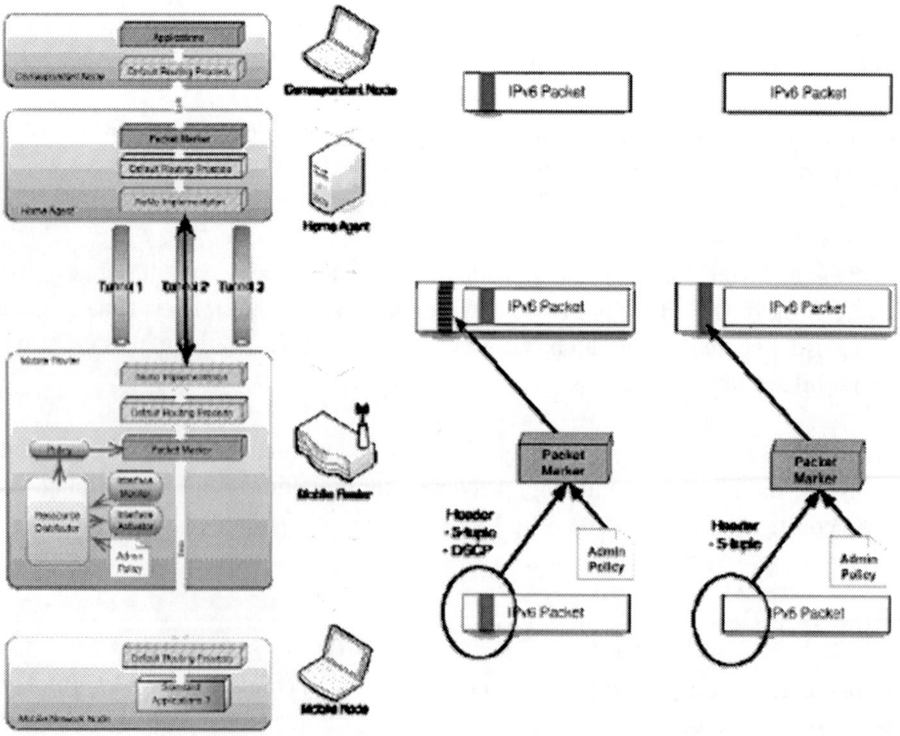

Figure 3.3. QoS management.

Service Differentiation

Diffserv (Differentiated Services) is a global traffic control mechanism that provides an interesting solution for traffic and network resources management. Diffserv is based on per-class resource allocations. In the Diffserv architecture, multi-field classifiers of edge routers use multiple values in the IPv6 header to classify packets and to mark them with a DSCP (DiffServ Code Points). Packets could be classified using the 5-tuple (SrcAddr, DestAddr, NextHdr, SrcPort, DstPort) but also with the previous DSCP and the IPv6 flow ID. Core network routers use only the DSCP to decide which behaviour to be applied to the packet.

Diffserv is simple, effective and commonly used but it also has several drawbacks in this context: First, the marking process is semantically different from what is expected in our architecture. Second, when marking the packet, we lose the initial value of the DSCP field that is used by applications to differentiate packets depending on their relative importance. This will prevent the end to end QoS classical process. Finally, the packet have to be unmarked just after getting out of the tunnel because the DSCP field's value could be interpreted wrongly in the network after the Home Agent (HA).

Simple Priority Mechanism

Some traffic can be automatically classified depending on their characteristics (5-tuple), anyway the application issuing a packet knows what the purpose of this packet is and then it can indicate setting up the DSCP which is the service level required for it.

Each packet contains the DSCP that tell to the mobile router which behaviour should be applied to the packet. The mobile router has dedicated queue for each PHB and applies a priority queuing between classes of services while other policies could be applied insides each class. By definition, EF (Expedited Forwarding) class of traffic is always served before all other pending traffic. This ensures that urgent packet will always get the Internet as soon as a connection is possible and whatever the Internet connectivity quality is.

Packet could be marked by the application or directly by the classifier at the mobile router. In the later case, the packet does not need to be actually marked and the packet sent in the MR-HA tunnel does not have a mark. Then, the QoS mechanism remains transparent to the rest of the Internet. In the former case, the packet is marked by the application. It carries the application semantic. In compensation, the mark has to be transparent to the rest of the Internet since all routers on the path between the HA and the CN (Correspondent Node) see the mark and could deal with it.

QoS and Nested Mobile Networks

In the case of nested NEMO the problem become more complex since the hosting mobile router cannot classify the traffic. Indeed, to classify traffic a router has to look at the packet header and use static rules to classify the packet regarding the value of given fields (source address, destination address, port source and destination, protocol number).

Unfortunately the hosting mobile router could not do that since it can only read the external header which is a IPv6 in IPv6 tunnel header between the nested mobile router and its HA. All information carried inside this tunnel is potentially hidden as the tunnel could be ciphered using IPSec. To solve this problem the first mobile router classifies the packets if they have not been classified yet by the issuing application. The diffserv mark is then copied inside the external tunnelling header. This solution works but it has a drawback: the mark is revealed to the Internet.

A mobile router could not know if it is the last mobile router before the Internet without a complex mechanism. Knowing that, it could decide to not mark the packet and even strip the mark before forwarding the packet in the last tunnel. Unfortunately this would lead to lose the mark carrying the application semantic if any.

We chose a simple mechanism to reduce the burden due to QoS field management in mobile router. First, a mobile router always classifies and marks the packet depending on static classification rules but saving mark set up by application. When marked packet arrives at the mobile router, it considers the mark and set the same mark in the external tunnel header. If the incoming packet does not have a mark, static classification rules are applied and then the mark is put on the external header too. This way, the mark appearing in the tunnelling header is always compliant with the mark seen in the carried packet.

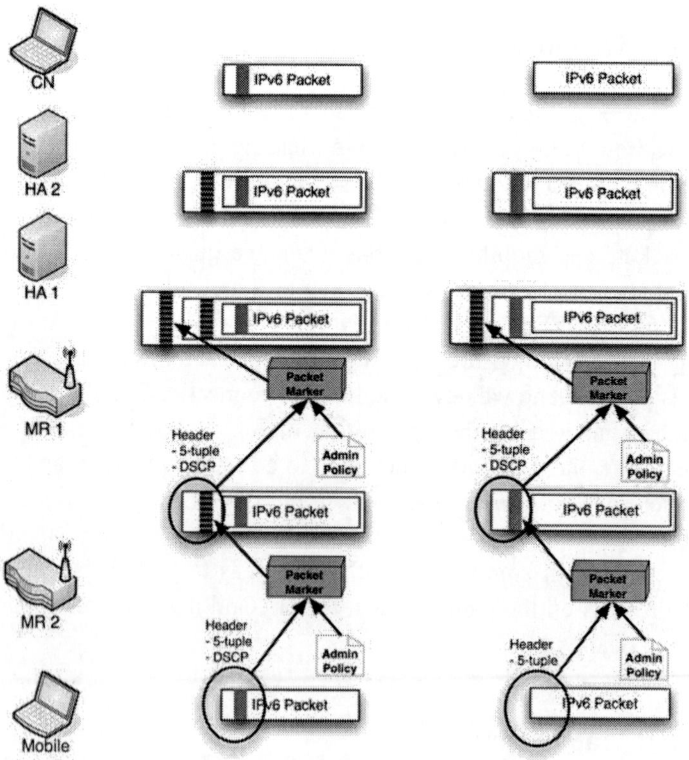

Figure 3.4. QoS management in nested case.

When the HA receive a tunnelled packet it will remove the external header and then the mark the mobile router has add. Router on the path between the HA and the CN will only see the mark set by the application on the original packet.

3.G. Improve Connectivity

For numerous raisons we have chosen to use IPv6 instead of IPv4. Unfortunately, for now there are still a lot of public access networks that only offer IPv4 services. In this section we review solution to use IPv6 protocol over IPv4 only accesses. There are two distinct problems to solve. The first one is related to the ability to take advantage of a wireless Internet access whatever it offers IPv4 or IPv6 connectivity. The second one is less important in our context. It refers to the ability to use IPv4 only services from our IPv6 only mobile network. The idea is to be able to communicate with IPv4 only server located in the IPv4 Internet without having to manage a complex Application–Level Gateway (ALG) inside the NEMO or inside the Home Network.

Dual Stack Mobile IP (cf. chapter 12 of this book) is currently being standardized at the IETF. It aims at solving the both problems. First it allows to register an IPv4 CoA with the HA when attaching to an IPv4 only access network. The Dual Stack Home agent will build an IPv4 tunnel when it has no way to do it with IPv6. The tunnel will carry IPv6 or IPv4 traffic indifferently. Using MCoA, the HA and the mobile router can have several tunnels opened. Few of them can operate in IPv4 while remainders use IPv6.

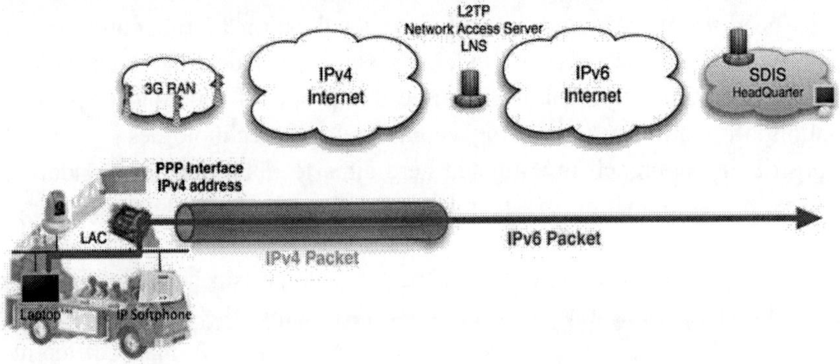

Figure 3.5. Using L2TP as IPv4/IPv6 transition tool.

NEMO has also been modified to be able to exchange IPv4 prefixes. This way it is possible to provide embarked node with an IPv4 address that allows it to reach the IPv4 Internet. Unfortunately this will not solve the question of prefix availability since a prefix should be allocated to each NEMO.

This solution is promising but no implementation existed when we designed our architecture. We study the possibility to use VPN client at the mobile router to build a tunnel between the Home Network or another IPv6 Internet Service provider and the mobile router. This tunnel could use whichever tunnelling techniques currently used to build VPN over IPv4 access. For example, Open VPN[6] client can be chosen. This solution allows to support NAT traversal if the tunnelling protocol is build over a transport level protocol.

In our design we have chosen on particular tunnelling techniques that is L2TP (Layer 2 Tunnelling Protocol). L2TP allows to build a PPP connection between the mobile router and

[6] http://openvpn.net/.

the IPv6 service provider; the home network in our example. This connection is build over IPv4 and use UDP as transport level protocol, which allows us to pass through NAT.

The main interest of this approach it that it builds a PPP interface in the mobile router that can be used just as normal communication interfaces. It is possible for example to configure it with an IPv4 address and an IPv6 one. Doing that will solve the two identified requirements. Moreover, since PPP is used, it is also possible to apply various well proofed compression techniques over the tunnelling reducing by the way the overhead due multiple encapsulations [71]. Some techniques have also been proposed [65] to reduce the overhead due to the NEMO and Mobile IP tunnelling but nothing has been still adopted at the IETF.

4. Improving Pedestrian Localisation

Intelligent Transports Systems are commonly associated with motorway applications. However, evolution of ITS, particularly when considering location-based transportation services, tent to consider pedestrian navigation and tracking as users' mobilities are changing as he become more and more mobile and ubiquitous. In pedestrian applications, ITS can be very useful to navigate in complex building, to obtain commercial information from surrounding shops or to simply get location-aware information. For a correct process, such applications need a precise outdoor and indoor localization (and a combination of both to avoid loss of service). This section addresses accurate pedestrian localisation, introduces some limitations intrinsic to technologies used and proposes a way to consider multi-modal transportation information systems using a combination of technologies.

The positioning approach investigated here aims to obtain accurate indoor and outdoor positioning with a combination of several location-based technologies. Sensors information and localization are transmitted between persons in operation and recorded in order to be used latter. In safety application particular infrastructure, which can be defaulting, can't be used. More the solution should not be expensive. Regarding the difficult issue of tracking pedestrians inside buildings (e.g. in and between metro stations), many different positioning approaches such as Local Positioning System (LPS) or GSM-based techniques have been developed and experimented [1]. The solution that illustrates this chapter is based on GPS receivers combined with inertial sensors and 3D indoor map-matching. In such approaches, GPS receivers are used when GPS signals are available, and MEMS (Micro Electro-Mechanical System) inertial measurement units otherwise. MEMS have been selected due to the fact that they are autonomous and don't need any infrastructure. These inertial sensors allow to estimate a relative localization and to derive travelled distance and the number and size of steps from accelerometers while the direction is rather provided by gyros and magnetometers. The main drawback of inertial system is that MEMS ambiguity introduces a location drift.

4.A. GPS localisation

Usually, geolocalisation is associated to Global Positioning System (GPS) that is a GNSS (Global Navigation Satellite System). In fact, GNSS is the most used system to obtain a precise localisation on each point of the Earth. The GPS has been developed by the United

States Department of Defence and is the only fully functional GNSS [2]. Utilizing a constellation of at least 24 satellites that transmit precise microwave signals, the system enables a GPS receiver to determine its location, speed, direction, time, date, altitude,… Many receivers relay position data using the NMEA 0183 protocol. Other similar systems are the Russian GLONASS, the upcoming European Galileo positioning system, the proposed COMPASS navigation system of China, and IRNSS of India.

GPS Principle

In a considerably simplified approach, each satellite is sending out signal with the following content: its identification, its position and the time reference. In addition to its own position, each satellite sends data about position of other satellites. Then, to determinate its position on earth, the GPS receiver compares the time when the signal was sent by the satellite with the time of the signal reception. From this time difference, the distance between receiver and satellite can be calculated. Finally, the localisation is calculated by triangulation. This means that at least four satellites are required to determinate a 3D position of the GPS receiver [3].

Receiver Characteristics

A microcontroller allows interpreting signals from GPS satellites and determinates the position of the GPS receiver. The SiRFstarIII chipset used seems interesting due to its ability to acquire and maintain a signal lock in urban or densely covered forest environments, and its faster Time to First Fix. SIRFstar III receiver announces a precision of less than 10m, which has been experimentally verified.

GPS Integrity

The position calculated by a GPS receiver requires the current time, the position of the satellite and the measured delay of received signal. The position accuracy is principally depending on the satellite position and signal delay. From theoretical parameters, the error using only the GPS C/A signal should be less than 3 meters. In practice different parameters deteriorate this resolution. The main phenomena, which disturb the GPS localisation, are presented here:

1) *Atmospheric conditions* which affect the speed of the GPS signals especially in the ionosphere. Using differential GPS or a bi-band receiver, which are more expensive, can compensate this effect.
2) *Multipath effects*: in this case, the radio signal reflects on buildings, canyon walls, ground, water… This multipath leads some delayed signal and cause inaccuracy.
3) *Environment perturbations*: this term regroups all perturbations from the receiver area as forest, high voltage line, wireless communication…

As example, Figure 4.1 illustrates some acquisition realized in free space (a) and under trees (b). In this case, the power of the received signal is reduced due to vegetation diffusion. This attenuation is reproduced at the location integrity.

Moreover, GPS localisation is unavailable for indoor environment. Indeed, the power of the GPS signal is too weak to pass through walls.

4.B. Complementary Locating Systems

Currently, GPS location is the most used to obtain a precise localization. But this system, as shown previously, presents some limitation and it's not always available which is annoying for pedestrian application. So, in this section different localization systems are briefly presented in order to justify our choice.

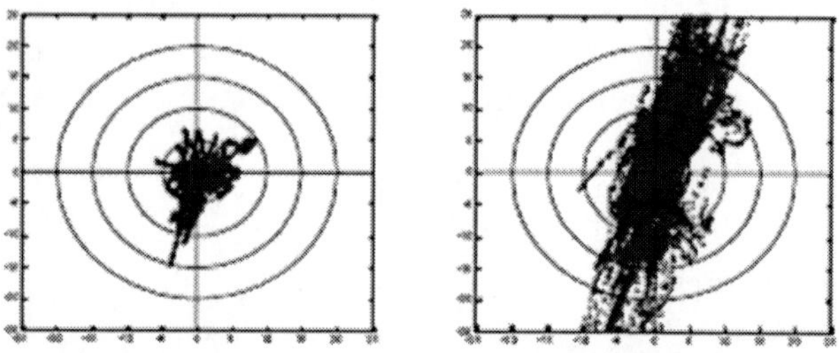

a- Free Space b- Under trees

Figure 4.1. Location integrity.

GPS Relayed

The principle consists in placing the relays in strategic places of the zone to cover to emit a replica with a greater power [4]. This technique is particularly useful for solving problems of attenuation (due to the passage walls or tree foliage), however it does not solve the problems concerning multipath. Moreover this method is particularly complex and expensive to set up because it requires the generation of perfectly synchronized signals.

GSM

The idea of the location by mobile phone [5] consists in using ground station, with directional antennas, for triangulation. The signal used comes from a channel of control, the BCCH (Broadcast Control Channel), which serves to transmit IDs cells adjacent to mobile stations with a constant power. The biggest problem is that the principle of localization requires the perfectly knowledge of the GSM transmitter position. More, the performances of this system depend of the influence of meteorological parameters. Regarding the localization resolution, it is still higher than ten meters.

RFID (Radio-Frequency Identification)

The principle is quite similar to the previous one. It supposes to equip the person of a device that transmits a radio wave with a known frequency. Then, the radio wave will be detected by a network of specific detectors [6]. This technique requires a particular infrastructure. This infrastructure can also be used for transmission of data. Concerning the location accuracy, it will be directly related to the type of signal used.

Sensors Network

Sensors are laid out in strategic paths and/or carried by different people [7]. Each sensor can communicate with each other's to construct a set of nodes. By processing information obtained by other sensors, a sensor can then be located in the network. The greater the number of sensors is, the better the resolution will be. This method is currently under development; it is not yet possible to quantify precisely the results obtained.

MEMS Localization

The localization from inertial measurement is fundamentally different from methods presented previously. Indeed, all previous methods are based on communications between transmitters and receivers. Thus, they depend on infrastructure more or less complex. With inertial measurement, information comes from various sensors (accelerometers, gyroscopes, magnetometers…) located on a person [8]. Miniaturization of sensors using MEMS brings together all the necessary sensors in inertial navigation equipment of small size. Then, the principle consists in collecting the information provided by all sensors to characterize the motion of a person (speed and direction). By knowing the origin of displacement, it is possible to carry out a relative localization.

The main drawback of this method is that the sensors are not perfect and generate errors. These measurement errors are transcribed directly on the localization, to be called drift. MEMS localization is particularly interesting in many application contexts (particularly for rescue services tracking) due to the fact that it is independent of infrastructure and it is always operational.

4.C. MEMS Localisation

In this part, MEMS localization technique is more detailed. Indeed, a precise localisation is needed for pedestrian ITS application in order to give or receive appropriate information (commercial, safety…). So, sensors used in our application are presented, and a first location algorithm is presented

Sensors and Principle

The used sensor is the inertial measurement unit MTx from Xsens Company. It gives 3D accelerations, 3D rates of turn (rates gyro) and 3D earth-magnetic fields. Accelerometers give the number and size of steps, the direction is provided by gyros and magnetometers. Then a relative localization is obtained from the initial position. The main difficulty of this localisation is that the localization derives due to sensors errors. So, the longer the way is, the more the drift will be important.

Localization

In the case of localization from MEMS, it is first necessary to characterize the walk of the person. This is done from information provided by the sensors of the inertial central. Thus, a study of the characteristics of the mechanics of walk is needed. A detailed study of this one was produced by Cavagna [9]. It results the movement of its barycentre can characterize motions of a person: Its hip. So, sensors are positioned on individual hip to characterize his walk without obstruct its displacement.

Initially, it is necessary to detect the steps of the person. For that, information provided by the accelerometers is particularly interesting. In fact, human walk creates a succession of acceleration related to the various steps of displacement. For a forward displacement, major accelerations are along the vertical axis; these accelerations are illustrated in Figure 4.2.

The number of steps is determinated from maxima or minima. Then the step size is obtained from empirical following expression [9]. Then the direction is directly determinated from the magnetometers. Indeed, those make it possible to directly obtain orientation of the sensor compared to the cardinal points.

$$L = \frac{0{,}3}{0{,}55} \times size \times (0{,}725 \times \frac{1}{time\_mi} - 0{,}495)$$

were *size* is the people height and *time_mi* the time between two minima.

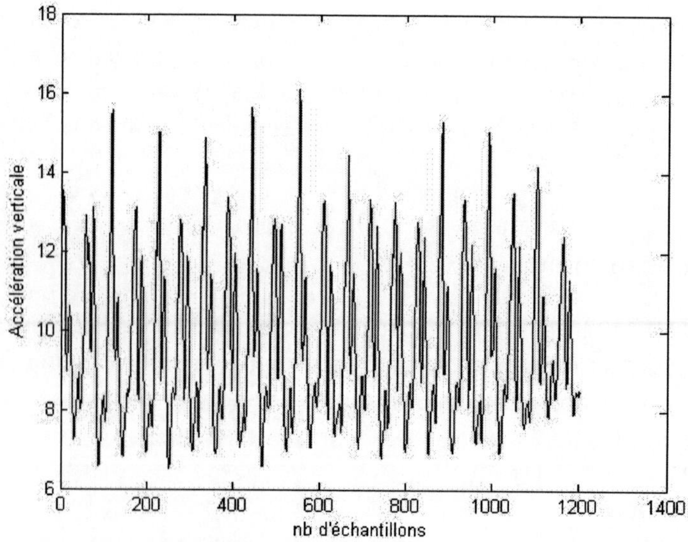

Figure 4.2. Recording of vertical accelerations during the walk.

The obtained localization is a relative localization, so it is necessary to precisely determine the initial position of the people. It can be given directly by the people on an interactively map or automatically from a valid GPS localization when it's available.

The direction is given by the magnetometers. Magnetometers sensors are very sensitive to their environment, so it is necessary to compensate - in part - their drift with the gyroscopes. For that, when the magnetometers indicate a change of direction, it checks if the vertical gyro has detected this change of direction. Otherwise, it is assumed that the change of direction was caused by the external environment (metal beam, electrical sheaths…) and is not taken into account.

4.D. GPS and MEMS Fusion

GPS and MEMS localization seems to be complementary. MEMS give a permanent localization but introduce drift and GPS allow obtaining precise localization when GPS signal are available. In this part, the approach to combine these two methods is presented.

Principle

The performance of GPS localization can be improved by using complementary sensors as inertial sensors. Inertial sensors provide information continuously (unlike GPS which depends on the received signal), so the best way to combine GPS and MEMS, is to use the GPS information, when it's available, to correct inertial localization. Thanks to hybridization with the GPS, it is possible to obtain good performance with low-cost sensors. The GPS is used to initialise the first localization and then to update accurate data (every second). In order to take into account the localization drift from inertial sensors, the estimation of localization is not deterministic as presented in the next section.

Fusion Algorithm

In our algorithm, the considered localization is not single but it's a set of localization named particles. The initial positions of particles are given by a valid GPS localization and the distribution by the GPS integrity. The number of used particles depends on the application. When inertial sensors detect a step of the person, all particles move. Each particle moves independently and differently. The step size is determined randomly from a normal distribution centred on the value L where L is the estimated size from the model in section 3.c. In the same way, direction of each particle is obtained from a random process with a normal distribution centred on the estimated direction from gyros and magnetometers. With these considerations, the area of valid localization becomes greater and greater very quickly. Then valid GPS localizations are used to reduce this area. When GPS localization is available, all particles are readjusted around this position according to the GPS integrity.

Results

Results obtained allow improving the GPS integrity in constrained environments (forest, urban environment…), and then the resolution is less than 5m when GPS is frequently available. Otherwise (very dense forest, indoor), localization is only obtained from inertial sensors and the position error become quickly very important. In fact, the longer the trajectory without available GPS is, the worse the integrity will be.

4.E. Indoor Map-Matching

In the case of localization in the forest or city, available GPS position can be expected occasionally to readjust particles position. Otherwise for localization within a building, getting available localization from GPS system is much less probable. It is therefore necessary to select another type of repositioning, the solution which will be presented in this section is the indoor map matching. This problem is more complex than the previous one and will be more developed.

Experimental Context

The environment used to the indoor experiments is illustrated in figure 4.3. This environment is particularly interesting because it is made up of offices of various sizes, collective offices and corridors. In order to verify the possibilities of localization indoor and their integrity, various circuits - more or less complex – have been followed within a building. Two circuits are presented Figure 4.3.

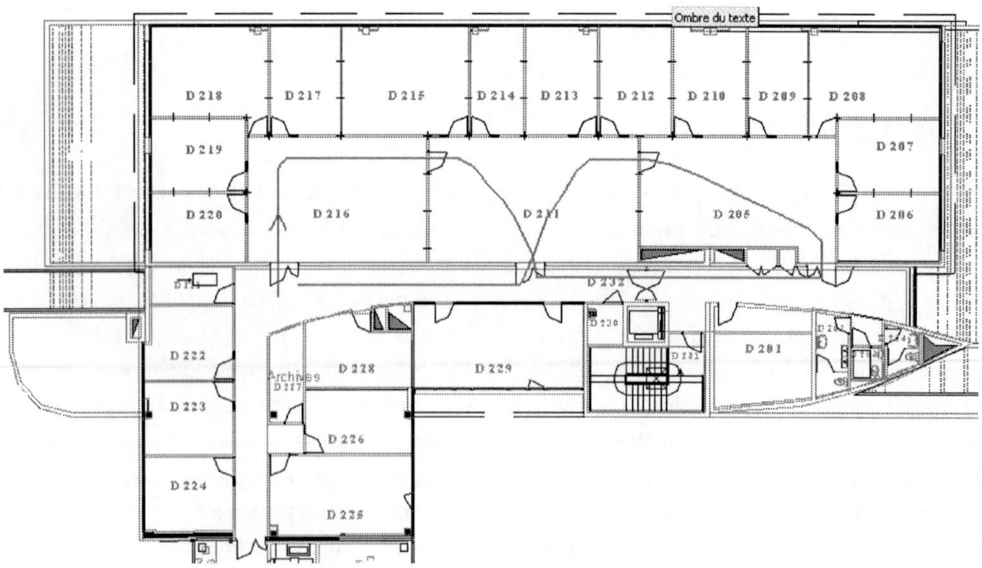

First circuit

Figure 4.3. Continued on next page.

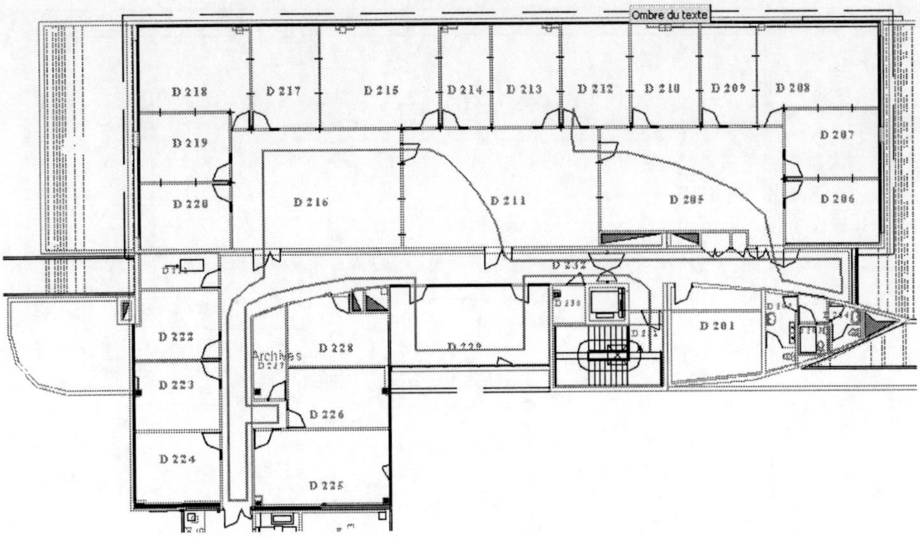

Second circuit

Figure 4.3. Circuits made in the building.

Figure 4.4 gives the trajectories obtained from inertial sensors. It can be seen that for all the ways, the general shape of the circuit can be found. For the first circuit, the arrival positions are much closed to the real positions; on the other hand the evolutions of the ways do not correspond to reality. The trajectories do not follow real directions and pass through walls to move from one room to another. Without a priori knowledge, it might be difficult to know constantly where the person is. In the same way, for the second circuit, the sensors drift accumulates and the final position is outside of the building.

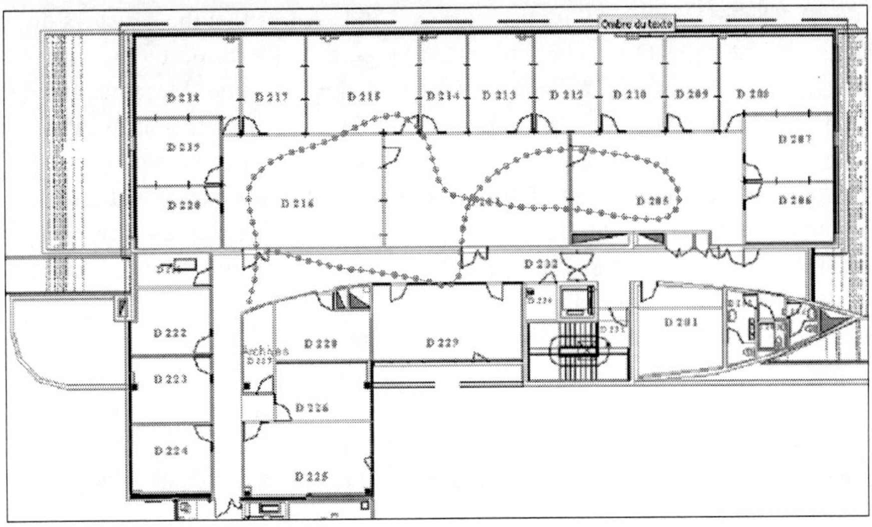

First circuit

Figure 4.4. Continued on next page.

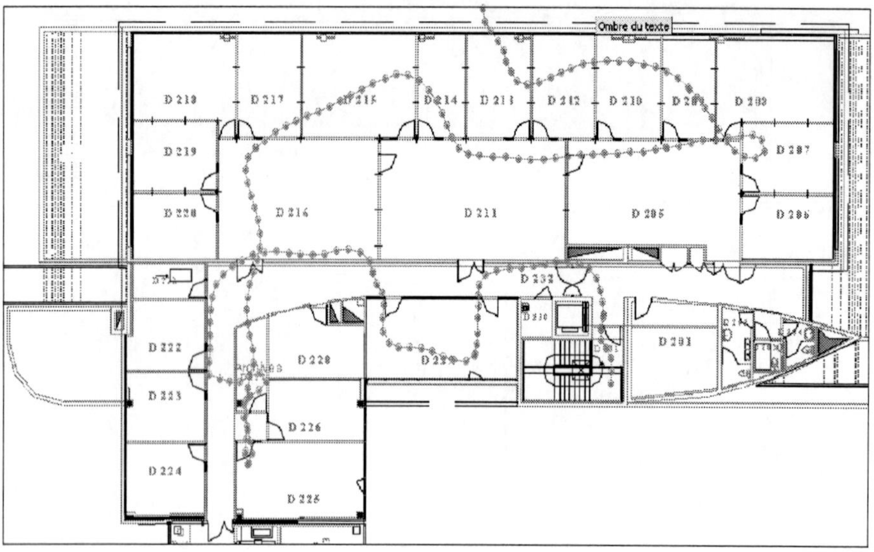

Second circuit

Figure 4.4. MEMS localisation.

In order to improve the localization, trajectories will be forced by the building infrastructure: the indoor map-matching.

Referencing Buildings

Map matching principles introduced in Section 2 is commonly used road navigation and path finding (personal navigation devices like Tom-Tom, ViaMichelin or Magellan include such features). The principle is to use a vector map and assume that the car stays on the road. Thus, the errors of localization can be corrected by putting the car on the nearest road when the GPS locates on the roof of a building or in the middle of a pond. Moreover, when a car is on a road, map-matching limits its upgrading capabilities to both directions of circulation.

In the case of the location of a person in a building, the problem is more complex. Indeed, each building has a particular infrastructure that might be formed corridor office more or less large meeting room, engine room… Furthermore, when person is in one piece, it can move in all directions. The only constraints used are the limits of the building and the passage of a person from one room to another through a door.

Some studies start to be interesting in indoor map matching [10] [11] [1], those are strongly inspired by road map matching: the buildings are broken down into knots (access doors to the building room) and stop connecting nodes. In our application a precise localization in each room is wanted, so this representation is not appropriate. More, for building with complex structures, this representation may be insufficient to take into account all possible trajectories.

It is therefore necessary to carry out a referencing of buildings showing the positions accessible by a person, and then available direction from each position is defined. The grid is an 8-connexcity orthogonal grid (Figure 4.5) with 0.5m steps.

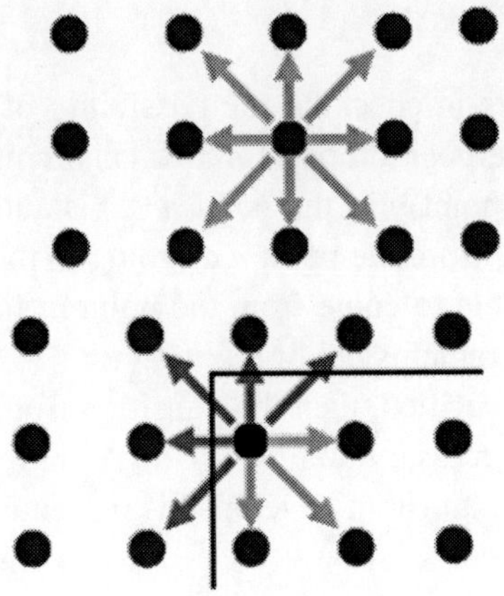

Figure 4.5. Illustration of the grid and the ability to move.

The step size (0.5m) was chosen as it allows discretizing all buildings; it is indeed rare to find corridors closer than 50cm. For each points of the grid, directions in which people can move are defined. For example, the configuration shown in Figure 4.5, where black lines represent a wall, for the black point it will be possible to move in the right directions, bottom-right and bottom (green arrows), however it is impossible to head in other directions: bottom-left, left, left, top, top, top-right (red arrows).

From these considerations, referencing of the considered building is realized (cf. Figure 4.6). From the plan it is possible to determine all the positions available to a person. They must then be determined for each position, which are the authorized directions.

Indoor Map-Matching Algorithm

The purpose of the indoor map matching algorithm is to check the credibility of the trajectory obtained from inertial sensors to suggest an alternative way if the latter is not possible. The first step of the algorithm is therefore to associate the localization obtained at the nearest point of the grid. For that, a simple calculation of distances between the point and positions located in the grid is necessary.

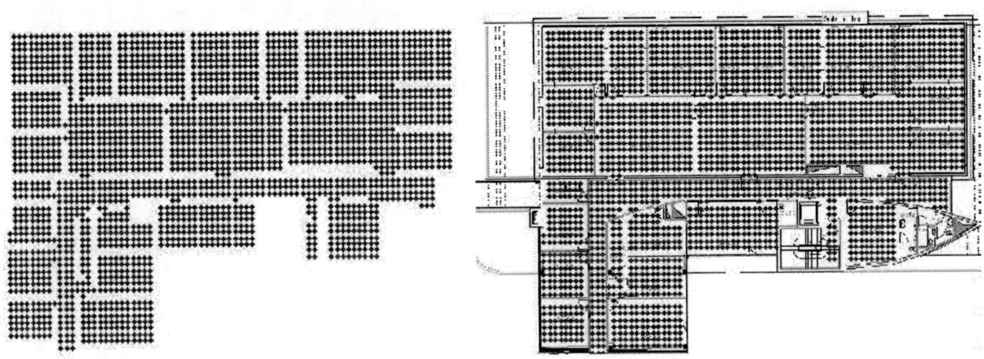

Figure 4.6. Accessible positions on the building.

Evolution Law

The evolution law consists in checking the possibility of upgrading between the current position point *n* and the future position point *n +1*. This requires the building referencing to check if *direction (n)* (of the point *n* to the point *n+1*) is authorized. That is to say that one verifies if it's possible starting from the point *n* of going in the direction of the point *n+1* and if for the point *n+1* it is possible to come from the point *n*. In addition, the distance between the point *n* and the point n *+1* remains below 1.5m.

If all these conditions are verified, then the transition from the point *n* to the point *n +1* is permitted. Otherwise, it is necessary to use an algorithm to find a new way to pursue trajectory in the building. The algorithm developed is presented in the following section.

Monte-Carlo Algorithm

The indoor map matching uses the method of Monte Carlo. This method is based on random processes, in order to find a solution to the problem studied: in our case the position of a person. The localization is obtained from a set of positions that evolve independently one from another.

The application is as follows: in first time the set of positions are placed uniformly in an area of $1m^2$ around the original position. Each position is called particle, the number of particles will determine the size of our system and will remain constant throughout the trajectory. To move from one position to the next one, the direction is determined randomly from a normal distribution centred on the direction given by the inertial sensors.

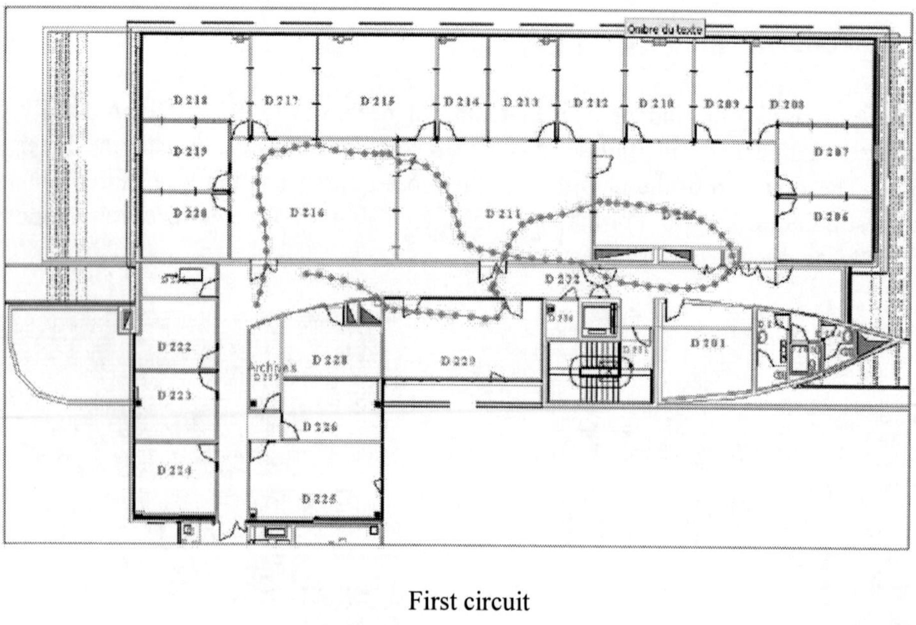

First circuit

Figure 4.7. Continued on next page.

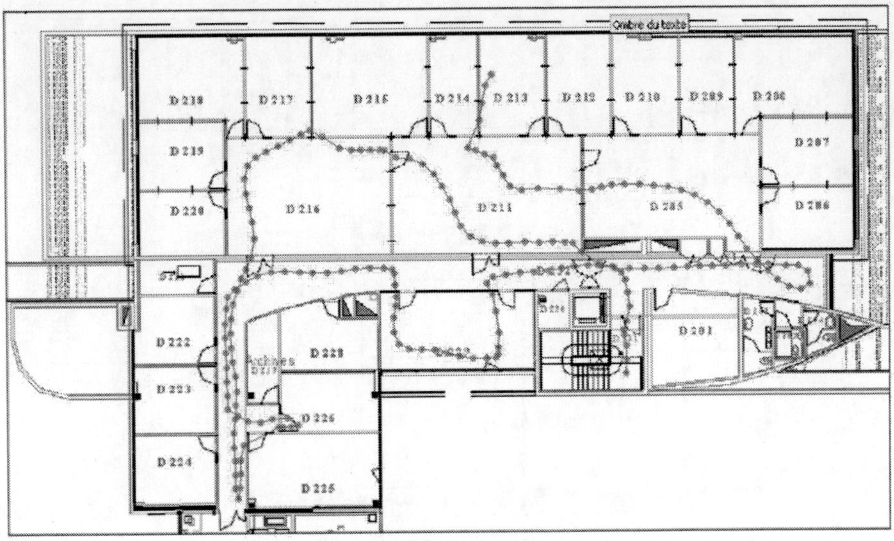

Second circuit

Figure 4.7. Localisation from Monte-Carlo algorithm.

Similarly, the step size is determined randomly from a normal distribution centred on the value obtained from the vertical accelerometer. Each particle will then follow the law of evolution described above. When a particle cannot move anymore, it is deleted and replaced by another one drawn randomly from among the valid particles. The results are shown in Figure 4.7. The trip represented in red corresponds to the average position of the particles. This average position takes into account positions that will be deleted in the next step, so the average position does not follow exactly the real trajectory and pass through walls. However, for each trip it is possible to find particles that follow the exact way and leading to the exact final position.

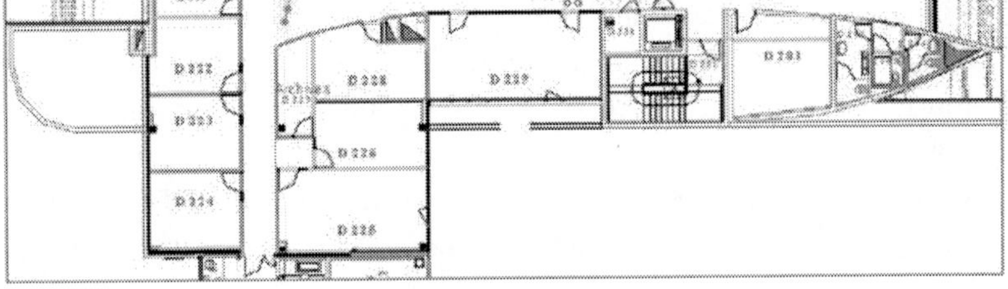

(a) Particles separation

Figure 4.8. Continued on next page.

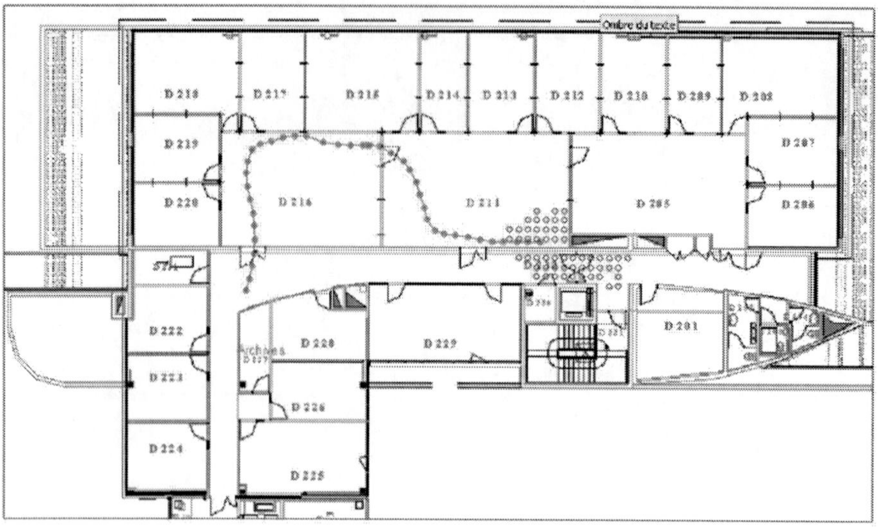

(b) Particles removal

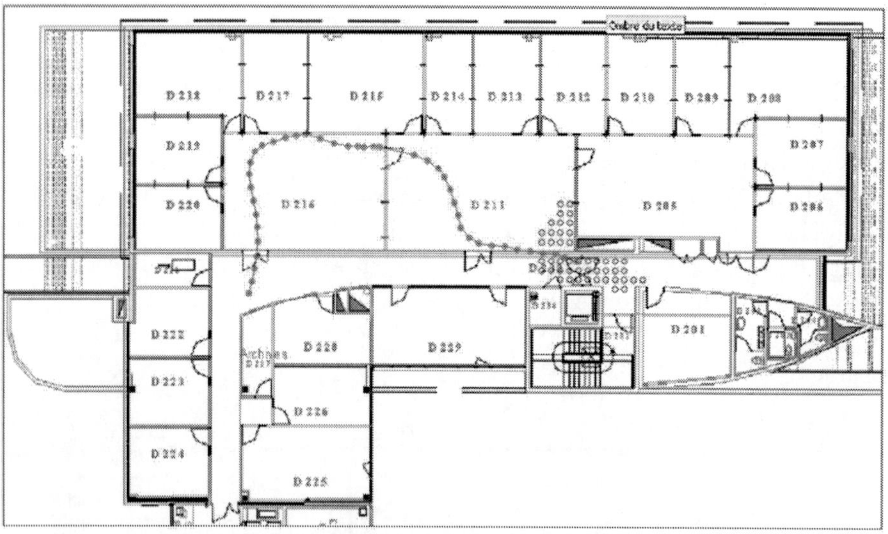

(c) Recalage of the average value

Figure 4.8. Evolution of the average value when moving from the office to the corridor.

In order to more precisely explain the evolution of the mean position, let's get to the first circuit during the passage of the room D211 to the corridor. Some particles remain in the room D211 and the other go in the corridor (Figure 4.8a the blue points correspond to the particles). The average position remains in the room D211, but when particles remained in the room D211 arrives to the limit of the room D211 they are deleted and replaced by valid particles in the corridor (Figure 4.8-b) and average value will then be located in the corridor (Figure 4.8-c).

This algorithm allows obtaining good results in localization and trajectory. Moreover, the fact of knowing the number of particles make it possible to determine in advance the algorithmic complexity what is essential for a real-time application.

Conclusion

This chapter aimed at presenting ways to combine wireless communications, localization techniques and information systems for the design of location-based services. Principles and methods have been linked and illustrated through an application that supports the real-time following of rescue services. Beyond this global approach for the design of intelligent transportations systems, one of the main purpose of this chapter is to demonstrate that transportation systems will tend progressively to integrated multi-modal transportation systems able to cope with individual (pedestrian, vehicle) and collective (public transportation) mobilities in order to ensure a continuity of location-based services. Such approach is motivated by the increasing of mobile and ubiquitous users looking for location-based services such as path finding or navigation whatever their transportation mode.

The role of information systems for ITS is crucial as it brings services transparently to users. The information system (Section 2) designed to illustrate these principles is based on 4-tiers distributed architecture that take advantage of communication facilities (Section 3) to collect and process location-based information generated by several mobile objects (ships, firemen, vehicles). Such software platform generally stores data in spatio-temporal database, process fusion algorithms and provides trajectory analysis as well as transportation information and services. Visualisation outputs of these location-based data can be done either with Web-based and technologies or standardized exchanges languages according to user's context and needs.

Wireless communications acts as a key component in real-time intelligent transportation systems. Their role is to establish connection between sensors, information systems and users. Human and vehicles in motion require a communication architecture that provides transparently persistent network connectivity. The approach described in this chapter allows mobile networks to host any kind of IPv6, devices from simple sensors to multimedia terminals. They just have to take in charge normal IPv6 auto-configuration to be able to communicate with any correspondent on the Internet. The support of a well-known QoS mechanism allows applications to specify in each packet how it is important from an application point of view. If the application does not want or has not been designed to add this mark in the packet header, the mobile network's administrator can add static rule in the router to mark the packet regarding its header (the 5-tuple). Moreover, this mechanism still works with nested mobile networks. This architecture has also some drawbacks regarding the way applications can have an influence on the way the mobile router manages interfaces and routes the traffic. But a trade-off had to be found between the availability of standards and implementations and the need for unsupported functionalities. In this architecture the main concerns was to improve the availability of the network disregarding other aspects such as the adaption of application behaviours to the network conditions experienced by the mobile router or the reduction of network usage cost.

In terms of localisation in constrained environment, different systems allow to determinate a position, but each of them has some constraints and limitations (Section 4). In order to obtain a precise localisation whatever the environment, the use of hybrid systems based for instance on GPS receiver and inertial sensors brings solutions to multi-modal transportations and to accuracy issue. Such approach is particularly well suited in the case of indoor localisation where GPS positioning is impossible. This chapter also illustrates the role of map-matching algorithms in indoor localisation. An example of algorithm based on Monte-Carlo allows obtaining good results for localisation and trajectory. It also shows that multi-sensors fusion and data processing return accurate pedestrian localizations (better than 5m) whatever is the environment.

Acknowledgements

Research support from the French Brittany Regional Council, the Côtes d'Armor departmental council and the Finistère departmental council permits the development of the LoCoSS project (localisation and communication techniques for rescue services) that has been used as an illustrative application in this chapter.

References

[1] Gilléron, P.Y.; Spassov, I.; Merminod, B.; *European Journal of Navigation,* 2005, 3(3). Indoor Navigation Enhanced by Map-Matching.

[2] Parkinson, B.W. Spilker, J.J. *Global Positioning System, Theory and application,* American Institute of Aeronautics and Astronautics, 1996.

[3] Hofmann-Wellenhof, B.; Lichtenegger, H.; Collins, J.; *Global Positioning System, Theory and Practice*, Springer Verlag, 2001.

[4] Chen, Y; Kobayashi, H. In *IEEE Proc. Intelligent Transportation Systems.* System design and analysis of repeating GPS, Oakland, US, 2001, pp2-5.

[5] Varshavsky, A. In *Seventh IEEE Workshop on Mobile Computing Systems and Applications.* Are GSM Phones THE Solution for Localisation?, 2006, pp34-42.

[6] Hahnel, D., Burgard, W., Fox, D., Fishkin, K., Philipose, M., *IEEE ICAR,* 2004, 1015-1020. Mapping and localization with RFID technology.

[7] Chen, J.; Hudson, R.; Yao, K.; *IEEE Trans. Signal Processing.* 2002, 50(8), 1843-1854. Maximum-Likelihood Source Localization and Unknown Sensor Location Estimation for Wideband Signals in the Near-Field.

[8] Ladetto, Q. *Capteurs et algorithmes pour la localisation autonome en mode pédestre,* Ph D Thesis, EPFL, 2002.

[9] Cavagna, G.A. *J. Physiologie.* 1969, 61, 3-42. Travail mécanique dans la marche et la course.

[10] Kolodziej, K.W.; Hjelm, J. *Local Positioning Systems, LBS Applications and Services,* TaylorandFrancis Group, 2006.

[11] Cross, P.; Kealy, A.; Scott-Youg, S.; Leahy, F. *GPS World.* may 2002, Intelligent Navigation, Inertial Integration: Double Assistance for GPS.

[12] *The Geospatial Web: How Geobrowsers, Social Software and the Web 2.0 are Shaping the Network Society,* Arno Scharl, Klaus Tochtermann (Eds.), Advanced Information and Knowledge Processing Series, 2007, Springer, ISBN 1-84628-826-6

[13] Goodchild, M. F., GIS and Transportation: Status and Challenges, *GeoInformatica,* 2000, pp. 127-139.

[14] Thill, J. C., *GIS in Transportation Research*, Pergamon Press, 2001

[15] Taylor, G. and Blewitt, G., Intelligent Positioning: GIS-GPS, Unification Wiley, 2006, 200 pp.

[16] Bertolotto, M., O Hare, G. M. P., Strahan, R., Brophy, A., Martin, A. N. and McLoughin, E., 2002, Bus catcher: a context sensitive prototype system for public transportation users, in Proceedings of the 3rd International Conference on Web Information Systems Engineering Workshops (WISE 2002 Workshops), Huang, B. et al (eds.), IEEE Computer Society, Singapore, pp. 64-72.

[17] Bertrand, F., Bouju, A., Claramunt, C., Devogele, T., and Ray, C., Web architectures for monitoring and visualizing mobile objects in maritime contexts, In Proceedings of the 7th International Symposium on Web and Wireless Geographical Information Systems (W2GIS 2007), pages 94-105, G. Taylor and M. Ware (eds.), Springer-Verlag, LN series in Computer Science (LNCS 4857), Cardiff, UK, 2007, ISBN 978-3-540-76923-1

[18] Brinkhoff, T., 2000, Generating network-based moving objects, in Proceedings of the 12th International Conference on Scientific and Statistical Database Management, Günther, O. and Lenz, H.-J. (eds), IEEEE Computer Society, Berling, Germany, pp. 253-255.

[19] Cao, H., Wolfson, O., and Trajcevski, G., 2006, Spatio-temporal Data Reduction with Deterministic Error Bounds. *The VLDB Journal — The International Journal on Very Large Data Bases*, vol.15 num.3, pp. 211-228.

[20] Civilis A., Jensen C. and Pakalnis A., Techniques for Efficient Road-Network-Based Tracking of Moving Objects, *IEEE Transactions on Knowledge and Data Engineering archive,* Vol. 17(5), 2005, ISSN:1041-4347

[21] Douglas, D. and Peucker, T., Algorithms for the reduction of the number of points required to represent a digitized line or its caricature, *The Canadian Cartographer*, Vol 10(2), pp. 112-122, 1973

[22] Follin, J.-M., Bouju., A., Bertrand, F. and Boursier, P., 2003, Management of multi-resolution data in a mobile spatial information visualization system, in *Proceedings of the 2nd Workshop on Web and Wireless GIS, WISE workshops*, M. Bertolotto et al. (eds.), IEEE press, Singapore, pp. 92-99.

[23] Gardiner, K. and Carswell; J., 2003, Viewer-based directional querying for mobile applications, in *Proceedings of the 2nd Workshop on Web and Wireless GIS, WISE workshops*, M. Bertolotto et al. (eds.), IEEE press, Singapore, pp. 83-91.

[24] Guth, P., Oertel, O., Bénard, G. and Thibaud, R., 2003, Pocket panorama: 3D GIS on a handheld device, in *Proceedings of the 2nd Workshop on Web and Wireless GIS, WISE workshops*, M. Bertolotto et al. (eds.), IEEE press, Singapore, pp. 100-111.

[25] Ishikawa, Y., Tsukamoto, Y. and Kitawaga, H., 2003, Implementation and evaluation of an adaptive neighborhood information retrieval system for mobile users, in Proceedings of the 2nd Workshop on Web and Wireless GIS, WISE workshops, M. Bertolotto et al. (eds.), IEEE press, Singapore, pp. 25-33.

[26] Kollios, G., Gunopulos, D. and Tsotras, V. J., 1999, On indexing mobile objects, in Proceedings of the 18th ACM Symposium on Principles of Database Systems, ACM Press, Philadelphia, Pennsylvania, pp. 261-272.

[27] Kang, H.-Y., Kim, J.-S. and Li, K.-J., 2006, Indexing Moving Objects on Road Networks in P2P and Broadcasting Environments, in *Proceedings of the 6th International Symposium on Web and Wireless Geographical Information Systems*

(W2GIS 2006), J. Carswell and T. Tezuka (eds.), Springer-Verlag LNCS 4295, Hong Kong, pp. 227-236.

[28] Li, X., Claramunt, C., Ray, C., and Lin, H., 2006, A semantic-based approach to the representation of network-constrained trajectory data, in *Proceedings of the 12th International Symposium on Spatial Data Handling (SDH 2006),* W. Kainz, A. Reidl, and G. Elmes (eds.), Vienna, Austria, Springer-Verlag, pp. 451-464.

[29] Luaces, M. R., Brisaboa, N. R., Parama, J. R. and Rios Viqueira, J. R., 2004, A Generic Framework for GIS Applications, in *Proceedings of the 4th International Workshop on Web and Wireless Geographical Information Systems*, Kwon, Y. J. et al. (eds.), Springer-Verlag, LNCS 3428, Goyang, Korea, pp. 94-109.

[30] Meratnia, N., By, R. 2004, Spatiotemporal Compression Techniques for Moving Point Objects, Proceedings of EDBT, , LNCS Volume 2992/2004, Springer, pp. 765-782

[31] Doyle, J., Han, Q., Weakliam, J. Bertolotto, M. and Wilson, D. C., Developing Non-proprietary Personalized Maps for Web and Mobile Environments, in *Proceedings of the 4th International Workshop on Web and Wireless Geographical Information Systems*, Kwon, Y. J. *et al.* (eds.), Springer-Verlag, LNCS 3428, Goyang, Korea, pp. 181-194, 2004.

[32] Papadias, D., Zhang, J., Mamoulis, N. and Tao, Y., 2003, Query processing in spatial network databases, in *Proceedings of 29th International Conference on Very Large Data Bases,* Freytag, J. F. et al. (eds.), Morgan Kaufmann, Berlin, Germany, pp. 802-813.

[33] Peytchev, E. and Claramunt, C., 2001, Experiences in building decision support systems for traffic and transportation GIS , in *Proceedings of the 9th International ACM GIS Conference*, W. G. Aref (ed.), ACM Press, Atlanta, pp. 154-159.

[34] Potamias, M., Patroumpas, K. and Sellis, T., Sampling Trajectory Streams with Spatiotemporal Criteria, *Proceedings of 18th International Conference on Scientific and Statistical Database Management (SSDBM'06),* pp. 275-284

[35] Saltenis, S., Jensen, C.S., Leutenegger, S.T. and Lopez, M.A., 2000, Indexing the positions of continuously moving objects, in *Proceedings of ACM SIGMOD International Conference on Management of Data*, Chen. W., Naughton, F. and Bernstein, P. A. (eds.), ACM Press, Dallas, Texas, pp. 331-342.

[36] Sharifzadeh, M., Azmoodeh, F. and Shahadi, C., 2005, Change detection in time series data using wavelet footprints, in *Proceedings of the 9th International Symposium in Advances in Spatial and Temporal Databases*, Medeiros, C. B., Egenhofer, M. and Bertino, E. (eds.), LNCS 3633, Springer-Verlag, Angra dos Reis, Brazil, pp. 127-144.

[37] Teixeira de Almeida, V. and Güting, R.H., 2005, Supporting uncertainty in moving objects in network databases, In *Proceedings of the 13th annual ACMGIS International Workshop on Geographic Information Systems*, ACM Press, pp. 31-40.

[38] Trajcevski, G., Wolfson, O., Hinrichs, K. and Chamberlain S., Managing uncertainty in moving objects databases, ACM Transactions on Database Systems, 29(3), pp. 463-507, 2004

[39] Chen, S., Ray, C., Tan, J., Claramunt, C., Integrated Transportation GIS for the City of Guangzhou, China, In *Proceedings of the 6th International Conference on ITS Telecommunications (ITS-T 2006),* pages 894-897, W. Guangun, S. Komaki, F. Pingzhi and G. Landrac (eds.), Chengdu, China, ISBN 0-7803-9586-7, 2006

[40] Li X., Lin H., Zhao Y.B., A connectivity-based map matching algorithm. *Asian Journal of Geoinformatics* Vol 5, pp. 69-76, 2005

[41] Comblet, F., Peyret, F., Ray, C., Bonnin, J.-M. and Le Roux, Y.-M., Telecommunications and geo-localization for ITS applications : The LoCoSS project, In *Proceedings of the 7th International Conference on ITS Telecommunications (ITS-T 2007)*, U. Finger, M. Fujise, G. Wen, C. Bonnet, S. Komaki, M. Lenardi (eds.), pp. 268-273, Sophia Antipolis, France, ISBN 1-4244-1178-5, 2007

[42] S. Rooney, S., Gardiner, K. and Carswell, J.D., Wireless Positioning Techniques: A Developers Perspective, In *Proceedings of the 7th International Symposium on Web and Wireless Geographical Information Systems (W2GIS 2007),* G. Taylor and M. Ware (eds.), Springer-Verlag, LN series in Computer Science (LNCS 4857), Cardiff, UK, , 2007

[43] Botts, M., Robin., A., Davidson, J. and Simonis, I., OpenGIS® Sensor Web Enablement Architecture Document, OpenGIS® Discussion Paper n°OGC 06-021r1, Version: 1.0, 69 pp. , 2006

[44] Li, X., Claramunt, C., Ray C., and Lin H., A semantic-based approach to the representation of network-constrained trajectory data, In *Proceedings of the 12th International Symposium on Spatial Data Handling (SDH 2006),* pages 451-464, W. Kainz, A. Reidl, and G. Elmes (eds.), Springer-Verlag, Vienna, Austria, ISBN 3-540-35588-x, 2006

[45] Frentzos, R., Indexing moving objects on fixed networks. *Proceedings of the 8th Intl. Symp. on Spatial and Temporal Databases (SSTD),* Santorini Island, Greece, pp 289-305, 2003

[46] Guting, R.H., Almeida, V., Ding, Z., *Modeling and Querying Moving Objects in Networks*. Research Report. Fern Universitat Hagen, 2004

[47] Hadjieleftheriou, M., Kollios, G., Tsotras, V.J., Gunopulos, D., Efficient indexing of spatiotemporal objects. *Proceedings of the 8th International Conference on Extending Database Technology*, Prague, Springer-Verlag, pp 251-268, 2002

[48] Miller, H.J., Shaw, S.L., *Geographic Information Systems for Transportation: Principles and Applications*. Oxford University Press, New York, 2001

[49] Nascimento, M.A., Silva, J.R.O., Theodoridis, Y., Evaluation for access structures for discretely moving points. *Proceedings of the International Workshop on Spatio-Temporal Database Management (STDBM'99),* Edinburgh, Scotland, pp 171-188, 1999

[50] Saltenis, S., Jensen, C.S., Leutenegger, S.T., Lopez M.A., *Indexing the Positions of Continuously Moving Objects*. Research Report. TIMECENTER, 1999.

[51] Sheffi, Y., *Urban Transportation Networks: Equilibrium Analysis with Mathematical Programming Methods*. Prentice Hall, Englewood Cliffs, NJ, 1985

[52] Sistla, A.P., Wolfson, O., Chamberlain S., Dao, S., Modeling and querying moving objects. *Proceedings of the International Conference on Data Engineering*, Birmingham, UK, pp 422-432, 1997

[53] Tao, Y., Papadias, D., MV3R-tree: A spatiotemporal access method for timestamp and interval queries. *Proceedings of the 27th International Conference on Very Large Databases,* Roma, Italy, pp 431-440, 2001

[54] Theodoridis, Y., Vazirgiannis, M., Sellis, T., Spatio-temporal indexing for large multimedia applications. *Proceedings of the 3rd IEEE Conference on Multimedia Computing and Systems*, Hiroshima, Japan, pp 441-448, 1996

[55] Vazirgiannis, M., Wolfson, O., A spatiotemporal model and language for moving objects on road networks. In: Jensen CS, Schneider M, Seeger B and Tsotras VJ (eds) *Advances in Spatial and Temporal Databases*, Springer, Berlin, pp 20-35, 2001

[56] Wolfson, O., Moving objects information management: The database challenge. In *Proceedings of the 5th Workshop on Next Generation Information Technologies and Systems (NGITS'2002),* Caesarea, Israel, pp. 75-89, 2002

[57] L. Suciu, J.-M. Bonnin, K. Guillouard, and B. Stévant. Achieving "Always Best Connected" through extensive profile management. In *The 9th Intl. Conference on Personal Wireless Communications (PWC'04),* Lecture Notes in Computer Science, Delft, Pays Bas, September 2004. Springer-Verlag.

[58] Ian F. Akyildiz, Xudong Wang, and Weiling Wang, 2005, Wireless mesh networks: a survey, *Computer Networks*, **47**:445–487, January 2005.

[59] F. André, J.-M. Bonnin, B. Deniaud, K. Guillouard, N. Montavont, T. Noël, L. Suciu, 2003, Optimized support of multiple wireless interfaces within an ipv6 end-terminal. In *SOC'03 (Smart Objects Conference),* Grenoble, France, May 2003.

[60] C2C Communications Consortium. *Car-2-Car Communication Consortium -* Manifesto, 2007.

[61] Williams, B. (Ed) *The CALM handbook: Continuous communications with and between vehicles*. ISO/TC204/WG16 (work in progress v2.0), November 2005.

[62] Thierry Ernst. Le support des réseaux mobiles dans IPv6. *Revue Technique et Science Informatiques (TSI),* 25(5), Juillet 2006.

[63] D.S. Kim, C.S. Hong, and T. Suda. A terminal mobility management architecture for IPv4 and IPv6 environments. In *1st International Workshop on Broadband Convergence Networks (BCN'06),* April 2006.

[64] M. Mahamat-faki, J.-M. Bonnin, and R. Ben Rayana. Evaluating traffic flow control on multi-tunnelled mobile router. *Short paper in WNEPT06 (Workshop on Networking in Public Transport),* Waterloo, Canada, August 2006.

[65] A.Ainaburo, P. Rawat, L. Toutain, J.-M. Bonnin, and E.-K. Paik. IP tunnelling header compression. Internet Draft - Work in progress (draft-minaburo-rohc-nemo-01.txt), February 2006.

[66] N. Montavont, M. Kassi-Lahlou, and T. Noël. Description and evaluation of mobile IPv6 for multiple interfaces. In *Wireless Communication and Networking Conference (WCNC'04),* volume 1, pages 144–148, Atlanta Georgia, USA, March 2004.

[67] Nagesh Nandiraju, Deepti Nandiraju, Lakshmi Santhanam, Bing He, Junfang Wang, and Dharma P. Agrawal. Wireless mesh networks: Current challenges and future directions of web-in-the-sky. *IEEE Wireless Communications*, pages 79–89, August 2007.

[68] Q.-T. Nguyen-Vuong, N. Agoulmine, and Y. Ghamri-Doudane. Terminal-controlled mobility management in heterogeneous wireless networks. *IEEE Communications Magazine,* April 2007.

[69] IEEE P1609. IEEE p1609.1 standard for wireless access in vehicular environment (WAVE) - multi-channel operation. *IEEE draft Standard*, 2006.

[70] IEEE 802.11 WG. Draft IEEE 802.11p wireless access in vehicular environment (WAVE). *IEEE draft Standard*, 2006.

[71] Priyanka Rawat, Jean-Marie Bonnin, and Laurent Toutain. An end-2-end tunnel header compression solution for nested mobile networks. In *ICLAN'07 (International Conference on the Latest Advances in Networks),* Paris, France, December 2007.

[72] Pablo Rodrigez, Rajiv Chakravorty, Julian Chestereld, Ian Pratt, and Suman Baner- jee. MAR: A commuter router infrastructure for the mobile internet. In *ACM Mobisys* 2004, June 2004.

[73] L. Suciu, J.-M. Bonnin, K. Guillouard, and B. Stévant. Towards a highly adaptable user-centric terminal architecture. In *The 7th Intl. Symposium on Wireless Personal Multimedia Communications (WPMC'04),* Abano, Italie, September 2004.

[74] L. Suciu and J.-M. Bonnin. A survey of multi-criteria network selection algorithms. In *Global Mobile Congress (GMC'04),* Shanghai, Chine, October 2004.

[75] L. Suciu, J.-M. Bonnin, K. Guillouard, and T. Ernst. Multiple Network Interfaces Management for Mobile Routers. In *5th International Conference on ITS Telecommunications (ITST 2005),* Brest, France, June 2005.

In: Wireless Technologies in Intelligent Transportation... ISBN 978-1-60741-588-6
Editors: Ming-Tuo Zhou et al, pp. 267-296

Chapter 10

WIRELESS MESH NETWORKS FOR MARITIME INTELLIGENT TRANSPORTATION SYSTEMS

***Ming-Tuo Zhou*[1,*], *Hiroshi Harada*[1], *Peng-Yong Kong*[2] and *J.S. Pathmasuntharama*[2]**

[1]Wireless Communications Laboratory,
National Institute of Information and Communications Technology
20 Science Park Road, #01-09A/10 TeleTech Park, Singapore 117674
[2] Institute for Infocomm Research
1 Fusionopolis Way, #21-01 Connexis, South Tower, Singapore 138632

Abstract

The conventional maritime communications are mainly based on satellite and HF / VHF / UHF networks that are either expensive or low-speed. With the development of economy and technologies, it needs better communication means in maritime Intelligent Transportation Systems as well as in systems providing general communication services like voice call and Internet surfing. This chapter introduces high-speed and low-cost wireless network technologies for maritime communications, on topics of maritime communication environment, analysis, evaluation and new developments of WiMAX mesh MAC and routing protocols for maritime communications.

1. Introduction

There are many reports of maritime disasters all around the world every year. Unfortunately, more than 80 percent of ship accidents are caused by human errors. It requires improvements and new developments of maritime Intelligent Transportation Systems (ITSs). A maritime ITS can reduce human errors through provision of automatic indicators and warnings. In addition, it can further aids to reduce ship accidents by providing hydrographic and navigational information and risks to vessels for safe and secure navigation, and by observing and managing vessel traffic from shore/coastal facilities. However, current maritime communication systems are mainly based on satellite networks and high frequency (HF)/very high frequency (VHF)/ultra-high frequency (UHF) systems, which are

*E-mail address: zhou.mingtuo@ieee.org

either expensive or low-speed. It requires to develop high-speed and low-cost communication means. In addition to being demanded for ITS applications, cost-effective broadband maritime communications also are needed by maritime users including ship crew, ship passengers, fleet managers and port operators for general communications, such as Internet surfing, file downloading, and fleet and ports management, and so on.

Broadband wireless technologies are actively studied for use on land. Developments include 3G networks, WiMAX networks, Wireless LANs, and so on. Recently, there are some projects aim to explore these technologies for maritime communications. Among of them, one promising technology is WiMAX mesh network that allows multi-hop communications. Compared to cellular networks, WiMAX mesh network is a better choice in maritime environment since it is difficult to setup cellular infrastructure in waters far from shore. A mesh network is capable of routing packets through multiple paths, self-organizing and self-healing, thus it is robust and suitable for sea communications in which radio links may experience frequent break due to sea wave movement and occlusion.

A maritime mesh network shares some common points of a vehicular ad hoc network (VANET) on land, such as in mobility, geographically constrained topology, rich power supply, and readily available location information. However, it differs from a VANET in communication environments. The properties of radio signal propagation in sea, ship movement due to sea wave, and ship inter-arrival time and speed play important roles in nodes connectivity and links quality in a maritime mesh network. Hence, designs and findings of WiMAX mesh MAC protocols and ad-hoc routing protocols that are primarily developed for on-land applications cannot be directly used in sea. Instead, analysis and new developments are necessary.

In following part of this chapter, section 2 presents current maritime communication systems and new maritime communication systems in development. Section 3 introduces maritime environment including radio channel model, sea wave movement model, and ship mobility model. Section 4 is introduction of WiMAX mesh MAC protocol, mainly on topics of MAC management messages scheduling and bandwidth requests and grants. In Section 5, analysis, considerations, and new developments of WiMAX mesh MAC protocols and routing protocols for maritime communications are presented. The chapter is concluded in Section 6.

2. Maritime Communication Systems

Since the invention of radio many years ago, myriad types of maritime wireless communication systems have been developed and applied, primarily for safety and secondarily for navigation and operational efficiency. Currently, a ship is required by law and international agreements to be equipped with a radio station depending on its size, purpose, or destination. For example, a large passenger or cargo ship that travels on the open sea must install a radio station for long distance radio communications. Passenger ships that travel along the coast are required to be able to communicate at a shorter range with coast stations. Smaller ships used for recreation (e.g., sailing, diving, sport fishing, water skiing) are not required to have radio station installed but may be so equipped by choice.

A shipboard radio station includes all the transmitting and receiving equipment installed on-board a ship for communications over water. Conventional maritime radio stations are mainly based on high frequency (HF), very high frequency (VHF), and satellite radios. With development of wireless and information technologies in recent years, new communication systems are envisaged and developed for maritime communications, and new applications that are hindered by limitations of traditional technologies are emerging. This section briefly gives reviews of the legacy maritime communication systems and introduces new systems under development for maritime applications.

2.1. Legacy Maritime Communication Systems

2.1.1. HF, VHF and UHF Systems

The following maritime communication systems use high frequency (HF), very high frequency (VHF), and/or ultra high frequency (UHF) radio frequencies.

- **Maritime Mobile amateur radio networks**
  A number of Maritime Mobile amateur radio networks are operated in HF range by licensed volunteers. The purposes of Maritime Mobile amateur radio networks include handling traffic from maritime mobiles, assisting in incidents of vessel distress and medical emergencies, acting as weather beacon for ships, and so on.

- **HF Marine Radiotelephone**
  HF Marine Radiotelephone first became common in the 1930s, and was extensively used for communications to ships and aircraft over water. In its typical operations, a ship calls a shore station, where the shore station's marine operator connects the caller ship to the public switched telephone network. Single-sideband (SSB) is used as the HF bands are crowded with many users.

- **HF Marine Radiotelegraph**
  HF Marine Radiotelegraph is used to communicate by means of Morse code facsimile, narrow-band direct-printing, or any coding and decoding technique for printed text over radio. Morse radiotelegraph has ceased operation worldwide in February 1 1999.

- **Marine VHF radio**
  Marine VHF radio operates in range between 156 and 174 MHz and is installed on all large ships and most motorized small craft. A marine VHF set is only allowed to operate in standard channels. The majority of channels have two frequency assignments, and use frequency modulation and duplex communications. Marine VHF radio has a number of applications:

  - *Marine VHF Radiotelephone* is used for voice communications with other ships and coast stations over short distances.
  - *Automatic Identification System* (AIS) is used for exchange of data in ship-to-ship communications and in communications with shore-based facilities. Information exchange includes identification, position, course, and speed. The purpose of AIS is to help identify vessels; assist in target tracking; and achieve situation awareness.

  - *International distress, safety and calling*
  - *Weather radio broadcast*
  - *Government, coast guard, or navy use*

- **Digital Selective Calling** (DSC)
  DSC is a part of Global Maritime Distress Safety System (GMDSS). It is a paging system used to automate distress alerts sent over terrestrial VHF, medium frequency (MF) and HF marine radio systems. A DSC distress alert, which consists of a preformatted distress message, is used to initiate emergency communication with ships and rescue coordination centers.

- **Marine Radar**
  Marine Radar is used for navigating, direction-finding, locating positions, and ship traffic control. It can be operated in HF range such as with coastal radar systems, and beyond UHF like 8-12 GHz.

2.1.2. Satellite Systems

Since 1970s, satellites are launched and many of them are communication satellites for both users on land and at sea. Modern communication satellites are equipped with digital transponders that receive, process and retransmit digital signals.

A typical satellite system for communications at sea is Inmarsat. It is originally founded in 1979 for the purpose of establishing satellite communications for the maritime community. Currently, Inmarsat's network can provide voice calls, data tracking systems, Internet surfing, distress and safety services. Its global maritime distress and safety services (GMDSS) to ships and aircraft are operated as a public service. It also can provide mobile Integrated Services Digital Network (ISDN) services, and most recently, IP-based services at up to 492 kbit/s via the Broadband Global Area Network (BGAN).

Specially for maritime use, Inmarsat has infrastructures of Fleet Phone, Fleet and Fleet Broadband (FB). Fleet Phone provides telephony service for master and crew on small fishing vessels, leisure craft and regional merchant fleets. Fleet is a family of networks that includes the Inmarsat-Fleet77, Inmarsat-Fleet55 and Inmarsat-Fleet33. Fleet provides mobile voice and data communication services, e-mail and secure Internet access for maritime industry. Features of Fleet is summarized in Table 1 below. Fleet Broadband provides IP-based services including standard IP (up to 432 kbps), streaming IP (up to 256 kbps), voice, Fax, ISDN (64 kbps), and text messages.

Another common maritime system that uses satellites is the Emergency Position Indicating Radio Beacon (EPIRB). An EPIRB is a compact, buoyant, self-contained radio transmitter designed for marine use when a ship is in distress. It is capable of continuously transmitting distinctive distress signal over at least 48 hours. The distress signal marking the location is received by satellite and is relayed to a Rescue Coordination Center.

2.2. Emerging Maritime Communication Systems

Unlike in the case of terrestrial environment, there is a lack of infrastructure for high bit-rate communications at sea. As a result, the current maritime communications must rely

Table 1. Inmarsat Fleet features [1]

Service	Fleet77	Fleet55	Fleet33
Coverage	Global -all services except 128 kbps (spot-beam)	Global -all services	Global -all services
Voice	Global /Digital	Global /Digital	Global /Digital
Data -Circuit Switched	64kbps ISDN 128kbps ISDN	64kbps ISDN (Euro standard)	9.6kbps dialup data
Data -Packet Mode	MPDS*	MPDS*	MPDS*
FAX	2.4bps (option) 9.6kbps (option) 64kbps G4 Fax	9.6kbps (option) 64kbps G4 Fax	9.6kbps
GMDSS -Distress & Safety	Voice - IMO A888(21)	Not Applicable	Not Applicable

*: MPDS – Mobile Packet Data Service

on conventional HF, VHF and UHF radios for ship-to-shore communications near coast waters and low-bandwidth and expensive Satellite communications for long-range ship-to-ship/shore communications. However, there is a general and increasing need for better communication solutions with the development of economy and technologies. The opportunities connected to maritime communication systems include:

- Need for better security, surveillance and environment – there is low or completely no bandwidth in areas that satellites have no coverage
- Need for higher efficiency and better work conditions for crew on-board – better and cheaper communication; availability of high bandwidth at low cost may accelerate the creation of new innovative applications in improving efficiency and work conditions on-board a ship
- Need for eNavigation – highly developed communication means are required
- New and better services for passengers, both on-board and on the coast
- New business opportunities for maritime industry

For the purpose to meet the above needs, a number of projects to develop new maritime communication systems have been launched recently. These emerging systems will extend and optimize the terrestrial high-speed communications to sea, and new information systems that integrate various existing and new maritime applications will be developed based on the networks.

2.2.1. TRITON Project

TRI-media Telematic Oceanographic Network (TRITON) project is a cooperation between the National Institute of Information and Communications Technology (NICT) Japan and the Institute for Infocomm Research (I^2R) Singapore. It was launched at 2006 and spans 5 years. TRITON is to develop a system that will serve multiple applications such as general voice and data communications, video surveillance, piracy prevention, pollution detection, navigational safety, etc. The project however does not address the development of these applications but merely provides the infrastructure and interfaces to ease the development of these applications. Some of these newly envisaged applications actually demand a strict network quality of service such as significantly higher bandwidth and lower delays.

The system envisaged in TRITON is a wireless multi-hop mesh network formed by neighboring ships, marine beacons, and buoys. The multi-hop mesh network will be connected to the terrestrial networks via land stations that are placed on shore regularly. In areas that there is no sufficient coverage or in deep sea that there is no relay to shore, satellites will be used to complement/supplement the network. The TRITON network is based on IEEE std 802.16-2004 [2], i.e., WiMAX mesh operation mode, and the operation frequencies is at 5.8 GHz, a candidate frequency of WiMAX network. Operating at other frequencies such as 2.4 GHz, and VHF frequencies that ITU recently claimed for International Mobile Telecommunications (IMT) is also possible.

Researches carried out with TRITON include:

- Radio channel modeling in sea
- Ship mobility modeling
- Development of maritime mesh network simulator
- Analysis of mesh connectivity
- Routing protocol development
- Scheduler development
- Middleware development

2.2.2. WISEPORT Project

WISEPORT (WIreless-broadband-access for SEaPORT) is the first project of Infocomm@SeaPort programme – a collaboration between the Maritime and Port Authority of Singapore (MPA) and the Infocomm Development Authority of Singapore (IDA) that is launched at September 2007 and will last 3 years. Infocomm@SeaPort programme aims to enhance communications in the port community, improve operations management, and increase the port community's handling capacity.

WISEPORT seeks to provide a mobile wireless broadband network within 15 km from Singapore's southern coastline. The infrastructure built up is Mobile WiMAX network based on IEEE standard 802.16-2005. It creates the first mobile WiMAX ready seaport in the world. During free trial period, March 2008 to March 2009, it provides 512 kbps mobile wireless access for a single subscriber. The rates in following operations can be 512 kbps, 1 Mbps, up to 5 Mbps.

WISEPORT benefits the maritime industry in a variety of important ways. It includes improving operational and business efficiencies, enhancing human communications and opening doors to new business opportunities that were previously hindered by high satellite communication costs. Envisaged WISEPORT's applications include:

- Electronic bay loading plans for container vessels
- Dynamic international chart data synchronization
- Automated ship-chandling and bunkering services
- Video surveillance of territorial waters
- Internet-based services
- IP telephone

2.2.3. MARCOM Project

The MarCom project is funded by industry and the Norwegian Research Council, starts in 2007 and spans for four years. It aims to provide wireless broadband communications for coast, Polar Regions, offshore and sea farming. It also works with the issues from IMO like eNavigation. More than 20 members of R&D institutes, universities and colleges, public authorities and industry from Europe and Asia work together for this joint initiative. Targets of MarCom for maritime industry are [3]:

- **Ferry monitoring:** to monitor installations on-board ferries in domestic trade
- **Pilotage of fairways and lighthouses:** navigation, remote monitoring and partly ship operation
- **Integrated operations:** a more complete cooperation among the actors involved in offshore operation in the oil industry
- **High-speed craft operations:** technical maintenance, navigation, infotainment, and ship operation
- **Mesh networking and hand over:** a multi-hop mesh network amongst ships, marine beacons and buoys
- **Onboard WLAN solution**

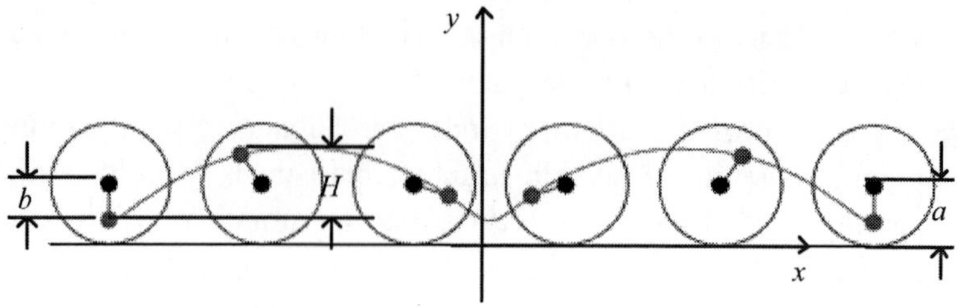

Figure 1. Illustration of modeling sea wave movement as Rrochoid waves.

- **High north operations:** monitoring ship operations and environment, security and territorial control, metrology, and safety and rescue operations in the High North
- **International shipping:** eNavigation, ship operation, and emergency management

3. Maritime Communication Environment

3.1. Sea States and Wave Movement Model

The sea wave movement continuously changes the orientation and height of an antenna installed on a ship, resulting in unstable antenna gain and the received radio signal power. It is necessary to characterize the sea wave movement and its effect on the radio link quality. The roughness of sea surface is described by so called sea states. A sea state includes parameters of significant sea wave height, average sea wave length, and average sea wave period. Signification sea wave height is the mean wave height of the one third highest waves. A detailed sea state table can be found at reference [4].

A general model of sea wave movement is Trochoid wave that is created by tracing the path of a point inside a circle. The parametric equations for a trochoid wave are

$$\begin{aligned} x &= a\phi - b\sin\phi \\ y &= a - b\cos\phi \end{aligned}, \tag{1}$$

where a is the radius of the circle and b is the distance of the point to the center of the circle. From Fig. 1, we know

$$\begin{aligned} a &= \lambda/2\pi \\ b &= H/2 \end{aligned}. \tag{2}$$

For a given location x and a time instance t the angle ϕ is calculated by

$$\phi = 2\pi\left(\frac{x}{\lambda} - \frac{t}{T}\right). \tag{3}$$

In above, λ is the average sea wave length, T is the average sea wave period, and H is the significant sea wave height of a given sea state.

3.2. Radio Channel Model in Maritime Environment

Similar to case on land, radio channel properties in sea are closely related to the propagation environment. Maritime radio environment can be generally categorized into three types [5]:

- wide sea passages and open sea environments
- near-coast regions with cliffs or hills
- ports, harbors and narrow sea passages with intense mobility of the environment and urban/suburband onshore surroundings

This section presents properties of path loss and delay spread of radio signals in above environments.

3.2.1. Path Loss Property

In a wide sea passage or an open sea environment with sea wave height less than 1 meter, the radio channel at GHz frequencies can be modeled as a two-ray channel [6] with a direct line-of-sight (LOS) path and a sea-reflected path. This is verified by intensive measurement campaigns for maritime mobile communications at frequencies 1.9 GHz [7], 2.5 GHz [5], and 5.8 GHz [8, 9]. The upper frequency that the two-ray channel model applicable can be calculated by good conductor condition $f << \sigma/2\pi\varepsilon_0\varepsilon_R$, where σ is conductivity, ε_0 and ε_R are permittivity and relative permittivity of sea water, respectively. With parameters of $\sigma = 5S/m$ and $\varepsilon_R = 81$ for sea water, the upper frequency is about 7 GHz.

With two-ray channel model, the large scale path loss can be calculated by

$$\frac{P_r}{P_t} = (\frac{\lambda_r}{4\pi d})^2 \times 4\sin^2(\frac{2\pi h_t h_r}{\lambda_r d}), \tag{4}$$

where λ_r is the radio wavelength, d is the distance between transmitter and receiver, h_t and h_r are antenna height of transmitter and receiver, respectively, and G_T and G_R are gain of transmitter and receiver antenna, respectively. There exist fading notches with depth more than 20 dB along the distance d because of destructive interference between the direct path and the sea-reflected path. In practice, due to the movement of sea wave, the effective height and gain of both transmitter and receiver antenna change with time and location, and then the fading notches vary with time and locations.

In near-coast regions with cliffs or hills, other than the strong direct connection and the sea-reflected path, reflections from rocks or hilly terrain on the coastline can be received. As the direct LOS path is still guaranteed in this case, the channel shows Ricean fading characteristic. The absolute reflection coefficients from the land reflections is typically well below 0.7, hence the fading depths are usually below 10 dB and then for prediction of large scale path loss, two-ray channel model is till applicable.

In the third type of maritime environment, plenty of reflections from other ships bodies, surround buildings, and hilly terrain could be received. It is more likely that the land-reflected paths become dominant when the direct path is severely attenuated by the sea-reflected path. A log-distance model is used to predict the large scale path loss in this type of environment. With this model, the received signal power P_r is written as follow [6]

$$P_r(d) = P_r(d_0) - 10nlog(d/d_0) + X_\sigma, \tag{5}$$

where d_0 is a reference distance, n is attenuation factor, and X_σ is a zero mean log-normally distributed random variable with standard deviation of σ. n represents the attenuation rate of the radio signal along the separation between transmitter and receiver, and σ indicates the shadow effect. These two parameters can be determined using linear regression of the path loss values measured against the log of normalized distance (d/d_0) in a minimum mean square error (MMSE) manner [10]. Reported values of n and σ are listed in Table 2.

Table 2. Reported attenuation factor and standard shadow deviation [8, 5]

Frequency	Reference distance	n	σ
1.9 GHz	405 m	3.4	3.95 dB
5.8 GHz	10 m	2.39	8.4 dB

When LOS is not dominant in the third environment, beyond a break point defined by $d_A = 20h_th_r/\lambda_r$ the signal attenuates with an increased attenuation factor. As reported in [8], n becomes 6.12 when distance is more than about 12 km, with $h_t = 4m$ and $h_r = 8m$.

In both second and third type of maritime environments, if the direct path is blocked by ship, platform, island, or peninsula between the transmitter and receiver, the connection is NLOS. In this case, the received signal power usually drops 10 to 20 dB and the radio channel shows Rayleigh fading characteristic. Again, the large scale path loss is predicated by the above log-distance model. Reported n and σ are listed in Table 3.

Table 3. Reported attenuation factor and standard shadow deviation in case of non-LOS [8, 5]

Frequency	Reference distance	n	σ
1.9 GHz	1025 m	3.606	4.5 dB
5.8 GHz	10 m	6.12	8.4 dB

To overcome the systematic fading notches occur with two-ray channel model and NLOS conditions in maritime environments, antenna diversity is recommended to be applied. In addition, a link margin of 10 to 20 dB is necessary.

3.2.2. Delay Spread Property

In above two-ray channel model for wide sea passages and open sea environments, the propagation delay between the direct path and the sea-reflected path, i.e., the delay spread

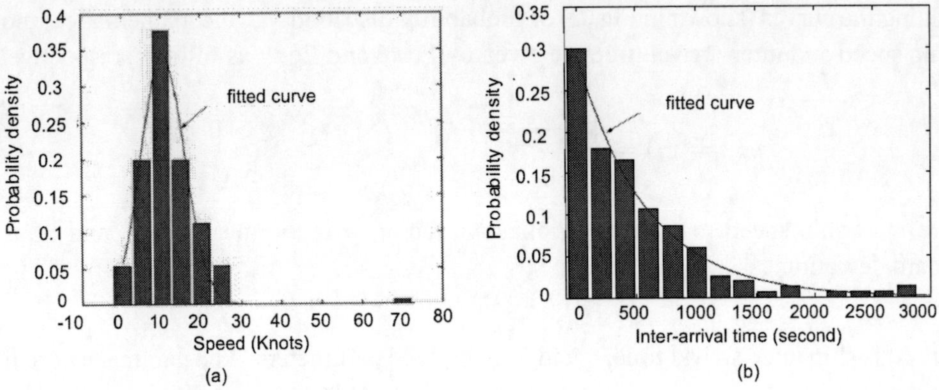

Figure 2. Distribution of (a) ship speed; (b) ship inter-arrival time crossing vertical line of longitude 140.15 in Singapore seaport.

is less than 1 ns, and hence the coherence bandwidth is more than 1 GHz [7].

In the second environment, there are several echoes reflected from the land objects. Although the radio channel is still modeled as two-ray, the system performance could be degraded by signal fading and intersymbol interference. Table 4 presents measured excess delay and relative power of the echoes at 1.9 GHz.

Table 4. Measured power excess delay and relative power of echoes [5].

Echo path	1	2	3	4
Excess delay (μsec)	0.1-0.82	2.2-2.8	3.5-5.5	6.1-6.7
Relative power (dB)	-31	-31	-33	-32

In the third maritime environment, much more reflections could be received than the second case, and then usually the mean excess delay is larger. In property of RMS delay spread, the two cases are comparable. However, when the direct path is blocked, i.e., in case of NLOS, the delay parameter values increase remarkably [5].

3.3. Ship Mobility Model

Mobility pattern of ships is critical in planning a mesh maritime communication network in a specified region. It includes information of statistical distribution of ship speed and ship inter-arrival time. Clearly there is no a uniform ship mobility pattern worldwide. It needs to analyze the location information of local ships. In this section, an example of ship mobility pattern in Singapore seaport is presented.

AIS information of ships in Singapore seaport that is provided by Singapore MPA is used in analysis of ship mobility pattern. It was focused on analyzing speed and inter-arrival time of all ships crossing the vertical line at longitude 140.15 from both directions.

By fitting the curves (shown in Fig. 2) of probability distributions, the mathematical models of ship speed and inter-arrival time are given by Eq. 6 and Eq. 7 as follows, respectively.

$$p\left(x=v_i\right)=\begin{cases}\frac{c}{\sqrt{2\pi\sigma_s^2}}e^{-(x-\mu)^2/\left(2\sigma_s^2\right)} & x>0\\ 0 & x\leq 0\end{cases} \tag{6}$$

where v_i is ship speed, c is the multiplicative factor, μ is the mean value, and σ_s is the standard deviation.

$$p\left(x=\lambda_i\right)=a\times b^x \tag{7}$$

where λ_i is ship inter-arrival time, a and b are derived parameters. The parameters are listed in Table 5.

Table 5. Ship mobility parameters [11].

Parameter	Westbound shipping lane	Eastbound shipping lane
a	0.292064	0.284495
b	0.998286	0.998324
c	4.763146	4.354804
μ	10.38378	10.800865
σ_s	5.488477	5.189037

3.4. Mesh Connectivity in Maritime Environment

In maritime environment, due to the large distance between communication devices, packets typically are forwarded through multiple hops before reaching a gateway to the terrestrial network infrastructure. If needed, a ship could function as a relay station for routing traffic to its final destination. However, ships move in an autonomous way according to instructions of the schedule and traffic management. This leads to dynamical change of connectivity of mobile ship users. It is important to maintain the best connectivity for providing ship mobile users a given level of QoS in terms of ensuring continuity of communication, failure backup and load balance.

In analysis of mesh connectivity in a specified maritime region, location information of ships during an observation period is a prerequisite. AIS data is one of the sources of such information. By assuming a radio transmission range, two ship nodes are considered connected if they are within this range each other. This section illustrates connectivity of ship nodes in Singapore seaport for maritime mesh network design.

The studied region is shown in Fig 3. The research focused on analyzing connectivity of 44 ship nodes traced in the region to a Base Station (BS) on shore. A ship node is said connected to the BS if and only if it has at least one one-hop, two-hop or three-hop route to

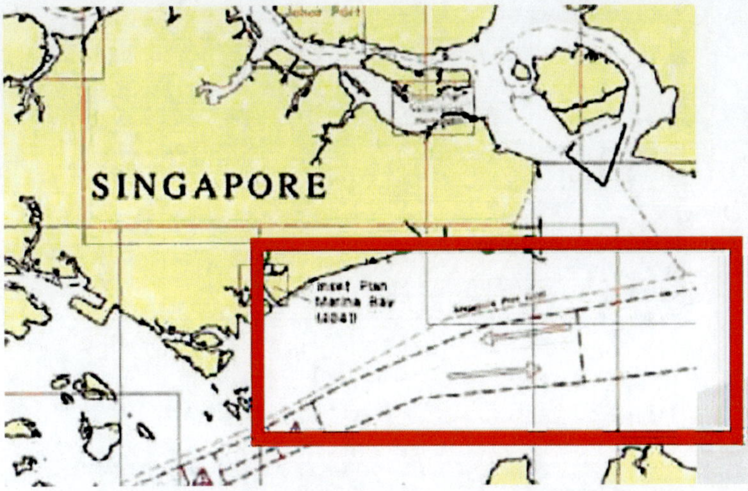

Figure 3. The region of Singapore seaport studied.

the BS. By varying ranges of BS-to-node and node-to-node, different connectivity can be obtained. Figure 4 shows the mesh connectivity of ship nodes and the number of routes to the BS, within in the 4-hour observation period, and with assumed BS-to-node range of 10 km and node-to-node range of 8 km.

4. 802.16-2004 Mesh MAC Protocol

WiMAX (Worldwide Interoperability for Microwave Access) is based on the IEEE std 802.16 and is believed an enabling technology for last mile wireless broadband access. Two operation modes are defined with 802.16-2004 standard: point-to-multiple point (PMP) and Mesh. With IEEE 802.16-2004, WiMAX mesh mode operates in frequency range of 2 to 11 GHz and initially it is designed for range extension in communities with sparse users. In TRITON and MarCom project it is used for maritime mesh communications. This section introduces MAC protocol of this standard.

4.1. Frame Structure

Figure 5 shows the frame structure of the IEEE std 802.16-2004 mesh operation mode. A single mesh frame consists of a control subframe and a data subframe. A control subframe has MSH-CTRL-LEN slots, and each slot duration is 7 OFDM symbols. During each control slot, a MAC management message can be transmitted. There are two types of control subframe: network control subframe and scheduling control subframe. Frames with a network control subframe occur periodically, with a period of Scheduling Frames. The first slot of a network control subframe is used for network entry, and other slots are for network configuration. Within a scheduling control subframe, the last MSH-DSCH-NUM slots are allocated for transmission of distributed scheduling messages, while the first MSH-DSCH-CTRL - MSH-DSCH-NUM slots are for transmission of centralized scheduling messages.

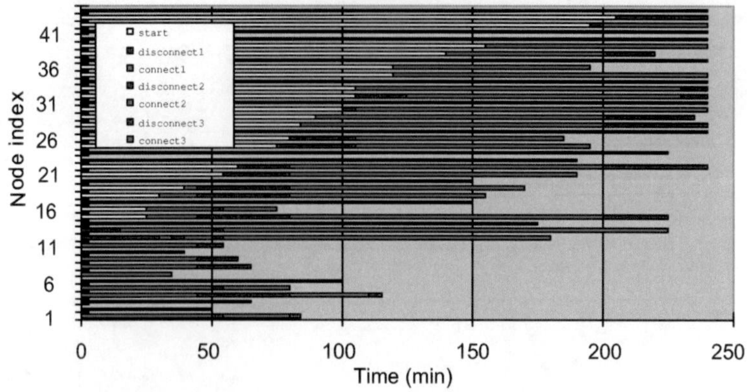

(a)

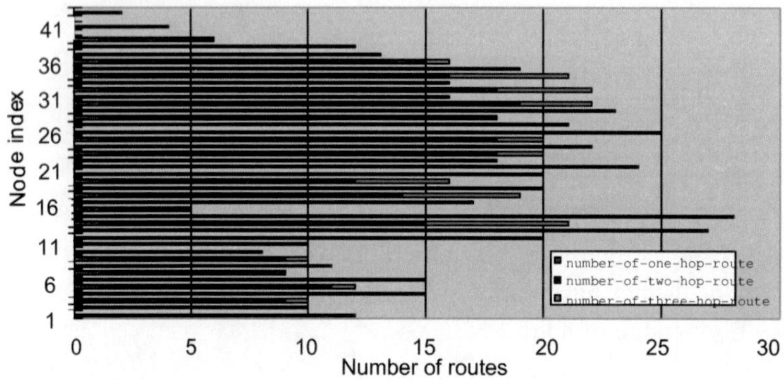

(b)

Figure 4. Mesh connectivity of ship nodes in Singapore seaport within 4-hour observation period with assumption of BS-to-node range of 10km and node-to-node range of 8km: (a) connectivity of ship nodes to on-shore BS; (b) route redundancy of ship nodes to on-shore BS.

The data subframe is divided into 256 minislots. This part is used for transmission of bursts of data. Also distributed scheduling message can be transmitted in data subframe.

4.2. Network Configuration

802.16-2004 mesh mode configures the network by broadcasting NCFG messages. A NCFG message contains physical information, time-synchronization information, schedule information of next NCFG transmission, neighbor's information, BS information, network parameters, and network entry information.

NCFG physical information includes its transmission power, logical antenna, network base channel and information of channels used in the network. Eight antenna directions and 16 logical channels can be supported in WiMAX mesh network. Network base channel indicates the logical channel number used to broadcast schedule control information.

Time-synchronization information includes frame number, network control slot number

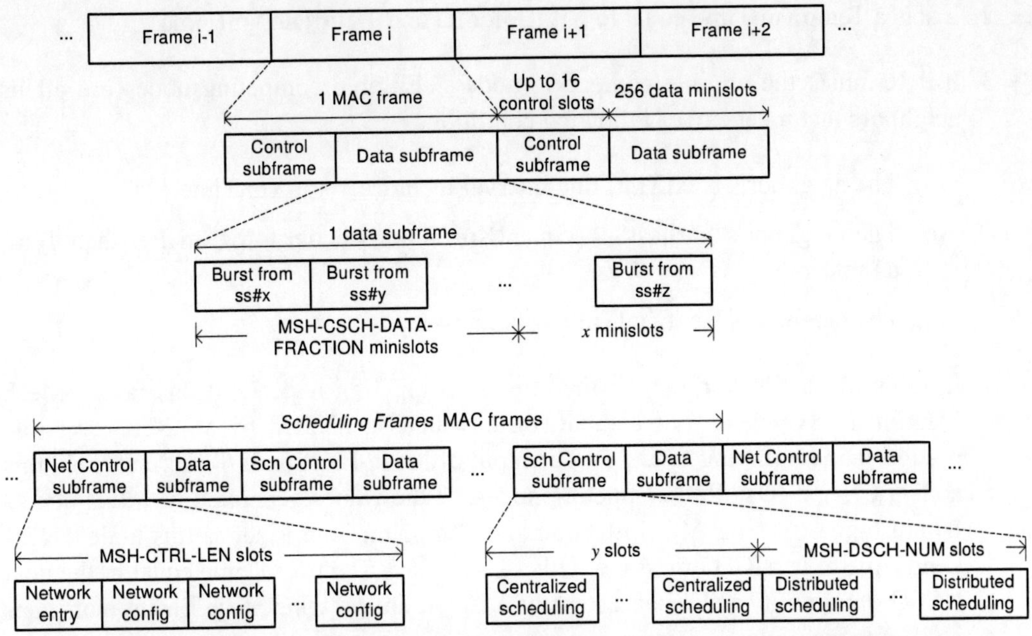

Figure 5. Frame structure of IEEE std 802.16-2004 Mesh operation mode.

in one frame and synchronization hop count. As control slot number in one frame, frame duration, and duration of control subframe are fixed, nodes can determine the start time of the frame. This is called coarse synchronization. Fine synchronization is achieved through PHY functions. A mesh node will keep synchronization to nodes with lower synchronization hop count, or if counts are the same, to the node with smaller node ID.

WiMAX mesh nodes schedule NCFG messages among two-hop neighbors in a distributed manner based on exchange of schedule information of next NCFG transmission. Upon reception of an NCFG message from an one-hop neighbor, a WiMAX mesh node get information NextXmtTime of this neighbor, their common one-hop neighbors and two-hop neighbors. NextXmtTime is the next NCFG eligibility interval given as follow

$$(2^{XmtHoldOffExponent}.NextXmtMx, 2^{XmtHoldOffExponent}.(NextXmtMx+1)], \quad (8)$$

where NextXmtMx and XmtHoldOffExponent are schedule information of the next NCFG message. A mesh node determines its next NCFG transmission time right before the current NCFG sending time using following procedure:

1. For each neighbor, it adds the node's NextXmtTime to the node's XmtHoldoffTime to arrive at the node's EarliestSubsequentXmtTime. XmtHoldoffTime is the number of NCFG transmission opportunities after NextXmtTime that this node is not eligible to transmit NCFG message. It is given by

$$XmtHoldoffTime = 2^{(XmtHoldoffExponent+4)}. \quad (9)$$

2. It sets a TempXmtTime equal to XmtHoldoffTime + current XmtTime.

3. It determines the eligible competing nodes. Eligible competing nodes are all its neighbors that meet any of following conditions

   (a) The neighbor's NextXmtTime interval includes TempXmtTime

   (b) The neighbor's EarliestSubsequentXmtTime is equal to or smaller than TempXmtTime

   (c) The neighbor's NextXmtTime is unknown

4. It holds Mesh Election among the eligible competing nodes and this node. Mesh Election is a procedure of comparison of values generated by a 32-bit reversible pseudo-random mixing function. The input of Mesh Election is TempXmtTime, this node's ID, and IDs of all competing nodes. If the values generated with this node's ID are biggest it wins Mesh Election and TempXmtTime is set as this node's NextXmtTime of its NCFG message. Otherwise, it sets TempXmtTime equal to the next NCFG opportunity and repeats steps of 3 and 4, until it wins Mesh Election or it has no competing nodes.

With above algorim, WiMAX mesh mode allows only one node to transmit in each NCFG eligible opportunity among two-hop neighbors.

Neighbor information contains physical connection information and schedule information (neighbor NextXmtMx and XmtHoldoffExponent) for scheduling NCFG transmission among two-hop neighbors. Also it contains information for logical connections like link quality, burst profile employed, and transmission power.

BS information includes hop count to the BS and energy/bit needed to reach the BS through this node. These information indicate nearness to the BS and allow better capacity tradeoff.

Network parameters report basic parameters of the network such as control subframe length, number of DSCH opportunities in schedule control subframe, scheduling frames, number of burst profiles, logical frequency channels and so on.

Network entry information is used to open, reject, or acknowledge a new node joining the network.

In addition, unlike to transmit schedule control information in a network base channel, a mesh node broadcasts NCFG messages with all available frequency channels in a cycled way. This is to ensure that all nearby nodes have chance to receive NCFG messages.

4.3. Network Entry

Following procedures are used for a new WiMAX mesh node to join the network

1. *Scan for active network and establish coarse synchronization with the network*
   At this stage, the mesh node continuously scan all possible frequency channel to receive NCFG message, from which it can achieve coarse synchronization with the network and acquire network parameters.

2. *Obtaining network parameters*
   The node continuously scans channels until it at least receives an NCFG message from the same node twice and it receives at least one NCFG message with Network Descriptor subfield that contains network parameters. It then selects one potential sponsoring node and synchronizes itself to this node.

3. *Open sponsor channel*
   The node sends NENT:Request message to the sponsor node in a control slot for network entry. Upon receiving NENT:Request, the sponsor node can reject with NCFG:NetEntryReject or accept with NCFG:NetEntryOpen. The NCFG:NetEntryOpen contains new node's MAC address and a temporary schedule (called sponsor channel) for transmission of higher layer authentication and configurations.

4. *Basic capacity negotiation and perform node authorization, registration, IP connectivity establishment, time-of-day establishment, and operational parameters transfer*
   The node confirms the temporary schedule with an NENT:NetEntryAck and performs authentication and configuration with security sublayer and basic capacity negotiation with the sponsor node. When finish, it closes the network entry procedure with an NENT:NetEntryClose and the sponsor node replies an NENT:NetEntryAck. After this the new node is added into the network.

4.4. Data Scheduling

WiMAX mesh nodes schedule data transmission in two ways: distributed scheduling and centralized scheduling. The first CSCH Data Fraction percent of the 256 minislots are used for centralized scheduling and the rest (1 - CSCH Data Fract) percent of the 256 minislots are for distributed scheduling.

4.4.1. Distributed Scheduling

In this scheme, WiMAX mesh nodes schedule data transmission among two-hop neighbors. A three-way handshake procedure is used by exchanging DSCH messages. First, the transmitting node sends a DSCH:Request containing demand level in terms of number of minislots and number of frames it needs. Second, upon reception of DSCH:Request, the receiving node sends back a DSCH:Grant containing location, range and frame persistence of minislots granted. Third, the transmitting node sends DSCH:Grant confirmation. By doing so, all the one-hop neighbors are acknowledged. The medium status in the allocated minislots are marked unavailable by the transmitting and the receiving node, as well as by their one-hop neighbors. And the availabilities of medium are informed among two-hop neighbors by exchanging DSCH:Availability, thereby near-collision-free transmission among two-hop neighbors can be achieved.

There are two manners for transmission of DSCH messages: coordinated distributed scheduling (CDS) and un-coordinated distributed (UCDS) scheduling. With CDS, DSCH messages are transmitted in control slots and the transmission timing uses the same distributed algorithm of NCFG. And then the transmission of DSCH messages in CDS are collision-free among two-hop neighbors. With UCDS, DSCH messages are transmitted in data subframe in a contention based way, making the handshake procedure susceptible to collisions.

4.4.2. Centralized Scheduling

Centralized scheduling allocates resource through scheduling tree, a topology structure with the Base Station of a cell as root. The BS broadcasts CSCF messages containing information of children nodes of each mesh node. Upon receiving CSCF message, all one-hop neighbors of the BS rebroadcast CSCF messages. And CSCF messages are rebroadcasted again by higher hop count mesh nodes. Mesh nodes have no child node do not broadcast CSCF messages. By this way, all nodes in the cell know the scheduling tree. The transmission timing of CSCF messages uses a breath-first algorithm. Based on knowledge of scheduling tree and the time receiving the last CSCF message, a node with lower hop count, or if same hop counts the one with smaller node ID transmits CSCF first.

CSCH messages are used for resource request and grant. To request resource, a mesh node sends a CSCH:Reqest to its parent node, together with resource requests from all its children nodes if it has. Transmission timing of CSCH:Request also uses breath-first algorithm, however, with a reversed sequence of CSCF messages. Finally, the BS knows request from each mesh node and it assigns resource of data minislots in one or two frames. If available data minislots are not sufficient to fulfill all requests, the BS scales down proportionally the sizes of all requests. Distribution of assignments is through CSCH:Grant, with a transmission sequence same as CSCF. The bandwidth of grant entries are in a CSCF scheduling tree order, then upon reception of a CSCH:Grant, a mesh node with request can identify its grant easily. And based on scheduled burst profile, it translates the assignment into time duration and it arranges the corresponding data transmission according to the order of CSCF scheduling tree. By this way, collision-free transmission of data packets can be achieved.

5. Maritime Mesh Networks

Due to the difficulty to set up cellular infrastructure in sea, mesh network is a better choice for ship-to-ship/shore communications, especially in waters relatively far away from shore. Advantages of wireless mesh networks are listed as follows.

- *Easy coverage extension*
  All mesh nodes can act as wireless router and can relay traffic within its coverage. The network coverage can be easily extended by installing new nodes.

- *Increased capacity*
  The network capacity can be increased by simultaneous transmission over multiple hops away, as that does in distributed scheduling of WiMAX mesh mode.

- *Path resilience and self-healing*
  If one path is failed, alternative path may be available as each mesh node can forward data packets. And a failed path, like the one occluded by sea wave temporally, may be

recovered by the nodes involved in the path themselves without the help of a higher-level node like BS in cellular networks.

- *Low cost*
  In a mesh network, there is no need a conventional BS as that used in cellular network. Usually, the cost of a conventional BS is much higher than a mesh node.

- *Easy and low-cost deployment*
  In a conventional cellular network, installation of nodes needs careful planning, and in some cases directional antennas and alignments are required. Wireless mesh network adopts omnidirectional antenna and it can be easily installed by users.

TRITON and MarCom Projects both employ WiMAX mesh network for ship-to-ship/shore communications.

A maritime mesh network is similar to a vehicular ad-hoc network (VANET) on land. In both cases, data packets are relayed over multiple hops before reaching destinations. They shares following common characteristics [12].

1. *Mobility*
   In vehicular ad-hoc network, vehicle movement is restricted by the road directions and traffic regulations. In busy narrow navigation channels, such as the Strait of Malacca, the Strait of Dover, the Strait of Singapore and the Turkish Straits, ship movement is restricted by shipping lane directions and traffic regulations. Ships can only make a turn at some designated areas. Within a shipping lane, ships are bounded by maximum and minimum speed.

2. *Geographically constrained topology*
   Consider each car as a network node, roads limit the network topology of VANETs to some straight lines with small-angle curves. The shipping lanes have the same effect of roads in limiting the network topology consider each ship as a network node.

3. *Rich power supply*
   In traditional mobile ad-hoc network (MANET), nodes are power limited and their lifetime depends on their batteries. On the contrary, cars and ships can provide continuous power to their computing and communication devices.

4. *Readily available position information*
   Increasingly, cars are equipped with accurate positioning systems such as the Global Positioning System (GPS) and GALILEO. Ships are similarly equipped with GPS and GALILEO to achieve accurate positioning, and are required to regularly announce their positions through the Automatic Identification System (AIS).

While terrestrial VANET for cars is actively researched, the findings are not directly applicable to a ship-to-ship multi-hop mesh network. This is partly because ships have a different mobility pattern than cars. For example, compared to a car, a 30,000 tons ship takes a much longer time to make a turn, as well as to accelerate and decelerate to a new steady state speed. Also, different from the static earth ground, sea surface is rough and random. The random movements of sea surface can lead to time-varying received signal quality as well as wave occlusions that break communication links.

This section presents considerations, designs and research results of maritime WiMAX mesh networking.

5.1. Considerations for Maritime WiMAX Mesh Networking

WiMAX mesh network is initially designed for range extension in communities with sparse users and the designs are optimized for static nodes. Maritime communication environment is quite different from land in terms of such as user's mobility characteristics, radio channel, and connectivity of nodes. To apply WiMAX mesh mode for maritime communications, some considerations and improvements are necessary.

- **Distributed Scheduling vs. Centralized Scheduling**
  Centralized scheduling employs scheduling tree for resource allocation. Upon change of network topology, acknowledgement across the whole cell network is needed by broadcasting CSCF and CSCH messages. In another hand, because of sea wave movement and movement of ships, a maritime mesh network is highly dynamic and there are frequent changes of parent-child relationships. This may leads to flush of CSCF and CSCH messages to reflect the frequent changes of a maritime mesh network. In addition, in some areas like seaports, the ship density is possibly high and then the number of nodes in the coverage could be more than tens or hundreds, making sizes of CSCF and CSCH messages possibly larger than the capacity of a single control slot. Moreover, in a time duration scheduled by standard centralized scheduling protocol, only one transmission is allowed across the cell network and then the capacity of the network is relatively low. These facts make centralized scheduling not suitable for maritime mesh networking.

  On the contrary, distributed scheduling allocates resource in a distributed manager. Change of a part of network topology has no impact on MAC management signaling of nodes several hops away and then the network is more adaptable to the dynamic maritime network topology. Another advantage is spatial reuse that is allowed in distributed scheduling and then the network capacity can be greatly increased by concurrent transmissions. Based on the above considerations, distributed scheduling is better in maritime WiMAX mesh networking.

- **Supporting to Mobility**
  Considerations and designs to support mobility in maritime mesh network is needed as the IEEE 802.16-2004 mesh only supports static nodes. It requires the network can dynamically update the neighbor list and corresponding parameters such as hops to a neighbor, estimated propagation delay, neighbor transmission power, transmission antenna, and burst profile. Further more, transmitting-receiving ship pairs that are several hops away may move close to each other and may have collisions in data transmission due to previous allocations with distributed scheduling. Mechanism of detection of minislots conflictions is required. Moreover, fast detection of link break due to sea wave occlusions and changes of ship locations is necessary by which the network can change routes fast to achieve better packets delay and link quality.

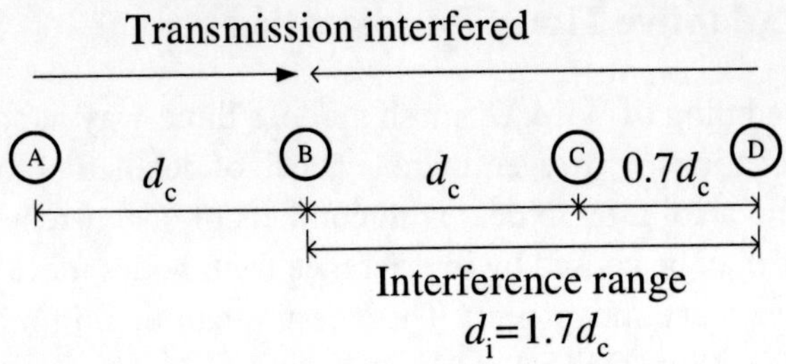

Figure 6. Illustration of interference from node beyond two-hop neighbors, d_c: communication range, d_i: interference range.

- **Packets Retransmission**
  In maritime environment the communication link quality is affected by sea wave movement, ship speed, and propagation impairments. In a mild sea condition and careful designs, the probability to have good link is high. However, following good link duration is a long bad link duration like several seconds because of the long period of sea waves. In this case, all retransmissions for a failed packet over the same link in the long bad link duration may also fail. Increasing the maximum number of retransmission before discarding a packet is not effective to improve the performance, and thus retransmission is not a good solution for link failures in maritime mesh networking. Instead, communication diversity can be used and based on this consideration, an aggregated-path routing approach is presented in [12].

- **Interference Range**
  In distributed scheduling protocol of WiMAX mesh network, it is assumed that the probability of interference beyond two-hop neighbors are sufficient low and then it can be ignored. However, this may be not true in open sea maritime environment, because the signal attenuation rate of a two-ray channel model is slower than that of a urban or sub-urban radio channel model. If interference range is close to or more than two hops and schedules are negotiated within two-hop range, transmission of MAC management messages and data packets will suffer interference from neighbors more than two hops away.

Figure 6 illustrates interference beyond two-hop neighbors. In the figure, communication distance d_c is the maximum distance that two nodes can communicate with each other with certain transmission power and modulation-coding scheme. Then node B and C are one-hop and two-hop neighbor of node A, respectively. It has been proved that with 2-way channel model, the interference range d_i is about 1.7 times of communication distance d_c, i.e., $d_i = 1.7d_c$ [13]. This indicates a three-hop neighbor node D (of node A), which is $1.7d_c$ away from node B may introduce interference at node B when A transmits MAC management messages and data packets. Then scheduling transmission among three hops neighbors are necessary in maritime environment.

5.2. Distributed Adaptive Time Slot Allocation

In distributed scheduling of WiMAX mesh mode, a three-way handshake process is defined for resource allocations. However, there is a lack of detailed algorithm to allocate the resource. It also has no a solution to deal with conflicts of concurrent transmissions. Conflicts of transmissions may be caused by interference from nodes more than two hops away, ships mobility, and sea wave movement. These issues can be minimized by a distributed adaptive time slots allocation (DATSA) [14].

The following method is used in DATSA to manage and allocate minislots of data subframe in distributed scheduling protocol. Each node keeps track of the availability of its and one-hop neighbor's minislots in data subframe. A granter marks all minislots it granted "busy" and a requester marks all minislots it accepted also "busy". One hop neighbors of the granter and the requester mark their corresponding minislots "busy" upon reception of DSCH:Grant from the granter and DSCH:Grant Confirmation from the requester. Other minislots in data subframe are marked as "free". No states "available for reception only" or "available for transmission only" is allowed as the interference range may close to or larger than two hops range in maritime environment. Further, each mesh node broadcast their minislots availability using DSCH:Availability that defined in the protocol, and each mesh node also keeps track of the minislots availability of its one-hop neighbors. When grant or accept minislots, a granter or a requester checks record of its own minislots availability as well as the one of its one-hop neighbors. If a mesh node is "busy" or any one of its one-hop neighbors is "busy" in a minislots range, it cannot grant or accept this minislots range. By doing so, minislots in use by a two-hop neighbor will not be used by this node, and then interference of concurrent transmissions can be minimized.

There is a delay between the time a granter node receives a DSCH:Request and the time this node sends its next DSCH message with corresponding grant. This delay is a random variable due to the Mesh Election algorithm for scheduling DSCH messages. In DATSA, the grant is generated right before transmitting the DSCH:Grant. This helps in avoiding conflict arises when two neighboring granter nodes allocated an overlapped minislots due to unawareness of each other's allocation. Similarly, a requester checks for minislots availability just right before sending its DSCH:Grant Confirmation.

As mentioned above, DATSA negotiates resource allocation among three hop neighbors. However, in a case that there is no common one-hop neighbor between two pairs of granter-requester two-hops away, conflicts may happen due to the lack of help from common one-hop neighbor. DATSA has a mechanism to solve this issue. A requester (i.e., receiver) continuously monitors the data packets reception. If there is successful reception during a number of consecutive frames, the requester sends a DSCH:Grant Confirmation with cancel information to the granter (i.e., transmitter). Upon reception of a DSCH:Grant Confirmation with cancel information, a granter releases the corresponding minislots and reply with a DSCH:Grant with cancel information. One-hop neighbors also release the corresponding minislots when receive a DSCH:Grant Confirmation with cancel information or a DSCH:Grant with cancel information. This detection mechanism is also helpful in detecting deterioration of link quality due to random sea wave movement and ship mobility that leads to zero data packets reception.

When a cancelation of minislots happens, DATSA interprets this as a detected allocation conflict. In following certain number of frames, in process of resource allocations, the same minislots range will be not used.

5.3. Scheduling Among Three-Hop Range

As discussed in section 5.1., radio signal in open sea maritime environment follows 2-ray model and then transmission of a WiMAX mesh node in sea may be interfered by three-hop neighbors if scheduling of transmission is only within two-hop neighbors. It is necessary to schedule transmission of MAC management messages and data packets within three-hop range.

To schedule NCFG and DSCH messages among three-hop neighbors, a node needs to consider next transmission time of neighbors within three hops. NCFG Nbr Physical IE carries information of NextXmtMx and XmtHoldoffExponent of one- and two-hop neighbors, thus next transmission time information is distributed among three-hop neighbors with the standard design. However, in standard WiMAX mesh network, a DSCH message only carries NextXmtMx and XmtHoldoffExponent of itself and one-hop neighbors. To enable scheduling DSCH among three-hop neighbors, a DSCH message is modified and carries next time information of itself, one-hop and two-hop neighbors. By doing so, in example shown in Fig. 6, node A knows DSCH and NCFG messages next transmission time information of node D through node B, then it can avoid to use the same next transmission time by including node D in the process of mesh election.

With coordinated distributed scheduling for maritime mesh communications, data scheduling is in a distributed manner. Scheduling data packets transmission among three-hop neighbors can be achieved by using the above Distributed Adaptive Time Slot Allocation. With DATSA, availability of a mesh node its own and its one-hop neighbors are tracked. If a data time range is used by an one-hop neighbor, a node marks itself unavailable during the time range and it broadcasts unavailability of this time range to its one-hop neighbors. As results, a node will not transmit during a time range that is using by a two-hop neighbor and can avoid introducing interference to transmission and reception between a two-hop and a three-hop neighbor. In example shown in Fig. 6, if node A is sending data to node B, node C will be notified by node B when node B confirms DSCH:Grant from node A, and then node C will mark the time range not available and broadcasts this information. Finally Node D will not use the time range as its one-hop neighbor node C is not available. As a result, node D will not introduce interference to transmission of node A to node B.

5.4. Comparison of Routing Protocols

There are a number of routing protocols have been developed for multi-hop and mesh wireless networks in recent years. Roughly they can be classified as proactive and reactive. A proactive routing protocol actively discovers and maintains their routes by periodically distributing routing tables throughout the network. On the contrary, a reactive routing protocol only finds and maintains routes on demand. Routing protocols also can be recognized as single-path one and multi-path one, depending on the number of routing paths to a destination node the protocol maintains. This section presents performance comparison of three routing protocols in maritime mesh networking: OLSR (Optimized Link State Routing), AODV (Ad-Hoc On Demand Distance Vector Routing), and AOMDV (Ad-Hoc On

Demand Multipath Distance Vector Routing), as representative of proactive, reactive, and multi-path routing protocols, respectively.

5.4.1. Mechanism of OLSR, AODV, and AOMDV

The above three routing protocols have different approaches to find and maintain routes. OLSR periodically sends two types of routing control messages, namely HELLO and TC. Hello messages contain one-hop neighborhood information, and are broadcasted so that each node knows information of its one-hop and two-hop neighbors. Each node selects a set of one-hop neighbors through which the node can reach all its two-hop neighbors. The selected neighbors are called Multi-Point Relay (MPR). MPR is announced by HELLO messages, then each node knows its MPR selectors, i.e., nodes that have selected them as MPR. Each node announces its MPR selectors by sending TC which is forwarded by MPR throughout the network. Upon receiving the TC, each node can build a topology table consists of entries that each of them indicates the last hop to reach a destination node. Based on the topology table, a proactive routing table can be built.

Different from OLSR, AODV setups a route only when a traffic flow has been started. The source node floods the network with route request which will be rebroadcasted by intermediate nodes. Eventually the route request will reach the destination node if there is a route between them. Upon receiving the first route request, the destination node will send a unicast route reply along the path traversed by the received route request. When the route reply reaches the source node, a route is considered setup and data packets can be forwarded. To limit routing overhead, an intermediate node will rebroadcast a route request only if it is not a duplicate copy, and the Time-To-Live (TTL) value in the IP header of the route request is not zero after being decreased by one.

AOMDV has inherited the route request-reply mechanism with a few modifications. First, the *Hopcount* field of AODV routing table is replaced by *advertised hopcount* that represents the maximum value of hop count among multiple routes that it has to the destination. A node will only accept alternative route with a lower hopcount compared to the advertised hopcount. Second, a field called *first hop* is added to route request to ensure link disjointness. All nodes that hear the route request from source node directly have to provide its IP address as the first hop in the route request. At the intermediate nodes, duplicate copies of route request are not discarded but their first hop is checked. The duplicate route request will be forwarded if its first hop has not been see before. Third, instead of replying only to the first received route request, AOMDV sends up to three route replies, one to each route request received from different neighbors of the destination node.

5.4.2. Performance Comparison

Table 6 lists an example of performance achieved with the three routing protocols for maritime mesh communications [15].

The first metric compared is initial packet delay. This metric gives an indication of the route discovery latency. As expected, OLSR has the lowest initial packet delay because it has setup routes before the traffic sources start generating packets. However, this low initial packet delay comes with high cost, as indicated by the second metric comparison – routing overhead. The routing overhead is measured in terms of routing packets. Table 6 shows

Table 6. Performance comparison of routing protocols for maritime mesh networking [15]

Metric	Sea State	OLSR	AODV	AOMDV
Initial packet delay (second)	3.0	0.56	1.51	2.20
	6.0	0.93	2.52	2.61
Average packet delay (second)	3.0	0.075	0.028	0.032
	6.0	0.246	0.078	0.046
Routing overhead (packets)	3.0	7901	283	212
	6.0	7698	1233	256
Delivery ratio	3.0	0.968	0.976	0.983
	6.0	0.685	0.883	0.892

the significantly higher routing control overhead incurred in OLSR as it periodically sends HELLO and TC messages.

The higher overhead in OLSR has negative effect on its other performance metrics like average packet delay. OLSR routes are highly node-joint since many nodes share a same node as their MPRs. This results in highly congested hotspot. In addition, OLSR suffers from frequent route changes as there are always multiple routes with equal hopcount to a destination. Each time OLSR switches to another route, the WiMAX mesh MAC layer has to perform resource reservation through the three-way handshake process and this leads to higher packet delay.

In higher sea state like sea state 6 in the example, OLSR is even worse in average packet delay. In higher sea state, a ship is subject to a larger scale of movement and this affects link quality and causes route break. OLSR needs longer time to discover a new route compared to other two, as well as compared to itself in lower sea state.

As shown in Table 6, AODV has lower initial packet delay compared to AOMDV. This is because in AODV, intermediate nodes rebroadcast less routing packets, and then there are less congestions in route discovery process. Thus in AODV route discovery can be faster as routing packets can be forwarded faster. In AOMDV, intermediate nodes may rebroadcast more route request because a duplicate route request may also be rebroadcasted if its first hop has not be seen.

Despite higher routing packets traffic, AOMDV has a lower routing overhead compared to AODV. This is because AOMDV has less number of route discovery because it finds multiple routes in a single route discovery. When a route fails, AODV needs to perform a new route discovery while AOMDV only needs to switch to a backup route. In higher sea state, this is more clearer, as in higher sea state, more frequent route failures happen due to larger scale of ship movement.

In lower sea state, AODV and AOMDV have similar performance in average packet

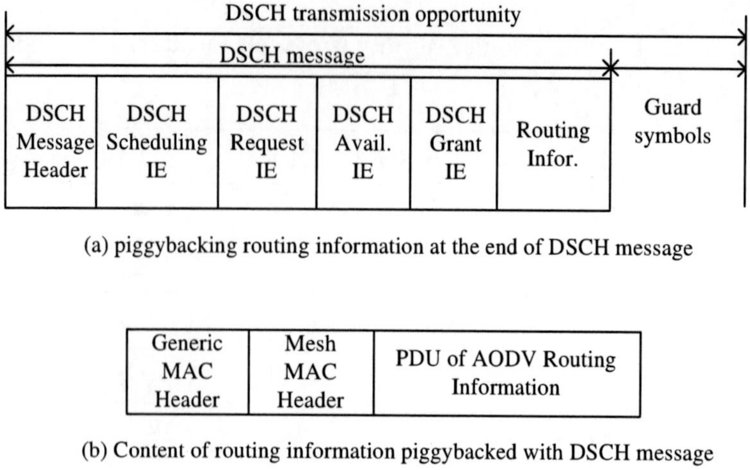

Figure 7. Illustration of routing information delivery by using DSCH message.

delay. However, in higher sea state, compared to AODV, AOMDV has lower average packet delay. The reason is same as above, in higher sea state, more frequent route failures happen and AODV needs more frequency route discovery.

In performance of packet delivery ratio, AODV and AOMDV have no clear difference. This is because same large-enough buffer size is used in both cases, and during period of route discovery with AODV and AOMDV, packets are buffered. In lower sea state, OLSR has a little bit lower delivery ratio than the other two. While in higher sea state, delivery ratio of OLSR is much lower than AODV and AOMDV. This is because in higher sea state, much more frequent route discovery happen in OLSR and the discovery process is long, leading to drop of packets due to limited size of buffer.

5.5. Routing Information Delivery by MAC Messages

As presented in above section, OLSR, AODV, and AOMDV all need to broadcast and unicast routing packets for route discovery and maintenance. Conventionally, routing packets need to use minislots in data subframes for distribution among neighbors in WiMAX mesh networks. To do so, a three-way handshaking process is needed for each routing packet when WiMAX mesh network uses distributed scheduling scheme. This is relatively time-consuming as it needs to perform resource request, grant, and grant confirmation with several around of exchange of DSCH information. A faster way is to deliver routing packets using DSCH messages. By this approach, routing packets are distributed in control subframes so it is not necessary to use the three-way handshaking process, and then both time and resource can be saved.

Figure 7 illustrates mechanism piggy-backing AODV routing packet using DSCH messages. One AODV routing packet is piggy-backed on a DSCH message PDU, right after the DSCH GrantIEs. As the number of DSCH IEs can be controlled, the whole DSCH message PDU with routing information still can be within the regularized seven-symbol duration. The routing packet can be AODV route request and route reply, etc.

The method of delivering routing information by piggy-backing on DSCH messages shows better performance than conventional method, in terms of initial packet delay, average packet delay and delivery ratio, especially in higher sea state. It is because by piggy-backing routing packets with DSCH message, faster route discovery and more available resource in data subframes can be achieved. In higher sea state that more frequent route discovery is required the advantage is clearer.

5.6. MAC-Based Routing Protocol

Routing protocol is important in ensuring connectivity and reliable packet delivery. However, none of the existing routing protocols is ideal for maritime wireless mesh networks. In Triton project, a new routing protocol called MRPT (MAC-based Routing Protocol for TRITON) is proposed [12]. In WiMAX mesh MAC protocol, control messages are periodically transmitted in dedicated control slots in a collision-free manner. It is proposed to piggy-back routing information onto the NCFG messages. The MRPT routing protocol establishes routes in both directions, from the ships to land station, and from the land station to ships.

5.6.1. Routing from Ship to Land Station

In this direction, the NCFG message from land station is used to carry additional information and is flooded towards the ships. The NCFG is designed to carry routing information that includes a sequence number, hop count and other routing metric. To distinguish different land stations within a region, a land station identification is also included as a field in the NCFG messages.

There are several metrics that can be used to ensure loop-freeness. MRPT uses sequence number to ensure loop-freeness. Sequence number has been widely used in other routing protocols, such as in AODV, AOMDV and DSDV (Destination-Sequenced Distance-Vector Routing). In MRTP, the sequence number can only be increased by the land station and is incremented with each new NCFG message that is sent out by land station. In the original WiMAX mesh MAC protocol, the NCFG messages are broadcasted and collected by all nodes in maintaining the neighborhood information. This neighborhood information intrinsically maintains the two-hop topology information in relation to a node. Information related to the nodes more than two-hops away are not propagated in the NCFG broadcast message. MRPT takes a different approach. The additional fields for routing will be propagated in each NCFG message sent by the land station or the ships. When a node other than the land station receives an NCFG message from a neighbor, and this node has not identified its current next hop to land station, and the link to the neighbor is eligible for routing, then this node will choose this neighbor as its next hop to the land station immediately.

When a node other than the land station receives an NCFG message from its neighbor, and a current next hop to land station has already been chosen by this node, it updates its sequence number only when it receives a larger sequence number from its current next hop. This is different from the mechanism used in AODV or AOMDV whereby the route is updated whenever the sequence number is larger than the current one and old route information is given up. This update happens even if the messages are sent from nodes that are

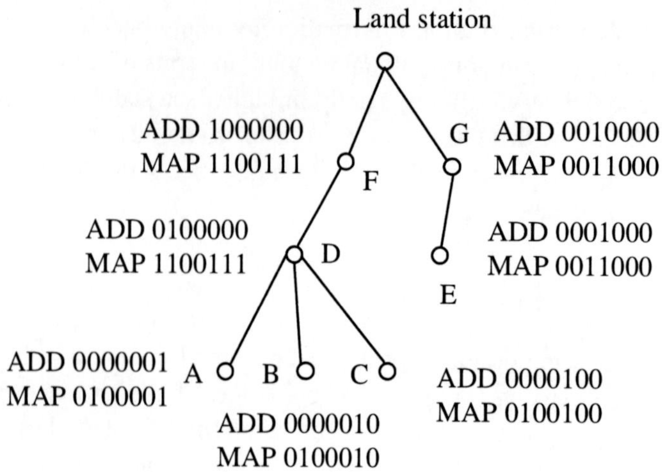

Figure 8. Bitmap address (ADD) assignment and bitmap encoding (MAP).

not the current next hop. MRPT does not adopt such a method as it may cause a node to frequently change its next hop and affect the performance in data forwarding because a stable routing structure is critical. The mechanism it adopted is the same as the one proposed in DSDV. In DSDV, a node updates its sequence number only if it receives a larger one from its parent node.

In the maritime mesh network scenario, the nodes are mobile and the network performance is rather sensitive to the stability of the route because any change in route may lead to renegotiation of the bandwidth. So, when routes are chosen, both routing cost and stability should be considered. MRPT uses hop count as the routing cost. Also, single path routing may not be stable and can be easily affected by the topology changes. Hence, MRPT includes schemes that allow each node to maintain multiple routing paths.

5.6.2. Routing from Land Station to Ship

The route setup procedure from land station to ships is different from that of ship to land station, because it is a one-to-many routing. MRPT proposes using a bitmap that efficiently encodes the logical routing tree for the routes formed from ships to land station. Then, by piggy-backing the bitmap in NCFG, the land station can be informed of the routing tree from each ship to its children. Choosing symmetric path allows MRPT to provide similar quality of service in both directions.

In the bitmap scheme, each ship is assigned a unique bitmap address (ADD) during the new node network entry procedure. As illustrated in Fig. 8, the ADD is N bits, where N is the maximum number of ships in the network at any given time. There is no need to allocate ADD to a land station. When there are multiple land stations, the land station identification can be included in the NCFG messages explicitly.

In Fig. 8, the solid lines indicate the chosen paths originating from ships to land station. Nodes A, B and C have chosen node D as their next hop for routes destined to land station. Because nodes A, B and C are one-hop away from node D, they are aware of node D's

ADD. Node A, B, and C then use an OR operation on the ADD of node D, which is their next hop node and broadcast this information in NCFG messages. When node D receives NCFG messages from node A, B, and C, respectively, it decodes the bitmap (MAP) and knows that nodes A, B, and C have chosen node D as the next hop forwarding node. Notice that node E may receive their NCFG messages from node C as well. However, node E is not indicated as the next hop to the land station in the bitmap of node C. In this case, node E does not add node C to be one of the children in the routing tree. This process repeats itself for the entire tree. Eventually, the land station receives one or several bitmaps indicating the location of all nodes in its subnet. Because the bitmaps are aggregated along the tree branches, the forwarding nodes along the way is able to indicate if a particular node is located under a certain routing tree branch. Using this technique, the land station is able to route packets directly to a ship by just decoding the bitmap along the tree branches.

6. Conclusion

Application of WiMAX mesh networking technology for broadband maritime wireless communications is introduced in this chapter. Compared with conventional HF/VHF marine wireless networks and satellite links, WiMAX mesh network can provide relatively low-cost and high-speed connections. Designing and planning a maritime WiMAX mesh network need to consider maritime environment such as radio channel model, sea wave movement, ships mobility, etc., and to develop or modify the current MAC standard and routing protocols.

References

[1] *http://www.inmarsat.com/Services/Maritime*

[2] IEEE Standard for Loacal and Metropolitan Area Networks Part 16: Air Interface for Fixed Broadband Wireless Access Systems, *IEEE std 802.16-2004*, October 2004

[3] *http://www.marcom.no*

[4] *http://www.eustis.army.mil/WEATHER/Weather_products/seastate.htm*

[5] Maliatsos K. N., Loulis P., Chronopoulos M., Constantinou P., Dallas P., Ikonomou M. (2006), Measurements and wideband channel characterization for over-the-sea propagation, *Proceeding of IEEE International Conference on Wireless and Mobile Computing, Networking and Communications* 2006, page: 237 - 244

[6] T. S. Rappaport (2002), *Wireless Communications Principles and Practice*, Upper Saddle River, NJ: Prentice-Hall Inc.

[7] Schmalenberger, R.M. and Edrich, M.G. (2000), Channel modelling for wideband data communication in a maritime mobile environment, *IEEE/AFCEA EUROCOMM 2000,* Information Systems for Enhanced Public Safety and Security, page: 150 - 154

[8] Joe, J., Hazra, S.K., Toh, S.H., Tan, M.W., Pathmasuntharam, J. S., Hoang, V.D., and Fujise, M. (2007), Path loss measurements in sea port for WiMAX, *Proceeding of Wireless Communications and Networking Conference* 2007, page: 1871 - 1876

[9] Zhou M. T., Joe J., Pathmasuntharam, J. S., and Fujise M. (2006), Characterization of Radio Path Loss in Seaport Environment for WiMAX Applications, *Proceeding of International Symposium on Antennas and Propagation* 2006, pp. 176

[10] Erceg, V., Greenstein, L.J., Tjandra, S.Y., Parkoff, S.R., Gupta, A., Kulic, B., Julius, A.A., and Bianchi, R. (1999), An empirically based path loss model for wireless channels in suburban environments, *IEEE Journal on Selected Areas in Communications*, vol. 17, no. 7, pp. 1205 - 1211

[11] Pathmasuntharam, J. S., Kong P. Y., Zhou M. T., Ge Y., Wang H. G., Ang C. W., Su W., and Harada H. (2008), TRITON: High Speed Maritime Mesh Networks, *Proceeding of IEEE 19th International Symposium on Personal, Indoor and Mobile Radio Communications* 2008, page: 1 - 5

[12] Kong P. Y., Zhou M. T., Pathmasuntharam J. S. (2008), A Routing Approach for Inter-Ship Communications in Wireless Multi-Hop Networks, *Proceeding of the 8th International Conference on Intelligent Transport System Telecommunications* 2008, page: 89 -94

[13] Xu K., Gerla M., Bae S. (2003), Effectiveness of RTS/CTS Handshake in IEEE 802.11 based Ad Hoc Networks, *Elsevier Science Ad Hoc Network Journal,* vol. 1, no. 1, page: 107 - 123

[14] Kong P. Y., Wang H. G., Ge Y., Ang C. W., Pathmasuntharam J. S., Su W., Zhou M. T. and Harada H. (2009), Distributed Adaptive Time Slot Allocation for WiMAX Based Maritime Wireless Mesh Networks, *Proceeding of IEEE Wireless Communications & Networking Conference* 2009, in press

[15] Kong P. Y., Wang H. G., Ge Y., Ang C. W., Su W., Pathmasuntharam J. S., Zhou M. T. and Hoang V. D. (2008), *Proceeding of IEEE Wireless Communications & Networking Conference* 2009, page: 2170 - 2175

In: Wireless Technologies in Intelligent Transportation... ISBN 978-1-60741-588-6
Editors: Ming-Tuo Zhou et al, pp. 297-322

Chapter 11

Heterogeneous Networks and Mobile Management in Intelligent Transportation Systems

***Jean-Marie Bonnin** and Rayene Ben Rayana*†**
Institut TELECOM; TELECOM Bretagne
Université Européenne de Bretagne (UEB)

Abstract

Car industry is evolving in a way to provide smarter and safer cars. Nowadays, embedded electronics provide users with a safer and more enjoyable traveling experience than ever before. An inescapable step in this evolution is the connection of vehicles to the Internet which opens the way to an infinite number of enhancements. Basically, services related to Intelligent Transportation Systems (ITS) can be categorized into safety related and infotainment related services. Some of them use the Internet while others rely on car-to-car communications.

In this chapter we explain how vehicles can take advantage from wireless communication technology diversity using IPv6 mobility protocols and standards to provide embarked applications with a full continuous IPv6 connectivity. Consequently, the Internet flexibility could simplify the development of various services (from security to infotainment). The CALM architecture, designed at ISO, is described and a focus is made on the support of multiple heterogeneous communication interfaces.

After a brief review of the work done in standardization bodies and in the academic world, this chapter provides an analysis of what should be implemented inside a vehicle. It also points out several missing features in the IETF/IEEE standards. The chapter ends, giving some insights of what a "full-featured" heterogeneous networks and mobility management framework for ITS should be.

1. Introduction

Everywhere over the world, governments have the aim to drastically reduce the number of deaths and injuries due to road traffic accidents. The main factor explaining most of the

*E-mail address: jm.bonnin@telecom-bretagne.eu
†E-mail address: rayene.benrayana@telecom-bretagne.eu

accidents remains the human factor. Therefore, beside the necessary education of drivers, there is an effort to reduce the relative importance of the human factor in the equation.

Even if passive security allows improving safety in the car, it is more interesting to avoid accidents by assisting the driver or even acting on his behalf when necessary. One way to achieve that is to make cars smarter and safer using telecommunication capabilities. Usually, when one talks about Intelligent Transportation Systems (ITS) services, it is often referred to safety-related services that rely on car-to-car (or vehicle to vehicle) communications. Even if these kinds of services could lead to a major push in road safety through a new generation of usages (collision avoidance, electronic brake and waring light, emergency vehicle crossing, etc.), number of them require that an important proportion of vehicles be equipped before becoming efficient. Some of them also required an expensive roadside deployment that will last for a while.

Car manufacturers also try to improve traveling experience for drivers and passengers providing them with various kinds of in-car infotainment services. These services make also an intensive use of the Internet.

Another, important trigger for developing a "fully networked car"[1] is the growing interest in sustainable development issues. Indeed, organizing multimodal transportation requires a tight coupling between different transportation systems through centralized databases. This would help to provide travelers with up to date information helping them to reduce the trip time and the global energy consumption.

As a consequence, it became critical to provide in-vehicle applications with a fully operational Internet connectivity. This chapter studies the way to use multiple access technologies to provide a better Internet access. It also discusses the advantages and the drawbacks of making applications aware of mobility events.

1.1. Context

The car-to-car consortium[2] has identified more than 300 applications that could use Internet connectivity. Among them, traditional ITS related applications such as vehicle tracking and traffic assessment are obvious. Additionally, a new class of applications related to infotainment is going to be available in cars, some to assist the driver (navigation system), others to amuse passengers.

More and more ITS-related services rely on information and communication technologies to achieve efficient cooperation between various components involved in providing the service: mobile stations or terminals, roadside units, central systems or databases, and numerous small electronic devices (sensors and controllers). In this chapter we are more interested in services relying on Internet connectivity no matter if its connectivity uses car-to-car communication to reach the Internet or not.

To make all these services possible while on the move, it is necessary to provide embarked applications with a continuous Internet connectivity at a reasonably high data rate and at a reasonable cost.

For a long time, ITS related works assumed that a universal communication technology could be provided for all ITS-related usages. Unfortunately, there are no available commu-

[1] It is the name of a workshop organized each year by the ITU at the International Motor Show in Geneva

[2] http://www.car-to-car.org/

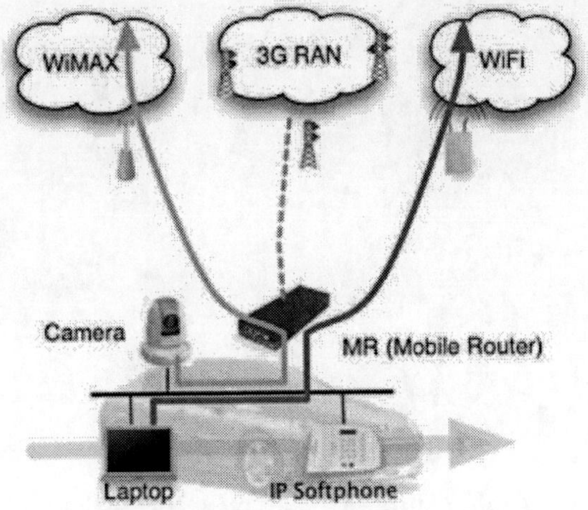

Figure 1. Heterogeneous access networks.

nication technologies that fulfill the requirements in terms of performance, cost, response time, availability and reliability. If many efforts have been spent on IEEE 802.11-based standards (IEEE 802.11p [27] and M5 [8]) this type of media would never be deployed world wide. Moreover, the current deployment costs are too high and law-related issues too complex to fully cover any country or district.

Fortunately, almost all needs could be covered using available public data network as stated in chapter 15. Nowadays, it is expected that various technologies will coexist allowing mobile devices equipped with multiple wireless communication interfaces to roam seamlessly between multiple heterogeneous access networks (see figure 1).

The use of heterogeneous wireless networks such as 3G, WLANs and WWANs has been the focus of numerous researches in an attempt to satisfy the variety of emerging applications that require different levels of QoS. Furthermore, the complementary nature of these technologies makes their diversity suitable to enhance the overall capacity and coverage. It provides better performances to users than any single technology deployment and can offer different levels of QoS and radio characteristics.

As already stated, embarked mobile devices will be diverse in terms of energy, memory, computation capability, etc. Therefore, all of them will not be able to manage their mobility as well as to embark multiple communication interfaces. This leads to introduce in the car a new device called Mobile Router. This Mobile Router (MR) will be in charge to manage networking issues on behalf of embarked equipments and applications. It can integrate multiple communication interfaces more easily than it could be done for every embarked device.

1.2. Standardization

Many stakeholders, spanning from public authorities to end users via manufacturers, are involved in ITS-related services and most of these services suppose cooperation between them. Interfaces between various equipments must be fully open to allow a well-balanced market development. Standardization bodies tackle with the definition of a communication

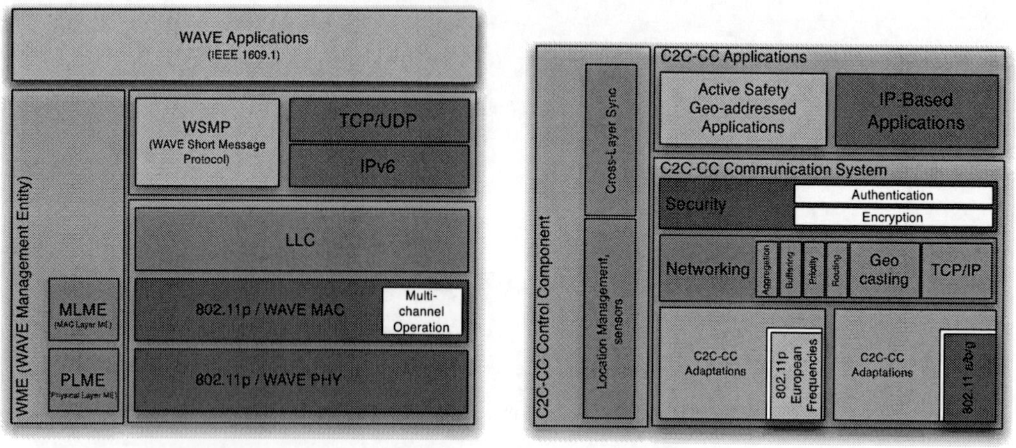

Overview of the WAVE Architecture (a) Overview of the C2C-CC Architecture (b)

Figure 2. Overview of proposed architectures.

architecture providing, among other features, Internet connectivity for vehicles.

The IEEE standardization body is designing the WAVE (Wireless Access in the Vehicular Environment) architecture [16] (see fig.(2 (a)) for a commercial deployment in 2010-2012. This architecture uses an IEEE 802.11 evolution, named IEEE 802.11p (DSRC), which allows in-car OBU (On Board Unit) to communicate with other cars in the range or with specialized RSU (Road Side Unit). A management plan and the capabilities to carefully manage multiple channels has been added to the legacy IEEE 802.11 standards. This US-centric initiative, which has been adapted to Europe in the C2C-CC (Car-to-Car Communication Consortium), is mainly devoted to car-to-car communications and to short range communications with RSU (e.g., toll gate). Recently, IPv6 has been chosen to provide Internet connectivity in WAVE architecture. WAVE has not been defined to allow heterogeneous networking and no routing features have been designed in WAVE OBU.

Just like the WAVE architecture, the C2C-CC architecture is mainly devoted to safety ITS-related applications. It uses WLAN technologies to provide both multi-hop car-to-car communications and car-to-infrastructure communications (see fig.(2 (b)). It is not really able to manage heterogeneous wireless accesses but it allows, beside the mandatory European version of the IEEE 802.11p, the use of public IEEE 802.11-based hotspots for Internet communication using IPv6. The C2C-CC middleware routes the traffic to the best suited interface encapsulating IPv6 packets in the specific C2C-CC frame format when it uses a car-to-car link to reach the Internet.

The CALM architecture (Communications Access for Land Mobiles) [29] (see fig.(3) is being designed by the working group 16 of the ISO technical committee 204. It aims at providing users with transparent continuous communications across multiple heterogeneous wireless access networks. The architecture is open and evolutionary to tackle with the difference between communication technologies and cars lifetimes. It has been explicitly designed to support heterogeneous networking allowing the integration of multiple access technologies (802.11-based, 3G, WiMAX, Millimeter Waves, etc.) through dedicated convergence layers.

To fulfill requirements identified in the C2C-CC manifesto [3], it integrates several network layers. The IP network layer, the one we are interested in, consists of IPv6, Mobile IPv6 and NEMO which have been defined at the IETF. Applications can be CALM aware or not. CALM Aware Applications interact with the management plan to give their requirements. The management plan controls the network layer to route applications flow to the interface that better meets requirements given by the application.

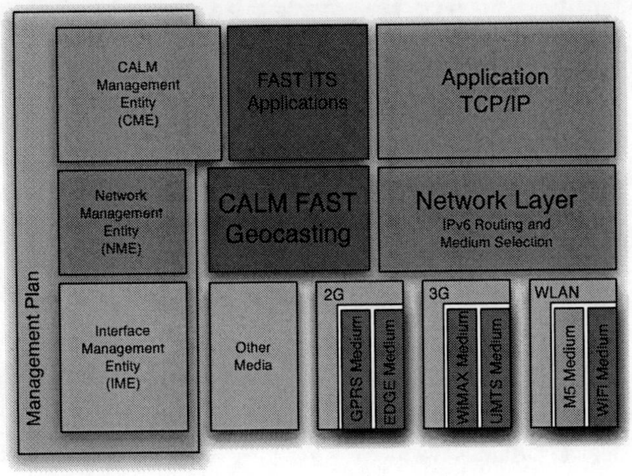

Figure 3. Overview of the CALM Architecture.

The CALM architecture is designed to be split into several devices within the on-board network. The CALM Modem (the MR) is responsible for interface management and connectivity continuity. All other on-board equipments only have to support IPv6 networking and implement a simple version of the CALM management plan [9]. It could be easy to integrate simple IPv6 devices as they are allowed to communicate as if they are on the fixed Internet. Various CALM deployment scenarios have been envisioned (see figure 4). The simpler one (4 (a)), consist of one single CALM devices that do not provide IPv6 connectivity to other on-board devices and do not allow evolution of the on-board network. The more complex one (4 (c)) allows a progressive deployment; a first CALM mobile router could be built-in the vehicle by the car manufacturer and devices could be added during the vehicle's lifetime. As vehicles have a longer lifetime than wireless communication technologies a new CALM router may provide new wireless interfaces and cooperate with other CALM devices to transparently enhance the Internet experience of the users. A first take of the architecture is currently being developed by the CVIS (Cooperative Vehicle-to-Infrastructure Systems) European project[3].

All these architectures provide Internet connectivity; Moreover, all of them have chosen IPv6 as a network layer since everybody in car industry is now convinced that there is no serious long term alternative for ITS other than IPv6. Anyway, only the CALM architecture explicitly deals with access network heterogeneity.

Heterogeneity support is tightly related to the support of multi-homing in the IETF world. Even if NEMO and Mobile IP were not initially designed for the simultaneous use

[3] http://www.cvisproject.org/

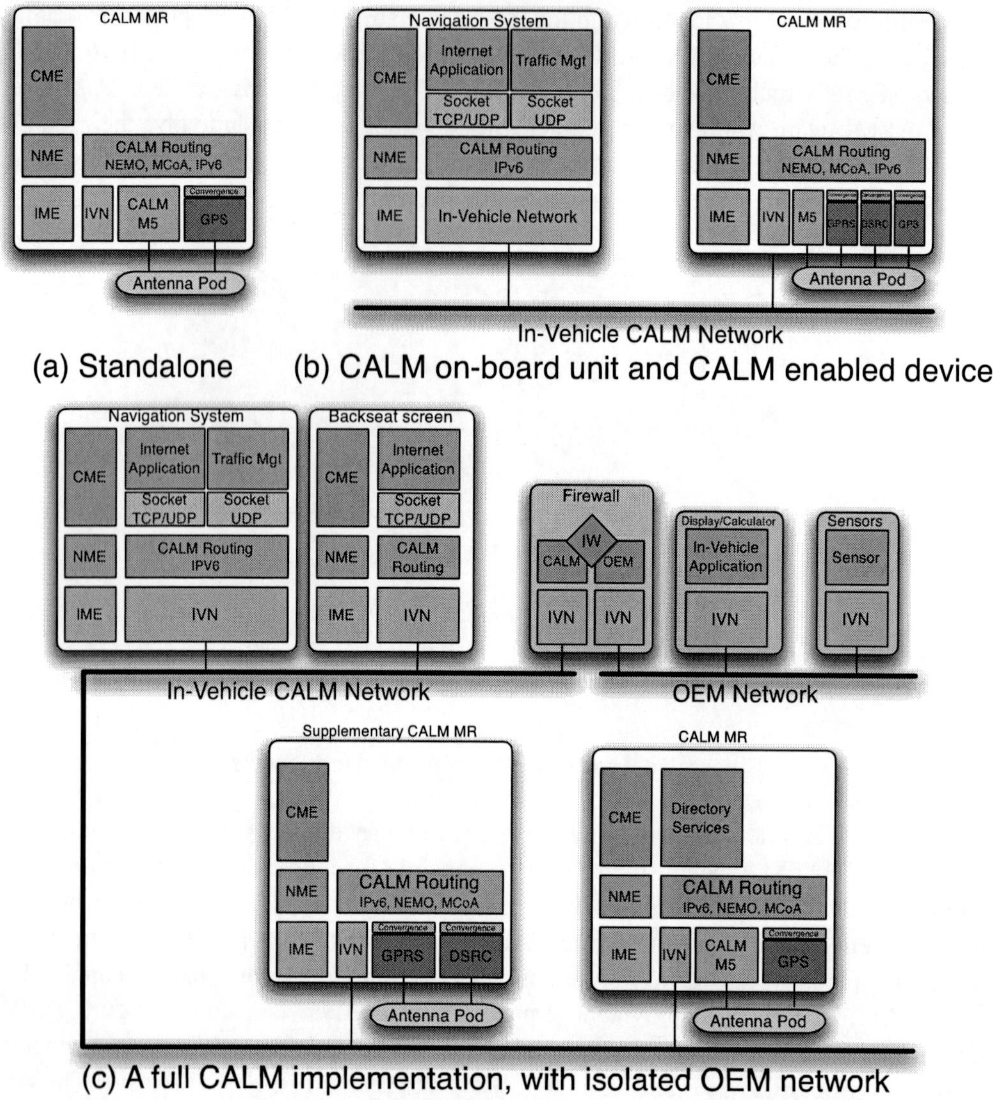

Figure 4. Various CALM deployment scenarios.

of multiple heterogeneous interfaces, many works are in progress in several IETF working groups. The edge multi-homing can be managed by end nodes using shim6 [15] or by intermediate entities through a comprehensive use of tunneling using MCoA [26]. The latter will be considered here since it is fully transparent to end nodes. Only the MR and the Home Agent (HA) are involved in the process. This fits well with the CALM architecture principles.

1.3. Focus of This Chapter

In the overall picture depicted in this introduction, the ISO CALM architecture seems to be the one that fits well with the requirements coming from automotive industry. The

ISO is responsible for designing the architecture and main features that should interact together. IETF, on its side, is in charge of designing and evolving networking protocol such as IPv6, NEMO and routing policy exchange. IEEE and all communication technology providers are responsible for the definition of their respective technologies that could be done independently. Anyway a collaboration could be launched to design the convergence layer in a first time and a fully integrated version of the technology in a second time. A stronger collaboration with the IEEE 802.21 [28] working group is in progress to integrate features provided by MIH (Media Independent Handover) inside the CALM architecture and probably a specific version of the convergence layer for the IEEE 802.21-compliant access technologies.

In chapter 15, a simple multi-homing solution for providing in-vehicle applications with full Internet connectivity is described. This solution is compliant with the main principles of the CALM architecture. The current chapter, describes the way the IP layer manages interface heterogeneity, it explores the missing parts that are not currently dealt with at the IETF and for which ISO does not provide a complete definition. All the decision part will probably never be standardized, as it is important to let battlefields to stakeholders to differentiate between them.

2. State of the Art

As explained before, the future of mobile networking is closely tightened to the ability to efficiently support heterogeneity. Heterogeneity in this context can be seen in terms of devices capabilities, application requirements, and network characteristics. In this chapter a focus is made on the ability to efficiently take advantage of the availability of heterogeneous access networks in the surrounding.

We can identify three different factors that are part of the equation we are trying to solve (see fig.5). First, administrators want a certain control over the network they operate, be it at the core network level (mobile network operators) or at the access network level (access network operator or ISP). Second, as the system is intended to be moving, the current environment (i.e., network availability, network load, financial cost, security, etc.) has to be taken into account. Third, applications have certain quality of service requirements to function properly (bandwidth, delay, etc.).

Several mobility management frameworks have been identified in the literature. Through the following paragraphs, a survey of several of them is provided to identify their differences and, most importantly, their common functional modules. This allows to design an exemplary architecture supporting all necessary functions. Some of these functions would require support that could not be provided by current research and standardization works.

2.1. Identified Solutions

2.1.1. The InternetCar Project

The InternetCar Project (1996-2002) is one of the pioneer projects proposing a network mobility solution for vehicles. According to [14], it is based upon two main technolo-

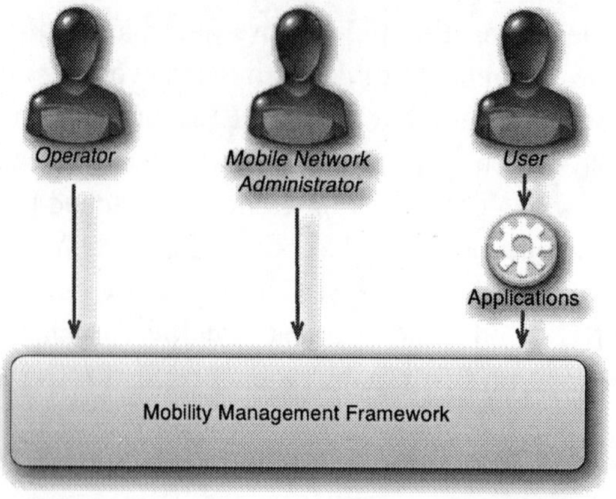

Figure 5. The Different Actors that Take Part in the Decision Process.

gies: *Interface Switching* and Prefix Scope Binding Update (PSBU). The first draft of RFC 3963 [4] and [6], state that *PBSU* is the first network mobility solution that have been proposed as an extension of the Mobile IPv6 solution [10]. It is very similar to NEMO Basic Support (see section 3.2.). The main difference relies on the management of the permanent address of the MR, which is assigned to the egress interface (external), and not to the ingress interface (internal).

The proposed solution uses a policy-based mechanism to choose the most suitable interface at a given moment called *Multiple Network Interface Support by Policy-Based Routing on Mobile IPv6*. This solution is similar to MCoA (see section 3.2.).

The project underlines the importance of multihoming to ensure sustainable connectivity. Unfortunately, it does not use these interfaces simultaneously to send data. The underlying technologies allow only a single interface to be used at a given time. Therefore, a per-flow routing policy cannot be enforced and dynamic load sharing cannot be achieved. Nevertheless, the project revealed several missing pieces to achieve an efficient network mobility framework [5] and was at the origin of several other solutions.

2.1.2. MAR: a Commuter Router Infrastructure for the Mobile Internet

In some works, the MR acts as an application or transport gateway to provide TCP flows or more generally application flows with some mediation functions [18]. It could for example deal with acknowledgement management to free the embarked device of it. The idea here, is to build a unique tunnel using multiple interfaces just as it could be done between router for the sake of load-balancing. The mobile router and a mobility access gateway manages the tunnel dealing with routing, acknowledgements and retransmissions to make them transparent to embarked devices and correspondent nodes.

As an example of this kind of solution, MAR is a network mobility management system developed in 2004. As stated in [20], MAR dynamically instantiates channels based on traffic demand, aggregates the bandwidth and dynamically shifts load from poor quality to better quality channels.

MAR is based upon the Mobile IPv4 protocol rather than Mobile IPv6. The technical terms used to describe the solutions are quite different from the terms used in the IPv6 world but the idea is almost the same. It uses a policy-based mechanism to exploit multiple interfaces at the same time. In this solution the mobile router also performs higher-level task taking care of packet loss and disordering to save TCP performances.

The results given in [20] show how GPRS interfaces provide variable throughput and blackouts. It also shows how the MAR architecture can combine three interfaces to provide a much reliable and effective virtual link.

Unfortunately this type of solution hardly rely on a gateway which is in charge of managing one of the two end points of the virtual link. This gateway has to manage complex states for all individual flows at the transport level as well as buffer associated to them (to keep packet in order). This makes really difficult the distribution of this mobile gateway function, while it is necessary for scalability and reliability purposes. It also has the same concerns than NAT (Network Address Translator) since it makes the deployment of new protocol (transport) more complex as it will involves routers and gateways. For these reasons, it looks preferable to make the mobility management transparent to the application and to manage flow in a very generic way, i.e. at the network level.

2.1.3. Ubique

The integration of various heterogeneous access networks in a ubiquitous wireless environment has been studied in several works (e.g., [21, 30]). The main idea of Ubique is to provide a unified platform architecture offering a seamless integration of heterogeneous technologies (e.g., [1, 32]). Previously proposed handoff solutions (e.g., [23, 31]) only try to keep the mobile user always connected. However, it is important to be able to use for each running applications the adequate technology among the current available technologies while providing the best tradeoff between bandwidth, bit error rate, one way delay and its induced cost (e.g., [7]).

Although various vertical handoff schemes have been recently proposed to offer seamless session continuity, there is a lack of mechanisms allowing a comprehensive network connectivity management while providing means to control some essential parameters such as financial costs, device energy consumption and service satisfaction. This mechanism should be able to take into account preferences or usage policies regarding the way user subscriptions and wireless interfaces could be used.

The Ubique is a Mobile IPv6 based framework designed to offer ubiquitous access to mobile users [24, 25]. The novelty of Ubique is that it approaches the problem differently. Rather than using handmade routing policies, Ubique proposes to generate these policies dynamically based on high-level parameters specified by the users, the administrator or the applications. In other words, instead of asking the system to send VoiP traffic on eth0, the user asks the system to ensure the delivery of the most important traffic while minimizing the costs.

In Ubique, the applications communicate directly with the framework to inform it that a new flow is to be initiated and specifies its requirements. The framework matches the flow requirements with the interfaces and networks that can be used. It ensures the respect of the administrator limitations and conveys the flow through the most suitable interface.

Ubique has been partially adapted to be implemented in a mobile router in [12] and takes benefits of the ability to allow simultaneous use of multiple interfaces. This is done by using multiple tunnels toward the Home Agent (i.e., MCoA)

2.2. Features Identified from The State of the Art

This survey of the solutions proposed to manage multiple interfaces in a mobile router and more generally in a mobile terminal, allows to identify the main features that have to be present in a mobile router. It is worth to note here, that a solution dedicated to mobile terminal could be generally adapted to the mobile router context but specific attention should be payed to adapt it to mobile network specificities.

2.2.1. Multiple Path Management

The main feature of a Mobile Network Management System is the ability to handle several paths simultaneously. According to [20], none of the wireless technologies available can offer the level of reliability desired for in-vehicle networking. Mobile Networks should be connected through multiple networks and technologies at the same time to be efficient.

2.2.2. Per Flow Path Selection

Settling multiple paths can be useful for backing-up traffic. In other words, when an interface is disconnected, another interface is ready to take over the flows. Multiple paths can also be used to increase reliability by duplicating traffic over several paths at the same time, but the most common usage will certainly be the per flow dispatching. This allows the user to select the most appropriate path or access network for each flow. It can be handled in a simple manner by providing a flow-interface correspondence table as in the MAR project. This can also be handled at a higher level as in Ubique.

Note that looking at the current CALM specification, the CALM management plan has to be involved in each flow routing decision and each time a new interface becomes available or disappears. This could lead to intolerable loss of packets. To avoid that, it is important to provide network layer with enough information to react timely to a change in the networking conditions. Of course this short-term decision should be done according to higher level constraints.

2.2.3. Advanced Interface Management

The formerly cited solutions agree on the simultaneous use of several network interfaces. However, they did not study the fact that these interfaces could be power consuming and that power resources can be limited in a mobile environment. Additionally, some connections are billed according to duration rather than the amount of data sent. It is not conceivable in this case to keep these interfaces running all the time. Thus, an advanced management system should be proposed to bring interfaces up when critical flows have to be sent and shut them down when the transmission is terminated.

2.2.4. Dynamic Network Selection and Switching

Nowadays, it is very common for network interface to have several wireless networks available at the same time. In the future, the number of these networks is expected to increase dramatically and it is obvious that the performance of a mobility management system can be improved by a comprehensive dynamic selection of the best network. A Mobile Router should be able to monitor the available wireless networks and evaluate, even roughly, the performance of each one.

2.2.5. Multiple MR

None of the above solutions deals with Multiple MRs. Nevertheless, MR Redundancy can be very interesting, particularly in trains where MRs can be geographically distant and take profit from a larger set of wireless networks.

Managing multiple MRs can also be useful in the case where a user have a hand-held device such as a mobile phone and that he wants to turns it into a mobile router each time he uses his car. This would add an additional egress interface to the whole system and improves the overall performance. However, since the car is already equipped with a mobile router, this requires a hot plug mechanism, which is quite more complex than the basic cold plug. Studying the different multiple MR mechanisms are out of the scope of this chapter. A detailed comparative survey of the available multiple MR solutions can be found in [11].

2.2.6. High-Level Decision Modules

As explained before, the Ubique architecture introduces a new way of taking into account the user preferences. Rather than applying a predefined set of routing rules, it generates them automatically using high-level (human friendly) objective definitions. Other representation of high level usage policy (or preferences) can be used (fuzzy logic, policy-based language, etc.) but it is important to allow the user to specify generic goals instead of describing specific situations. This would help in making the system more autonomous since it can handle situations not explicitly specified by the user. It is also more efficient if the decision is made using accurate decision algorithms dealing with the actual situation of the mobile router.

2.2.7. Support of Mobility Aware Applications

Applications requirements, in terms of bandwidth, delay, jitter, etc., are getting increasingly diverse. Furthermore, "Adaptive" applications may adapt their own behavior to resources available on the different access networks [1]. Adaptive applications and network architecture have to interact tightly to reach this level of interaction [17, 25]. This is more complex in a context where applications do not run on the device that manages interfaces.

It is then necessary to exchange information between the MR and embarked devices. Two operation modes could be envisioned. In fig.6.a, the MR announces available resources and each application decides on its own to adapt its behavior. In fig.6.b, the router perform a centralized resource management taking into account requirements declared by applications.

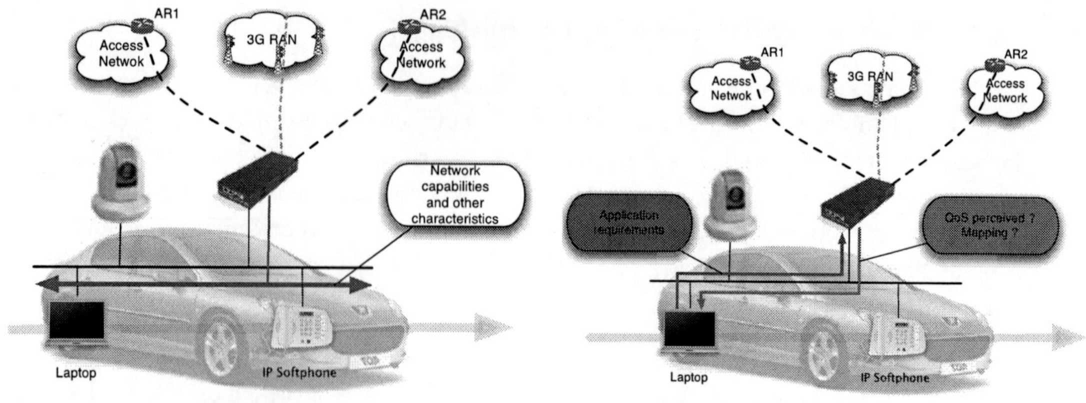

(a) (b)

Figure 6. Exchanges between MR and embarked devices.

It is necessary to offer a unified way to obtain mobility-related information for interoperability concerns. This will enable applications to participate in the decision process. These protocols have to be as maintainable and flexible as possible to foresee future technology evolutions and needs. Further details on mobility aware applications are available in section 4..

3. Proposal for a "Full-featured" Architecture

In the following, we introduce, through an example, the various elements that should compose a communication architecture for ITS. This architecture includes a high-level mobility management framework which is highly inspired by the architecture of the REMORA project[4] in which authors are involved in. As you can see in fig.7, modules are divided into three categories: Monitoring Modules, Decision Modules and Enforcement Modules, but the overall architecture is built on a network device which uses IPv6 and related mobility management schemes. This section includes also a survey of the underlying IPv6 network technologies that have been chosen for this architecture.

3.1. Architecture Overview

Monitoring Modules monitor the system environment to collect valuable data necessary to the decision process. Data collected may include vehicle speed, geolocalization, battery status, available wireless networks, etc.

Decision Modules have three types of input: First, they take the data collected by the Monitoring Modules. Second, they gather the preferences of the different actors of the system. This includes operators, system administrators and final users. Finally, the system needs flow requirements (which are static in this case since there is no communication with the applications yet). These allow a better dimensioning of the needed resource level and results into more accurate decisions. The resulting decision consists in two policies: The

[4]This project is a collaborative project supported by the ANR (French government).

Interface Management Policy and the Routing Policy. Once produced, these policies are delivered to be enforced. Later in this chapter, when we will introduce mobility aware applications, we will see that a third policy, called Application Policy, has to be produced to manage application adaptation.

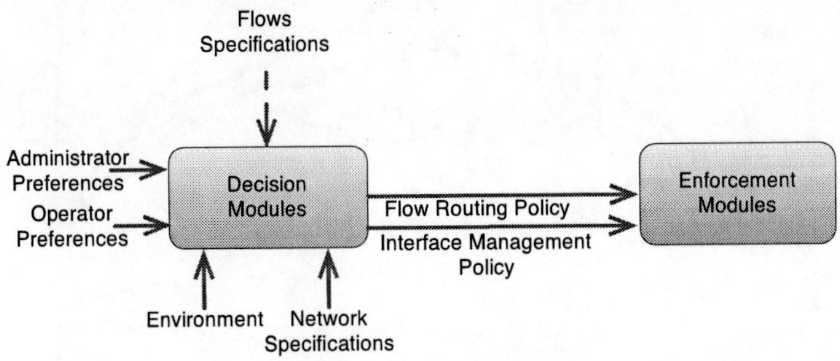

Figure 7. Overview of a Mobility Management Framework.

Enforcement Modules are in direct interaction with the network layer of the operating system's kernel. In several works (e.g., [1]) dedicated enforcement modules have been developed to fit project requirements. Unfortunately, these modules are hard to maintain since operating system kernels are constantly evolving. Consequently, such a framework should use extensively the high-level tools offered by the system. In this case, the role of the Enforcement Modules consists mainly in translating and adapting the policies produced by the Decision Modules to the current operating system tools. Therefore, the framework would be much easier to maintain and could be easily ported on multiple platforms.

In the following paragraphs, a manner to implement these modules is given as an example. A simplified view of the proposed system can be seen in fig.10. In this figure the classification of the modules as Monitoring, Decision and Enforcement is maintained and inter-module interaction is schematized by arrows. We are aware that this can be handled differently but what is important here is the presence of the different functionalities identified in the state of the art and the proposal of a design of the interactions that could take place between them.

3.2. Networking

IETF launched its first working group dealing with the support of mobile node at the end of 90s. Until recently the mobility management was considered only on an end-to-end basis with the support of a location server inside the network, namely the Home Agent (HA). The functions of the latter are two fold: first it has to track the location of the mobile node and to maintain an association between a permanent address depending on the home network (the Home Address or HoA) and a temporary address (the Care-of Address or CoA) depending on the actual location of the mobile node in the Internet. Each time the mobile node changes its attachment to the Internet (i.e. its CoA) it has to update the association on the HA. Second, it routes the traffic between the mobile node and the correspondent node for whom it should remain transparent. All the traffic travel inside a secured tunnel (IPsec) between the HA and the mobile node (MN). IETF also defined a routing optimization mode where the correspondent (CN) could be involved in the mobility management to use a more direct path between the MN and the CN and thus avoid to make a detour to the HA.

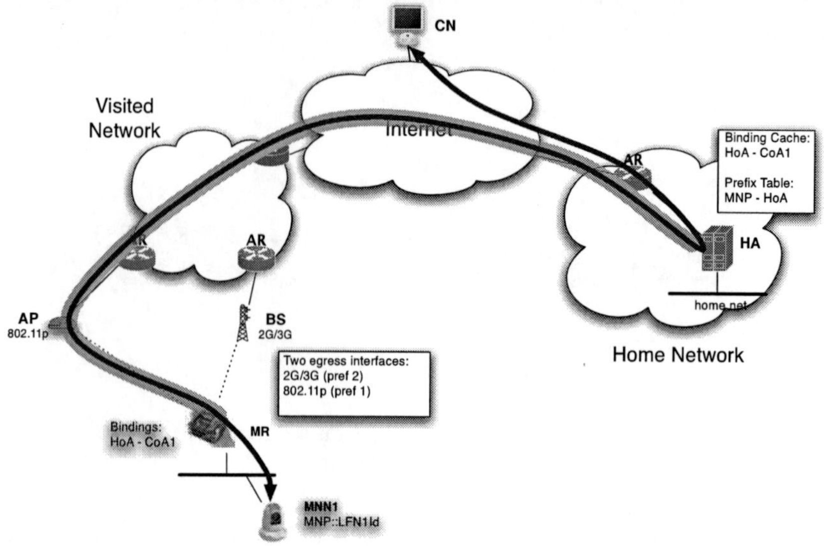

Figure 8. Example of a NEMO Basic Support Use Case.

In 2002, after initial discussion started two years before, a new working group devoted to network mobility was created and began to identify requirements and solutions to fulfill them. A simple solution based on a straightforward evolution of Mobile IPv6 has been defined, namely NEMO Basic Support [4]. This solution mainly introduces the Mobile Router (see fig.8) which is in charge to manage the IP mobility for all mobile networks attached to it. The mobile network gets a Mobile Network Prefix (MNP) depending on the home network. This prefix does not change while the mobile network moves. The MR behaves like an MN registering its locally acquired CoA with the Home Agent. It also announces the prefixes it uses for its mobile network. Then the HA is able to route all the traffic sent to an address belonging to one of the MNP declared during the tunnel establishment.

When the MR changes its point of attachment (and then its CoA) it only has to update the association on the HA. Mobile Network Nodes (MNN) continue to use the same address they have acquired during the normal IPv6 auto-configuration process. Although, they are not involved in the mobility management as long as they remain attached in the Mobile Network.

By default, the NEMO specifications do not deal with multi-homing. Usually the mobility management process considers only one interface to the Internet at a given time. If the mobile node or router has more than one interface attached to the Internet (and then more than one CoA) it has to choose one of these interfaces has primary interface. The latter will then be used for updating the binding at the HA. It could be useful to establish the layer-two connections (or attachments) with an access network before the current one becomes useless. Once this attachment has been done and a CoA has been configured, the MR has just to consider the new CoA as primary to perform a heterogeneous handover.

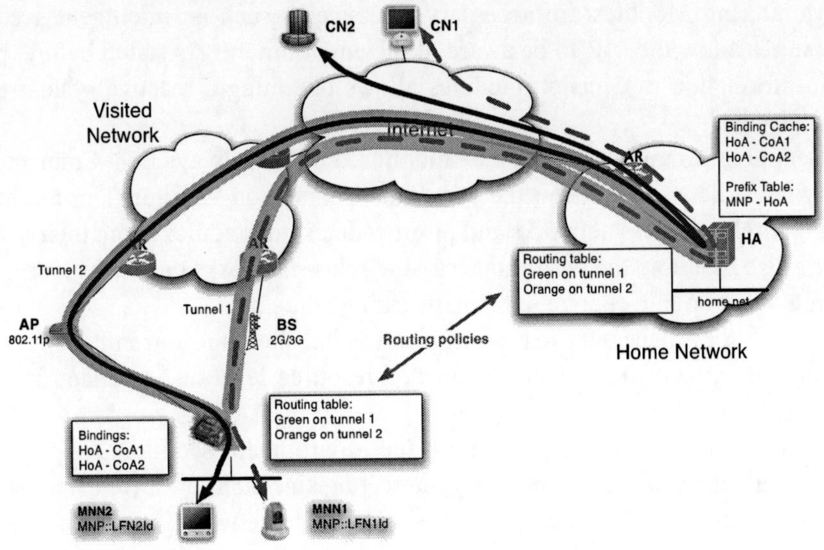

Figure 9. Example of a NEMO + MCoA Use Case.

A work in progress (draft MCoA [26]) also defines a way to register multiple CoA for one HoA at the HA. This way, the HA has several tunnels to reach the HA (see fig.9). Only one of these tunnels is used at a given time; having multiple tunnels established could only reduce the time required to reestablish the tunnel when the MR wants to perform a handover but the benefits would be very limited. The main interest of this solution comes with the ability to manage multiple tunnels simultaneously.

3.3. Monitoring

The system administrator, usually the vehicle's driver, cannot be asked to interact with the system while driving. This contrasts with its needs that are variable during a travel. In fact, the driver's expectations from the system depend closely on the context. For example, a user may privilege security and link reliability during working hours and prefer cost economy during his free hours. A user may want to differentiate the behavior depending on certain parameters such as power supply. For example, a user can ask to deactivate a particularly power consuming network interface if the vehicle's battery level is under a certain threshold. Finally, some particular locations can require a particular treatment. For example, it is expected that, in the near future, wireless devices will have to be shutdown in gas stations due to inflammation risks. Hence, the user will need to bring down the system's interfaces each time he stops at a gas station.

Of course, it is possible to define operating modes that can be selected by the driver when needed (e.g., "Work Mode", "Cost Economy Mode", "Low Power Mode", "Gas Station Mode", etc.), but, this requires an interaction with the driver, which can be annoying and dangerous. It is by far preferable to have a fully autonomous system that is configured using solely high-level parameters allowing the system to react in function of the current context.

The Monitoring Modules are necessary to conceive such an intelligent system. The sensing features allow the MR to be aware of the environment. As stated before, providing up-to-date information to decision modules allows obtaining a reactive system that takes accurate decisions.

An essential information that must be monitored to be "Always Best Connected" is the available wireless networks. This is the role of the Connection Monitor. This module listens passively to wireless access networks and informs decision modules about their availability. The module also monitors the performances of wireless networks currently in use, allowing the system to react upon a change in the networking context.

A Flow Monitor is necessary to provide information about the currently conveyed flows. It will allow the system to evaluate the needed resource level and to adapt itself to flow requirements.

The last necessary monitoring module is the Environment Monitor. Among the monitored parameters, the vehicle speed is very interesting since it allows, for example, to privilege WiFi networks when the vehicle is stationary and 3G networks when it is moving. The time and date parameters give the possibility to choose cheaper networks even if they have complex billing method (e.g., UMTS is cheaper from 8 p.m. to 6 a.m. and during weekends). It allows also differentiating the system behavior depending on whether it is work time (the company is billed) or free time (the user is billed). Another interesting parameter is the battery level. Monitoring this value allows the system to decide if it should limit its power consumption (e.g., when the vehicle engine is stopped) or if it can act regardless of power consumption. Additionally, this module should be highly configurable and should allow users to monitor almost any value that can be asked for through a simple script. Thus, the module can be adapted to usages that have not been foreseen by designers.

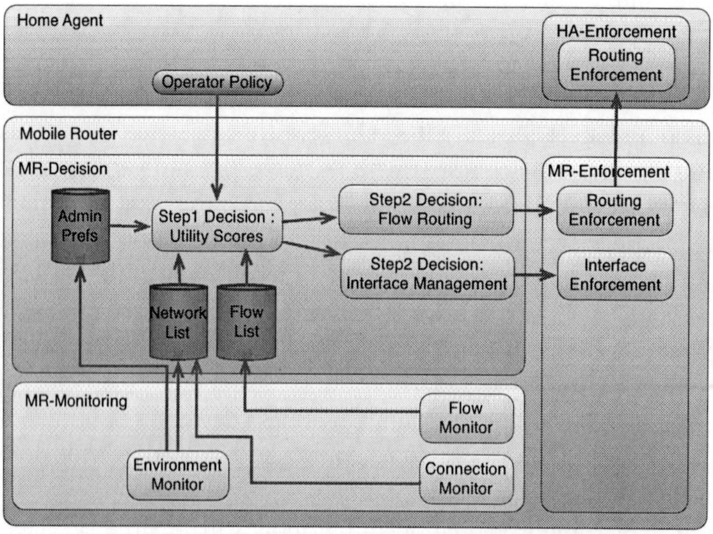

Figure 10. An Example of Implementation of a Mobility Management Framework.

3.4. Configuration

3.4.1. Administrator Preferences

As stated before, the administrator needs depend closely on the context. Therefore, the system configuration has to be decontextualized. In other words, instead of giving the value of a parameter, the administrator should rather give its different values in different contexts. To achieve this, various languages can be used. They have to express conditional statement in a human-like language to be easily modified by users. In fig.11, you can find an example of an administrator configuration file written as pseudo-algorithm.

```
Section Objectives:
Minimize Cost;
Minimize PowerConsumption;
Maximize Bandwidth;
Maximize ConnectionStability;
Maximize Security;
Minimize Jitter;
Maximize
OperatorSatisfaction;

Section Modes:
CostEconomy
{
    Cost = 100
    Bandwidth = 20
    PowerConsumption = 0
    ConnectionStability = 10
    OperatorSatisfaction = 20
}
MaxStability
{
    Cost = 10
    Bandwidth = 70
    PowerConsumption = 0
    ConnectionStability = 100
    OperatorSatisfaction = 20
}
MaxBandwidth
{
    Cost = 10
    Bandwidth = 100
    PowerConsumption = 0
    ConnectionStability = 70
    OperatorSatisfaction = 20
}
PowerEconomy
{
    Cost = 50
    Bandwidth = 20
    PowerConsumption = 100
    ConnectionStability = 20
    OperatorSatisfaction = 20
}

Section Pseudo Algorithm:
VehicleLocation in ('France')
    BatteryLevel in [20..100]
        DayOfWeek in ('Saturday', 'Sunday')
            RETURN CostEconomy
        DayOfWeek in other
            HourOfDay in [8..12, 14..18];
                VehicleSpeed in [0..30]
                    RETURN MaxBandwidth
                VehicleSpeed in other
                    RETURN MaxStability
            HourOfDay in other;
                RETURN CostEconomy
    BatteryLevel in [0..20];
        RETURN PowerEconomy
VehicleLocation in other;
        RETURN CostEconomy
```

Figure 11. An example of a decision process configuration file.

This file includes three sections. The first section specifies the objectives: It tells which are the parameters that the administrator needs to maximize or to minimize. The second section gathers the different operating modes of the system. As stated before, these operating modes could be changed directly by the user but, to minimize the interaction with the driver, it is preferable to let the system select the mode on its own. The operating modes indicate the weight of each "performance" parameter. The higher the weight is, the more important the parameter will be. Finally, the last and most interesting section is the pseudo algorithm that allows selecting the adequate mode in function of the context variables.

In this example, the user chose to give priority to cost when he is abroad. When he is at home (i.e. France), the user privileges power economy when battery level is low.

Otherwise, he checks if he is connecting during work hours in which his company is billed for the connection fees. If the company is billed, the user privileges bandwidth or stability depending on the vehicle speed. If the user himself is billed, he gives advantage to cost.

The pseudo algorithm allows expressing the administrator needs in a natural way and allows an infinite number of configuration possibilities depending on the needs of every user, whether they have been planned or not. The output of the pseudo algorithm is the operating mode (e.g., CostEconomy, MaxBandwidth, etc.).

3.4.2. Operator Preferences

The influence of the operator on the decision can be handled in several ways. The most viable solution seems to send a generic (i.e. for all the users) set of rules that associate for each flow type (i.e., port number) a list of mandatory networks, a list of preferred networks and a list of forbidden networks.

The administrator should have a way to limit the impact of the Operator Preferences on the decision. Experienced users desiring to profit from competition between operators have to minimize it. Alternatively, common users may follow their operator's recommendations. Note that, nothing have been specified in the CALM architecture to allow operator policies to be taken into account in the decision process nor to exchange such policies.

3.4.3. Network List

In order to choose among several networks, the system has to identify them and evaluate their performance and their costs. Thus, a Network List should contain an entry for each known network specifying its cost, its characteristics and a way to identify it and to connect to it. When a piece of information is missing (e.g., bandwidth), a default value is set depending on network's technology (e.g., usual bandwidth of a 802.11 network).

This list has to specify all the networks that the MR can connect to. As a consequence, this list can become heavy and it will not be easy for a user to maintain it.

As a first step, the Network List could be filled in by the administrator or downloaded from a foreign server (e.g., operator server, network card manufacturer website, etc.). A second step would be the addition of a performance monitoring module that would update the performance parameters dynamically.

In a third step the list could be filled in automatically: When an unknown network is detected and if no other known network is available, the administrator is asked to add it in the Network List. This operation can be delayed to avoid bothering the driver of the vehicle or it can also be automated by a simple default action (e.g., "Always add" or "Always ignore").

A fourth step would be the contextual filtering of this list to reduce the number of entries considered in the decision process. This additional module can take into account current GPS positions and other clues to evaluate the probability of availability of known networks. This information could be learnt progressively during the mobile router's journeys. The Network List can also be decontextualized. This means that the user can specify characteristics (performance, cost, etc.) that depend on the context. The most obvious example are the billing schemes. In the example below (figure 12), the user is billed 0.3 Euro per MB during the day and connection is free from midnight to 6 A.M. When the vehicle is abroad, each MB sent or received is charged 1 Euro.

```
UMTS
{
    ConnectionScript = ``pppd ...''
    ConnectionStability = 80
    Security = High
    Download, Upload
    {
        Cost  =   CASE VehicleLocation in ('France')
                      CASE HourOfDay in [0..6]
                          0.0 Euro/MB
                      ELSE
                          0.3 Euro/MB
                  ELSE VehicleLocation
                      1 Euro/MB
                Bandwidth = 600kb/s
        Jitter = DEFAULT
        BitErrorRate = DEFAULT
        Delay = DEFAULT
    }
}
```

Figure 12. An example of a decontextualized network description.

3.4.4. Flow List

A Flow List should contain the types of flows to be conveyed. For each flow it is necessary to specify the manner to recognize belonging packets (port numbers, IP addresses, etc.). A Flow Monitor adds information about whether the flow is alive or not and evaluates the bandwidth currently used by each flow. The administrator may also add a priority value to certain flow types. For example, he can privilege a videoconference flow over a video streaming. This way, when resource level declines, the MR can drop video streaming packets in favor of videoconference packets.

In addition, the Flow List specifies the minimum requirements for each flow. A network that does not meet these requirements is simply not considered as a candidate in the routing decision. When there are more than one candidate network, the system has to elect the one that fits better flow needs. To orient the decision a flow could indicate in the Flow List the relative importance it gives to network parameters. For example, an FTP flow may prefer to increase the bandwidth while a VoIP flow may prefer getting more stable and reducing its round trip time.

3.5. Decision

The decision process have to take into account high level considerations such as operator preferences, all available networks and all flows to be routed. This could result in a heavy process that could not be triggered each time something happens.

In a mobility context, networks events (handover, disconnection, etc.) are frequent and require fast reaction that cannot be afforded while taking into account an important number of parameters. To overcome this, the decision process should be split in two steps:

The first step consists in evaluating the matching degree of flows with the networks stored in the Network List while respecting the restrictions and the preferences of the operator, the administrator and the flows. To achieve this, a utility score could be calculated for each network-flow tuple. The greater the score is, the more compatible the tuple is.

The second step has two tasks: The first task is to route the flows through the different tunnels available. The second task, which is highly correlated with the first, consists in bringing up and down interfaces as needed and looking for "better" networks.

3.5.1. First Step: Utility Scores

Within the Network Grader (see figure 10), the information provided by monitoring modules is combined with the Flow List, the Network List, the operator preferences and the administrator preferences. Weights provided by the different inputs allow to obtain a utility score for each network-flow tuple. The score is computed not only regarding the networks currently in use, but also the networks that are potentially reachable. The output of this module is a list of flows and for each flow, a list of the compatible networks and their respective utility scores.

3.5.2. Second Step: Flow Routing

This task consists in distributing flows over available interfaces while sharing resources. Priority holding flows are served first and the other ones can be dropped if needed. The objective of this task is to maximize overall flow satisfaction regarding flows priorities. Simultaneously, the system tries to find alternative access networks that can fit better aggregated flow requirements. In addition, some interfaces can be brought down to save energy and money. The decision's second step module filters the output of the Network Grader by eliminating the non-connected networks. The obtained list is a set of FlowIDs (Flow IDentification) with a list of zero or more NetworkIDs-scores (Network Identification) attached to each FlowID.

The flow profiles that do not have any NetworkID attached are simply discarded later. The flow profiles that have a single NetworkID attached are marked to be conveyed through the corresponding tunnel if the tunnels resources are sufficient. The flow profiles that have several NetworkIDs attached (e.g., Flow1) can be handled in several manners:

An obvious and simplistic way to choose one of the NetworkIDs proposed is to choose the first one (with the highest score). The problem with this solution is that it does not guarantee a satisfactory load sharing. For an overall satisfaction increase, it is wise to study the second and third tunnel choices of each flow depending on the priority of the flows and on the difference in the scores.

To make such a decision, the module needs an approximation of the bandwidth needed by each flow as well as an approximation of the bandwidth currently available for each tunnel. Both values should be given by monitoring modules.

3.5.3. Second Step: Interface Management

The Interface Management module is in charge of continuously looking for better and cheaper networks. To evaluate the need for a network, it uses the network-flow tuples scores previously described and computes the score of each combination of reachable networks. The best combination is chosen and the module generates a policy that reflects this decision. Of course, hysteresis is necessary to avoid redirecting flows back and forth on the interfaces.

Additionally, this module is responsible for cost and power mastering. It shutdowns interfaces when they are not really needed: This is the case when there is no critical flows that require the interface and the non critical flows can be dispatched on the other interfaces.

The result obtained is a list of interface-network tuple. The indicated interfaces should be activated if needed and connected to the corresponding network. The other interfaces have to be deactivated. Once this policy enforced, the Flow Routing Decision Module will adapt itself and redirect the flows conveniently.

3.6. Enforcement

The enforcement modules should be comparatively simple. They just translate the decision modules policies into system commands. The Interface Enforcement module operation consists in calling interface configuration system commands such as *ifconfig, iwconfig and pppd* in the case of Unix-based systems. The module configures the interfaces and reports whether the operation was successful or not.

The Routing Enforcement module has to be enforced on both the MR and the HA. This ensures to have the entire communication between a CN and an MNN (back and forth flows) to be conveyed into the same tunnel. To achieve this goal, two major solutions are available: *Flow Binding* [22] and *Policy Data Set for Flow Distribution + MCoA* [13,26].

The main difference between these solutions is that Flow Binding handles the policy exchange at the level of the NEMO implementation by adding a sub-option to the Binding Update message whereas *Policy Data Set for Flow Distribution + MCoA* handles the exchange at the application level. The policy exchange, which is completely independent in this case, can be achieved using HTTP/SOAP as stated in the draft or any other protocol.

This solution could be adopted if flexibility is important, it offers to the user the ability to create and exchange complex policies while being very easy to use for research purposes.

4. Mobility-Aware Application Management

The architecture described above aims to optimize the satisfaction of the operator, the administrator and the applications through high-level decision modules. Nevertheless, the application satisfaction is somewhat basic since it is based upon static flows descriptions. A static description implies that the applications needs do not vary through time. It also implies that all applications using the same port numbers must necessarily have the same requirements, which is wrong. Applications requirements vary through time and depend closely on available resources as well as final user preferences.

Flow management can be handled in a more intelligent way at the application level starting from the statement: “Who knows application requirements better than the application itself?”. And, by “application”, we mean not only the applications but also the user behind this application since their interaction is not defined.

In the standard IP model, applications do not have to be aware of network status. TCP has the task to share the available bandwidth among the different contenders using the same link. This was satisfying in the wired networks since link conditions were relatively stable but, in a mobility context, conditions can vary widely and can no longer be efficient (see [20]).

Applications requirements, in terms of bandwidth, delay, jitter, etc., are very different. Some critical applications need few bandwidth but constant connectivity (e.g., medical monitoring). Some other applications, such as VoIP, need a certain amount of bandwidth and are very sensitive to delay and variation in packet arrival times. Such applications may prefer not to operate if their requirements are not fulfilled. FTP-like applications just need as much bandwidth as available.

Furthermore, there are more complex cases where application needs depend on the resources available on the network. For example, a mail client can wait to be over a high speed network to download the mail attachments instead of doing it over GPRS. A video-conference application can also reduce its needs by transmitting only voice and blocking video in the case of serious network conditions degradation.

More generally, applications and network architecture have to interact to be more reactive to network changes and to distribute the available bandwidth in a clever manner.

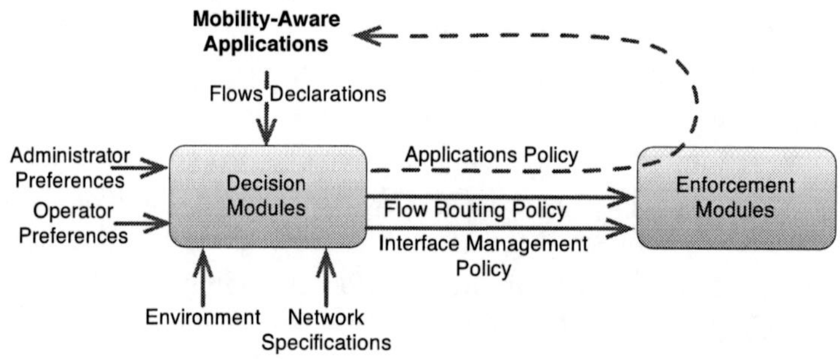

Figure 13. Overview of a Mobility Management Framework with Mobility Aware Applications.

To reach this level of interaction, we need Mobility Aware Applications (See fig.13). These are high-level applications that have network awareness features allowing them to participate in the network's resources management and benefit from information given by the MR.

Making applications aware of mobility context and offering them the possibility to interact with the decision modules has already been discussed in the literature (see [1,17,25]). These solutions have shown an effective improvement of the global bandwidth as well as a better dispatching of the flows on the different network interfaces. It is wise to bring these benefits to network mobility. In fact, the cited solutions treat cases where applications, network interfaces and decision modules are on the same physical device and where it is relatively easy for applications to monitor network status and react in consequence. In a mobile network, network interfaces and decision modules are on the MR while applications are on MNNs. This raises several issues such as communication reliability and security.

Applications can take part in the decision at several levels: First, applications can declare their flows to the MR to help it to dimension its needs and allocate more resources if necessary. These are called declarative mobility aware applications. Alternatively, applications can receive recommendations from the MR to adapt their requirements to the available resources. These are called adaptive mobility aware applications. Of course, the ideal situation is to have all the applications that are adaptive and declarative at the same time but this ultimate goal is not realistic: small-sized applications and devices will not afford such complex modules. Thus, solutions have to be proposed to deal with this heterogeneity. A more in-depth study of mobile applications issues is proposed in [2].

5. Conclusion

The recent boom in wireless networks technologies and the diversification of the connected devices opened the way to several network-related services. It is stated that, in the very near future, users will be permanently connected to the Internet no matter where they are. Moreover, mobile devices are getting more and more diverse. Nowadays, it is easy to find in the market portable game devices including a wireless interface to offer to the user the opportunity to play online. One can easily imagine that, in the near future, the same features will be available on other portable devices such as multimedia players, camcorders and GPS Navigation systems. Later, mobile sensors will be able to provide sensitive information in real-time (e.g., medical monitoring of a patient health state).

Current wireless access technologies do not allow these small-sized devices to have more than one or two wireless interfaces due to energy consumption, size and cost issues. Therefore, these devices would not benefit from the diversity offered by the various network technologies. They also will not be able to manage mobility and maintain an ubiquitous access to the Internet. However, considering the problem otherwise, one can notice that these devices are often used in environments such as personal vehicles and public transportation systems. Those environments can manage the ubiquitous access issues on behalf of the devices and provide them with a stable and easy-to-use access network (e.g., Wi-Fi or Bluetooth). Mobile devices have only to manage handovers that occur when moving from an environment to another, which is less frequent.

Buildings having a wired Internet connection can easily provide such a service to the mobile devices located in their surroundings. This is commonly the case for the industrial and the scholar institution buildings and increasingly common in private accommodations and hotels. However, to provide the same service into vehicles, several aspects have to be studied in order to handle a seamless mobility through multiple access networks. A first step was the design of the NEMO (NEtwork MObility) Basic Support protocol by the IETF. NEMO's approach introduces the notion of Mobile Routers (MR), which will be part of modern vehicles and manage all complexities related to multi-interfaces and seamless mobility management. The main benefit behind multi-interface handling is to benefit from technology diversity and heterogeneous networks coexistence.

In fact, operators propose Wi-Fi hotspots within urban areas. These offer a connection quality that is suitable for most usages but remains very limited in terms of space. Thus, actual Wi-Fi hotspots are not suitable for moving users. To compensate for this weakness, several cellular networks architectures have been proposed. These networks have been developed to offer ubiquitous connectivity almost everywhere. Unfortunately, they have the disadvantage of providing relatively low bandwidth to the final user comparatively to their important cost.

In addition, real experiments conducted over production networks (e.g., GPRS, UMTS, CDMA2000) indicate that such links are currently plagued with several problems such as high and variable round trip times, patterns of burst of packet losses, frequent link outages, and significantly lower bandwidth than originally claimed. In other words, it seems that there is currently no wireless technology can offer reliability level required for providing seamless connectivity to mobile vehicles. Several authors stated that the most viable solution to reach this level must exploit wireless networks co-existence and diversity. A node should consider all reachable wireless networks as a unique heterogeneous network.

In a heterogeneous network it is possible to use a combination of several networks, each of which being optimized for some particular service. This results in a system that delivers each service through the network that is most efficient for that service. Access networks may also be combined to increase the available capacity. Different access networks might also be used for uplink and downlink traffic. This can be useful for user applications like web browsing and e-mail, which in many cases are asymmetrical in nature causing more downlink than uplink bandwidth.

ISO is currently designing the CALM architecture that aims at providing continuous access to embarked applications and devices. A full featured management plan has been envisioned to allow the simultaneous use of multiple network interfaces while taking into account application needs. This chapter described how such management could be implemented and which functions it should include to take full advantages of technology diversity.

Further works have to be made to develop a fully operational CALM architecture. It is first necessary to provide a way to distribute decision and routing operation when several mobile routers are present in the mobile network. Therefore, the management plan should be able to interact smoothly with routing optimization [19] and other networking layers such as CALM FAST and Geocasting (see figure 3).

References

[1] F. André, J.-M. Bonnin, B. Deniaud, K. Guillouard, N. Montavont, T. Noël, and L. Suciu. Optimized Support of Multiple Wireless Interfaces within an IPv6 End-terminal. In *SOC'03 (Smart Objects Conference)*, Grenoble, France, May 2003.

[2] R. Ben Rayana and J.M. Bonnin. Mobility aware application manager for mobile networks. In *ITST'08 (Intelligent Transportation Systems and Telecommunication)*, Phuket, Thailand, October 2008.

[3] C2C Communications Consortium. Car-2-Car Communication Consortium - Manifesto, 2007.

[4] V. Devarapalli, R. Wakikawa, A. Petrescu, and P. Thubert. Network Mobility (NEMO) Basic Support Protocol. *RFC* 3963 (Proposed Standard), January 2005.

[5] T. Ernst and K. Uehara. Connecting Automobiles to the Internet. *ITST: 3rd International Workshop on ITS Telecommunications*, 2002.

[6] T. Ernst, K. Uehara, and K. Mitsuya. Network mobility from the InternetCAR perspective. In *Advanced Information Networking and Applications, 2003. AINA 2003. 17th International Conference on*, pages 19–25, March 27–29, 2003.

[7] E. Gustafsson and A. Jonsson. Always Best Connected. *IEEE Wireless Communications*, February 2003.

[8] ISO/TC204/WG16. CALM M5. **ISO** (work in progess), 2007.

[9] ISO/TC204/WG16. CALM Management. **ISO** (work in progess), 2007.

[10] D. Johnson, C. Perkins, and J. Arkko. Mobility support in IPv6. **RFC** 3775 (Proposed Standard), June 2004.

[11] R. Kuntz, J. Montavont, and T. Noel. Multiple Mobile Routers in NEMO: How Neighbor Discovery Can Assist Default Router Selection. *IEEE International Symposium on Personal Indoor and Mobile Radio Communications (PIMRC)*, 2008.

[12] M. Mahamat-Faki, J.-M. Bonnin, and R. Ben Rayana. Traffic Flow Control on Multi-Tunneled Mobile Router. In *short paper in WNEPT06 (Workshop on Networking in Public Transport)*, Waterloo, Canada, August 2006.

[13] K. Mitsuya, K. Tsakada, R. Wakikawa, and R. Kuntz. A policy data set for flow distribution. *Draft IETF* (Work in progress) draft-mitsuya-monami6-flow-distribution-policy-04.txt, August 2007.

[14] K. Mitsuya, K. Uehara, and J. Murai. The In-vehicle Router System to support Network Mobility. *International Conference on Information Networking (ICOIN), South Korea, Feb*, 2003.

[15] Eric Nordmark and Marcel Bagnulo. Shim6: Level 3 Multihoming Shim Protocol for IPv6. Internet Draft (work in progress).

[16] IEEE P1609. IEEE P1609.1 Standard for Wireless Access in Vehicular Environment (WAVE) - Multi-Channel Operation. IEEE draft Standard, 2006.

[17] Emilio Garcia Pedro M. Ruiz. *Improving User-Perceived QoS In Mobile and Wireless IP Networks Using Real-Time Adaptive Multimedia Applications.*

[18] Eranga Perera, Vijay Sivaraman, and Aruna Seneviratne. Survey on network mobility. *ACM SIGMOBILE Review*, 2004.

[19] C.E. Perkins and D.B. Johnson. Route Optimization for Mobile IP. *Cluster Computing*, 1(2):161–176, 1998.

[20] Pablo Rodriguez, Rajiv Chakravorty, Julian Chesterfield, Ian Pratt, and Suman Banerjee. MAR: A Commuter Router Infrastructure for the Mobile Internet. In *MobiSys '04: Proceedings of the 2nd international conference on Mobile systems, applications, and services*, pages 217–230, New York, NY, USA, 2004. ACM.

[21] Farhan Siddiqui and Sherali Zeadally. Mobility Management Across Hybrid Wireless Networks: Trends and Challenges. *Computer Communications*, 29(9):1363–1385, 2006.

[22] H. Soliman, N. Montavont, N. Fikouras, and K. Kuladinithi. Flow Bindings in Mobile IPv6 and Nemo Basic Support. Draft IETF (Work in progress) draft-ietf-mext-flow-binding-00.txt, May 2008.

[23] E. Stevens-Navarro and V.W.S. Wong. Comparison between vertical handoff decision algorithms for heterogeneous wireless networks. In *IEEE Vehicular Technology Conference (VTC)*, Melbourne/Australia, May 2006.

[24] Lucian Suciu, J.-M. Bonnin, K. Guillouard, and B. Stévant. Achieving "Always Best Connected" Through Extensive Profile Management. In *The 9th Intl. Conference on Personal Wireless Communications (PWC'04)*, Lecture Notes in Computer Science, Delft, Pays Bas, September 2004. Springer Verlag.

[25] Lucian Suciu, J.-M. Bonnin, K. Guillouard, and B. Stévant. Towards a highly adaptable user-centric terminal architecture. In *The 7th Intl. Symposium on Wireless Personal Multimedia Communications (WPMC'04)*, Abano, Italie, September 2004.

[26] R. Wakikawa, K. Uehara, T. Ernst, and K. Nagamix. Multiple Care-of Address Registration. Draft IETF (Work in progress) draft-wakikawa-mobileip-multiplecoa-05.txt, February 2006.

[27] IEEE 802.11 WG. Draft IEEE 802.11p: Wireless Access in Vehicular Environment (WAVE). IEEE draft Standard, 2006.

[28] IEEE 802.21 WG. Local and Metropolitan Area Networks: Media Independent Handover Services (IEEE P802.21/D00.04). IEEE draft Standard, December 2005.

[29] Bod (Ed.) Williams. *The CALM Handbook: Continuous Communications with and between vehicles*. ISO/TC204/WG16 (work in progess v2.0), November 2005.

[30] M. Williams. Directions in Media Independent Handover. *IEICE Transactions on Fundamentals of Electronics, Communications and Computer Sciences - Special Section on Multi-dimensional Mobile Information Networks*, July 2005.

[31] B. Xing and N. Venkatasubramanian. Multi-Constraint Dynamic Access Selection in Always Best Connected Networks. In *IEEE MobiQuitous'05*, pages 56–64, San Diego/CA/USA, July 2005.

[32] W. Zhang, J. Jaehnert, and K. Dolzer. Design and evaluation of a handover decision strategy for 4th generation mobile networks. In *57th Semiannual Vehicular Technology Conference (VTC'2003-Spring)*, April 2003.

In: Wireless Technologies in Intelligent Transportation… ISBN: 978-1-60741-588-6
Editors: Ming-Tuo Zhou et al, pp. 323-360

Chapter 12

ROUTING AND DATA DISSEMINATION IN INTELLIGENT TRANSPORTATION SYSTEM NETWORKS

Péter Laborczi[1,a], Attila Török[1,b], Miklós Máté[2,c] and Rolland Vida[2,d]

[1]Institute for Applied Telecommunication Technologies
Bay Zoltan Foundation for Applied Research, Budapest, Hungary
[2]Budapest University of Technology and Economics

Abstract

Today, cars become more and more intelligent. They are equipped with hundreds of sensors and microcomputers that are able to measure speed, tire pressure, brake temperature, to detect raindrops on the windshield, and to help the driver to react in function of these measurements. The next step researchers started to work on was related to car-to-car communication: how could a car disseminate relevant information to other cars, in order to increase road safety. A vehicular network can be considered as a special case of a wireless ad hoc network. However, the traditional ad hoc routing algorithms might not work properly, or should at least be modified according to the specificities of such an environment, mainly the high speeds and the fast changing topology. In this chapter we present first the shortcomings of traditional communication solutions in ITS scenarios, and describe then ITS specific routing and information spreading techniques, based on partial, directed flooding or geocasting (i.e., location-based multicasting). At the end of the chapter we also analyze hybrid architectures that involve road side infrastructural elements to help the communication between the mobile vehicles.

---

[a] E-mail address: laborczi@ikti.hu.
[b] E-mail address: torok@ikti.hu.
[c] E-mail address: mate@tmit.bme.hu.
[d] E-mail address: vida@tmit.bme.hu.

Introduction

Intelligent Transportation Systems (ITS) are intended to ease the everyday life of drivers by reducing the risk of accidents, improving safety, increasing road capacity and reducing traffic jams. Many car-manufacturers are currently focusing on ITS applications where cars are acting autonomously based on the readings of their on-board sensors (Intelligent Vehicle Initiative concept). However, despite the promising applications and the imminent market penetration, these systems are doomed to relative inefficiency because of the absence of shared knowledge and group reasoning.

A more advanced vision, called Automated Highway System (AHS), introduced communication between cars as well. Successful tests were carried out at the beginning of the '90s already. However, due to the lack of proper communication infrastructures and protocols, this direction was neglected for years. A boost for AHS-like solutions was given by the emergence of the mobile ad hoc and opportunistic networking paradigms. On the other hand, ITS applications made ad hoc networking techniques to become a reality, representing the "killer application" the ad hoc research community has lacked for years.

Due to their inherent characteristics, viable communication is harder to support in ITS scenarios than in conventional ad hoc networks. Vehicles are usually moving much faster than traditional ad hoc nodes; moreover, a vehicular network might be very heterogeneous in terms of node density, becoming fragmented in many cases. Reliability is also compromised due to the usually high interference in urban scenarios. Thus, there was a need to reconsider the ad hoc networking protocols, and to use new concepts that fit better the specificities of ITS applications.

The first part of this chapter presents therefore the evolution of communication protocols in traditional ad hoc networks, and outlines their shortcomings in ITS scenarios. To better understand the specific requirements of these scenarios, we then present in more details some concrete ITS applications like spreading road condition information, or providing adaptive cruise control. We describe then ITS specific routing and information spreading techniques that can cope with the above mentioned requirements. Traditional routing is in general not acceptable, as routing paths change far too often due to the high mobility, and cannot be maintained up to date without generating unacceptably high signaling traffic. Therefore, more efficient information spreading techniques have to be applied, based on partial, directed flooding or geocasting (i.e., location-based multicasting). These solutions are important also because in many ITS applications there is no single destination node, but rather a destination area; e.g., all the vehicles that are in a specific area have to be alerted about changing road conditions or imminent traffic jams. Finally, many ITS applications build on a hybrid architecture, with road side infrastructural elements that, among other tasks, might serve as landmark nodes and help the communication between the mobile vehicles. Special techniques, borrowed from the field of opportunistic networking, can be used here to address the challenges of such hybrid systems. The last subsection of the chapter addresses these architectures.

Routing and Information Spreading in Traditional Mobile Ad Hoc Networks

With the advances of wireless communication technology, low-cost and powerful wireless transceivers are widely used in mobile applications. Mobile networks have attracted significant interests in recent years because of their improved flexibility and reduced costs. Compared to wired networks, mobile networks have unique characteristics. In mobile networks, node mobility may cause frequent network topology changes, which are rare in wired networks. In contrast to the stable link capacity of wired networks, wireless link capacity continually varies because of the impacts from transmission power, receiver sensitivity, noise, fading and interference. Additionally, wireless mobile networks have a high error rate, power restrictions and bandwidth limitations.

One of the most popular method to distinguish mobile ad hoc network routing protocols is based on how routing information is acquired and maintained by mobile nodes. Using this method, mobile ad hoc network routing protocols can be divided into proactive routing, reactive routing and hybrid routing.

Proactive Routing Protocols

Proactive routing protocols, also called table-driven solutions, are based on the nodes of a mobile ad hoc network continuously evaluating routes to all other reachable nodes, and attempting to maintain consistent, up-to-date routing information. By doing so, a source node can get a routing path immediately if it needs one. All nodes need to maintain a consistent view of the network topology. When the topology changes, updates are propagated throughout the entire network to notify the change. Thus, routes are proactively maintained, regardless of whether data traffic exists or not; the overhead to maintain up-to-date network topology information is high

The Wireless Routing Protocol (WRP) [34] is a proactive routing protocol that uses an improved Bellman-Ford Distance Vector routing algorithm Each mobile node maintains a distance table, a routing table, a link-cost table and a message retransmission list. An entry in the *routing table* contains the distance to a destination node, the predecessor and the successor along the paths to the destination, and a tag to identify its state, i.e., is it a simple path, a loop or invalid. An entry to the *link-cost table* is maintained for each neighbor, and contains the cost of the link connecting to the neighbor, and the number of timeouts since an error-free message was last received from that neighbor. The mobile nodes exchange routing tables with their neighbors using update messages, sent either periodically or whenever link state changes occur. Finally, the *retransmission list* contains information about which neighbor has not acknowledged an update message. If needed, the update message will be retransmitted to those neighbors. In WRP, a node checks the consistency of its neighbors after detecting any link change. A consistency check helps to eliminate loops and speed up convergence. The major shortcoming of WRP is that it needs considerable memory and computing resources to maintain several tables. Moreover, as a proactive routing protocol, it has a limited scalability and is not suitable for large mobile ad hoc networks.

The Destination Sequence Distance Vector (DSDV) [42] protocol is also based on the traditional Bellman-Ford algorithm, but its mechanisms to improve routing performance are quite different from those of WRP. An entry in a DSDV routing table stores the next hop towards a destination, the cost of the path leading to that destination, and a destination sequence number. The sequence numbers are used to distinguish stale routes from fresh ones and avoid the formation of loops. Route updates are transmitted to the neighbors periodically, but if significant changes occur, they can trigger an immediate update message as well. DSDV has two ways of sending these routing table updates. In case of a "full dump" update the entire routing table is included in the update message(s). As opposed to this, an incremental update contains only those entries that have been changed since the last update was sent.

The Fisheye State Routing (FSR) [41] is a proactive unicast routing protocol based on a Link State routing algorithm with reduced overhead to maintain network topology information. FSR maintains accurate distance and path quality information about the immediate neighboring nodes, and it progressively reduces the level of details as the distance increases. To reduce the size of link state update messages, different update periods are used for different entries in the routing table. Link state updates corresponding to nodes within a smaller scope are propagated with higher frequency.

Reactive Routing Protocols

Reactive routing protocols, also called *on-demand* solutions, are based on routing paths being searched only when needed. Compared to the proactive solutions, these protocols are much more scalable, as less control overhead is generated. However, in reactive routing protocols source nodes may experience long delays due to route searching, before they can forward data packets. DSR and AODV are the two most known protocols in this family, so in the next paragraphs we present them briefly.

Dynamic Source Routing (DSR) [17] is a reactive unicast routing protocol that uses a source routing algorithm. Each data packet contains complete routing information to reach its dissemination. Additionally, each node uses caches to maintain route information that it has learned The operation of the protocol can be split in two major phases: the route discovery phase and the route maintenance phase. When a source wants to send a packet, it first consults its route cache. If the required route is available, the source includes the routing information inside the data packet before sending it. Otherwise, it initiates a route discovery operation by broadcasting route request packets. These contain addresses of both the source and the destination, and a unique number to identify the request. Upon receiving a route request packet, a node checks its route cache. If it doesn't have routing information for the requested destination, it appends its own address to the route record field stored in the request, and forwards it to its neighbors. If a node receives a request that has processed before, it drops that request, reducing thus the communication overhead. If the route request packet reaches the destination, or an intermediate node that has a route to the destination, a route reply packet is generated. This reply will contain the addresses of nodes the route request packet has traversed, concatenated with the route in the intermediate node's route cache, if necessary. DSR generates an increased traffic overhead by including complete routing information in each data packet, which is an even more serious drawback in a wireless environment with scarce communication resources.

As opposed to DSR, the Ad Hoc On-demand Distance Vector Routing (AODV) protocol [43] maintains information in the routing tables of nodes, but only for active paths, i.e., paths through which there is an ongoing communication. Every mobile node keeps a next-hop routing table, which contains the destinations to which it currently has a route. A routing table entry expires if it has not been used or reactivated for a given expiration time. When a source wants to send packets to the destination, but has no available route, it initiates a route discovery, by broadcasting a route request packet. Sequence numbers are associated to the requests, to ensure loop-free operation. The route discovery process is based on an expanding ring search algorithm; the requests start with a small TTL (Time-To-Live) value, and if the destination is not found, the TTL is increased. Each node maintains also a cache, to keep track of the requests it has received, and the return paths to the originators of those requests. When the route request reaches the destination, or a node that has a route to the destination, a route reply packet is forwarded back to the source. Sequence numbers are also used to ensure the freshness of the information; thus, the reply packet is sent back to the source only if the destination sequence number is equal to or greater than the one specified in the request. AODV uses only symmetric links; thus, the route reply follows the reverse path of the respective route request. Upon receiving the reply packet, each intermediate node along the route updates its next-hop table entries with respect to the destination node.

Hybrid, Hierarchical Protocols

Proactive routing protocols maintain up-to-date routing information for each possible destination node; thus, routes can be provided immediately to a source, but at a cost of a continuously high signaling cost. Reactive protocols maintain up-to-date information only about active paths; thus, signaling cost is significantly reduced, but at a cost of high initiation delay when targeting a destination node the source does not have a route for. Hybrid, hierarchical routing protocols were therefore proposed to combine the advantages of both proactive and reactive approaches, and overcome their shortcomings. The main idea is to apply the proper proactive and reactive routing approach at different hierarchical levels, respectively.

The Zone Routing Protocol (ZRP) [13] is based on the network being divided into routing zones according to distances between mobile nodes. Given a hop distance *d* and a node *N*, all nodes within hop distance at most *d* from *N* belong to the routing zone of *N*. Different routing approaches are then used for inter-zone and intra-zone packets. To reach a destination in its zone, the source uses a proactive routing protocol called Intra-zone Routing Protocol (IARP); link state information is maintained for each possible destination inside the zone, routes being thus immediately available when needed. When targeting a destination outside the source's zone, the reactive Inter-zone Routing Protocol (IERP) is used; if there's no active path to the destination, a route discovery has to be initiated. Most of the existing proactive and reactive routing schemes can be used as IARP and IERP respectively.

Another similar solution, called the Hybrid Ad Hoc Routing Protocol is proposed in [37]; a two-level zone based hierarchical network structure is defined, and depending on whether the source and the destination are in the same zone or not, intra- and inter-zone routing is applied. The protocol relies on the routing tables constructed by the Distributed Dynamic Routing protocol (DDR) [38]. Nodes in DDR periodically exchange topology messages with their neighbors. First, each node elects his preferred neighbor, as the node that has the most neighbors. Then, a forest is constructed by connecting each node to its preferred neighbor. Each tree of the forest forms a zone, the network being thus divided into a set of non-overlapping dynamic zones. Each mobile node keeps routing information for all other nodes in the same zone. The nodes belonging to different zones but within the direct transmission range of a node are defined as gateway nodes. They have the responsibility of forwarding packets to neighboring zones. The advantage of DDR is that it does not rely on a static zone map to perform routing. However, the nodes that were selected as preferred neighbors by many other nodes might become performance bottlenecks, both in terms of radio resources and battery lifetime.

Position-Based Routing in Ad Hoc Networks

Today more and more wireless communication devices (PDAs, 3G phones, etc) are equipped with GPS, being thus able to provide quite precise position information. There are thus more and more ad hoc routing solutions that besides topology information try to make use of geographic node positions as well. We call these approaches position-based, or geographic routing solutions.

The main idea is that the source sends the data towards a given geographic location, instead of a network address. If each node is aware of its own location, and the source is aware of the location of the destination node, then the message can be routed to the destination without any knowledge of the network topology or a-priori route discovery.

Most of the position-based unicast routing strategies rely on two techniques: *greedy forwarding* and *face routing*. Greedy forwarding tries to take the message closer to the destination in each step, selecting the most suitable next hop node based only on local information. The most suitable neighbor can be the one which minimizes the distance to the destination in each step, but we can consider another notion of progress as well, namely the projected distance on the source-destination-line, or the minimum angle between the neighbor and the destination. Greedy forwarding can lead into a dead end if at a certain point there is no neighbor closer to the destination. Then, *face routing* is applied to recover from that situation and find a path to another node, where greedy forwarding can be resumed. The most known routing strategies from this category are Greedy Face Greedy (GFG) [3], Greedy Perimeter Stateless Routing (GPSR) [18], and the GOAFR+ family of protocols [23].

There are a large number of other routing protocols for mobile ad hoc networks. Flow oriented routing protocols find a route on demand, by following present flows. In link stability based routing protocols control overhead is reduced by using the stability of links instead of hop numbers as metric for routing path selection. Power aware routing protocols, designed mainly for wireless sensor networks, aim at distributing as much as possible the relaying load of nodes, and prolonging thus the network lifetime. Finally, there are many multicast routing protocols as well defined for mobile ad hoc networks. However, the goal of this section was not to give a complete overview of all the existing ad hoc network routing solutions; there are many excellent surveys presenting such an analysis [1][27]. We intended rather to present the basic ideas that these routing solutions rely on, in order to be able to emphasize in the following sections why these approaches are not suitable for a vehicular environment.

Short Introduction to ITS Applications

ITS applications can be grouped in three major areas, based on their primary goals: increasing road safety, improving transport efficiency, and providing Internet access while on the move.

Driver safety and accident prevention has been recognized as an important issue to focus on by the automotive industry, governments and the research community. The basic research challenge is to identify critical situations, and broadcast the related information to other possibly involved or interested vehicles, alerting the driver or, if both technically and legally possible, intervene in the vehicle control unit. The first approaches for observing the environment were based only on local observations, made by the vehicles themselves, like the detection of other vehicles through a radar system, or the recognition of lane separation lines by on-board cameras. An enhanced service is then based on sharing these observations among the vehicles via wireless communication. By doing so, vehicles are informed about accidents, road works and road conditions, they are aware of each other's positions, and can even cooperatively accelerate and decelerate in a convoy, using an adaptive cruise control mechanism. In fact, many accidents are due to the lack of cooperation between drivers. Cooperative driving is thus an important concept that could form the basis of services such as curve warning, lane merging warning, turn conflict warning, etc. [46]

If a vehicle reduces its speed significantly, e.g., because of an accident, it can broadcast its location to its neighbor vehicles, which will relay the emergency message further. Consequently, vehicles behind the critical situation will alert their drivers before they can physically see the accident. Furthermore, data transmitted from a roadside base station to a vehicle can warn a driver that it is not safe to enter an intersection. Safety critical applications require an extremely reliable communication with low delays, typically bellow 100 milliseconds.

Traffic optimization is another application area which has specific requirements towards the ad-hoc network. With the increase in the amount of vehicles around the globe, traffic delays continue to grow, wasting more than 40 hours each year for peak time travelers. Many research papers [12][19[pointed out that a significant reduction of traffic jams can be achieved through the use of vehicular ad-hoc networks (VANETs). Vehicles could serve as data collectors and transmit the traffic condition information in the vehicular network. Transportation companies could utilize this information to actively relieve traffic congestion. Practically, vehicles could detect if the number of neighboring vehicles is too large and/or their average speed is too low, and then relay this information to vehicles approaching the location. To make it work better, the information could be relayed by vehicles moving in the opposite direction as well; by doing so, it may be propagated faster to the vehicles approaching the congested location. Thus, they will have enough time to choose alternative routes. Vehicles can also collect other sensor based data, e.g., about the weather, the road surface, the construction zones, railway crossings, or emergency vehicle signals [46]. In contrast to safety related applications, in traffic optimization neither reliability nor delay is a critical parameter.

Besides multi-hop vehicle-to-vehicle communications, future VANETs will also comprise roadside gateways to the Internet. Through these gateways, passengers sitting in the vehicles that pass by could get connected to the Internet in an economic but time-restricted manner [2]. This opens the door for a large array of possible applications; for example, department stores can disseminate sale advertisements to vehicles within vicinity to attract consumers as well as providing dining and parking information. The need for reliability and delay depends on the specific application.

All of these applications rely on an ad-hoc network that has significantly different characteristics than traditional networks analyzed in the previous subsection. The most important differences we have to mention are the high mobility (high relative speed) of the vehicles and the large geographical extent of the network. In most of the cases, a vehicular ad-hoc network is not completely connected, but consists of clusters of vehicles in communication range. Although the movements of the vehicles are constrained by the road layout, the topology of the network changes very rapidly [35]. If we establish a route between a source and a destination node (i.e., vehicle), that route will quite often break even before data transfer can start. As a result, many well studied structures for data dissemination, such as trees, clusters, grids, are extremely hard to set up and maintain, even if network partitioning or merging can usually be predicted based on the movements of nodes. Another problem is that conventional broadcast mechanisms for data dissemination may lead to broadcast storms [36], as network node density is usually quite high in urban areas and extremely dense during rush hours or traffic jam [56]. Consequently, it is not possible to set up and maintain routes in the traditional sense; special information spreading techniques are needed instead.

Although information dissemination from a server to clients has been thoroughly studied in the research community (e.g., [33]), the above mentioned unique characteristics of VANETs bring to the surface new research challenges and the need for defining new metrics to characterize these networks. One such metric is the connectivity degree of the network. This depends mostly on two factors: the range of the wireless link and the fraction of participating vehicles (penetration rate). The lack of connectivity raises questions about whether the vehicular ad hoc network can effectively disseminate traffic and safety information, or not. In order to determine this, it is important to analyze the requirements of the application, and answer questions, like *what is the effective communication bandwidth the network must support* or *what is the useful dissemination radius for the information*? Traffic information is useful at a certain location if it helps drivers in reducing their commuting time [12]. Safety information on the other hand is useful if it reduces the probability of accidents.

A number of EU, US and Japanese projects are dealing with applications of vehicle-to-vehicle (V2V) and vehicle-to-infrastructure (V2I) communication. For example, the Fleetnet project [57] focused on emergency braking notification, distribution of traffic data and providing "Internet on the Road". The project titled Communications for eSafety (COMeSafety) [58] and the Car-to-Car Communication Consortium (C2C-CC) [59] prepare the standardization process with respect to technologies for V2V and V2I communications.

California PATH (Partners for Advanced Transit and Highways) [60], established in 1986, is a multi-disciplinary program whose mission is to develop solutions to the problems of surface transportation systems through cutting edge research. The program comprises several research projects. Among them, the California Connected Traveler Program aims at empowering travelers with real-time traffic, transit and road safety information. With alerts sent directly to drivers as road conditions change, motorists will be able to choose the quickest and safest routes, integrate with public transit, and adjust their routes to avoid developing hazards and congestion.

A further project is Cooperative Vehicles and Road Infrastructure for Road Safety (SAFEspot) that deals with V2V and V2I communication-based safety solutions. The Cooperative Vehicle-Infrastructure Systems (CVIS) project [61] creates a unified technical solution allowing all vehicles and infrastructure elements to communicate in a continuous and transparent way with each other in the above mentioned three applications: road safety, efficiency and Internet access.

Routing in ITS Networks

As previously outlined, vehicular ad hoc networks (VANET) exhibit different characteristics compared to conventional mobile ad hoc networks (MANET), because of the highly dynamic topology, the special movement constrains, and the frequent disconnections respectively. Due to the high speed of vehicles, the topology and connectivity are constantly changing, especially when the vehicle density is low. In addition, in city scenarios the buildings will block the line of sight between vehicles; thus, direct communication is not always possible.

Because of the dynamic nature of VANET topologies, finding and maintaining routes in an ITS application is very challenging. On the other hand, certain advantages can be also derived from the specific ITS context. For example, vehicles are moving along predefined trajectories (road segments, highways), which implies a possible successful prediction of the vehicle's future position. This section will outline some special routing techniques, which are able to cope with the additional difficulties encountered in urban ITS scenarios.

Conventional position-based routing techniques used in MANETs are based on a local behavior; the packet forwarding decision is based on the local knowledge about the forwarder and the neighbor node's coordinates. Thus, it avoids the excessive overhead of topology discovery and route maintenance. Although position-based routing can achieve high packet delivery rates in MANETs, it performs poorly in networks where the frequent topology changes induce topology holes. These topology holes will cause the greedy forwarding scheme applied by the position-based routing to fail, since the packet can reach a host that has no neighbor closer than itself to the destination. To recover from such a delivery failure situation, the protocols might apply techniques like the planar graph face transversal method used in GPSR (Greedy Perimeter Stateless Routing). However, this method still presents some problems, such as the possibility of obtaining longer paths with higher delays, the possibility of having routing loops due to mobility, or having packets forwarded in the wrong direction.

To overcome the problems of position-based routing protocols, various techniques have been proposed, usually taking into consideration some additional information in the routing process, e.g., the information extracted from the digital map of the navigation system.

In [48] and [29] the authors propose to use spatial awareness in the geographic forwarding scheme (Figure 1) by constructing a Geographic Source Route (GSR), consisting of a list of intermediate road vertices derived from a spatial model. The spatial model is constructed from the available digital road maps and describes with high-level abstractions the spatial environment in which the vehicles are located in. The GSR is used as a source routing header in the data packets. The forwarding nodes map their neighbor's position into the graph model, and then choose the one with the shortest path along the GSR as the next hop. Recovery methods are also proposed to avoid cases when there are no suitable neighbors on the GSR. In such cases, switching to greedy forwarding, packet buffering, or GSR re-computation is proposed.

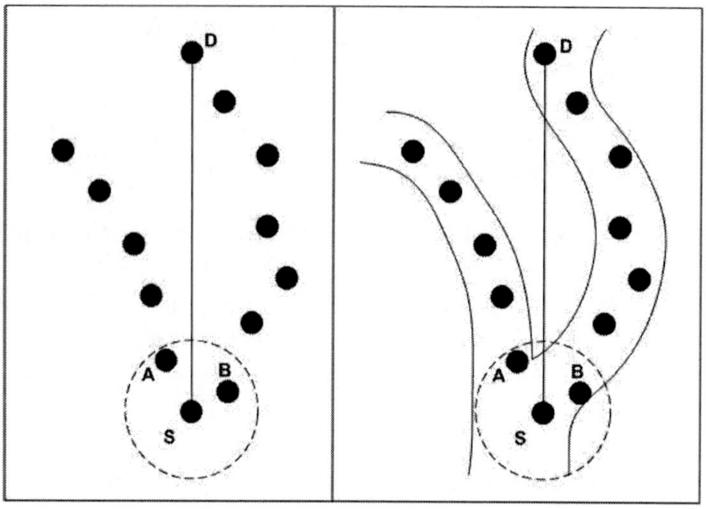

Figure 1. Spatial awareness for geographic forwarding using digital road maps.

Lochert et al. [30] also proposed a solution called GPCR (Greedy Perimeter Coordinator Routing), where the availability of digital maps is not assumed. As long as the packet is traversing a road segment, a restricted greedy algorithm can be applied; actual routing decisions have to be taken only at the road junctions. Therefore, the goal is to forward the packet until a junction is met, where a coordinator node is designated. It is very important to switch off from greedy mode in the intersection, since by using the greedy forwarding strategy the packet could take a wrong path. This coordinator node will help to find the alternative path to the destination. However, in certain cases a repair strategy still has to be used to get out from a local minimum. This can happen when there is no neighbor node that is closer to the destination than the forwarding node itself. In such a case the right-hand rule is applied to select a street which the packet should follow.

The next step in the evolution of VANET routing protocols was reached with the introduction of traffic pattern awareness. In the A-STAR (Anchor-based Street and Traffic Aware Routing) protocol [28] both static (e.g., route of city buses) and dynamic (e.g., traffic monitoring results) traffic information is used to rate the anchor paths with highest delivery probability. A local recovery strategy is also proposed, which is claimed to be more efficient than the one used in GSR. When a packet is routed to a local maximum, a new anchor path is computed and the packet is saved by traversing this new path. Then, the street where the local maximum occurred is marked as temporarily "out of service", in order to prevent other packets from entering in the same void area. This information is distributed in the network, and the "out of service" streets are not used for anchor computations for a certain period.

Simulation results show that A-STAR solution presents better results than GSR and GPSR, since due to traffic awareness better paths are selected during the routing procedure.

In [55] the authors propose a protocol called VADD (Vehicle-Assisted Data Delivery), which uses the idea of carry and forward (data mules) combined with traffic pattern aware road selection. The scope of VADD is to select a forwarding path with the smallest packet delivery delay. Therefore, if it is possible, the wireless channel is used for transmission, instead of the carry and forward technique. Using a delay model, a stochastic method is proposed to estimate the data delivery delay for the selection of the next intersection. Using this method the packet can be forwarded to the neighboring vehicle with the smallest delivery delay. However, it is not trivial to determine the next hop among all the available neighboring nodes in order to ensure the packet's pre-computed direction.

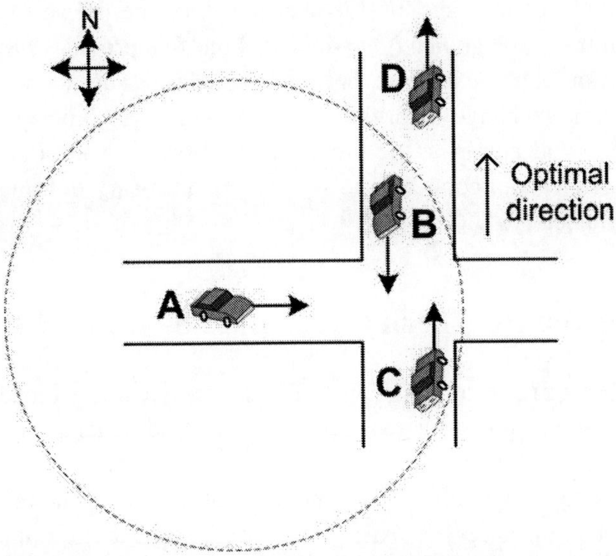

Figure 2. Selection of next vehicle in the intersection mode of VADD.

In Figure 2 vehicle A has a packet to forward towards the optimal direction, North. At this time two neighbor nodes are available: B (moving south) and C (moving north). One choice is to select B, since it is geographically closer to the destination and provides connection possibility with D. Another choice is to select C, because C is moving north. These choices lead to the proposal of several forwarding strategies, such as Location First Probe (L-VADD), Direction First Probe (D-VADD), etc. L-VADD tries to find the closest contact towards the preferred forwarding direction as the next hop; however, L-VADD may result in routing loops. D-VADD selects the contacts moving towards the selected direction, and then selects as the next hop the one closest to that direction. It is also proved that D-VADD is free from routing loops at intersection areas. Extensive simulation results show that VADD outperforms existing solutions (GPSR, epidemic routing) in terms of packet delivery ratio, data delivery delay and traffic overhead.

Wu et al. [52] present a protocol that combines opportunistic, geographical, and trajectory based forwarding. A forwarding trajectory is specified extending from the source to the destination location, along which the messages are routed. The opportunistic forwarding is used to determine which nodes will participate in the message forwarding, while the geographical strategy is applied when the destination area is reached. The forwarding trajectory is selected based on a formula which takes into account the static road information (distances and traffic conditions). During the message forwarding phase a message holder is designated, called the "message head". The message head is the vehicle closest to the destination area along the forwarding trajectory. This means that the message is "stucked" at this node, since there is no communication possibility to forward the message further along the route. Only the message head tries to pass the message to other vehicles that are closer to the destination. If such a vehicle is found, the message head role will change. To avoid loosing of messages due to the disappearance of the message head, it is allowed to a group of vehicles near the message head to actively forward the message instead of the message head vehicle only.

There is of course a tradeoff between data reliability and message overhead. The information concerning the message head is exchanged among the vehicles in the messages themselves, through the *message head pair* information; this provides information about the message head location. Data exchange between vehicles assumes the knowledge of the neighbor set. The data exchange algorithm is defined so as to quickly propagate the new messages, but to stop and eliminate the propagation of obsolete information. Messages are stored until the vehicle cannot contribute to suppress unnecessary transmissions anymore, as it deviates from its trajectory.

Information Spreading Techniques in ITS Networks

One specific characteristic of VANETs is that in many cases a broadcast- or multicast-like information spreading solution is required instead of traditional routing between two end nodes, as the messages don't have a single destination node, but a group of nodes in a certain area. These solutions have to carefully take into account the use of the radio resources. In an ad hoc network no centralized service exists to control the time division multiple access to the channel. Since radio transmission is inherently broadcasting, transmissions without access control result in interference with other transmissions, decreasing the utilization of the radio channel. The participating nodes should therefore apply a distributed algorithm to detect channel availability and enforce reliable channel sharing.

Broadcasting in mobile ad hoc networks is deeply studied in [36]. Two key issues have to be taken into account. First, the broadcast is spontaneous, i.e., due to the lack of synchronization and a-priori topology information any node can initiate a broadcast at any time. Second, the broadcast is also unreliable, as no acknowledgment is sent on a broadcast packet; this is to avoid serious medium access contention that would lead to inefficient channel utilization. Blind flooding, i.e., when all nodes rebroadcast the message unconditionally after receiving it for the first time, has an additional problem: the rebroadcasts are highly redundant, because most neighbors of the sender already received the message. These factors collectively are referred to as the *broadcast storm* problem.

Specialized protocols were developed to address the problems emerging in VANETs and to take advantage of the positional information and the predictable vehicle movements. The data dissemination algorithm chosen to be used in a VANET must ensure fast and reliable delivery of safety messages. The vehicles participating in the communication are expected to be equipped with navigational devices that provide positional information, and a digital map, which is very useful aid to the routing algorithms, enabling *directed broadcast* as well.

This section will present the most interesting and promising ideas in the field of directed broadcast and emergency message dissemination in VANETs.

Directional Broadcast

The sole goal of some protocols is to choose the farthest vehicle in the direction of the broadcast (i.e., the direction of the destination area) as a next forwarder of the received message by taking advantage of the position information. Three protocols of this family are presented here.

Position-Based Forwarding

The plain, neighbor-aware Position-Based Forwarding (PBF) [20] works as follows: the forwarding node F checks its neighbor table for neighbors being geometrically closer to the destination area than itself. Out of these potential forwarders one minimizing the remaining distance is selected by the so-called Most Forward (progress) within Radius (MFR) heuristic. It is not guaranteed that the MFR strategy finds a route in a given scenario, because it greedily minimizes the remaining distance; however, this is not a problem for cars traveling on the same highway.

Neighbors are detected using periodic beacons. Accuracy can always be increased with higher beaconing rates, but in this scenario estimating the current position using previous knowledge of position and/or movement, called Dead-Reckoning, seems to be a promising improvement due to the one-dimensionality of the network layout. Direction can be modeled as same or opposite, and even the speed of a vehicle can be assumed to remain stable within a few seconds.

The authors of [20] study two next hop selection algorithms based on dead-reckoning and one simple defensive neighbor selection approach to detect possibly unavailable neighbors. The main goal of these algorithms is to reduce the probability that a selected next hop has already left the communication area of the sending node. The Active Anticipation algorithm requires speed and direction to be contained in the beacons. On the other hand the Passive Anticipation algorithm computes the driving direction and speed of a node out of the position values from two consecutive beacons.

The Defensive Neighbor Selection scheme gives a worst case estimation of B leaving the radio range of node A using its previous distance (dAB), the time passed since the last beacon (tB), and the maximum allowed speed (vmax). Node B is considered unavailable if dAB+tB2 vmax is greater than the communication range. The factor of 2 is chosen because in PBF it is not differentiated whether the neighboring node moves in the same direction as node A, or the opposite one.

Smart Broadcast Protocol

The Smart Broadcast Protocol, a position aware algorithm, was developed to be implemented on top of an IEEE 802.11-like system [9]. Transmissions are assumed to be correctly received within a given area *A* in the direction of the dissemination; this area is effectively strip-shaped if the nodes are vehicles on a highway. *A* is partitioned to *N* adjacent and non-overlapping sectors $S_0, .. S_{N-1}$, where S_0 is the sector the farthest away from the originator. Nodes in each sector use a contention window of size

$$W_j=w_j+CW_j,\ CW_j=\{0,1, \dots ,cw_j-1\}, j=0, \dots ,N-2,$$

where $w_0=0$ and $w_{j+1}=w_j+cw_j+1$. The backoff time of the nodes in sector *j* is independent and has uniform probability distribution function

$$\varphi_j(h)=1/cw_j,\ h \in \{0, cw_j-1\}.$$

Nodes follow the IEEE 802.11 CSMA/CA protocol during the backoff phase. Upon receiving a copy of a queued packet, the node checks if it is farther from the previous hop than the repeater. If the repeater is closer to the source, then the sectorization is restarted using the repeater as previous hop. Otherwise, the rebroadcast procedure is aborted and the message is dropped.

To increase the reliability and robustness of the message propagation, the nodes initiate a backoff of length W_N after broadcasting a message. The message is rebroadcast again if a valid rebroadcast is not overheard within the backoff period.

Contention-Based Forwarding

The Contention-Based Forwarding (CBF) protocol [10] is a directed flooding scheme which was originally developed for sensor networks, and later adapted to VANET environments. CBF is a position based greedy forwarding scheme that performs well in highway scenarios.

Nodes that use the CBF protocol set a timer upon receiving a packet to schedule the retransmission. Every CBF data packet contains the position of the node that has just forwarded the packet (last hop), the ID and position of the final destination, and the ID of the packet. The timeout value is calculated based on the progress the node provides towards the packet's destination, which is defined as

$$p_i = dist(l, d) - dist(i, d),$$

where *dist* is the Euclidean distance, and *l*, *d*, *i* are the positions of the last hop, the destination and the actual node, respectively. The timer is set to

$$\tau\left(1-\frac{p_i}{p_{\max}}\right),0\le p_i<p_{\max}$$

$$\infty, otherwise$$

$$t=\{$$

where pmax is the radio range and τ is the maximum allowed forwarding delay. If t is infinite, or the node overhears another node transmitting a packet with the same ID, the packet is dropped. After successful forwarding the nodes store the IDs of the transmitted packets to avoid duplicates.

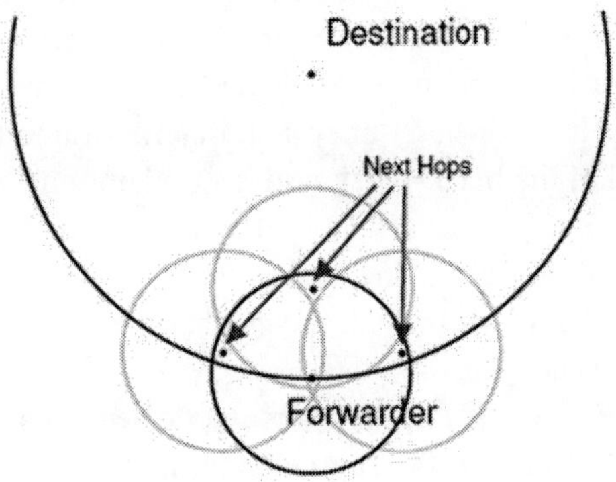

Figure 3. Packet duplication without area based suppression in CBF.

In Figure 3 three nodes are able to forward the message at the same time, as they are out of each other's radio range. To avoid creating duplicates in scenarios like this, CBF applies an area based suppression scheme besides the timer based forwarder selection. The next hop node is only allowed to be inside a reduced area, called suppression area. The optimal shape of the suppression area is a Reuleaux triangle, according to [11].

Reliable Dissemination

The MAC protocol of the widely known and used IEEE 802.11 [14] protocol stack only supports reliable one hop packet transfer in unicast mode, by a four-way handshake (RTS/CTS/DATA/ACK), called virtual carrier sense. In ITS networks, however, emergency signaling might require a reliable broadcast medium. Reliable broadcast protocols exist for mobile ad hoc networks; they usually adapt the four-way handshake of IEEE 802.11 for broadcast transmissions. Such protocols might be adapted to VANETs as well, e.g., [54] suggests using Batch-Mode Multicast MAC (BMMM), where the source sends a Ready for Acknowledgment (RACK) frame after the DATA frame to control the acknowledgments of the receivers. This might give good reception reliability, but the complicated ACK scheme can seriously reduce channel utilization; thus, a little less reliable protocol may give much better performance.

The Urban Multihop Broadcast (UMB) protocol [21] assigns the task of forwarding and acknowledging the received packet to the farthest away node only, without any a-priori topology information. The UMB protocol uses Request to Broadcast (RTB) and Clear to Broadcast (CTB) frames instead of the original RTS/CTS handshake of the IEEE 802.11 protocol, because the broadcast DATA frame has more than one destination. In an RTB frame the source node includes its position and the intended broadcast direction. If the source wants to disseminate the message in more than one direction, a new RTB frame should be generated for each direction. When the nodes in the direction of the dissemination receive this RTB frame, they compute their distance to the source node. Based on this distance, they send an energy burst (channel jamming signal) called black-burst. The black-burst method is used by many authors for different purposes in wireless networks (e.g., time synchronization); here, it is used to select the farthest away node for retransmission. The receivers set the length of the black-burst signals proportional to their distance to the source. Since the positions of the receiver nodes are unique, using the position information to determine the length of the black-burst most likely yields to a single node that is elected as being the farthest away.

Nodes start to send their black-burst after a short inter-frame time (SIFS), upon hearing the RTB frame. At the end of the black-burst, nodes stop jamming and listen to the channel. If one finds the channel empty, it means that its black-burst was the longest and it is now responsible for replying with a CTB frame after a duration called CTBTIME, where SIFS < CTBTIME < DIFS. If the channel is busy, it means that there are some other vehicles farther away, and the CTB packet is not sent.

After successfully receiving a CTB, the source node sends its broadcast packet as shown in Figure 4. In this broadcast packet, the source node includes the ID of the node which has sent the CTB, making that node responsible for sending an ACK to the source and forwarding the broadcast packet. This ACK frame ensures the reliability of the packet dissemination in the desired direction, just like in the original IEEE 802.11 MAC. Although all other nodes between the source and the ACK sender receive and process the broadcast packet, they do not rebroadcast or acknowledge it.

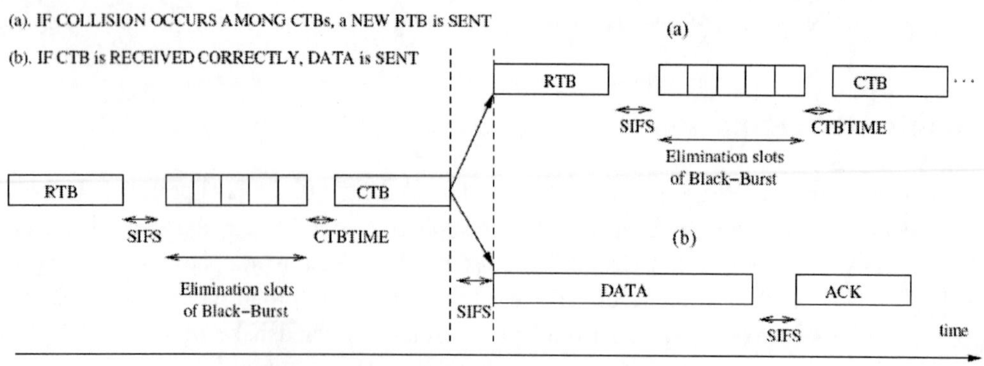

Figure 4. Virtual carrier sense in UMB.

At the intersections, the UMB protocol suggests installing repeaters to forward the packet to all road segments. Mobile nodes detect the intersection areas by using the digital map, and deliver the packets to the repeaters with the original IEEE 802.11 unicast forwarding. After receiving the broadcast packet, the repeater initiates new directional broadcasts in all road segments other than the one the message was received from.

The Ad hoc Multihop Broadcast (AMB) protocol from the same authors [22] requires no fixed repeater nodes in intersections; ordinary vehicles are elected instead to temporarily act as repeaters. A new intersection broadcast mechanism is proposed to find the vehicle closest to the intersection, as it is considered a good candidate for branching the packet, because it is likely that vehicles closer to the intersection have a better coverage of the other road segments.

The new intersection broadcast mechanism is composed of two phases. The first phase is choosing a *hunter* vehicle which tries to select the closest vehicle to the intersection. A vehicle becomes hunter if the next intersection is within its radio range. In the second phase the hunter selects the vehicle closest to the intersection by sending an Intersection RTB (I-RTB) frame; the receivers reply with a black-burst, according to their distance to the intersection. Unlike the normal RTB, where the farthest vehicle from the source sends the longest black-burst, when vehicles receive an I-RTB, the vehicle closest to the intersection sends the longest black-burst. If the repeater selection of the hunter vehicle is unsuccessful, the hunter vehicle itself becomes the repeater.

Emergency Message Propagation

The most important applications of a VANET are the ones related to safety, e.g., cooperative collision avoidance. These require multicast-like data dissemination, as the messages do not have a certain destination node, nor do they need to reach every single node in the network. Geocasting is a multicasting scheme commonly used in position-aware ad hoc networks, where the target of the message is a geographic area, called geocast area. Flooding is used inside the geocast area, and multiple methods exist to reach the area with directed flooding if the source is outside the area. The following two protocols are special geocast schemes that define the geocast area.

Localized Urban Dissemination

The key observation behind the Localized Urban Dissemination protocol (LUD) [32] is that in urban scenarios, i.e. in downtown, the message flow is forced to follow the road network, because the buildings block the radio propagation. The geocast area is not described by the source as a polygon, but is obtained following a multi-step distributed decision process by the vehicles. The forwarders in junctions decide if the next road segment belongs to the region or not, by estimating the probability of vehicles coming on that road segment being interested in the message.

The packet dissemination algorithm must ensure localization, as emergency messages are only useful in a certain area around the source. LUD achieves this using a scheme called gossiping, which means the messages are rebroadcast with probability p at each node. This value p is carried in the packet header, and acts like a stochastic Time-To-Live (TTL) boundary. If $p<1$, the messages cannot reach infinite distances from the originator node, as the rebroadcasts are independent Bernoulli trials; thus, the number of hops a packet can survive follows a geometric distribution, and the expected value of the hop number is $\mathbf{E}(h)=1/(1-p)<\infty$.

With a constant p rebroadcast probability, the shape of the geocast area is determined by the topology of the road network only. There are certain areas, however, where the message is not useful, because the vehicles are not likely to go to the alert location from there (e.g., due to one-way streets). These areas can be filtered out in a stochastic way if the rebroadcast probability p is changed in junctions according to the a-priori knowledge of the usual behavior of the vehicles. This information may be guessed based on the ranks of the roads, or a Steering Probability Database (SPDB) might be built based on a traffic observation service, and downloaded into the navigational devices for better accuracy.

Changing the rebroadcast probability by the deciders is the core part of the protocol, as it defines the large scale behavior of the data dissemination. There seems to be no optimal solution for this; many different schemes can be applied, each of these schemes having its advantages and requirements. The authors in [32] propose two schemes, the first being memoryless, and the second being slightly directed.

The first scheme gives a formal computation to set p based on two events:

- A_i: The message reaches the ith decider
- B_i: The ith decider goes to the source of the message

The fundamental equation of this scheme is:

$$\mathbf{P}(A_i) = C\,\mathbf{P}(B_i),$$

where C is a constant chosen by the source to control the distance the message should reach. The probability of both event A and event B can be expressed with formulas if the model of the road traffic is simplified enough. Because both $\mathbf{P}(A_i)$ and $\mathbf{P}(B_i)$ follow a geometric distribution, the resulting scheme becomes memoryless, and we obtain:

$$p_i = \sqrt[h_i]{q_i(1-s_i)},$$

where q_i is the steering probability for the decider to follow the path of the message back to the source, s_i is the stop probability for the next road section (to model non-infinite paths), and h_i is an estimation of the number of hops needed to reach the next junction on the road the decider came from. This scheme is fairly simple, especially if the root calculation is optimized numerically, but it yields to a very strange geocast area due to multipath propagation.

The second scheme is an effort to counter the effects of multipath propagation. In the first scheme the receipt probability in a certain position is proportional to the number of different paths leading there from the source. The straight road the source goes on is an important target of the dissemination, but it is only reachable on a single path (the message is not allowed to approach the source), so the alert is not propagated there. Using heuristics, this second scheme can increase the receipt probability on that road at the expense of loosing the formal description of the decision.

This protocol can be adapted to highway scenarios as well; the message flow is localized with the gossiping, but no decision schemes are needed as the intersections are very rare. The flooding controlled by gossiping is expected to provide fast alert propagation along the road, but its reliability is questionable without acknowledged forwarding.

Bidirectional Perimeter-Based Propagation

Another approach to emergency message propagation is the Bidirectional Perimeter-based Propagation (BiPP) protocol [47], which builds a Regional Alert System (RAS) by notifying all vehicles inside the alert region, as long as the alert is active. Cars can only enter the alert region if they cross its boundary, called perimeter. Instead of having periodic broadcasts throughout the entire region, it is sufficient to broadcast where roads cross the perimeter.

The alert is represented as a tuple of the form *(location, start_time, duration, safety, operate).* The *start_time* and *duration* fields indicate the timeframe within which the alert is active. The *safety* field gives the desired radius of the alert; when a car moves to within the safety distance, it should have "heard'" about this alert. Outside the *operate* radius the alert is not disseminated.

If the scenario is a single road, there is a token held by a vehicle between the *safety* and *operate* distances, in each direction from the alert. The authors call these left and right tokens, and these two tokens are handled separately. The cars holding the tokens are responsible for notifying all inbound vehicles about the alert, by periodically broadcasting the message. The number of the broadcasters is reduced by eliminating duplicate tokens; if a token holder receives a broadcast for the same token, it drops its token.

If the vehicle holding the token is inside the safety radius, it broadcasts frequently in order to have the token reach the safety radius as fast as possible. In the perimeter area the token is kept close to the safety radius to prevent the accidental passing of the operation radius. Outside the operation radius the token is dropped, and a new one must be instantiated if the alert is still active.

Theoretical Bounds of Dissemination Efficiency

The main goal of [44] is to investigate the fundamental tradeoff between *safety-level* and *emergency message resource wastage* encountered in hazard warning applications. The dependence between 1-hop and multi-hop time-constrained reliability is studied under various message dissemination strategies for a given network topology, channel characteristics, and interference model. The study uses a probabilistic approach, and finds time constraints on hazard warning reception. That means the probability $P\ d,t,p$ that a vehicle located at distance d from the hazard warning initiator receives correctly the warning within time t is estimated, where p is a parameter modeling 1-hop reliability.

A set of n equally spaced cars on a straight line road are represented as a string S_n of bits, where the least significant bit (the rightmost one) corresponds to the initiator, and the i-th bit to the i-th car in the cloud. The bits encode whether a certain car has received the token (bit value is 1), or not (bit value is 0). A car holding the token is a 1-node, and a node that has not yet received the token is a 0-node. The value of the bit string $S_n(t)$ changes in rounds as the token is propagated, where t is the round index.

Initially only the initiator holds the token, $S_n(0)=0...01$. During the first round, the initiator transmits the message, and all nodes within r_T positions from it become a 1-node with probability p. In later rounds multiple nodes transmit at the same time, resulting in interference. This effect is modeled by the r_I interference radius: a 0-node becomes 1-node with probability p only if a transmitter is within r_T and no other transmitters are within $r_I \geq r_T$ range.

At round k some bits are 1, some become 1, and some remain 0 in $S_n(k)$, depending on the positions of the transmitters and the radio interference (modeled by the reception probability p). It is obvious that there can be several 1 chains in any round, separated by at most r_T long 0 chains (called internal 0-nodes).

Fast backward propagation is needed, because the information contained in the emergency message has a very limited lifetime, and reliable propagation is needed because a single vehicle missing the warning can become extremely dangerous for the other vehicles. The goal of the analysis is to estimate the probability P that bit d turns into 1 within a certain number t of rounds. This probability greatly depends on the strategy used to select the transmitter nodes at each round. The study examines the propagation under controlled conditions in order to understand what is the achievable safety level versus the resource demand, i.e., the number of transmissions required to fully cover an area.

The strategy used to select the transmitters set at each round is called *Global*, because it is a centralized strategy that uses global knowledge: the set of transmitters for rounds k and $k+1$ if $r_I=r_T$, or k, $k+1$, $k+2$ if $r_T < r_I \leq 2r_T$ is computed by knowing the complete status of $S_n(t)$ at time k-1.

In the first round of this stage the leftmost 1-node is scheduled to transmit, and some other 1-nodes are selected greedily to avoid interference. Nodes within r_T distance from the selected transmitters are marked as covered, the other nodes are marked as uncovered. In the second round the consecutive uncovered nodes are grouped, and a 1-node is scheduled to transmit in round two for each group. The third round is similar to the second, but it is only executed if $r_T < r_I \leq 2r_T$.

The *Global* strategy is not a realistic dissemination protocol, as it requires global knowledge of $S_n(k\text{-}1)$, but it is very interesting, because it is within constant factor from the optimal propagation strategy. The *Idealized* strategy, which is considered to be optimal, is a stochastic process that at every round *i)* turns every internal 0-node into a 1-node independently, with probability p, and *ii)* every node i such that $k < i \leq k+r_T$ (node k is the leftmost 1-node) is turned into a 1-node with probability p. This strategy is called *Idealized*, because both *i)* and *ii)* cannot be satisfied for all possible $S_n(k\text{-}1)$ due to the interference among concurrent transmissions.

The following approximation factor is used to compare the effectiveness of the strategies: $E(\mathbf{k}(t))$ is the expected value, $\mathbf{k}(t)$ denoting the position of the leftmost 1-node under the *Idealized* strategy. A strategy S is said to be within a factor (α, β) from the optimal, if and only if *i)* at each round, to all internal 0-nodes it gives a chance of at least p/α of turning into a

1-node, and *ii)* $E_S(\mathbf{k}(t)) \geq E_S(\mathbf{k}(t)) / \beta$. This way, both the speed of the backward propagation and the completeness of the coverage are observed.

It can be proven that the *Global* strategy is within a factor $\left(p/\left(1-\sqrt{1-p}\right),2\right)$ from optimal, if $r_I = r_T$, and within a factor $\left(p/\left(1-\sqrt[3]{1-p}\right),3\right)$, if $r_T < r_I \leq 2r_T$. Strategy *ImGlobal* is a modification of strategy *Global*, where the leftmost 1-node is scheduled for transmission in every round to ensure fast backward propagation. This strategy is within a factor $(1-(1-p)^{(1/h)},1)$ from optimal, where h=2, if $r_I = r_T$, and h=3, if $r_T < r_I \leq 2r_T$. This means that the backward propagation of strategy *ImGlobal* is optimal, and its coverage ratio is better than that of strategy *Global*.

Using the above described dissemination strategies time-constrained reception probability can be investigated, and bounds can be given to the probability $P(d,t,p)$ that a node at distance d from the originator correctly receives the token within t rounds. Using limit theorems and simulations, the following conclusions can be drawn:

1) the beneficial effect of increased 1-hop reliability tends to decrease as the distance from the emergency message initiator increases;
2) the relative benefit on multi-hop reliability of having high 1-hop reliability tends to decrease as car density increases;
3) the dissemination strategy has a major impact on multi-hop reliability.

The result of the analysis shows that emergency message dissemination strategies have to be vehicle density-aware as well, in order to ensure optimal backward propagation.

Traffic Information Spreading in ITS Networks

Traffic and Travel Information (TTI) systems [5][24] are used to inform the driver about traffic situations. Efficient and timely propagation of traffic information is essential in such systems. The communication can take place between vehicle and vehicle (ad hoc network), or vehicles and infrastructure (centralized network) [3][25]. In the case of centralized TTI infrastructure [24][3], a traffic information centre collects, maintains, and supplies all traffic information, e.g., by using a cellular mobile technology (GPRS). The information centre has a global view of the road network, but there is an extra cost of processing and communication.

Therefore, the research community is currently focusing more on distributed traffic information dissemination [7][26][45][50], where inter-vehicular communication is applied. The vehicles share the traffic information among each other by using a local wireless technology, e.g., IEEE 802.11b or DSRC (Dedicated Short Range Communications). The distributed approach has only the installation cost of the on-board unit, without permanent communication costs. Traffic information spreading in these networks is achieved by means of a broadcasting/flooding mechanism [12][50]. To overcome network fragmentation, the vehicles usually maintain and carry a copy of the TTI. This information is disseminated along the road segments, while the vehicle is traveling around the road network. The frequency of subsequent transmissions will control the quality of the TTI reports, in terms of delay and accuracy. If the frequency of TTI transmissions is high, the time necessary for the information to reach the outer bounds of the geographic area is lower. The accuracy of TTI also varies in function of the amount of communication involved in the travel information gathering and transmission. Frequent information exchange leads to a more accurate picture about the traffic situation. Unfortunately, frequent information inter-transmissions lead to superfluous TTI dissemination, which consumes more bandwidth, overwhelms the network and causes frequent collisions.

Broadcast storms and superfluous forwarding can be reduced by using adaptivity in the broadcasting mechanisms [50][45]. Adaptivity can be introduced by controlling the frequency of information exchange (timely manner) or limiting the dissemination only to areas where the TTI is really necessary (spatial manner).

In the following subsection we present inter-vehicular communication protocols, which limit the superfluous TTI dissemination, either in a timely or a spatial manner. It will turn out that the spatial limitation of TTI information spreading should employ also similar routing techniques like the ones presented in the previous section.

Timely Limitation of Information Spreading

Inherently, there is a compromise between using a short inter-transmission interval, which will reduce the TTI distribution time for the cost of larger bandwidth per vehicle, and choosing a larger interval to reduce bandwidth requirements but increasing the TTI propagation delay. The following protocols introduce some levels of adaptivity in the frequency of TTI dissemination. The scope of dynamic adaptation is the reduction of the amount of required bandwidth per vehicle. The TTI is sensed by individual vehicles, then locally analyzed and combined with TTI received from other vehicles. The distribution of TTI is done by one-hop broadcasting, adaptively modifying the interval between subsequent transmissions.

The authors in [12] propose two mechanisms to disseminate information about traffic conditions. In the *Naive Scheme* vehicles transmit traffic reports at a pre-defined maximum rate of dissemination. At every dissemination period the vehicle examines its database for the travel times on the recently traveled road segments. From these, it selects the one with the most significant difference compared to the expected travel time of the corresponding road segment. The *Smart Scheme* disseminates traffic information only if it considers something 'interesting' to report. There is also a periodic transmission with a pre-defined maximum transmission rate, but contrarily to the Naive Scheme, the traffic report is only transmitted when its difference to the expected travel time exceeds a certain threshold. Upon receiving traffic information, the travel time estimates are updated considering both the new and the old values of the travel times. In both schemes, the vehicles disseminate traffic reports about a road segment in a geographic area centered on the respective segment. The size of this area is determined by the traffic application, and is used as a parameter in the evaluation of the protocols.

Wischhof et al. introduce a similar adaptive broadcast mechanism for TTI distribution in [50]. Here, adaptivity is introduced by the *Provoked Broadcast* scheme, where the broadcasting interval is dynamically altered in function of the observed events. Upon a data packet reception, its content is examined and compared with the local TTI database to determine the type of adaptation. From the received message for each road segment weights are calculated; these indicate the difference between the previously stored and the newly received knowledge (timestamp, TTI value). The calculated difference is compared to predefined threshold values, and if it exceeds the respective threshold, then the corresponding weight is increased with a constant value. Then, a total weight is calculated as a sum of the weight values of the distinct road segments (TTIs).

A provocation and a mollification threshold is also used to determine if a provoking or mollifying event has occurred (Figure 5). When the total weight of the message is less than the mollification weight, an increase of the remaining transmission time will occur, while receiving a message with a weight larger than the provocation weight will cause a decrease in the transmission time. Thus, if two vehicles are sharing significantly different knowledge, the threshold mechanism will provoke the receiver to rebroadcast the newly received message faster, with a reduced interval. An increase in the transmission interval (mollification) occurs when the received message has a very similar view of the traffic situation.

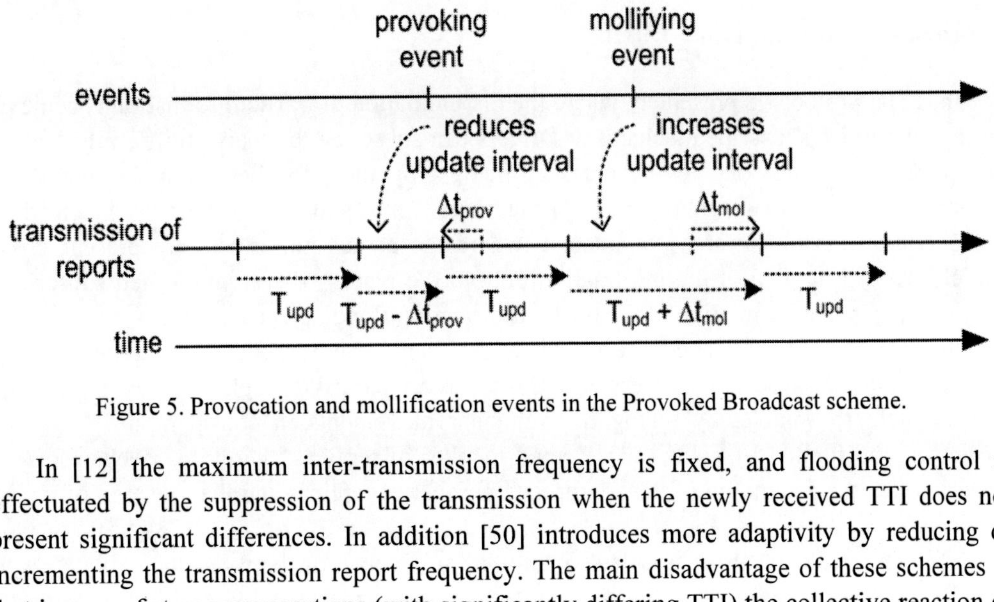

Figure 5. Provocation and mollification events in the Provoked Broadcast scheme.

In [12] the maximum inter-transmission frequency is fixed, and flooding control is effectuated by the suppression of the transmission when the newly received TTI does not present significant differences. In addition [50] introduces more adaptivity by reducing or incrementing the transmission report frequency. The main disadvantage of these schemes is that in case of strong provocations (with significantly differing TTI) the collective reaction of receivers can still lead to frequent collisions. There is also a difficulty in fine tuning the system parameters (e.g. the provocation and mollification thresholds), since the schemes strongly depend on the characteristics of the distributed TTI.

Spatial Limitation of Information Spreading

Superfluous TTI dissemination can be alleviated even more efficiently by controlling the dissemination, not only by reducing its rate, but also limiting its spatial coverage. Since the vehicles are moving along road segments, and these constitute predefined routes, this information can be used to propagate the TTI to the domain of interest, exclusively. This *domain of interest* is a set of places where information from a traffic jam on a certain road influences the route choice of the driver. At these places, the vehicles are still able to change their routes without a drastic deterioration in their travel time. This leads to spatial context-awareness in TTI infrastructures, which represents an open research issue.

Recently, a large number of researchers [48][55][6] have started to use special context-aware information in vehicular communication protocols. This information is extracted from the digital maps [48] or even from more specific context (e.g., time schedules of public transportation traffic [6]), which helps the forwarding vehicle to choose the best route for its packets. However, these protocols are concerned with information routing, not TTI dissemination.

For spatial limitation, TTI can be disseminated proactively, using *data-push model*, or based on a *data-pull model*, when the information is obtained on-demand. In the first case the data is usually broadcasted (disseminated) along the road segments, while in the second case the data is transported to the locations of interest and broadcasted only in case of demand. In both cases the question is how to control and limit the traffic information spreading only to the places where the respective information is useful.

Protocols Using the Data-Push Model

In [25] a protocol is presented, where the dissemination of traffic information is achieved through a publish/subscribe mechanism. Drivers can subscribe to receive traffic information; this information is carried to the areas of interest by the publishers, vehicles which are participating in the forwarding of the TTI messages. Thus, subscriptions are used to lead the broadcasted TTIs to the areas of interest, where the actual subscribers are present. Since vehicles carry a replica of the information only to the places where uninformed subscribers are present, a spatial limitation of the traffic information flooding is achieved.

In order to notify all the necessary subscribers, event replicas are generated and maintained in the subscriptions areas. These replicas are passed through periodic broadcasting from vehicle to vehicle, by taking the vehicular movements in consideration. Broadcast storms can be limited by using only a small number of replicas; therefore, mechanisms are presented for the generation of the appropriate number of replicas and the selection of replica carriers.

The next replica carrier is selected through an interaction between the replica owner (carrier) and its one-hop neighbors. A message is broadcasted by the carrier with the subscription topic, whereupon all one-hop neighbors respond with a message, indicating their subscription status, position and direction. The subscription status (*informed*, *interested* or *not interested*) will reflect the neighbor's interest in the respective notification. The replica owner will use the direction information to group the neighbor vehicles into clusters. From the cluster with the most uninformed subscribers will be selected randomly the next replica carrier, which moves towards the opposite direction. If all the neighbors are already informed, the cluster with the most subscribers will be selected.

An appropriate number of replicas are necessary to disseminate the information to the corresponding areas of interest with a high delivery ratio and a minimum message overhead. Replicas guided to areas where is no interest must be deleted, while in areas with a large number of uninformed subscribers a sufficient number of replicas must be created. Therefore, the replica owner will use the number of neighbor responses as an estimate of replica necessity in the respective area. The result of the last *k* polls will be used for statistics to adopt the number of the replicas to the subscriber demands. This method ensures that if replicas cannot find uninformed subscribers or no subscribers at all, they will be deleted; where there is a large flow of uninformed subscribers, additional replicas will be created.

However, due to the carrier selection and TTI replication mechanisms, it is not always guaranteed that the information carriers will meet their subscribers. The chance that a replica survives insensitivity, and meets proper subscribers, depends on the estimate of the replica's necessity, which is represented by the number of last k polls. Thus, the successful outcome of the protocol highly depends on the topological context and the fine tuning of the system.

In [45] an approach for information dissemination to a set of specific target areas is presented, where the structure of the road network is taken into consideration. Here the authors generate a *propagation function*, which encodes the destinations (target zones) and preferred routes for the message. The route that a message should take is driven by the directions of maximum decrease in the propagation function.

The message originator does not encode a predefined trajectory with the propagation function; rather, the route is the result of the function evaluation at each routing hop. The probabilistic function is carried by the data packets of routing protocols, similarly to the method presented in [48]. The propagation function can be viewed as a special type of source routing, taking a different approach than traditional source routing (DSR) or trajectory-based forwarding (TBF).

The computation and encoding of the propagation function is effectuated at the originator node, by using the additional information about the road network and traffic conditions. Considering averaged traffic values in different time frames could help to build an off-line library of the propagation functions. Replacement and combination of the propagation functions along the route is also possible.

The authors also present several protocols to explore in various settings the different impacts of the information encoded in the propagation function. In these protocols the message forwarding decisions are based on probabilistic choices, derived from the propagation function. The protocols use broadcast transmissions, and store and forward techniques (data mules) to achieve good performance in terms of dissemination coverage and message delivery ratio.

Since the preferred routes are encoded in the packets, this method is vulnerable to topology changes, which could lead to frequent failures in TTI delivery. On the other hand, there is no presented method to calculate the propagation function, i.e., the locations where the information should be propagated. Therefore, this protocol is not ready to be applied in urban scenarios.

Protocols Using the Data-Pull Model

In [7] an ad-hoc service infrastructure for vehicular networks is proposed to provide location-aware services, which distribute on-demand information describing road conditions and other geographic area related facilities. The infrastructure is based on the stateless Vehicular Information Transfer Protocol (VITP). VITP specifies the syntax and the semantics of messages exchanged between the components of the vehicular network, called *VITP peers*. These peers are running on devices which establish on-demand ad hoc communication groups, collect, store and combine information in order to resolve incoming requests.

The traffic information is obtained by a pull-based protocol, issuing a location-sensitive query along the route. The queries are routed towards their designated location of interest via the available network. Upon arrival to the destination area, the queries are intercepted and processed by the local vehicular infrastructure. Then, through on-the-fly collaboration, a reply message is generated and sent back to the source location. The responses can help drivers to adjust their path while driving to some destination on a given route. VITP also supports a push-based propagation of messages to disseminate various alerts about emergency situations. The service provisioning is based on an extended client-server computing model. The queries issued by the driver are responded by a set of vehicles collaborating to establish a virtual ad hoc server (VAHS), to resolve the incoming query and to send back the response. The VITP communication is based on a best-effort approach, as the VAHS peers do not have information about the other group members. The VAHS is identified by the query and its target location area.

The VITP request is transported between VAHS peers until its *return condition* is satisfied. The return condition defines when the VITP reply can be created; it is defined explicitly in the query and depends on the semantics of the query itself. Intermediary peers receiving VITP queries can respond with a reply if their caches contain matching information. This affects the return condition of the request; thus, the reply decision must be based on the semantics of the VITP message.

Unfortunately, there is no mechanism specified to calculate at what extent the queries, respectively the TTI reply/caching, should be propagated inside the road network. The authors present simulations, where the queries are propagated only to a randomly selected value of 400-800 meters inside the road segments. This means that in certain situations the information will not be received in time to calculate the proper bypass route (since it is coupled to the queries). Therefore, additional mechanisms are necessary to determine the critical points until when the queries must be propagated, i.e., from where the TTI information has to be gathered.

Calculation of the Domains of Interest

In [48] a different step is taken into the design of distributed TTI systems, by proposing an algorithm to calculate the domains of interest on a digital map. Based on this information, the dissemination protocol will be able to reduce superfluous forwarding, not only temporally, but also spatially. The SPatially-Aware Congestion Elimination algorithm (SPACE) is designed to determine the locations (domain of interests), where a possible event (e.g., traffic jam) on a certain road influences the route choice of the driver. Thus, a well defined area is given, where information about traffic jam is useful. This information could be used in the above presented protocols to fine tune their performance.

The algorithm simulates the impact of an event on a set of edges of the graph. For each affected optimal path p of the graph it is assumed that the travel time on edges of E_f decreases significantly. Then, the weights of the affected edges of E_f are increased, or these edges are temporarily excluded from the graph, and a new optimal path (the by-pass route) is calculated by running the shortest path algorithm in the temporary graph, resulting in path p^*. The output of the algorithm is an Impact Vector $I_{Ef}(e)$ that shows whether an event on edges of E_f has an impact on the route choice of vehicles traveling on edge e, and if yes, in what extent.

The value $I_{Ef}(e)$ expresses the average amount of vehicles on edge e, whose route choice is impacted by the knowledge about an event on roads of E_f.

By using SPACE, the broadcasting urge is also suppressed by the spatial context-awareness. Before broadcasting the message, the value of the Impact Vector is checked. If it is positive (and above a threshold), then the current road is part of the domain of interest of E_f ; otherwise, disseminating the TTI at this part of the road network is useless. However, it could happen that later the vehicle will cross a road from I_{Ef} . This can be checked by examining the vehicle's route, or the TTI can be also buffered until its lifetime expires and broadcasted only if a road from I_{Ef} is encountered. The outcome of SPACE can be calculated off-line, and stored in a database provided along with the digital map. Downloading the Impact Matrix is also possible, e.g., by using GPRS communication. Each vehicle receiving a TTI message examines the attached Impact Vector and reacts according to the context by calculating its route on-line.

A classical TTI dissemination protocol [12] is enhanced with the context-aware feature to show the advantages of SPACE. The simulation results show a significant improvement in case of SPACE in terms of channel utilization (nr. of information packets used), while the travel time of vehicles remained unaffected.

Unfortunately, in [48] there is no exact method proposed to set the threshold from where the impact factor of a road is negligible. Further studies are necessary, since the wrong adjustment of this parameter could lead to uninformed drivers, which could cause additional traffic jams in the network.

Infrastructure-Based and Hybrid Architectures

Up to this point, we have focused on infrastructure-less (ad-hoc) V2V networks, which offer a low-cost and easy to deploy solution for a number of localized applications. However, V2V communication has several drawbacks, as it is unreliable under specific circumstances and it has no access to external resources. Consequently, it is often needed to provide infrastructure based (fixed) V2I vehicular networks, instead of (or besides) the V2V communication. Infrastructure-based networks provide reliable broadband communication services, and can access online resources not available in vehicles. Moreover, research has begun to provide infrastructure-less and infrastructure-based communication at the same time. This hybrid architecture exploits the benefits of both V2V and V2I approaches, where the distributed, ad hoc vehicular network is supplemented by centralized, infrastructural elements.

Hybrid Wireless Technologies

A number of existing wireless technologies can be used to create vehicular networks of different architectures, including Wireless Wide Area Networks (WWAN), like 2G, 3G, or the developed 4G mobile technologies, Wireless Metropolitan Access Network (WMAN), like WiMAX, or Wireless Local Area Networks (WLAN), like WiFi. The communication links can be grouped in different ways. We present here the classification worked out in the CVIS project [7], where the following types of communication links have been defined:

1. *Localized Directed or Omni-directional Communication* - includes communications between two roadside units (I2I), or between vehicles and a roadside (V2I), or between two vehicles (V2V). Four WLAN technologies belong to this group:

    - *Microwaves (M5)* - supports directional and omni-directional communication and localized broadcast communication between moving objects, and between moving objects and roadside installations, with a minimum data rate of 6 Mbps, and covering a communication distance of several hundred meters. This technology is considered by the European car industry, in e.g., the C2C-CC project, with a preliminary focus on safety related applications, but not restricted to them only.
    - *Millimeter waves (MM)* - supports localized directed communication. MM allows for much higher data rates (in the order of hundreds of Mbps) in the range of several hundred meters.
    - *Microwaves CEN DSRC* - arguably the most successful ITS technology today, with wide spread usage all over the world. The main current use is for payment systems and access control.
    - *Infrared (IR)* - in accordance with the strong capabilities to "beam" information in dedicated directions, IR supplies directed communication services. This technology is widely spread in Japan these days. It provides 2-8 Mbit/s transfer-rates with a range up to 100 m.

2. *Wide-area Communication (WWAN)* includes communications based on cellular networks and using satellites.

    - *GSM/GPRS/UMTS/HSDPA*: High Speed Downlink Packet Access transmission protocol (HSDPA) supports transmission speed of 1.8 Mbps, while UMTS supports 384 kbps on 3G UMTS networks. If no UMTS is available, GPRS is a fall back solution with lower data rates.

WMAN networks are not defined in the CVIS project, but could be integrated if required. The design principle of such a hybrid system is that new air interfaces can easily be integrated into the platform. The communication subsystem of CVIS follows the series of draft standards under the acronym CALM (*Continuous communications Air interface for Long and Medium range*) [7]. CALM describes a hybrid mobile network based on the above mentioned various communication channels such as GPRS, UMTS, M5, MM, and IR. The research community has begun to work out and apply policy-based rules to select the optimal communications channel at any time and any place, for always-on and optimal connection.

Hybrid Wireless Architectures

The research on selecting the optimal communication channel from the technologies presented above aims at combining the two technologies and to exploit the high bandwidth of

WLAN and the wide coverage of WWAN. In [15] a classification is proposed for general hybrid networks. In [53] this classification is applied for vehicular networks:

- One-hop WWAN - In this case just the cellular based network is used, which covers large areas, but can serve a limited number of high-bandwidth users.
- *One-hop WLAN* - In this case high speed roadside access points are used (V2I) without V2V communication. This approach provides high throughput but limited coverage, i.e., network access within the range of the wireless link of the road-side unit.
- *Multi-hop WLAN* - This extends the coverage of one-hop WLAN with ad-hoc multi-hop V2V communication.
- *One-hop WWAN and one-hop WLAN* - This is a hybrid setup of one-hop WWAN and one-hop WLAN. WWAN covers the entire area, while WLAN access points are installed to specific places. Where both WWAN and WLAN are available, vehicles can choose to access either one of them, or both. Outside the WLAN coverage they can access only the WWAN.
- *One-hop WWAN and multi-hop WLAN* - This is a hybrid of one-hop WWAN and multi-hop WLAN. Vehicles can access the WLAN access point either directly or by V2V communication. If the next WLAN road-side unit is unavailable because the ad-hoc network is not connected, WWAN is the fall back network.

One-Hop WLAN

A specific infrastructure-based scenario, Internet access from moving vehicles on the road, was worked out in the Drive-thru Internet project [39]. This fits in the One-hop WLAN category of the above classification, and can be a building block of a hybrid architecture. The aim of the Drive-thru Internet project is to enable mobile users to automatically detect and access a wireless network while walking, driving, or otherwise passing through a hot-spot, even at high mobility, when only a short period of connectivity is available. In contrast to ubiquitous connectivity, nomadic users are expected to remain in a hot-spot area for a short period of time, and leave it before fully configuring their mobile devices or end running their respective Internet applications [40]. Asynchronous applications such as e-mail access and transmission as well as applications based on distributed object synchronization (such as calendars, offline file systems, data bases, and repositories for distributed authoring) are able to deal reasonably well with intermittent connectivity.

The concept is based on the Persistent Connectivity Management Protocol, which acts as a “session” protocol that offers persistent end-to-end communications even in the presence of interruptions. It provides the service of managing intermittent connectivity for transport layer endpoints. The Persistent Connectivity Management Protocol entities are Drive-thru clients and Drive-thru proxies – transport and application layer intermediaries that manage persistent connections on behalf of communicating nodes and their applications. Tests show that for vehicles traveling at 120 kmh, a minimum data volume of 30MB could be transmitted in virtually all test runs in a single pass [39].

Multi-hop WLAN

In [16] a MANET dispatch system for taxis is described, which is a Multi-hop WLAN system in the above classification. In this application a city taxi company's dispatch unit informs individual taxis about passenger pickups, assigns passengers to empty taxis, and passes on additional information such as directions and news about weather or traffic conditions. In contrast to WWAN based solutions, which are mostly used for such applications, this paper presents a multi-hop WLAN based system. On the other hand, the system is not a clear V2V network, since the taxi network has a central message source, although communication is achieved in an ad hoc fashion. This is a hybrid scenario in the sense that the system is a fully mobile ad hoc network, but there is a kind of centralization as well, in the sense that all dispatch information originates from one point, requiring that all nodes connect to it directly or indirectly to receive jobs.

The operation of the system is based on simple message exchanges. Taxis regularly send status updates to the dispatch unit, and the dispatch unit sends job requests and other information to taxis. Messages are always sent to the nearest reachable hop in order to minimize transmit power and interference. Given that the location of the dispatch unit and any taxi stands are known, the system can use one of the location aware routing protocols described earlier.

The average network coverage is one of the most important factors in such a system. The number of reachable vehicles increases as more nodes are available to act as relay points. On the other hand, the effect of traffic congestion on the efficiency is twofold. The average network coverage and the average and maximum outage times generally increase until a certain extent of traffic congestion. This is due to the fact that increased congestion slows down the mobility of nodes in high traffic areas, such that they are more likely to be in contact for the required duration, hence improving coverage. At the same time, nodes that are out of coverage and caught in traffic are more likely to experience longer periods of outage.

Above a given amount of cars on the streets, traffic congestion spreads increasingly throughout the entire map, such that all areas become highly congested rather than just central, high traffic areas. This subsequently results in a slight dip in average coverage and an increase in outage times. These results suggest that with light to medium congestion the dispatch system will be able to operate on average similarly, or even better than in zero congestion conditions. However, in conditions of very high congestion, coverage may suffer slightly and in any case congestion increases the risk of longer outages.

Another type of infrastructure based hybrid dissemination system is presented in [56]. Data is broadcast from an information source to many vehicles on the road, but there is no information flow in the other direction, from the vehicles to the center. The basic principle of the proposed data pouring strategy can be seen in Figure 6. Data are periodically broadcast from the road-side data centers to vehicles on the road, in the direction where the information is of interest. In the enhanced version, data poured from the data centers are buffered and rebroadcast at the intersections in order to reach vehicles that cross the road.

Two attributes are attached to the data item, originated from the data center: the dissemination zone which is defined as a rectangle area, and the expiration time which specifies the duration when the data item is valid. Vehicles outside of this dissemination zone discard the data to save bandwidth. Vehicles communicate with each other and with the data center through short range wireless (WLAN) channel.

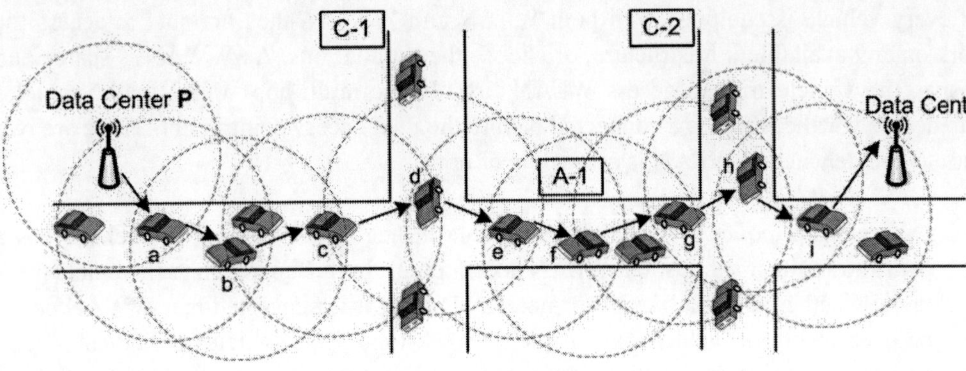

Figure 6. Data pouring strategy.

The data pouring scheme makes use of the partially predictable vehicle mobility limited by the road layout: it broadcasts the data to one or several roads, called axis roads. The axis roads are selected from those main roads going through the data center and which normally have higher vehicle traffic density than other roads. The data pouring scheme also delivers data to vehicles moving on the roads that intersect with the axis roads. However, it does not proactively push data to the crossing roads. Vehicles on the crossing roads will get the data when they reach the intersection; therefore, the dissemination zone of a data item only includes the axis roads.

To propagate the data to the desired road, the data item needs to be consecutively broadcasted along the road by other vehicles, as presented in Figure 6. By this way, data can be delivered between two data centers via V2V communication. To deal with the broadcast storm problem [36], each designated broadcasting vehicle selects one vehicle that is the farthest away in the data propagation direction from its neighbor list, and designates the selected neighbor as the next broadcast node by adding it to the packet header. After receiving the data, the designated forwarder vehicle rebroadcasts the data. By this way, the data are disseminated to vehicles on the axis roads.

An advanced version of this approach is called *Intersection data buffering and rebroadcasting*. This scheme follows the basic idea of the multi-hop WLAN data pouring, but additionally it relies on a simple device called relay and broadcast station to improve the dissemination capacity. This is a roadside unit that enables one-hop WLAN communication, a simple computing device with a small amount of memory and a wireless card. This unit is used to buffer data and rebroadcast them periodically. As a result, the data center does not need to ensure data delivery to the end user. Instead, it only transmits data to those relay and broadcast stations on the axis roads.

One-Hop WWAN and Multi-Hop WLAN

After introducing one-hop and multi-hop WLAN solutions, we show the results carried out on hybrid WWAN and WLAN architectures presented in [51]. The following hybrid architectures are considered: (1) One-hop WLAN, (2) Multi-hop WLAN, (3) One-hop WWAN and Multi-hop WLAN. In the last case different WLAN hop limits are considered, like 1, 2, 3 or 4, bounding the number of hops to the next WLAN access point.

If every vehicle is equipped with both WLAN and WWAN, they are able to access the network on any available infrastructure, or choose the optimal one. As WWAN is scarce and expensive, the vehicle tries to access WLAN first. In the multi-hop WLAN case, vehicles have to decide whether there is a connected path to the next access point or not. There are two policies to decide whether WLAN is accessible or not:

- *Deterministic policy* – A vehicle sends data through the WLAN if the vehicle has a path shorter than *m* hops to a WLAN access point. In this approach each vehicle must know whether such path exists or not. In [15][31] it is suggested that WWAN could be used to inform vehicles about the network topology. Vehicles periodically send updates about their locations to the center and the aggregated information is broadcasted to the vehicles via WWAN. Another approach is to have vehicles running a localized routing protocol. Due to the high mobility of vehicles, both approaches will generate significant overhead.
- *Statistical policy* – It is assumed that vehicles know the locations of WLAN access points and they receive statistics about vehicle traffic conditions broadcasted by a shared WWAN link. Vehicles use this information to compute the probability that they are connected to the closest WLAN access point in fewer than *m* hops using an analytical model [51]. By this way, vehicles only need to maintain one-hop neighborhood information. Using this statistical policy, in some cases (a) there will be no path to the access point although it is determined that there is a path to it, or (b) the access point would be accessible although it is determined that there exist no path to it. In the first case WWAN is used when the next hop is not reachable. In the second case WWAN is used by the source vehicle although WLAN could be used as well.

Wu [51] takes the following scenario to illustrate some results on a highway without intersection. The road length is 11 km, it is assumed that 1800 vehicles are present on the road, i.e., the number of instrumented vehicles is 1800 times the penetration ratio, WLAN radio range is 250 meters, data traffic rate per vehicle is 16 Kbps, and aggregated WWAN bandwidth is 6 Mbps.

It is important to see what the required WWAN capacity to serve communication requests of all vehicles is. The simulation results show that the WWAN one-hop architecture is normally not a feasible design option to meet the communication demands. The hybrid WWAN - WLAN architecture increases the system capacity and reduces the WWAN data traffic load as some data traffic is diverted to WLAN, especially with multi-hop forwarding. When the hop limit is 3, the average WWAN data traffic load is about 20% of the WWAN one-hop case.

It is important to study the number of WLAN access points needed for the hybrid architecture. As expected, there is a correlation between WWAN capacity, the number of WLAN access points, and the hop limit in providing the required capacity and coverage. The number of required WLAN access points was computed for various architectures. It was assumed that WLAN access points are placed at equal distances between neighbors.

Simulation results on the above mentioned 11 km long road show that the hybrid architecture can significantly reduce the number of required WLAN access points [51]. Multi-hop forwarding (even one more hop) can help to reduce the number further, lowering construction and maintenance costs. The hybrid architecture exhibits flexibility in adapting to different vehicle traffic conditions. Either the number of access points or the hop limit can be adjusted for adaptation. Let us consider an example when the penetration ratio is 0.4, it is assumed that there are 7 access points, and the hop limit is 2. Afterwards, vehicle traffic conditions change, and the instrumented vehicle density changes as well, i.e., the penetration ratio is increased from 0.4 to 0.6. In this example the system has two options: (1) it can notify vehicles to change the hop limit to 3, or (2) it can increase the number of access points to 9.

Tradeoff Between Cost and Performance

Designing a hybrid vehicular network, consisting of a fixed and an ad-hoc network is a tradeoff between cost and performance. In addition, the changing vehicle traffic condition over time must be considered as well. WWAN can provide continuous connectivity, but is not able to support a large number of instrumented vehicles. Pure WWAN is only feasible when bandwidth requirements are very limited. The alternative approach is to place a sufficient number of WLAN access points along the road providing an easy-to-deploy high data rate access. However, this needs dense access points, resulting in high installation and maintenance costs. The number of WLAN access points or the load of the WWAN network can be reduced by allowing multi-hop forwarding.

Multi-hop forwarding can increase WLAN coverage but may cause reduced usable channel capacity due to interferences. Therefore, the increased coverage may not translate into increased user throughput when the network is heavily loaded. However, the introduction of multi-hop WLAN provides the necessary flexibility required for adapting to varying vehicle traffic conditions [51].

A WLAN hop limit should be defined based on data traffic load, vehicle density, and other factors (e.g., delay and routing overhead). Data traffic load determines whether the reduced capacity caused by multi-hop forwarding becomes a limiting factor. Vehicle density decides whether there are significant benefits in terms of coverage extension. If carefully deployed, the hybrid WWAN - WLAN architecture can provide continuous connectivity, increase user throughput, and reduce the number of WLAN access points and communication cost.

Conclusions

Traditional routing solutions defined for mobile ad hoc networks are not suitable for a vehicular environment, due mainly to the high speed of the nodes that results in a highly dynamic network topology. Therefore, specific routing and information strategies have to be defined, keeping in mind the special requirements of such a network setup. In this chapter we tried to summarize the main ideas along which different approaches were defined, presenting solutions based on opportunistic or trajectory-based forwarding, directional broadcast or gossiping. At the end of this chapter we have to conclude that in spite of the numerous existing solutions, there are still several open issues that have to be addressed. However, taking into account the importance of ensuring efficient and reliable vehicular communications, the research community will most probably put significant effort in the forthcoming years as well to handle all those problems.

About the Authors

Dr. Péter Laborczi received his M.Sc. and Ph.D. degrees in Computer Science from the Budapest University of Technology and Economics, in 1999 and 2002, respectively. Between 2002 and 2004 he worked as visiting researcher at Arsenal Research in Vienna, Austria in the framework of an Intelligent Transportation Systems related European Union Marie Curie Postdoctoral research program. Currently he is a senior research fellow at the Institute for Applied Telecommunication Technologies (Bay Zoltan Foundation for Applied Research). He has been involved in several national and European research projects focusing on ITS networks and applications. His research interests include many aspects of ITS, e.g., routing, network optimization, and applications like Floating Car Data and Adaptive Cruise Control.

Mr. Attila Török received the M.Sc. degree from the Technical University of Timisoara, Romania, in 1999. Between 1999 and 2003 he attended the doctoral school of the Budapest University of Technology and Economics (BME), Budapest, Hungary. He also participated in several European research projects on the topic of wireless mobile ad hoc networks. During 2003, he worked as a visiting researcher at the Korean Electronics Technology Institute (KETI) in Seoul, South Korea. Currently he is working as a researcher at the Bay Zoltan Foundation for Applied Research on the topic of Intelligent Transportation Systems. His research interests are in the field of wireless ad-hoc and vehicular networking protocols. Besides research issues, he is also focusing on the practical aspects of introducing a Floating Car Data (FCD) system in Budapest. Mr. Török and Dr. Laborczi also teach the undergraduate course titled “Infocommunications in Transportation Systems” at BME.

Mr. Miklós Máté is a PhD student at the Budapest University of Technology and Economics, where he previously obtained his MSc degree in 2007. He is also employed as a part time researcher at Ericsson Hungary. His current research interests are related to efficient information dissemination techniques in vehicular ad hoc networks

Dr. Rolland Vida is an associate professor at the Budapest University of Technology and Economics (BME), Department of Telecommunications and Media Informatics. He obtained his B.Sc. and M.Sc. degrees in Computer Science from the Babes-Bolyai University, Cluj Napoca, Romania in 1996 and 1997 respectively, and his Ph.D. degree from the Université Pierre et Marie Curie, Paris, France, in 2002. He has recently managed or participated in several national and European research projects on the topic of wireless ad hoc networks and intelligent transportation systems. He has also served as a TPC Member for VTC 2007, the 65th IEEE Vehicular Technology Conference that was held in Dublin, Ireland, in April 2007. Dr. Vida has taught since 1996 several networking courses at universities in Romania, Slovakia and Hungary.

References

[1] M Abolhasan, T Wysocki, E Dutkiewicz, „A review of routing protocols for mobile ad hoc networks", *Elsevier Ad Hoc Networks Journal*, Vol. 2, No. 1., pp. 1-22. January 2004.

[2] M. Bechler, L. Wolf, O. Storz, W.J. Franz, "Efficient discovery of Internet gateways in future vehicular communication systems", in *Proceedings of the 57th IEEE Semiannual Vehicular Technology Conference*, VTC 2003-Spring, Vol.2, pp. 965-969, April 2003.

[3] P. Bose, P. Morin, I. Stojmenovic, J. Urrutia, "Routing with guaranteed delivery in ad hoc wireless networks", in *Proceedings of DIALM '99, the 3rd International Workshop on Discrete Algorithms and Methods for Mobile Computing and Communications,* Seattle, Washington, United States, pp. 48-55, 1999.

[4] Brickley, et al., "A Data Dissemination Strategy for Cooperative Vehicular Systems", in *Proceedings of the 65th IEEE Vehicular Technology Conference*, VTC-Spring 2007, Dublin, Ireland.

[5] N. Bulusu and S. Jha, "Wireless Sensor Networks", *Artech House,* 2005, ISBN 1-58053-867-3.

[6] J. Burgess, et al., "MaxProp: Routing for Vehicle-Based Disruption-Tolerant Networks", in *Proceedings of IEEE Infocom 2006,* Barcelona, Spain, April 2006.

[7] M. D. Dikaiakos, et al., "Location-Aware Services over Vehicular Ad-Hoc Networks using Car-to-Car Communication", in IEEE *Journal on Selected Areas in Communications*, Vol. 25, No. 8, October 2007.

[8] A. Eriksen et al, "D.CVIS.3.1 Reference Architecture", 2006.

[9] E. Fasolo, R. Furiato, A. Zanella. "Smart broadcast algorithm for inter-vehicular communications", in *Proceedings of WPMC'05, International Symposium on Wireless Personal Multimedia Communications,* Alborg, Denmark, September 2005.

[10] H. Fuessler, H. Hartenstein, J. Widmer, M. Mauve, and W. Effelsberg, "Contention-based forwarding for street scenarios", in *Proceedings of WIT 2004, the 1st International Workshop in Intelligent Transportation,* Hamburg, Germany, pp. 155-159, March 2004.

[11] H. Fuessler, J. Widmer, M. Kasemann, M. Mauve, and H. Hartenstein, "Contention-based Forwarding for Mobile Ad-hoc Networks", *Elsevier's Ad Hoc Networks*, vol. 1, pp. 351-369, November 2003.

[12] S. Goel, T. Imielinski, K. Ozbay, "Ascertaining viability of WiFi based vehicle-to-vehicle network for traffic information dissemination," *Proceedings of the 7th International IEEE Conference on Intelligent Transportation Systems*, pp. 1086-1091, October 2004.

[13] J. Haas, "A new routing protocol for the reconfigurable wireless networks", in *Proceedings of the 6th IEEE International Conference on Universal Personal Communications*, San Diego, CA, USA, pp. 562-566, October 1997.

[14] Z. J. Haas, J. Deng, and S. Tabrizi, "Collision-free Medium Access Control Scheme for Ad-hoc Networks", in *Proceedings of IEEE MILCOM* 1999.

[15] H.-Y. Hsieh, R. Sivakumar, "Internetworking WWANs and WLANs in Next Generation Wireless Data Networks", in *Proceedings of 3G Wireless and Beyond*, San Francisco, CA USA, May 2002.

[16] E. Huang, W. Hu, J. Crowcroft, I. Wassell, „Towards Commercial Mobile Ad Hoc Network Applications: a Radio Dispatch System", in *Proceedings of MobiHoc'05, the 6th ACM International Symposium on Mobile Ad Hoc Networking and Computing*, pp. 355-365, 2005.

[17] D. Johnson, D.A. Maltz, "Dynamic source routing in ad hoc wireless networks", in Mobile Computing (T. Imielinski and H. Korth, eds.), Kluwer Academic Publishing, 1996.

[18] B.N. Karp, H.T. Kung, "GPSR: Greedy Perimeter Stateless Routing for Wireless Networks", in Proceedings of MobiCom'00, the *6th Annual ACM/IEEE International Conference on Mobile Computing and Networking, pp.* 243-254, August 2000.

[19] H. Kim, D.J. Lovell, Y.-S. Kang, W. Kim, "Data Quantity for Reliable Traffic Information in Vehicular Ad Hoc Networks", in *Proceedings of Vehicular Technology Conference, VTC-2007 Fall*, pp.1401-1405, 2007.

[20] T. King, H. Fuessler, M. Transier, and W. Effelsberg, "Dead-reckoning for Position-based Forwarding on Highways", in *Proceedings of WIT'06, 3rd International Workshop on Intelligent Transportation,* Hamburg, Germany, pp. 199-204, March 2006.

[21] G. Korkmaz, E. Ekici, F. Özgüner, and Ü. Özgüner, "Urban Multi-hop Broadcast Protocol for Inter-vehicle Communication Systems", in *Proceedings of VANET '04, 1st ACM International Workshop on Vehicular Ad hoc Networks*, New York, NY, USA, pp. 76-85, 2004.

[22] G. Korkmaz, E. Ekici, and F. Özgüner, "An efficient fully ad-hoc multi-hop broadcast protocol for inter-vehicular communication systems", in *Proceedings of ICC'06, IEEE International Conference on Communications,* Istanbul, Turkey, pp. 423-428, June 2006.

[23] F. Kuhn, R. Wattenhofer, Y. Zhang, A. Zollinger, „Geometric Ad-hoc Routing: Of Theory and Practice", in *Proceedings of ACM PODC,* Boston, MA, USA, July 2003.

[24] P. Laborczi, et al., "Optimal Route Guidance Based on Floating Car Data", 10th World Conference on Transport Research, Istanbul, Turkey, July 2004.

[25] P. Laborczi, et al., "Vehicle-to-Vehicle Traffic Information System with Cooperative Route Guidance", in *Proceedings of the 13th World Congress on Intelligent Transport Systems*, London, United Kingdom, October 2006.

[26] A. Leontiadis, C. Mascolo, "Opportunistic Spatio-Temporal Dissemination System for Vehicular Networks", in *Proceedings of MobiOpp'07*, San Juan, Puerto Rico, USA, June 2007.

[27] C. Liu, J. Kaiser, „A Survey of Mobile Ad Hoc Network Routing Protocols", Technical Report, University of Magdeburg, 2005.

[28] G. Liu, B.-S. Lee, B.-C. Seet, C.H. Foh, K.J. Wong, and K.-K. Lee, "A routing strategy for metropolis vehicular communications", in *Proceedings of ICOIN'04, the International Conference on Information Networking,* Busan, Korea, pp. 134–143, February 2004.

[29] C. Lochert, et al., "A routing strategy for vehicular ad hoc networks in city environments" in *Proceedings of IEEE Intelligent Vehicles Symposium (IV2003),* pp. 156–161, June 2003.

[30] C. Lochert, et al., "Geographic routing in city scenarios", *ACM SIGMOBILE Mobile Computing and Communications Review (MC2R),* vol. 9, no. 1, pp. 69–72, January 2005.

[31] H. Luo, R. Ramjee, P. Sinha, L. Li, S. Lu, "UCAN: A Unified Cellular and Ad-Hoc Network Architecture", in *Proceedings of MobiCom'03, the Ninth Annual International Conference on Mobile Computing and Networking*, San Diego, USA, September 2003.

[32] M. Máté, R. Vida, "Probability-based information dissemination in urban environments", in *Proceedings of Eunice 2008*, Brest, France, September 2008.

[33] G. Muhl, A. Ulbrich, K. Herrmann, and T. Weis, "Disseminating Information to Mobile Clients Using Publish-Subscriber," *IEEE Internet Computing*, 2004.

[34] S. Murthy, J.J. Garcia-Luna-Aceves, "An Efficient Routing Protocol for Wireless Networks", ACM Mobile Networks and Applications *Journal, Special Issue on Routing in Mobile Communication Networks*, pp. 183-197, October 1996.

[35] V. Naumov, R. Baumann, T. Gross, "The Study of Inter-Vehicle Ad Hoc Networks on Realistic Vehicular Traces", in *Proceedings of ACM MobiHoc,* Florence, Italy, May 2006.

[36] S. Ni, Y. Tseng, Y. Chen, J. Sheu, "The Broadcast Storm Problem in a Mobile Ad hoc Network", in *Proceedings of ACM Mobicom,* pp. 151-162, Seattle, Washington, 1999.

[37] N. Nikaein, C. Bonnet, N. Nikaein, „HARP - Hybrid Ad Hoc Routing Protocol", in Proceedings of IST'2001, *International Symposium on Telecommunications*, Tehran, Iran, 2001.

[38] N. Nikaein, H. Laboid, C. Bonnet, „Distributed dynamic routing algorithm (DDR) for mobile ad hoc networks", in *Proceedings of MobiHOC'2000, First Annual Workshop on Mobile Ad Hoc Networking and Computing*, 2000.

[39] J. Ott, D. Kutscher, "A Disconnection-Tolerant Transport for Drive-thru Internet Environments", in *Proceedings of IEEE Infocom'05*, Miami, March 2005.

[40] J. Ott, D. Kutscher, "Why Seamless? Towards Exploiting WLAN-based Intermittent Connectivity on the Road," in *Proceedings of the TERENA Networking Conference*, TNC 2004, Rhodes, June 2004.

[41] G. Pei, M. Gerla, T.-W. Chen, "Fisheye State Routing in Mobile Ad Hoc Networks", in Proceedings of the ICDCS'2000 Workshops, Taipei, Taiwan, pp. D71-D78, April 2000.

[42] C.E. Perkins, P. Bhagwat, "Highly dynamic Destination-Sequenced Distance-Vector Routing (DSDV) for mobile computers", *ACM Computer Communication Review,* Vol. 24, No.4, pp. 234-244, October 1994.

[43] C.E. Perkins, E.M. Royer, "Ad hoc on demand Distance Vector routing, mobile computing systems and applications", in *Proceedings of the 2nd IEEE Workshop on Mobile Computing Systems and Applications*, New Orleans, LA, pp. 90-100, February 1999.

[44] G. Resta, P. Santi, and J. Simon, "Analysis of multi-hop emergency message propagation in vehicular ad hoc networks", in *Proceedings of MobiHoc '07, the 8th ACM International Symposium on Mobile Ad hoc Networking and Computing, New* York, NY, USA, pp. 140-149, 2007.

[45] D. Sormani, et al., "Towards Lightweight Information Dissemination in Inter-Vehicular Networks", in *Proceedings of ACM VANET'06*, Los Angeles, California, USA, September 2006.

[46] A. Stampoulis, Z. Chai, "Survey of Security in Vehicular Networks", *Project CPSC 534*.

[47] Q. Sun, H. Garcia-Molina, "Using ad-hoc inter-vehicle networks for regional alerts", Technical report, Department of Computer Science, Stanford University, 2005.

[48] J. Tian, L. Han, K. Rothermel, "Spatially Aware Packet Routing for Mobile Ad Hoc Inter-Vehicle Radio Networks", in *Proceedings of IEEE Intelligent Transportation Systems*, vol. 2, pp. 1546-1551, Shanghai, China, October 2003.

[49] A.Torok, P. Laborczi, G. Gerhath, "Constrained Dissemination of Traffic Information in Vehicular Ad Hoc Networks", accepted to the *68th IEEE Vehicular Technology Conference, VTC2008-Fall*, Calgary, Canada, September 2008.

[50] L. Wischhof, et al., "Adaptive Broadcast for Travel and Traffic Information Distribution Based on Inter-Vehicle Communication", in *Proceedings of IEEE Intelligent Vehicles Symposium*, June 2003.

[51] H. Wu, "Analysis and Design of Vehicular Networks", PhD Dissertation, Georgia Institute of Technology, 2005.

[52] H. Wu, et al., "MDDV: a Mobility-centric Data Dissemination Algorithm for Vehicular Networks", in *Proceedings of ACM VANET'04, 1st ACM International Workshop on Vehicular Ad Hoc Networks*, Philadelphia, USA, October 2004.

[53] H. Wu, R. Fujimoto, M. Hunter, R. Guensler, "An Architecture Study of Infrastructure-base Vehicular Networks", in *Proceedings of MSWiM'05, the 8th ACM International Symposium on Modeling, Analysis and Simulation of Wireless and Mobile Systems,* Montreal, Canada, October 2005.

[54] R.-M. Yadumurthy, Adithya C. H., M. Sadashivaiah, and R. Makanaboyina, "Reliable mac broadcast protocol in directional and omni-directional transmissions for vehicular ad hoc networks", in *Proceedings of VANET'05, the 2nd ACM International Workshop on Vehicular Ad hoc Networks*, New York, NY, USA, pp. 10-19, 2005.

[55] J. Zhao, G. Cao, "VADD: Vehicle-Assisted Data Delivery in Vehicular Ad Hoc Networks", in *Proceedings of IEEE Infocom 2006*, Barcelona, Spain. April 2006.

[56] J. Zhao, Y. Zhang, G. Cao, "Data Pouring and Buffering on the Road: A New Data Dissemination Paradigm for Vehicular Ad Hoc Networks," *IEEE Transactions on Vehicular Technology,* vol.56, no.6, pp.3266-3277, November 2007.

[57] The FleetNet project: http://www.et2.tu-harburg.de/fleetnet/

[58] The COMeSafety project: http://www.comesafety.org/

[59] Car-to-Car Communication Consortium: http://www.car-to-car.org/

[60] The PATH project: http://www.path.berkeley.edu/

[61] The CVIS project: http://www.cvisproject.org/

In: Wireless Technologies in Intelligent Transportation… ISBN: 978-1-60741-588-6
Editors: Ming-Tuo Zhou et al, pp. 361-392

Chapter 13

ROAD TRAFFIC ESTIMATION USING CELLULAR NETWORK SIGNALING IN INTELLIGENT TRANSPORTATION SYSTEMS

David Gundlegård and Johan M. Karlsson
Department of Science and Technology
Linköping University, 601 74 Norrköping, Sweden

Abstract

In the area of Intelligent Transportation Systems the introduction of wireless communications is reshaping the information distribution concept, and is one of the most important enabling technologies. The distribution of real-time traffic information, scheduling and route-guidance information is helping the transportation management systems in their strive to optimize the system. The communication required to transfer all this information is rather expensive in terms of transmission power, use of the scarce resources of frequencies and also the building of an infrastructure to support the transceivers. By using information that already exists and is exchanged within the infrastructures of the GSM and UMTS networks, a lot of the resource problems are solved. The information that could be extracted from these cellular networks could be used to obtain accurate road traffic information to support real-time traffic information. In this way the cellular networks not only becomes the means to distribute information but also a source of road traffic information.

From the analysis made it is obvious that the potential of retrieving valuable road traffic information from cellular systems in a cost efficient way, i.e. by using already existing signaling data, is very high. It has however not been clear what to expect from these systems in terms of accuracy, availability and coverage. In this chapter the basics for this is laid out and discussed in detail. A practical trial has also been performed and the results show clearly the potential as well as the differences in using the GSM compared to the UMTS network. The advantages and drawbacks are discussed and backed up by real measurements in two different road environments. The main advantages of using the existing signaling data, i.e., passive monitoring compared to active monitoring where the terminal sends extra data is discussed and could be summarized in three components, no user acceptance is necessary, no extra signaling is necessary and it does not drain the terminal battery.

In the future it is likely that vehicles need to communicate more frequently with each other and with some kind of traffic control centre. This traffic will also be very useful in order to estimate road traffic information using the signaling information obtained from the cellular system. However, the enhanced communication systems will also change traffic patterns in the cellular networks which will affect the potential of estimating road traffic from cellular systems. The evolvement indicates that the terminals will be in active state almost constantly, and hence the updating information will be more frequent and the information more accurate.

Abbreviations

A-GPS	Assisted GPS
AOA	Angle of Arrival
BSC	Base Station Controller
BTS	Base Transceiver Station
CDMA	Carrier Division Multiple Access
CGI	Cell Global Identity
CN	Core Network
DCH	Dedicated Channel
DOA	Direction of Arrival
E-OTD	Enhanced Observed Time Difference
FACH	Forward Access CHannel
FCC	Federal Communications Commission
FCD	Floating Car Data
GPRS	General Packet Radio Service
GPS	Global Positioning System
GSM	Global System for Mobile communication
HLR	Home Location Register
ITS	Intelligent Transportation Systems
LA	Location Area
LMU	Location Measurement Units
MAHO	Mobile Assisted Handover
MEHO	Mobile Evaluated Handover
MM	Mobility Management
O/D	Origin-Destination
OTDOA	Observed Time Difference of Arrival
PCH	Paging Channel
RA	Routing Area
RNC	Radio Network Controller
RRM	Radio Resource Management
RSS	Received Signal Strength
RTT	Round Trip Time
RXLEV	Received Signal Strength
RXQUAL	Received Signal Quality
SACCH	Slow Associated Control Channel
SFN	System Frame Number
SGSN	Serving GPRS Support Node
SIR	Signal to Interference Ratio
TA	Timing Advance

TDMA	Time Division Multiple Access
TDOA	Time Difference of Arrival
TEMS	Telecommunications Management System
TOA	Time of Arrival
TTFF	Time To First Fix
UMTS	Universal Mobile Telecommunications System
URA	UTRAN Registration Area
UTRAN	UMTS Terrestrial Radio Access Network
WCDMA	Wideband-CDMA

1. Introduction

Wireless communication is one of the most important enabling technologies for Intelligent Transportation Systems. By using different kinds of wireless network technologies it is possible to distribute real-time traffic information, scheduling, route-guidance information etc. All wireless communication networks require signaling in order to support mobility management and radio resource management. This signaling information can in cellular networks, such as GSM and UMTS, also be used to estimate road traffic state information. This section describes the cellular networks potential as a source of road traffic information instead of means to distribute it.

Although the technology of using cellular networks to estimate road traffic information has been subject for analysis for quite some time, it is still far from being mature. It is not clear what to expect from these systems in terms of accuracy, availability and coverage. The potential of the system is although quite clear; it is possible to retrieve a lot of traffic data in a cost efficient way, i.e. by using existing signaling data without the need to invest time and resources in a sensor infrastructure. This is essential, especially in terms of using the existing scarce resource of frequencies and cost of building a new infrastructure.

Evolving communication systems will change communication patterns, and both the design of the communication system and the communication pattern of the users will affect the potential of estimating road traffic information from cellular systems. In order to realise a safer and more efficient transportation system, it is likely that vehicles need to communicate more frequently with each other and with some kind of traffic control centre. This traffic will also be very useful in order to estimate road traffic information in combination with the signaling information obtained from the cellular networks. Increasing the communication to and from vehicles will generate signaling traffic that in turn can be used to improve the traffic information available. Once there is an improvement in traffic information, there will be more communication to and from vehicles. Following this positive spiral, shown in *Figure 1*, is a potential path towards a large improvement in available real-time road traffic information.

Two fundamental characteristics are important for the development of road traffic information from cellular networks. First of all is the spatial correlation between road network and cellular network, i.e. where the vehicle tracking is challenging due to dense road network, the cellular location data has higher quality due to much smaller cells. Secondly the correlation in time between congestion and when people use their cell phones is high, i.e. when there is a lot of road traffic, there is also a lot of cellular traffic available. Outside the cities the road networks and cellular networks are sparser, see *Figure 2*. Furthermore, the requirements of QoS by the mobile operator are increasing, which means that the operator will have to keep track of the cell phones in more detail in the future. A drawback of this technology is that the objective of the mobile operator is to minimize the signaling traffic generated by cell phones, in order to use the network more efficient. However, this objective might shift slightly if the road traffic information becomes commercially demanded by subscribers of the operator.

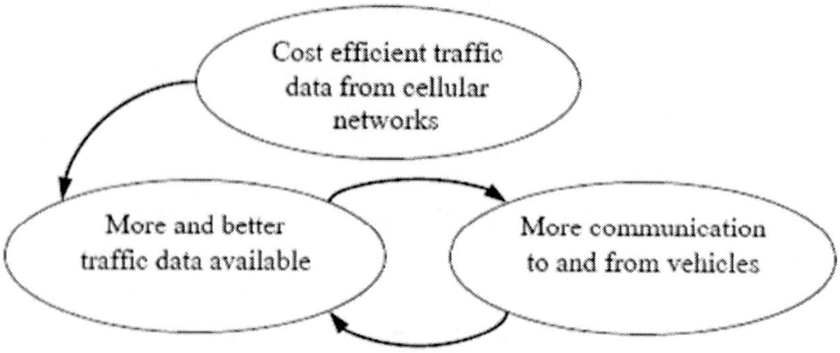

Figure 1. The positive spiral of available road traffic information from cellular networks.

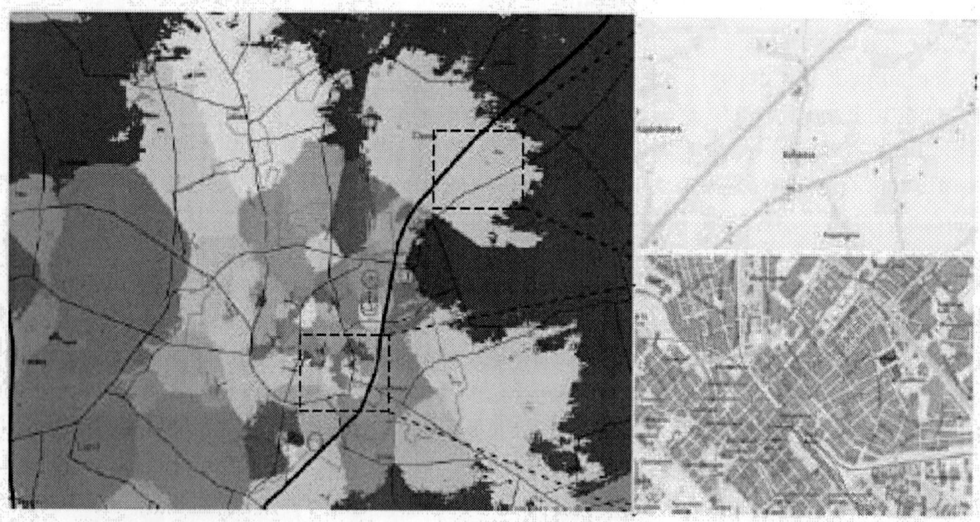

Figure 2. Illustration of a cellular network in Sweden, where it can be seen that the cells are small where the road network is dense and sparse outside the cities.

2. Background

The general concept of public system architectures for mobile communication systems is a cellular network concept. The cellular networks are built based upon a number of small geographical areas, cells, covering a larger area. Each cell consists of a base station transmitting the information within the limited area. For analytical purposes these areas are usually hexagonal cells, which totally covers an area and are rather similar to real transmission patterns. *Figure 3* shows a road segment and the cells covering the area as an example of how a road segment crosses several cells in its path.

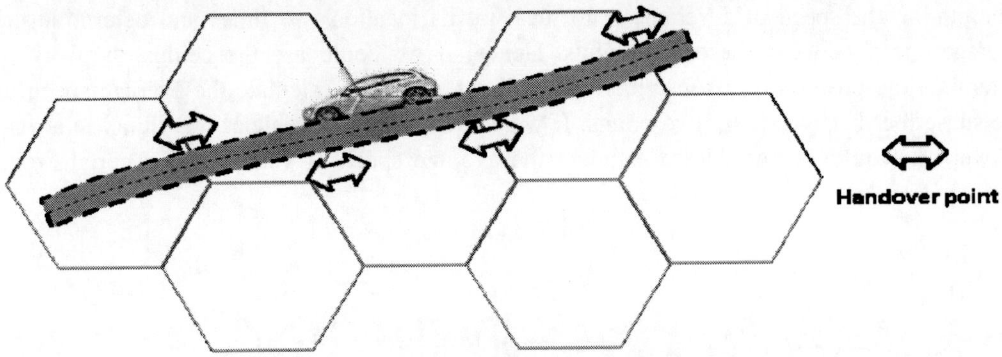

Figure 3. Travelling on a road segment involves a large number of cell border crossings. The handover between the cells are used to estimate traffic information and can be seen as virtual detectors.

The system architectures for both GSM and UMTS are rather similar in their concept. An overview of the system architectures are to be found in *Figure 4*. A GSM Base Transceiver Station (BTS) holds the transmit and receive equipment for one or more cells. It constitutes the interface between the network provider and the mobile phone. The Base Station Controller (BSC) administers the transmit and receive resources of the connected base stations. There are two categories of channels, signaling and traffic channels. Both the signaling channels and the traffic channel (handling the actual payload) are processed here. The concept for UMTS is rather similar, the base station is here called Node B and the device controlling a number of these cells is denoted Radio Network Controller (RNC). *Figure 4* also provides the interfaces between different sections of the architecture. The interface between the BTS and BSC in GSM is denoted A_{bis}, and the corresponding interface in UMTS is denoted I_{ub}. The interface between the GSM systems radio parts and the core network (CN) side is denoted A and the corresponding interface for UMTS is denoted I_u.

In the core network we find the devices that connect to other networks. If the communication should go through the circuit switched network side, it uses the Mobile Switching Centre (MSC). The MSC carries out all the duties of an ordinary wireline network switch, such as processing, finding a path and supplementary services. It is also the link between the wireless networks and the wireline network. If the communication is more data oriented it would go through the packet switched network side and hence through the Serving GPRS Support Node (SGSN). In *Figure 4* both devices are connected to the "network", it should be pointed out that this is just a simplification and is in reality a number of different networks with different network features and service level agreements.

The purpose of a cellular communication system is to offer mobile communication to subscribers of the system. In order to do this the mobile operator has to keep track of where in the network a certain subscriber is located. The location is used to reach the subscriber if it has incoming data transfers and to assign the subscriber to the most appropriate radio base station. To be able to handle the mobile subscribers the operator store a number of details of each user, this is stored in the Home Location Register (HLR). Information provided by the

HLR is for example if the subscriber is active or not, and if active in which part of the network the subscriber is situated at the moment.

The location within the network can with knowledge of the network structure be transformed to a position of the subscriber, i.e. coordinates. The basic operation of determining the speed of a vehicle is to measure its location two times and determine the average speed between these two points. Hence, if we could use the cellular network to determine the position of a subscriber at two times we could calculate the average speed of the subscriber between these two points. If we could do this for vehicles travelling on a road network it would be possible to estimate the average speed of vehicles on a certain road section.

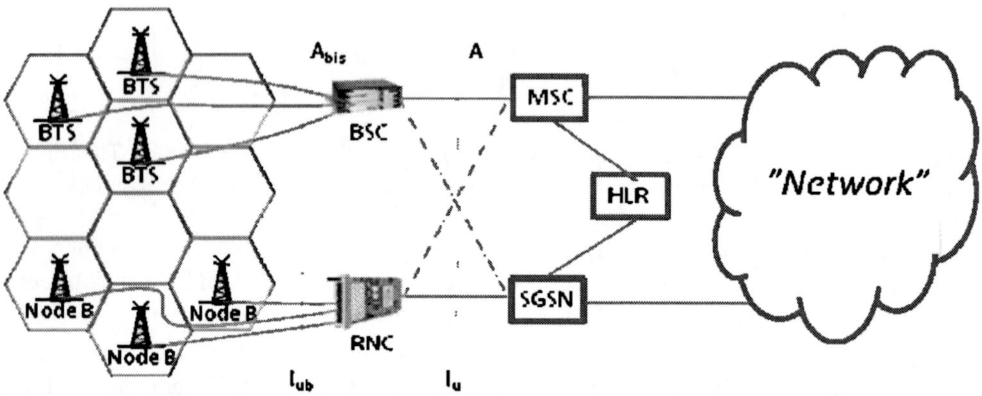

Figure 4. The cellular network architecture for the GSM (upper part) and UMTS (lower part) and their connection to the core network through the A and Iu interface, respectively.

Estimating travel times by analysing traffic from mobile phones has been subject for research since mid 1990´s. This was at a time when telematics services were supposed to expand rapidly within a few years. The cellular networks gave a possibility to communicate with the vehicles, and now it was possible to use it as a source of traffic information too. The foreseen rush of telematics services failed to come through and the willingness to pay for such services was much lower than expected. With a rather slow telematics market the interest in traffic information generated by cellular networks was reduced. One of the major problems with the telematics market was the lack of quality traffic information and this is still one of the main issues. Generating traffic information from cellular networks could be an important step in increasing the efficiency in road traffic information systems.

The task of determining a detailed position of a user in a cellular network without using dedicated positioning equipment, e.g. a GPS receiver, is challenging. Even with the most sophisticated positioning methods available today and perfect conditions the accuracy will likely not be better than 20 meters. To achieve this relatively high accuracy it is important to keep in mind that scarce bandwidth resources are used every time a position is calculated. However, the possibilities in estimating road traffic information from cellular networks roots from the assumptions that we are only interested of positions on the road network, several measurements are available for each user and a lot of users are travelling a given road section. The first assumption makes map matching possible, which increases the accuracy significantly [1]. The second assumption gives the possibility to determine a specific route for a user. The third assumption gives statistical credibility to the data.

One of the main advantages using cellular networks for positioning is the possibility to choose the positioning points. The chosen points should be unique and stable over time in order for the algorithm to perform well. The algorithm could choose a handover point, i.e., a position where the mobile is switched from one cell to the next as it moves along a path, or a point where a certain cell has a dominating received transmission power. These positioning points are relatively easy to detect and hence a good choice to base a speed or route calculation on. However, there will not always be one of these good behaving positioning points at each end of a road section, in that case a mapping between the measured road section and the target road section is needed. The mapping could be solved by adjusting the target road sections to the available cellular data, creating new positioning points using measurement reports or simply extrapolating measured road sections to target road sections. *Section 4* elaborates this issue further.

3. Cellular Network Positioning

The availability of positioning data for cellular mobile users is different depending on network, the state of the terminal, the users' preferences and the geographical location. In this section we will elucidate some of the main properties of positioning in cellular networks. It will be based on objective facts and its merits and drawbacks will be derived on in later sections. More details in GSM and UMTS specific characteristics can be found in the referenced 3GPP standards [2-7]. For an extensive treatment of general positioning technologies and methods, see e.g. [8-10].

A standard classification of cellular positioning methods is to relate the method to which extent the network and the terminal is involved in the positioning process. The different categories are network-based, handset-based and handset-assisted.

- *Network-based positioning.* Positioning within this category rely on measurements and calculations made by nodes in the cellular network. This implies that all positioning related functionality is located in the network and that all the processing is carried out by the network. An important result of this characteristic is that the network-based positioning methods supports legacy terminals, i.e. no changes has to be made to existing terminals. Network-based positioning implies multilateral positioning, i.e. the terminal transmits data that is received by multiple network nodes. Hence, a drawback of network-based positioning is that the terminal needs to be in active mode in order for a position to be calculated. Since the cellular network determines the position of the terminal, it also falls into the category of remote positioning.
- *Handset-based positioning.* These kind of positioning methods rely on positioning measurements and calculations made by the terminal. This implies that terminals are required to be updated with positioning functionality that is not supported in legacy terminals. The main advantage with handset-based positioning is that the terminal does not need to transmit in order to calculate a position, instead unilateral positioning is used, i.e. the terminal receives signals sent by multiple network nodes

and uses these for positioning. Since the terminal calculates its own position, handset-based positioning can also be referred to as a self positioning method.

- *Handset-assisted positioning.* Sometimes it is more convenient to use a unilateral approach and let the terminal report measurements to the network, where the position is calculated. If standardised measurements are used, e.g. signal strength or timing advance, legacy terminals can be used for positioning and additional functionality is not necessary in the radio access part of the network. Like network-based positioning, this is regarded as a remote positioning method.

It should be noted that remote positioning can be performed with handset-based methods by letting the terminal transmit the calculated position to the network, this is referred to as indirect remote positioning. Using the same idea, self positioning can be performed with network-based positioning, and this is referred to as indirect self positioning.

After making a distinction between where and how the measurements are performed, we will now focus on the methods used to make the positioning estimate. The more information provided the better the accuracy of the positioning estimate. The most trivial methods are based on the knowledge that the target is within a certain area. For example, once a base station register a terminal we know it is within the radio coverage of that base station (proximity sensing) or if it passes by certain radio device (signpost positioning) we know that it passed by a known position at a specific time. The signpost positioning requires a dedicated infrastructure, the denser the more accurate estimates. Another rather rough estimate is to use dead reckoning, used already by Columbus at sea. By approximating the direction and speed and apply this to the last known position we make an extrapolation which becomes the positioning estimate. This method is also known as inertial navigation. The next step is to use lateration, i.e. the range or range difference including two or more measurements. In circular lateration the distance to a reference position is known and hence we could draw a circle around this position. If we do that using one more known distance from another position we will get another circle with two intersections and hence two possible locations. The addition of a third measurement gives a unique position estimation in two dimensions. If we are only able to calculate the difference in time of arrivals of the same signal at several points we obtain a hyperbolic. The crossing of two or more hyperbolics becomes the positioning estimate (requires at least three reference stations). The advantage of using time differences is that no synchronisation between the terminal and reference station is required, on the other hand synchronisation between reference stations is required instead. These methods are also denoted Time of Arrival (TOA) and Time Difference of Arrival (TDOA), respectively.

Another method to estimate the position is to use angulation. This method is only applicable if either side is equipped with antenna arrays and hence able to detect from which approximate direction the signal is arriving. With several such sites, the terminals position estimate is restricted to a line that crosses both the target and the base station. The intersection of several of these lines becomes the positioning estimate. This method is sometimes also denoted Angle of Arrival (AOA) and Direction of Arrival (DOA). Pattern matching, or fingerprinting, is a potential solution where some other methods fails due to shadowing or gives poor results for other reasons. The methods are based upon the fact that there exists a database with known signal characteristics from different places, usually laid out as a grid. The measured signal is compared to the database and the most equal, by some definition, entry becomes the positioning estimate.

In addition to these methods several other methods based upon range calculation by time or received signal strength (RSS) are to be found. They are, however, mostly special cases of the methods mentioned is this section. Another category of methods is hybrid approaches where the methods mentioned are used in different combinations.

There are a number of the positioning methods mentioned that are more or less standardised and in use today. For example the CGI-TA (Cell Global Identity-Timing Advance) gives information in which cell the terminal is situated and how far from the base station, i.e., a circle on which the terminal could be at any point. In the case the cell is divided into sectors we will also receive in which sector of the cell the terminal is situated. The precision of this method is coarse, however, all information already exist in the network. The corresponding method in UMTS is Cell ID-RTT (Round Trip Time). Another used method is Enhanced Observed Time Difference (E-OTD), which is based on unilateral TDOA positioning. E-OTD requires the involved three base stations to be synchronized and additional hardware in the network as well as updated cell phones. The UMTS method corresponding to E-OTD is called Observed Time Difference of Arrival (OTDOA). The multilateral versions of E-OTD and OTDOA are called Uplink TDOA (U-TDOA) in both GSM and UMTS. The TDOA-based positioning methods typically have better accuracy than e.g. CGI-TA, however, more processing, updated cell phones and new hardware is required. Also Assisted GPS (A-GPS) is standardised in GSM and UMTS for high precision positioning, these standards rely on a GPS receiver in the cell phone. More information regarding these methods can be found in e.g. [6-7].

While knowing all these methods the next questions would be the performance metrics of the position estimate. There exist several criteria for evaluating the quality of the performed estimate. Obviously, most important is accuracy and precision of the estimates. These two terms should not be mixed up or used interchangeably as often could be seen. The accuracy defines how far away from the real position the estimate is, while the precision defines how close a number of estimates is to their mean value. This means, as an example, that the precision could be high but with very low accuracy if all estimates are close to each other but far away from the real position.

An often forgotten, but very essential, performance metric category is the practical ones while comparing the different methods. The *overhead* of performing positioning could almost entirely be divided into signaling and computational overhead, where the former reflects the information that has to be sent over the scarce air interface resource while the latter refers to the processing power to calculate the positioning estimate. The *coverage* is the definition in which environments the positioning estimates should be able to be delivered, such as indoor, rural or urban environment. Some services require positioning estimates in all environments while other is restricted to specific environments. Other practical performance metrics are *power consumption*, especially on the terminal side which has limited battery resources, *latency*, i.e., the time it takes to go through the entire positioning process until an estimate is delivered. The first time this is done this is referred to as Time To First Fix (TTFF), which could be more or less essential depending on the service. Finally, the *cost* also have to be taken into account, the cost could be divided into infrastructure cost to install the system and operational cost to run the system.

Positioning in cellular networks is limited to the information that is available to collect from the network. The GSM network was designed without positioning in mind, which makes the positioning related information limited. The positioning related information that is available in GSM and UMTS is described in the coming sections. A very good overview can also be found in [9].

Location Data in GSM

The backbone in an efficient travel time estimation system is analysis of signaling data generated by users in busy state, i.e. during voice calls or data sessions. This signaling data generated by busy state terminals is in GSM handled by Radio Resource Management (RRM) algorithms located in the radio access network. Complementary data can be obtained from positioning functions in the network (active monitoring) or signaling data generated by idle state terminals. Signaling data of idle state terminals is handled by Mobility Management (MM) algorithms located in the core network.

RRM is only active when the terminal is in busy state and an important task of RRM is to initiate handover. The Base Station Controller (BSC) is responsible for the handover decision and use information from measurement reports sent by the terminal and the current Base Transceiver Station (BTS). This information is very useful in the process of tracking a terminal. The terminal and the BTS repeatedly send information about received signal strength (RXLEV) and signal quality in bit error rate (RXQUAL). The fields are 6 bits long and correspond to a resolution of 64 discrete values. The terminal measures the signal quality and strength on the downlink and the BTS measures the signal quality and strength on the uplink. Based on the neighbouring list that is broadcasted by the BTS, the terminal tunes in to neighbouring cells and measures the signal strength. From the terminal, measurement reports are sent on the Slow Associated Control Channel (SACCH) once every 480 ms, the BTS adds the uplink measures and forwards a measurement result to the BSC.

Due to the propagation delay from the terminal to the BTS the terminal has to start its transmission earlier in order to avoid interference on adjacent timeslots. When the terminal shall start its transmission the delay is calculated in the BTS and the terminal is informed via a timing advance (TA) value that is sent on the SACCH to the terminal. The TA field is 6 bits long and corresponds to a resolution of 550 m. The TA value can also be used by the BSC in the handover decision and is included in the measurement report from the terminal. The BSC can use the TA value to roughly estimate the terminals velocity and, if a hierarchical cell structure is used, assign highly mobile terminals to a cell on a higher level. The TA value is also important for the BSC to complement the signal strength measurements in order to determine to which cell the terminal should be handed over.

When the terminal is in idle mode, i.e. powered on but not used for voice calls, data sessions or signaling, MM algorithms in the core network keep track of in which part of the network the terminal is located. The location information of a terminal in idle mode is much sparser and has a resolution of Location Area (LA), which consist of a configurable number of cells. The mobile terminal sends an LA update message when it detects a new LA identity broadcasted by the currently strongest BTS. During the LA update the terminal goes into busy state and more location information can be retrieved during a short period of time. A detailed description of GSM MM and RRM can be found in e.g. [11]. The relation between standardised location data reports and magnitudes of sampling distances are shown in *Figure 5.*

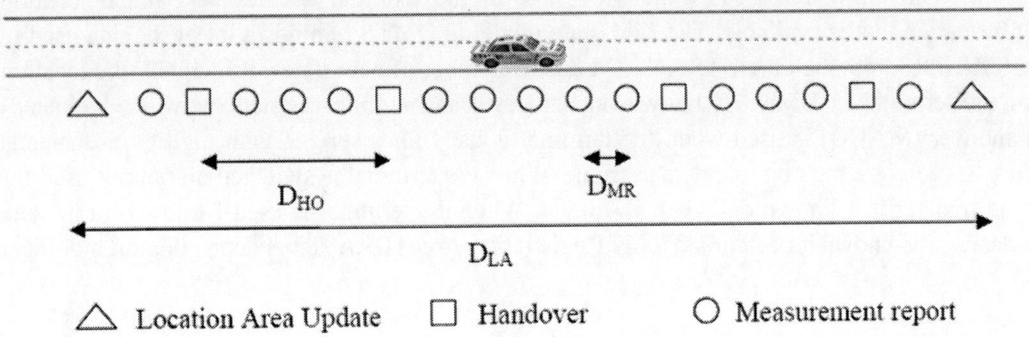

Figure 5. Distance between location data reports in GSM. D_{LA} is the distance between location area updates, magnitude from several km up to several tens of km. D_{HO} is the distance between handovers, magnitude from several hundreds of meters up to several kilometers (several tens of km in rural areas). D_{MR} is the distance between measurement reports, magnitude from several meters up to several hundreds of meters.

A GPRS-attached terminal does also generate location data, the information is however slightly different from circuit switched GSM data. When the terminal is attached to the GPRS network it can be in two mobility management states, stand-by and ready state. When the terminal is in ready state it can send and receive user data. A major difference between GPRS ready state, compared to circuit switched busy state, is that the terminal itself is responsible for which BTS to communicate with (mobile evaluated handover) [5]. The terminal listens to neighbouring cells during packet transfer and decides if it should stick with the current BTS or change to a better one. In ready state cell identity, TA-value and signal strength to the serving cell is useful location parameters. Since mobile evaluated handover is used in default mode, the terminal does not report signal strength to surrounding cells in ready state and this information cannot be used for tracking the terminal. However, the network can instruct the terminal to send measurement reports if necessary. In stand-by state the terminal is connected to the GPRS network, but is unable to send user data. In stand-by state the terminal only performs Routing Area (RA) updates. A RA comprises one or more cells and is comparable to, but not the same as LA, see *Figure 6* for a schematic view.

3G Differentiating Characteristics

As for GSM networks, the mobility of terminals in UMTS networks is handled by MM and RRM functions. MM and RRM are implemented in a similar manner in both systems, there are however a couple of fundamental differences. In UMTS RRM is solely handled by the UMTS Radio Access Network (UTRAN), this is achieved by connecting the Radio Network Controllers (RNC) with each other. Another important difference between the systems is that the support for Quality of Service (QoS) for different service classes in UMTS calls for more adaptive MM. This is solved by implementing MM functions not only in the core network, but also in UTRAN. More information regarding UMTS MM and RRM can be found in e.g. [12-13].

In both GSM and UMTS the MM state of the terminal decides how much location information that is available. The MM state model in UMTS reminds a lot of the one used in GSM/GPRS, although the UTRAN MM adds a number of new states. Principally the location of the terminal in UMTS is known on cell level and mobile assisted (network evaluated) handover (MAHO) is used when the terminal is used for a service with high QoS demands (e.g. speech or a high bit rate data session). When the terminal is switched on but not used for data transfer it is known on LA or RA level. When the terminal is used for low bit rate data transfer it is known on cell or UTRAN Registration Area (URA) level depending on mobility.

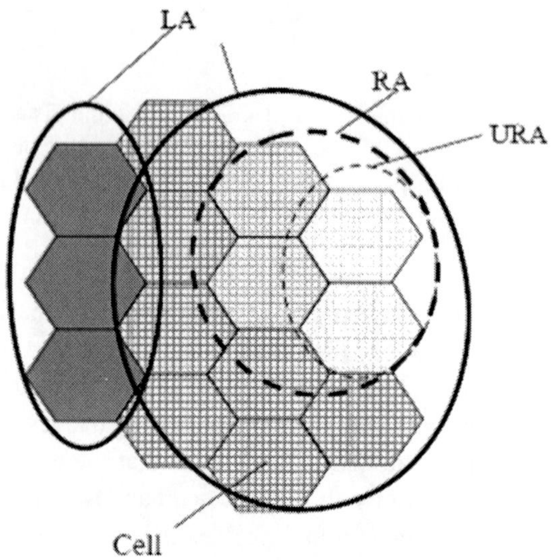

Figure 6. Relation between Location Area (LA), Routing Area (RA), UTRAN Registration Area (URA) and cell.

The location of the terminal is known in most detail when the terminal is used for circuit switched services or high speed data services, i.e. when the UTRAN MM state is Cell DCH (Dedicated Channel). In this state MAHO is used, which means that the terminal continuously reports data of the radio connection that can be used to locate the terminal. This state is similar to busy state of circuit switched GSM. The RRM of the two systems are however quite different, which leads to a number of important differences in available information. The main differentiating characteristics of UMTS RRM compared to GSM are:

- Handover control
- Time alignment
- Power control

These functions will affect the available information of a connection. The most important difference in handover control is the use of soft handover in UMTS. This means that the terminal is connected to several base stations at the same time and the location of the terminal can be determined in more detail. The terminal is expected to be in soft handover during 20-40% of the time [13]. Another difference in handover control is how the network makes the handover decision, i.e. the characteristics of the measurement reports that are sent from the terminal and the base station to the RNC.

In UMTS the periodic measurement report interval is configurable between 0.25 and 64 seconds, depending on radio environment and the state of the mobile terminal [14]. The frequency of event triggered reports are dependent of the frequency of actual events, e.g. a new radio link addition to the active set, but also on the operator configurable parameters time-to-trigger, hysteresis and offset value. More detailed information on UMTS measurement reports can be found in e.g. [3-4, 14]. Signal strength and quality of serving base station(s) are similar to the ones in GSM. The maximum number of surrounding base stations that can be measured is increased from 6 in GSM to 32 in UMTS. A TA value is not calculated in UMTS (WCDMA) networks since it is not a TDMA based system, but other time alignment measurements are available, e.g. round trip time and time difference between base stations [4].

The soft handover used in UMTS means that a terminal can be connected to several base stations simultaneously, whereas in GSM the terminal is only connected to one base station at the time. To track a vehicle, both measurement reports containing radio parameters and handover points can be used. When it comes to calculating travel times it is important to have two accurate estimations of the vehicle's position in order to make a good estimation of the travel time between those points. The handover points in GSM are a good candidate to estimate those positions. However, in UMTS the terminal will not change from one base station to another, instead radio links will be added to and removed from the terminals active set. As shown in [15], soft handover points in UMTS seem to be very promising to use for travel time estimation.

A potential problem with travel time estimation is the use of cell breathing, which probably will be more utilised in UMTS than in GSM. In cell breathing the size of the cells can change dynamically depending on the capacity need of different areas. An important tool to determine travel time is to measure the time between handover points, and if cell breathing is used the handover points will change with time. It will hence be more difficult to predict the handover points, but the pilot power of the base stations is known and can be used as input for the predictions.

An important function of the Code Division Multiple Access (CDMA) based systems is fast power control. In UMTS the inner loop power control makes it possible to adjust the terminal's power level 1500 times per second [13]. This can be compared to approximately two times per second in GSM. This means that the power level between the terminal and its serving base station(s) is measured at least 1500 times per second, and hence a massive collection of data is available. This data can be used to locate the terminal relative the connected base station(s) and be useful in the travel time estimation. However, a potential problem is that the inner loop power control is performed in the base station, which means that the information is not available in the RNC where it is viable to collect it. The RNC is though responsible for the outer loop power control, i.e. to control the target signal to interference ratio (SIR), which has a frequency of 10-100 Hz [13].

Reports to support handover and power control will be useful in order to locate the terminal according to relative received power level from different base stations. Power levels might not be the most efficient measurement to use when a terminal shall be located; this is due to fast and shadow fading. More often time differences are used to calculate the position of a terminal, which leads us to the time alignment comparison of UMTS and GSM. As described, time alignment is important in GSM and is managed with the TA-value calculated by the BTS and sent to the terminal. Since UMTS is a CDMA based system, the time alignment is not needed in order to avoid co-channel interference and is not implemented. However, similar time-alignment measurements are also available in UMTS.

During soft handover it is important to minimize the buffer needed in the terminal to combine the signal from the base stations. To do this, the terminal measure the time difference between the base stations and send this information to the network, which compensates for this by time alignment of the base station signals. The time difference is measured in terms of time difference between the system frame number (SFN) of different cells and is often referred to as the SFN-SFN time difference. If we know the real-time difference between the base stations, which can be measured by base stations or location measurement units (LMUs), we can narrow down the position of the terminal relative the base stations (cf. TDOA positioning). Another possibility is to use the round-trip time (RTT) measurements calculated by UTRAN. Reference [4] states that the RTT should have a measurement period of 100 ms and an accuracy of ± 0.5 chips. One chip accuracy in time correspond to approximately 80 m. Reference [16] claims however that it is possible to measure RTT with the accuracy of 1/16 of a chip, which corresponds to approximately 5 m.

So far terminals in Cell DCH state have been discussed. However, terminals in Cell Forward Access Channel (FACH) and Cell Paging Channel (PCH) state will be very useful in travel time estimations since the terminal performs cell updates in both of these states. Depending on the size configuration of URA, also terminals in URA PCH state might produce useful information. Terminals in these three states are characterised by having small amounts of data with low QoS demands to transfer [12]. If the URA is configured to be large, the information from URA updates can be used for the same purpose as RA or LA updates, e.g. as input for O-D matrix estimations or traffic flow measurements over large areas.

Table 1. Location data relevant for road traffic information estimation in GSM and UMTS networks

	GSM	UMTS
Synchronization level [bit length in m]	1108 m	78 m
Time alignment	TA	RTT/SFN-SFN
Cell size	< 35 km	< 10 km
Registration areas	LA/RA	LA/RA/URA
Measurement report interval	Periodic 0.48 s	Periodic 0.25-64 s Event-triggered
Max number of cell measurements	6	32
Power control frequency	2.1 Hz	Inner loop: 1500 Hz Outer loop: 10-100 Hz

Another fundamental difference between UMTS and GSM is the physical layer implementation. The characteristics of the modulation and wide spectrum of UMTS make it more suitable for positioning [8]. These characteristics may also affect the possibility to estimate the speed of the terminal according to the reception properties of the signal. This type of speed estimation can be useful in estimating travel times; it will however depend on implementation of the technique in the cellular networks. Different solutions to do this are described in [17-19].

The physical layer implementation also calls for another crucial characteristic of UMTS compared to GSM, in general significantly smaller cells. A denser network of base stations makes the location accuracy better, both in terms of active and passive monitoring as described in the following sections.

4. Estimation Approaches

Commercial companies have contributed to a large part of the development in estimating road traffic information from cellular networks in recent years. This means that few research publications are available describing the estimation approaches. Good exceptions are the work of e.g. Zhi-Jun Qiu et al. at University of Madison Wisconsin, see e.g. [20-24], and Bruce Hellinga et al. at University of Waterloo, see e.g. [25-27]. In order to estimate road traffic information from cellular networks the following basic steps can be included:

1) Location data collection
2) Terminal classification
3) Map matching
4) Route determination
5) Traffic state calculation

The *location data collection* phase involves how to gather the relevant data from the cellular network. The available data is described in *section 3*. The *terminal classification* focus on determining which of the mobile terminals that are located on the road network and in which type of transport mode they are. The *map matching* phase includes how to connect the location data to the road network, whereas the *route determination* process is used to determine the route of the vehicle, consisting of several road segments. The last step is to perform the *traffic state calculation*. Depending on the traffic information that is of interest, e.g. long term traffic flow estimation, origin-destination (O/D) estimation, travel time estimation and incident detection, different approaches and location data will be used. The steps can be carried out in a different order, in combination with each other and there can also be iteration between the different steps.

Location Data Collection

Two different approaches can be distinguished when collecting location data from the cellular networks. The first approach is based on the Federal Communications Commission (FCC) mandate that all mobile phones should be possible to locate with certain accuracy. A similar agreement exists in the EU countries and the operators active in EU. This implies that mobile phones can be located periodically and hence an average speed can be determined for the vehicle. This approach has the drawback that it generates extra traffic in the network and might be more vulnerable to privacy issues. The second approach relies on monitoring the generated traffic without trying to explicitly locate any of the mobile phones. The first approach is referred to as *active monitoring* whereas the second is referred to as *passive monitoring*.

Active monitoring can use any of the standardized positioning technologies described in *section 3*, e.g. CGI-TA or E-OTD. In passive monitoring it is possible to use all the data generated by the terminal to track a vehicle, this data is also described in detail in *section 3*. The problem with the passive monitoring approach is that terminals only generate detailed location information in busy state. This reduces the number of available probes significantly compared to all mobile phones that are switched on. The hybrid approach is based on passive monitoring with the possibility to complement with active monitoring upon certain criteria, e.g. large variations in estimations and light network load.

The possibility of estimating road traffic information with the help of cellular networks is well known, although the technique is not widely used today. Several commercial solutions are available, most of them situated in the U.S.[1]. Commercial companies, public organizations and universities have carried out field tests and simulations in order to evaluate the possibilities of the technology. A number of tests have been carried out in the U.S., but field tests have also been carried out in for example Austria, Canada, China, Finland, France, Germany, Israel, the Netherlands and Spain, see e.g. [28-35].

The Cellular Applied to ITS tracking And Location (CAPITAL) project was one of the first attempts to exploit cellular data to extract traffic information. The operational project started in 1994 in Virginia and ran for 27 months. The system used TOA together with AOA positioning to actively monitor different subscribers. The solution is based on active monitoring and it was unable to extract any useful information [28, 36]. Since then a lot of experience has been made in field tests and simulations and a number of projects have reported promising results.

The projects following CAPITAL have taken many different approaches to extract information from the cellular networks. Early papers in the area are mainly focused on the active monitoring approach. Also the developed simulation models are based on this approach. Reference [37-39] use simulation to evaluate the impact of system parameters, e.g. sampling interval and positioning error, in active monitoring based systems. Also [40] is an active monitoring based simulation model, and the focus is to evaluate a segment based approach to estimate travel times.

A number of papers assume configurations to the cellular network in order to generate more detailed signaling traffic [41-45]. These systems will be able to estimate the traffic conditions better than standard passive monitoring systems, but it is doubtful that the signaling configurations will be implemented in a commercial communication system. A couple of field tests has been carried out where cell phones are altered in order to send location and speed data with regular intervals, it is important to distinguish these tests since they require user acceptance and special software and is better categorized in the area of regular floating car data (FCD). It is possible that also active monitoring will require user acceptance since arbitrary switched on cell phones can be tracked and the tracking actually drains the terminal battery.

Due to an increasing number of mobile terminals on the roads that generate useful data, a wish to minimize network load and better tracking algorithms, the passive monitoring

[1] AirSage, Applied Generics, CellInt, IntelliOne, ITIS Holdings, Trafficcast.

approach has gained popularity. Several ways of passively collecting signaling data from the network are proposed. The first one is based on analysis of billing information sent from the core network. This approach is used in [31, 46] and makes it easy to collect the data typically in a single interface for the whole network. The billing information is not as detailed as the information available in other parts of the network and therefore systems are proposed using either the A- or A_{bis}-interface of the GSM network. If the A-interface is used, fewer installations have to be made for a certain geographic area. On the other hand the available location data is not as detailed as in the A_{bis}-interface. Basically the difference is that in the A-interface only handover and location area updates can be used, whereas in the Abis-interface also measurement reports can be used to estimate the location of the terminal. Passive monitoring via A- or A_{bis} interface is used in e.g. [32, 47, 48].

The natural extension to use either active or passive monitoring is to combine these two approaches; this is also suggested in recent commercial systems [49-51]. It is, however, unclear from present publications if it has been evaluated in a field test. This approach makes it possible to gather more information when it is most useful without putting unnecessary load on the network.

Terminal classification

Large amounts of signaling data is generated in a cellular network. However, we are only interested in tracking the terminals that are travelling on the road network. Furthermore the objective is often to estimate the travel time as it will be experienced by as many users as possible. This means that we are not interested in tracking motorcycles, bicycles, buses and taxis in separate bus lanes etc. since their travel times are not applicable to most of the road users. Another problem is that multiple cell phones in a single vehicle or bus will bias the travel time estimation.

First of all there is a need to filter out users that are not related to the road network, e.g. stationary users and pedestrians. This step can be performed in combination with the traffic state calculation. Stationary users can be filtered out by ignoring terminals that has a speed, as determined by e.g. handovers, below for example 6 km/h [50]. This means that terminals not moving fast enough will never be considered in the traffic state estimation. A similar approach is described in [22], with the difference that outliers are filtered in relation to the average speed in the previous time period. An interesting problem arises during congestion when vehicles are travelling at a lower speed than the chosen threshold. However, a straightforward way of solving this is to consider terminals that recently have been registered for a higher speed as valid probes, although they currently might be travelling at a lower speed than the threshold. Using this approach we will sometimes classify valid terminals as pedestrians during severe congestions. However, these severe congestions are most likely best detected using dedicated algorithms and possibly using other type of data.

As mentioned, buses and taxis can bias the estimations if they travel in a separate lane. Also motorcycles, bicycles and even trains can be a problem in some cases. Taxis, motorcycles and bicycles are very difficult to detect unless we have many samples in a time interval. In [52] a promising method to classify buses and possibly also trains using their timetables is described.

Map Matching and Route Determination

Since the location data from the cellular network is relatively coarse, the map matching step is quite important. Map matching and route determination is often combined. Depending on whether active or passive monitoring is used the map matching approach will differ. If active monitoring is used, due to the relatively slow sampling rate, an approach similar to map matching for regular FCD is used, i.e. variants of point-to-curve matching. Since the location accuracy is quite low, using previous locations and connections between different sections is more important compared to GPS-based systems. Map matching for active monitoring systems is described in detail in e.g. [27, 29]. If passive monitoring is used the sampling rate (from measurement reports) will typically be much higher, which means that some kind of curve-to-curve matching might be interesting. A difference with passive monitoring is also that the terminals are located spatially uniform, i.e. at the same place every time a travel time is calculated, which means that the map matching task is more predictable.

If the task of map matching involves how to match one or more location samples to a road section, the route detection task is to determine the order of sections that has been traversed. This can be done using the terminal locations in combination with the network structure and section decisions. Useful input to the route determination is also the measurement reports, since pattern matching techniques can be used to determine the most likely section or route. The tracking of cell phones using measurement reports in combination with Markov Models and Kalman filtering are discussed further in [53, 54].

When determining the route and matching positioning measurements to points on the road, a radio map including handover locations and signal strengths is needed. This map can be estimated using propagation models, test drives or a combination. Although test drives require more work, most likely they will also improve the accuracy of the radio maps and hence improve travel time estimations.

Traffic State Calculation

Once the section and/or route has been determined, the traffic information can be calculated. For O/D-matrix and traffic flow estimation the route determination is unnecessary, in that case map matching is the last step before state estimation. Handovers and measurement reports are not really useful in O/D-estimations since the number of active terminals is so few. However, location area and periodic updates are very useful since they are performed by all cell phones in a network. As described in *section 3*, all cell phones report their location when a telephone call is in progress, when an SMS is sent, when it is used for data transfer, when the location area is changed and also periodically (e.g. every fourth hour). This information is very useful in the O/D-matrix estimation process. If a cell phone is tracked during a longer period of time (at least several days) a good estimation of the travel behaviour can be made. Details of O/D-matrix and traffic flow estimation can be found in e.g. [55, 56].

In order to estimate travel times, two accurate locations with suitable separation are needed. These locations are straightforward using active monitoring, since it is the periodically collected position estimates. In passive monitoring a lot more information is typically available from the active terminals, and we can choose to estimate the position of the terminals at locations that are suitable from both a travel time and positioning accuracy point of view. As illustrated in *Figure 7,* the location can be estimated using handovers or by analysing measurement reports and defining proprietary location triggers. The potential gain in using proprietary triggers is that the handovers are not optimised to estimate positions, but instead optimising cellular network performance. The handovers can for example be a function of network load or interference, which can give a bias to the predicted handover location. The drawback of using a proprietary location trigger is that a lot more processing is needed.

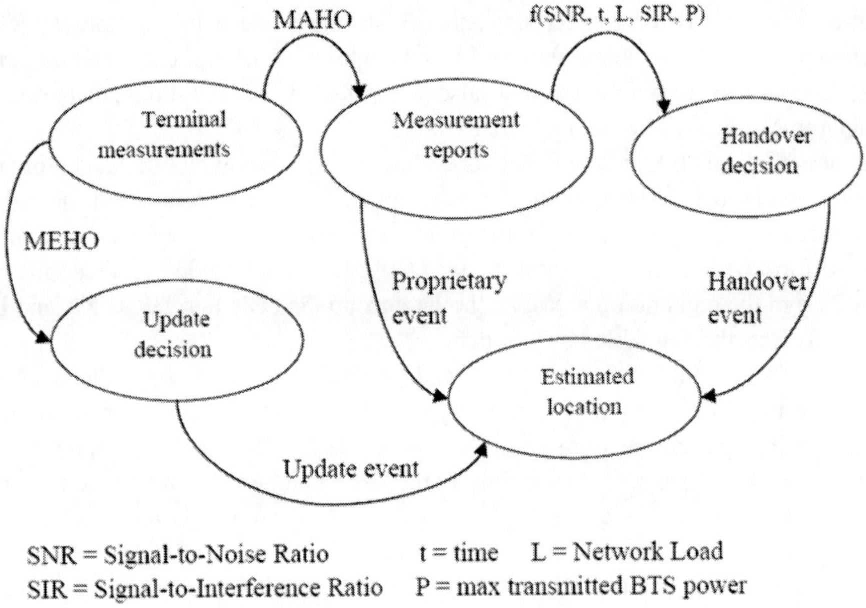

Figure 7. Different approaches to position estimation using measurement reports for travel time estimation.

Whether active or passive monitoring is used, the travel time between estimated locations needs to be mapped to the target links, where travel times are supposed to be reported to travellers. In active monitoring it is straightforward to give weights to the measurement in relation to the distance travelled on the target link [57]. As mentioned previously, in passive monitoring it could be solved by adjusting the target road sections to the available cellular data, creating new positioning points using measurement reports or simply extrapolating measured road sections to target road sections. Another possibility is that a target road section consists of several estimated positions, and in this case the estimated travel times should be weighted in relation to their length compared to the total length of the section.

There are only a limited number of traffic state estimation models for passive monitoring available in the literature. The traffic state is in [21] estimated using handovers and both a first-order model including average speed and a second-order model also incorporating estimated traffic flow. The estimated state is then filtered using a particle filter. In [22] handovers and a first-order model in combination with a Kalman filter are used to estimate average speed.

Incident detection can be carried out using travel times, traffic flow, or a combination of the two estimations. However, as first described in [37], there is also a strong correlation between congestion and the number of phone calls, since people tend to start making phone calls when entering a queue.

5. Network Characteristics and Data Quality

The quality of the road traffic information estimated from cellular networks is dependent on a number of aspects related to the geographic location of the implementation. The potential quality depends on the cellular network structure and configuration as well as the road network structure together with traffic characteristics of both the cellular and road network. The actual performance will also be affected by the implementation of the positioning, tracking and filtering methods.

The structure of the cellular network determines the available data for position estimation. Smaller cells give generally less ambiguity in position estimations and also more handover locations that can be used for travel time estimation. The route determination is also easier to perform with more cells within range. A configuration made by the mobile operator is how to design the registration areas, i.e. how to group the cells into URA, RA and LA, and this will also affect the available location data.

Correspondingly, the structure of the road network will also affect the challenges in extracting useful information. One of the major challenges today is to estimate high quality travel times using cellular networks in urban environments. The challenge using this kind of system in an urban environment lies mainly in the large number of non-vehicle terminals and possible routes. Furthermore, the mean speed of vehicle terminals is not much higher than non-vehicle terminals, especially in the case of congestion. The travel times are also exposed to large variations due to waiting times in signalized intersections. An interesting approach to overcome the problems of large variations in urban environments is described in [25].

The traffic characteristics of the cellular network are naturally important for the estimation quality. The road traffic characteristics are affecting indirectly, i.e. few cars means few cell phones and hence few measurements. This is not a large issue since it is most often no traffic related problems when the traffic flow is low. The number of calls in the cellular network is however a key indicator of the quality. Since the measurements are relatively noisy, it is important to have several samples in a time period in order to achieve high quality estimations. The relation between number of samples and estimation quality is discussed in detail in [24]. Also the lengths of the phone calls are of great importance, since the cell phone needs to pass several location points (e.g. handovers) in order for a travel time to be estimated.

The traffic information quality is also to a large extent affected by the location accuracy, which is discussed further in the remaining part of this section.

Location Experiments

As described earlier, travel time estimation require at least two reasonably accurate location estimates. The accuracy of these locations also determines the accuracy of the travel times. In order to estimate these locations, measurements of channel characteristics such as signal strength or propagation time are needed. These measurements can be obtained from the access network during terminal transmission, from measurement reports made by the mobile terminal or both. The time interval between measurement reports is crucial for the positioning accuracy and can be used to define a lower bound on accuracy. Whether the terminal location is estimated based on handovers or on raw measurement reports, the event is based on the measurement reports. The potential error introduced by the measurement report interval corresponds to half of the actual measurement report period. In GSM the measurement report interval is 480 ms, which corresponds to a location error of 2 meters driving at 30 km/h and 8 meters driving at 120 km/h. As described previously, the measurement reports in UMTS can be event-triggered or periodic. If the measurement reports are event-triggered, there is theoretically no error addition. However, the measurement report will most likely be delayed to an integer number of the shortest periodic reporting interval. The shortest periodic time interval is 250 ms and this corresponds to 1.0 and 4.3 meters for 30 and 120 km/h, respectively. In UMTS the periodic interval is configurable by the operator between 0.25 and 64 seconds, which means that the quality will depend on operator configuration. The error introduced by the measurement report seems in most cases negligible, hence other factors such as signal variations due to e.g. multipath propagation and handover inconsistency due to interference contributes to a larger extent to the location error. To get an idea of the potential of road traffic estimations from cellular networks an empirical study of the handover location error in GSM and UMTS is performed. For a more detailed discussion of the results, see [15].

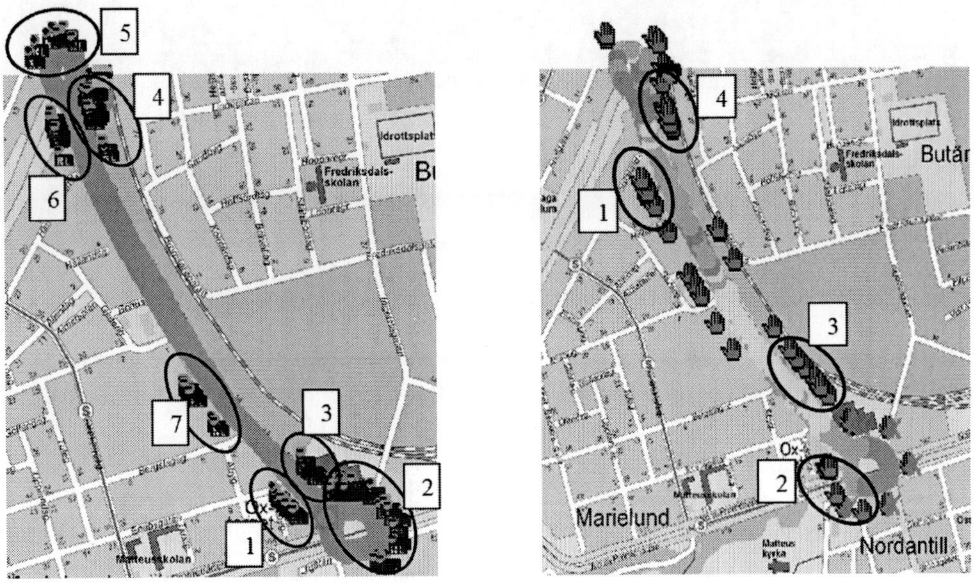

Figure 8. Handover locations in UMTS (left) and GSM (right).

The aim of the location experiments is to compare the location accuracy of handover points in GSM and UMTS. The first experiments were carried out on a 900 meters long street segment in a “sparse” urban environment. The segment was driven fifteen times back and forth with test equipment for GSM and UMTS (Ericsson TEMS Investigation 7.0) and a GPS receiver. Signaling data was collected from a GSM terminal and a UMTS terminal simultaneously, both with ongoing telephone calls. For comparison another test run was made in a different environment. The second test was performed in a suburban highway environment with less complex GSM cellular structure. The handover locations of the sparse urban environment are shown in *Figure 8*.

Figure 8 indicates quite large differences in the behaviour of the GSM and UMTS networks regarding handovers. In the suburban highway environment several UMTS and GSM handover zones were detected. Three of these handover zones for both GSM and UMTS were analysed in detail using seven test rounds on a 1.5 km road segment.

The percentage of times a specific handover is completed in the same handover zone, driving the same route several times, is referred to as consistency. The consistency of handover points will affect the travel time estimations. If a predicted handover does not occur, it will feed the system with no data or corrupt data, which will affect also an averaged calculated travel time. The consistency of the different handover points for GSM and UMTS are shown in *Figure 9*.

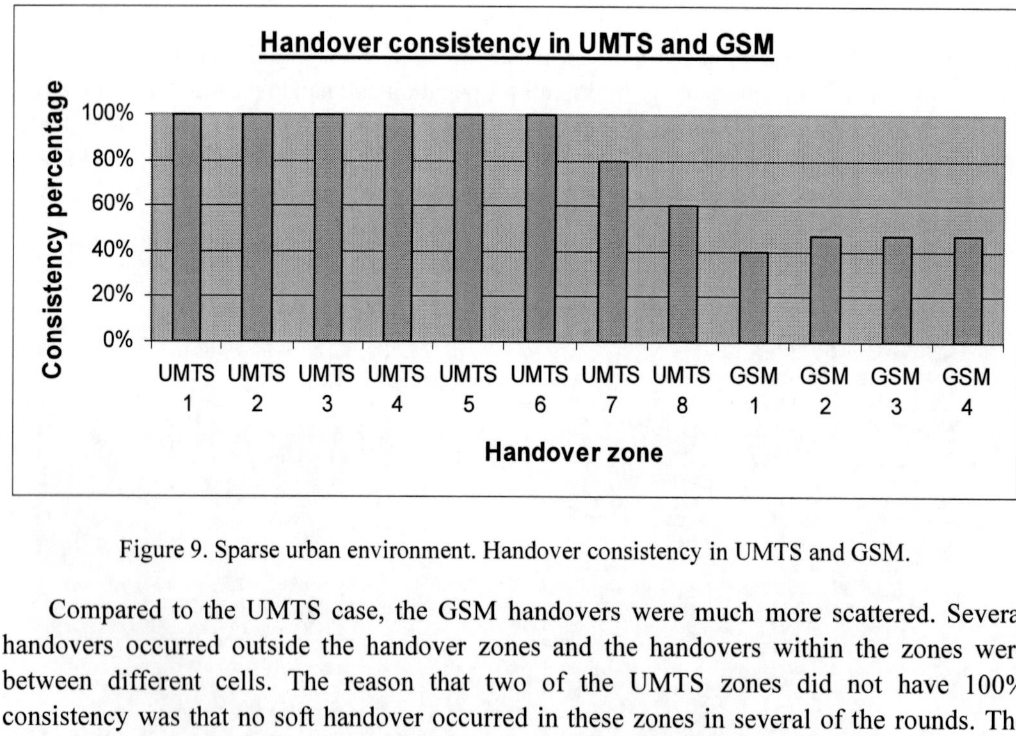

Figure 9. Sparse urban environment. Handover consistency in UMTS and GSM.

Compared to the UMTS case, the GSM handovers were much more scattered. Several handovers occurred outside the handover zones and the handovers within the zones were between different cells. The reason that two of the UMTS zones did not have 100% consistency was that no soft handover occurred in these zones in several of the rounds. The GSM inconsistency on the other hand, was due to handover between different cells in the same handover point. An explanation of the behavior of the UMTS and GSM networks can be that the UMTS network has more free capacity than the GSM network, and there is a possibility that the inconsistency of the GSM handovers are due to cell capacity limitations. However, the test runs were driven at two different days and times and it is not likely that the network was congested at both occasions. Since the results were similar for GSM in both test runs, a congested network does not seem to be a good explanation. A more plausible explanation might be the different cell structures of GSM and UMTS in the test area. Due to relatively few subscribers in the UMTS network, a hierarchical cell structure has not yet been deployed, as often is the case for the GSM network. A hierarchical cell structure might explain some of the scattered handovers and the fact that several different handovers were detected in the same handover zone.

The handover location accuracy is assessed using data from the 15 rounds to calculate an average handover point and measure the deviation from this point in the different rounds. The deviation is measured from the average point of all handovers within a handover zone. It should be noted that since the GSM handovers were very inconsistent, handovers between different cells will be grouped to the same average point, which will affect the GSM handover location accuracy measurements negatively. The handovers in UMTS are defined as the first or last radio link addition depending on handover zone. Missing handovers in a zone are ignored in the accuracy measurements, i.e. the consistency and accuracy diagrams should be assessed and analysed independently. All handover positions are mapped to the one-dimensional space of the traversed road segment. The positioning error of the GPS receiver is included in all measurements.

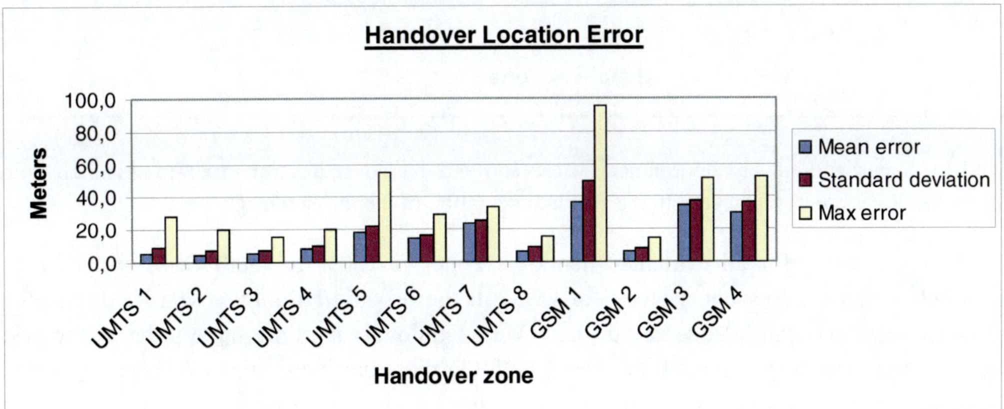

Figure 10. Sparse urban environment. Mean and max handover location error (in deviation from average handover point) together with the standard deviation of the measurements.

Figure 10 shows that the accuracy is quite good for a majority of the handover zones in both GSM and UMTS. The mean error is below 20 meters for more or less all of the UMTS handover zones, and below 40 meters for the GSM zones. However, a large variation in accuracy between different handover zones can also be seen, as well as a large difference between mean and max error. The large max error of handover zone 4 in UMTS was due to an outlier in one of the rounds. The UMTS accuracy seems in general much better than for GSM, which, as mentioned previously, partly could be explained by different cell structure and traffic pattern in the area. It should be noted that it is possible that the soft handover technique used in UMTS might render much more stable handover points compared to GSM, but it is not clear from the experiments what will happen if the traffic pattern and cell structure of the UMTS network changes. In the measurements the true value of the average handover point for the data set is used to calculate the location error. This value will not be known in a travel time estimation system, but instead estimated using test runs or coverage maps.

The previous hypothesis that the low accuracy and consistency of the GSM handovers is due to a complex hierarchical cell structure is evaluated with a second test in a different road and radio environment. This environment has most likely a flat GSM cell structure and the road is a highway with speed limit 90 km/h. In this environment it is also possible to evaluate how the vehicle speed affects the handover accuracy. It is possible that a higher speed gives a larger handover zone due to larger movement between measurement reports. The results of the test runs are shown in *Figure 11*.

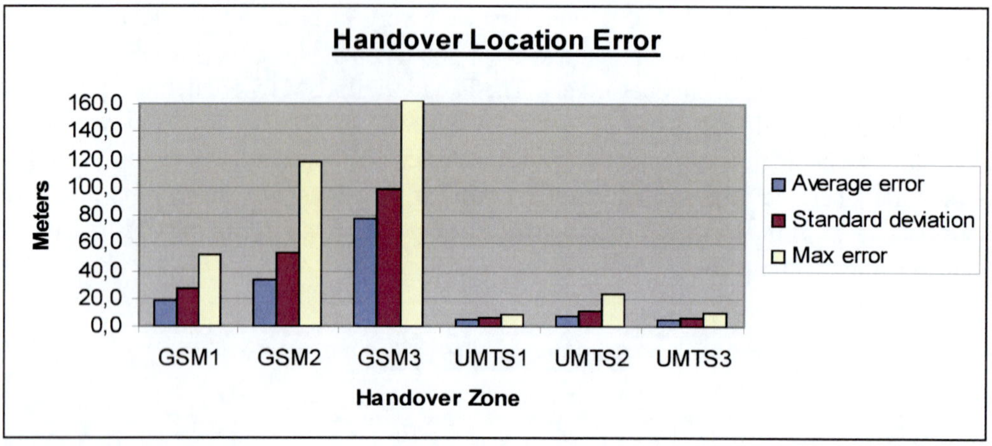

Figure 11. Suburban highway environment. Mean and max handover location error (in deviation from average handover point) together with the standard deviation of the measurements.

The consistency for all evaluated handover zones is 100% for both GSM and UMTS. This implies that the low consistency for GSM in the urban environment can be due to the cell structure. The location accuracy of the UMTS handovers is in the same order as the best ones from the urban environment, i.e. very good. On the other hand, the GSM measurements have a large variation where one handover zone has an average error of 20 meters where as another one has an average error of almost 80 meters. From the results we can expect that the vehicle speed does not have a dramatic effect on the accuracy, the handover consistency can be largely affected by cell structure and the UMTS handover location accuracy seems to have much higher location accuracy.

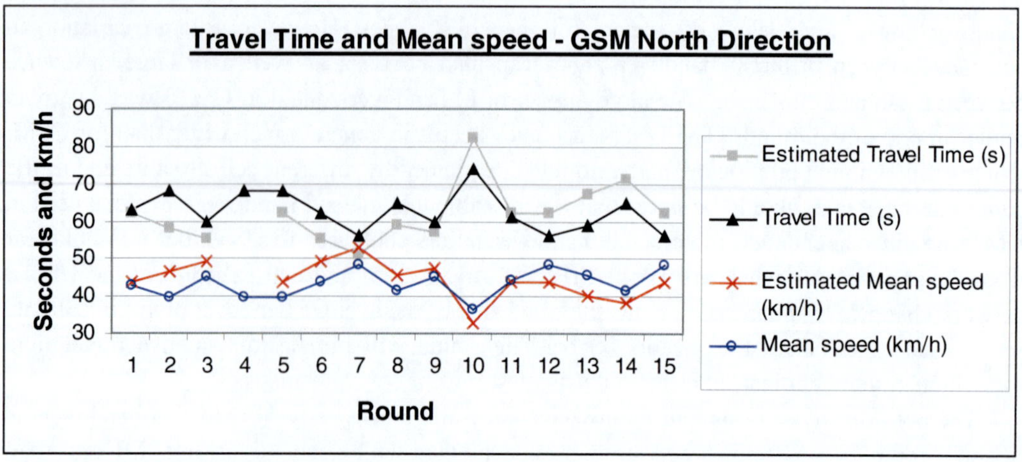

Figure 12. Travel times and mean speed estimated with GSM handover zones 3 and 4.

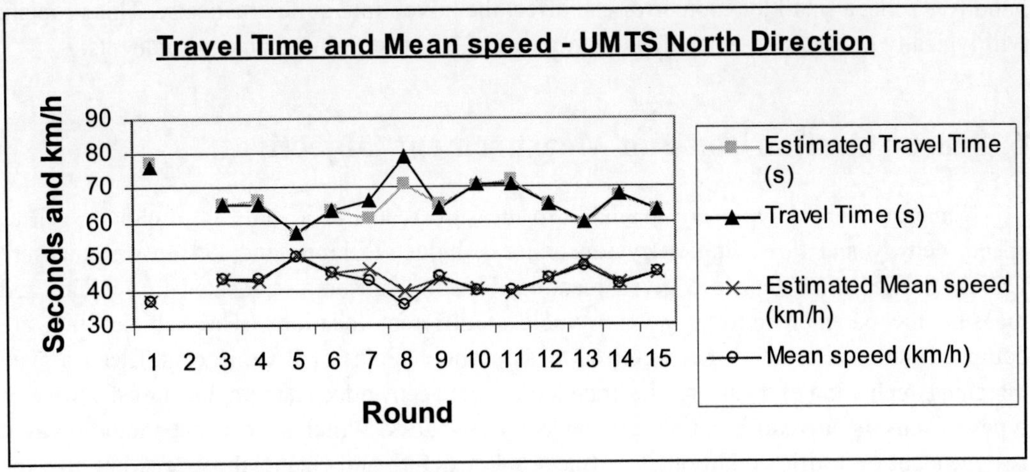

Figure 13. Travel times and mean speed estimated with UMTS handover zones 3 and 4.

Figures 12 and *13* show the actual travel time and the estimated travel time together with actual mean speed and estimated mean speed for the street segment traversed in the test runs. The travel times and mean speeds are calculated between the average handover points in the GSM network (handover zones 3 and 4) and the UMTS network (handover zones 3 and 4). The actual travel time is calculated from the TEMS log file between the average handover points for all rounds. The estimated travel time is calculated from the TEMS log file between the actual handover points in the current round.

It is important to mention that the travel times are calculated under the assumption that all vehicles are tracked to the correct road segment and that it is identified as a vehicle. In a real system, these tasks are typically very challenging, especially in an urban environment.

As can be seen in *Figure 13*, the estimated UMTS travel times follow the actual travel times extremely well. However, as can be seen in *Figure 12*, also the estimated GSM travel times follow the actual travel times in a good way. Four main factors affecting travel time accuracy are handover location accuracy, handover consistency, number of available probes in a time interval and road segment length. Handover location accuracy and consistency have been discussed, but also the number of available probes and the road segment length are important for travel time accuracy, interestingly they are also correlated. Increasing the road segment length will make the handover location error less significant and yield better travel time accuracy. More vehicle probes will give a better accuracy due to the possibility of averaging the travel time value between several probes in a travel time reporting interval. However, increasing the road segment length will also decrease the number of ongoing telephone calls or data sessions that completes the whole segment, which gives less probes for averaging. Shorter road segments for travel times also lead to more detailed traffic information, which might be useful in for example urban environments.

The road segment lengths (distance between average handover points) of the experiments were 750 and 785 meters for GSM and UMTS, respectively. The experiments indicate that it is possible to make good travel time estimations from cellular networks for this segment length, even shorter segment lengths are possible depending on radio environment, number of probes and accuracy requirements. Another factor that will affect the travel time accuracy is the average vehicle speed in the handover zones. If a vehicle is travelling slowly in a handover zone, a small location error can affect the travel time quite drastically. This problem will typically be common when deploying this kind of systems in urban environments.

6. Sensor Comparison and Measurement Objectives

Numerous sensor types are available to measure road traffic state information such as speed, density and flow. Stationary sensors, e.g. inductive loops and IR sensors, measure vehicle traffic parameters in a given location. Floating sensors are located in vehicles and measure the parameters for a given vehicle at different locations. The vehicles that are equipped with sensors are often referred to as probe vehicles or floating cars. License plate matching technologies measure the travel time between video camera locations. Different types of sensors have different advantages and drawbacks. Which sensor that should be used is dependent on traffic conditions, road network structure and financial aspects, but also the main application of data is relevant for this choice.

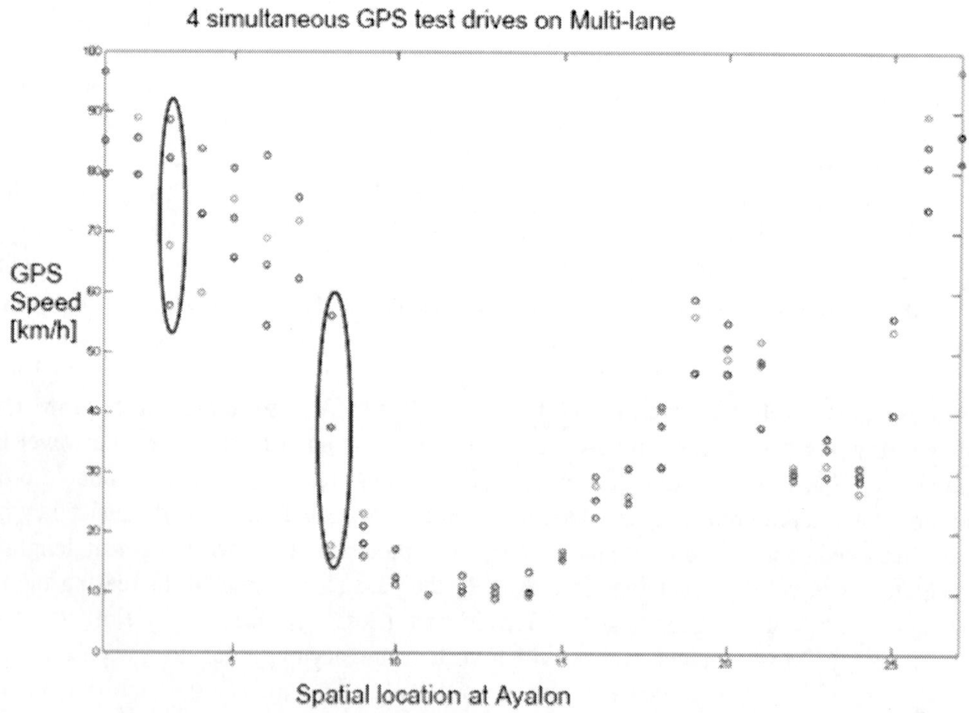

Figure 14. Variability of individual travel times travelling the same multi-lane road section simultaneously (Source: ITIS Traffic Services Ltd).

The performance of different systems for traffic information depends on a number of factors related to the measurement procedure and the number of sensors, but also on how the performance metrics are defined. Ideally we would want to know our own travel time given that we start our travel at a certain time in the near future. However, that is a quite challenging task, and the aim is currently to report the historical travel time dating e.g. five minutes back. Travel times fluctuate due to individual driving patterns and it is not obvious if we want to know the lowest, average or the highest travel time in the reporting interval. Speed fluctuations due to driver behaviour decreases when the road gets more congested and the incident or slowdown detection is another important application. However, even in a congested situation the travel time between different lanes can differ significantly. *Figure 14* shows a good example of the variability in travel times measured by GPS equipped vehicles travelling the same route simultaneously. The test was performed by ITIS Traffic Services Ltd on Ayalon freeway in Tel-Aviv.

A common performance metric is the time for a system to detect an incident. The definition of an incident varies, e.g. 40% slowdown, and affects the performance. For stationary sensors the time to detect an incident depends on the time it takes for the incident to propagate to the sensor. The time for the incident to propagate to the sensor depends on the average speed of the vehicle, which typically is very low during an incident. In this case the sensor spacing is the crucial factor and a dense network of sensors is related to large investment and maintenance costs. For floating sensors the time to detect an incident depends on the time for the sensors to propagate to the incident. This time is highly correlated to the number of vehicles with sensors (penetration) and the reporting interval of the sensors. Using signaling data from cellular networks gives a potentially high number of floating sensors with a very short reporting interval. The issue in these kinds of systems is the relatively low location accuracy discussed in previous sections. Since the time to detect incidents is dependent on sensor spacing for stationary sensors and sampling interval for floating sensors, ideally we would like to increase the number of sensors and decrease the sampling interval in congested areas. For stationary sensors this is rather difficult. For regular FCD systems and active monitoring systems this is achieved by the increase in number of vehicles in congested areas, and it is also quite easy to modify sampling intervals of these probes depending on the traffic conditions. Passive monitoring systems also benefit from the increased number of vehicles, but also from the fact that cell phones are more often used during congestion. As previously described, hybrid monitoring systems can combine the advantages of passive and active monitoring systems.

Another important performance metric is the travel time or average speed deviation relative some kind of ground truth. As previously explained and as can be seen in *figure 14*, it is not straightforward to determine the ground truth. Travel times from cellular networks are often evaluated against loop detectors, which measure the average speed at certain locations and can only be used to estimate the travel time. It is also quite common to use GPS equipped vehicles to measure experienced travel times for individual cars, which might vary significantly between different drivers and lanes. Although it might be costly, a good way of evaluating travel time accuracy is to use license plate matching cameras.

In the comparison between different sensors, it should be noted that all floating sensors rely on stochastic elements, i.e. that floating cars are travelling in all areas with a minimum frequency and that cell phones are used in vehicles. Leaving out effects of extreme weather conditions and severe congestions, stationary sensors are more deterministic in nature and have a vehicle penetration of almost 100 %. Considering this, the best results are most likely obtained when information is merged from several sensors with different characteristics.

7. Conclusion

Since the failure of the pioneering CAPITAL project in 1994 the estimation technologies have been improved tremendously. A couple of years ago succesful field trials started showing up in different parts of the world with different technology providers and since then the number of field trials and large scale implementations have increased dramatically. Recently three extensive reports have been published summarising the different projects and field trials that has been performed within the field of estimating road traffic information from cellular networks [31-33]. The conclusions from these reports are quite similar; the technology has large potential although more research is needed in order to use it as efficiently as possible. The technology has shown very good results on highways during light and medium congested traffic. However, during severe congestion the results are varying between different vendors and different trial areas. Promising results have been shown in arterials and even urban environments, the results in these environments are though quite few and very varying. Problems have also been identified during the morning rush, since people are reluctant to use the cell phone early in the morning.

Detailed analysis of specific deployments of commercial systems have been made recently in e.g. [58, 59]. However, raw signaling data results and network structure are not shown which makes it difficult to make general conclusions of future possible performance. In order to do that the cause of degradation has to be isolated, e.g. location inaccuracy, wrong route determination or simply to few samples in a time interval. Simulation models can also be a good way of analysing expected performance of these systems, both in general and for specific geographic locations. Several simulation studies have been performed based on the active monitoring approach. Since recent systems are focused on passive monitoring, there is a demand for passive monitoring simulation models that models the cellular network dynamics in greater detail.

Future systems will be feeded with more information from data communications, rather than telephone calls. This part of the research area is surprisingly unexplored and development is needed in order to utilise the increase in data traffic. The shift towards UMTS will also affect the performance in these systems. In the coming years there will be many countries with a combination of UMTS and GSM terminals. Most likely the share of UMTS terminals will grow. In Sweden, for example, the number of new UMTS terminals has passed the number of new GSM terminals. Hence, using also UMTS terminals for travel time estimation will increase the number of floating sensors in the system, which is an important factor when it comes to both travel time accuracy and time to incident detection. Increasing the number of vehicle samples in a travel time reporting interval is very important for accuracy, especially since the system produces relatively noisy measurements.

The higher data rate and shorter delay together with dynamic measurement reporting in UMTS makes the network react much faster to changes in the radio environment, which affects the location accuracy of network events. This, in combination with the soft handover principle that makes a radio link addition or removal without large delay, might be the reason to the much better UMTS location accuracy in the evaluated tests. Generally speaking there is also a better synchronization between base stations and mobile terminals in UMTS and (eventually) a more dense radio network which gives a potentially higher location resolution.

This is independent of whether handovers or something else is used to determine specific locations of the mobile terminal on the road.

The higher location accuracy in the UMTS network can be used to make the travel time accuracy better or maintaining the relative accuracy while making the travel time segments shorter. Shorter travel time segments are necessary in e.g. urban environments and are also useful when detecting incidents.

Relying on people using their cell phones in order to retrieve road traffic information is not always a perfect solution. It is likely that hybrid approach solutions will gain in popularity if a good way of selecting the cell phones to actively monitor is presented. The UMTS network utilising event-triggered measurement reports gives a new dimension to the hybrid monitoring approach. The location data available from a cellular network is highly dependent on network configurations. So far there has been little incentive for the operator to adjust the cellular network to perform better in road traffic estimation. This might change as soon as the demand for traffic information is high enough. An interesting analysis of this topic can be found in [60].

Most vehicles today have at least one cell phone onboard and a lot of valuable data is possible to extract from the cellular networks. If it is used in combination with other sensors or as a standalone technology will differ between different deployments. However, it would be a major waste not to use it at all.

References

[1] Gustafsson, F. and Gunnarsson. F. (2005). Mobile positioning using wireless networks: possibilities and fundamental limitations based on available wireless network measurements. *Signal Processing Magazine, IEEE vol. 22, Issue 4*, pp. 41- 53.

[2] 3GPP (2005). *Requirements for support of radio resource management*, TS 25.133, v. 7.2.0.

[3] 3GPP (2007). *Radio Resource Control (RRC); Protocol Specification.* TS 25.331, v. 8.0.0

[4] 3GPP (2005). *Requirements for support of radio resource management.* TS 25.133, v. 7.2.0.

[5] 3GPP (2002). *General Packet Radio Service (GPRS); Service description; Stage 2.* TS 03.60, v. 9.7.0.

[6] 3GPP (2007). *Radio Access Network; Functional stage 2 description of Location Services (LCS) in GERAN.* TS 43.059, v. 8.0.0.

[7] 3GPP (2007). *Stage 2 functional specification of User Equipment (UE) positioning in UTRAN.* TS 25.305, v. 8.0.0.

[8] Drane, C. and Rizos, C. (1998). *Positioning Systems in Intelligent Transportation Systems*. Norwood: Artech House.

[9] Küpper, A. (2005). *Location-Based Services: fundamentals and operation.* West Sussex: Wiley.

[10] Bensky, A. (2008). *Wireless Positioning Technologies and Applications.* Norwood: Artech House.

[11] Heine, G. (1999). *GSM Networks: Protocols, Terminology, and Implementation.* Boston: Artech House.

[12] Kaaranen, H., Ahtiainen, A., Laitinen, L., Naghian, S. and Niemi, V. (2001). *UMTS Networks – Architecture, Mobility and Services.* West Sussex: John Wiley and Sons, Ltd.

[13] Holma, H. and Toskala, A. (2002). *WCDMA for UMTS – Radio access For Third Generation Mobile Communications* (Second Edition). West Sussex: John Wiley and Sons, Ltd.

[14] Hiltunen, K., Binucci, N. and Bergström, J. (2000). *Comparison Between the Periodic and Event-Triggered Intra-Frequency Handover Measurement Reporting in WCDMA.* Proceedings of WCNC.

[15] Gundlegård, D. and Karlsson, J. M. Handover Location Accuracy for Travel Time Estimation in GSM and UMTS. To appear in *IET Intelligent Transport Systems*

[16] Borkowski, J., Niemelä, J. and Lempiäinen, J. (2004). *Performance of CellID+RTT Hybrid Positioning Method for UMTS Radio Networks.* Proceedings of The Fifth European Wireless Conference.

[17] Zhao, L. and Mark, J. W. Mobile Speed estimation Based on Average Fade Slope Duration. *IEEE Transactions on Communications,* Vol. 52, No. 12, pp. 2066-2069

[18] Kirsch, M. and Berens, F. (2002). *Mobile Speed Estimation for 3G Mobile Radio Systems using the Normalized Autocovariance Function.* International Zurich Seminar on Broadband Communications.

[19] Xiao, C., Mann, K. D. and Olivier, J. C. (2001). Mobile Speed estimation for TDMA-based Hierarchical Cellular Systems. *IEEE Transactions on Vehicular Technology,* Vol. 50, No. 4, pp. 981-991

[20] Jin, J., Qiu, Z. and Ran, B. (2006), *Intelligent Route-based Speed Estimation using Timing Advance,* Proceedings of IEEE ITS Conference.

[21] Cheng, P., Qiu, Z. and Ran, B. (2006), *Particle Filter Based Traffic State Estimation Using Cell Phone Network Data,* Proceedings of IEEE ITS Conference.

[22] Qiu, Z. and Ran, B. (2008), *Kalman Filtering Applied to Network-based Cellular Probe Traffic Monitoring,* Proceedings of TRB.

[23] Qiu, Z., Cheng, P. and Ran, B. (2007). Investigate the feasibility of traffic speed estimation using cell phones as probes. *Int. J. Services Operations and Informatics,* Vol. 2, No. 1, pp. 53-64.

[24] Qiu, Z. and Ran, B. (2007). *Sample Size Requirement for Cellular Probe Traffic Monitoring System*, Proceedings of ITSWC.

[25] Hellinga, B. et. al, (2008). Decomposing travel times measured by probe-based traffic monitoring systems to individual road segments. *Transportation Research Part C,* Volume 16, Issue 6, pp. 768-782

[26] Hellinga, B., Fu, L. and Takada, H. (2005). *Traffic Network Condition Monitoring via Mobile Phone Location Referencing – An Emerging Opportunity*, Proceedings of 6:th Transportation Specialty Conference

[27] Takada, H., Hellinga, B. and Fu, L. (2006). *Link Travel Time Estimation based on Mobile Phone Location Tracking,* , Proceedings of ITSWC.

[28] Yim, Y. (2003). *The State of Cellular Probes*. California PATH Research Report.

[29] Fontaine, M.D. and Smith, B.L. (2004). *Improving the Effectiveness of Traffic Monitoring based on Wireless Location Technology.* Final report Virginia Transportation Research Council.

[30] Rose, G. *Mobile Phones as Traffic Probes*. (2004). Proceedings of Conference of Australian Institutes of Transport Research.

[31] Sauret, P. (2003). *Pilot Project for Travel Time Measurements with Mobile Phones in Spain.* 10th ITS World Congress.

[32] Karhumäki, T. (2002). *The Utilisation of GSM-Network in Travel Time Monitoring.* ITS Workshop on Road Monitoring.

[33] Subbarao, W.V. et. Al. (2007). *Travel Time Estimation using Cell Phones (TTECP) for Highways and Roadways.* Final report prepared for Florida Department of Transportation.

[34] Virginia Transportation Research Council (2005). *Probe-based Traffic Monitoring State-of-the-Practice Report,* NCHRP 70-01.

[35] Hellinga, B. and Izadpanah, P. (2007). *An Opportunity Assessment of Wireless Monitoring of Network-Wied Road Traffic Conditions.* Final report prepared for Ministry of Transportation of Ontario.

[36] University of Maryland Transportation Studies Center (1997). *Final Evaluation Report for the CAPITAL-ITS Operational Test and Demonstration Program.* University of Maryland, College Park.

[37] Ygnace, J.L. and Drane, C. (2001). *Cellular Telecommunication and Transportation Convergence*. Proceedings of IEEE Intelligent Transportation Systems Conference.

[38] Fontaine, M. and Smith, B. (2004). *Impact of System Parameters on the Accuracy of Wireless Location Technology-based Traffic Monitoring System.* Proceedings of IEEE Intelligent Transportation Systems Conference.

[39] Cayford, R. and Johnsson, V. (2003). *Operational Parameters Affecting the Use of Anonymous Cell Phone Tracking for Generating Traffic Information.* Proceedings of Transport Research Board Annual Meeting.

[40] Hsiao, M. and Chang, J. (2005). *Segment based Traffic Information Estimation Method Using Cellular Network Data.* Proceedings of IEEE Conference on Intelligent Transportation Systems.

[41] Bolla, R: and Davoli, F. (2000). *Road Traffic Estimation from Location Tracking Data in the Mobile Cellular Network.* Proceedings of Wireless Communications and Networking Conference.

[42] Saraydar, C., Tekinay, S. and Choi, W.J. (2004). *Efficient Vehicular Traffic Monitoring using Mobility Management in Cellular Networks.* Proceedings of International Conference on Networking, Sensing and Control.

[43] Linauer, M. and Leihs, D. (2003). *Generating Floating Car Data by using GSM-network.* 10th ITS World Congress.

[44] Schneider, W and Mrakotsky, E. (2005). *Mobile Phones as a Basis for Traffic State Information.* Proceedings of IEEE Conference on Intelligent Transportation Systems.

[45] Myr, D. (2003). *Traffic Information Gathering via Cellular Phone Networks for Intelligent Transportation Systems.* U.S. Patent 6 577 946, June 10.

[46] White, J., Quick, J. and Philippou, P. (2004). *The Use of Mobile Phone Location Data for Traffic Information.* Proceedings of 12th IEE International Conference on Road Transport Information and Control.

[47] Alger, M., Quartier, F. and Lang, T. (2003). *Generation of Road Traffic Information from GSM Signaling Data. Proceedings of* 10th ITS World Congress.

[48] Rutten, B., de Wolff, P. and van der Vlist, M. (2004). *GSM as the Source for Traffic Information.* Proceedings of 11th ITS World Congress.

[49] Feldman, I., Trinker, A., Meltzer, Y., Eshpar, A. and Lotem, A. (2003). *Method and System for Modeling and processing Vehicular Traffic Data and Information and Applying Therefor.* U.S. Patent 6 587 781, July 1.

[50] Kaplan, J. and Avni, O. (2005). *Method for Monitoring Cellular Communication and System therefore.* U.S. Patent 6 947 835, September 20.

[51] Smith, C.W., Wilkinson, C., Carlson, K., Wright, M.P. and Sangal, R. (2005). *System and Method for Providing Traffic Information using Operational Data of a Wireless Network.* U.S. Patent 6 842 620, January 11.

[52] Schollmeyer, R. and Wiltschko, T (2007), *Classification of Public Transport Vehicles using Cellular Mobile Radio Data.* Proceedings of ITS Europe.

[53] Mangold, S. and Kyriazakos, S. (1999). *Applying Pattern Recognition Techniques based on Hidden Markov Models for Vehicular Positioning Location in Cellular Networks.* Proceedings of VTC '99, Amsterdam.

[54] Hellebrandt, M. and Mathar, R. (1999). Location Tracking of Mobiles in Cellular Radio Networks. *IEEE Transactions on Vehicular Technology, Vol. 48, No. 5*, pp. 1558-1562

[55] Wideberg, P., Caceres, N. and Benitez, F.G. (2007). Deriving Origin-Destination Data from a Mobile Phone Network: *IET Intelligent Transport Systems, 2007, Vol. 1, Issue 1.* pp. 15–26

[56] Bekhor, S., Hirsh, M., Nimre, S. and Feldman, Israel (2008). *Identifying Spatial and Temporal Congestion Characteristics using Passive Mobile Phone Data*, Proceedings of TRB 2008.

[57] Fontaine, M.D., Jakkala, A.P. and Smith, B.L. (2007). *Probe Sampling Strategies for Traffic Monitoring Systems based on Wireless Location Tehnology.* Final Report Virginia Transportation Research Council.

[58] Bar-Gera, H. (2007). Evaluation of a cellular phone-based system for measurements of traffic speeds and travel times: A case study from Israel. *Transportation Research Part C,* Vol. 15. pp. 380-391

[59] Liu, H.X., et. al, (2008). *Evaluation of Cell Phone Traffic Data in Minnesota,* Proceedings of TRB.

[60] Hsiao, W. C. and Chang, S.K. (2006). *The Optimal Location Update Strategy of Cellular Network Based Traffic Information System,* Proceedings of IEEE Intelligent Transportation Systems Conference.

INDEX

A

B

C

D

E

F

G

H

I

K

L

M

V

W

X